Geophysical Monograph Series

Including

IUGG Volumes

Maurice Ewing Volumes

Mineral Physics Volumes

GEOPHYSICAL MONOGRAPH SERIES

Geophysical Monograph Volumes

1 Antarctica in the International Geophysical Year *A. P. Crary, L. M. Gould, E. O. Hulburt, Hugh Odishaw, and Waldo E. Smith (Eds.)*
2 Geophysics and the IGY *Hugh Odishaw and Stanley Ruttenberg (Eds.)*
3 Atmospheric Chemistry of Chlorine and Sulfur Compounds *James P. Lodge, Jr. (Ed.)*
4 Contemporary Geodesy *Charles A. Whitten and Kenneth H. Drummond (Eds.)*
5 Physics of Precipitation *Helmut Weickmann (Ed.)*
6 The Crust of the Pacific Basin *Gordon A. Macdonald and Hisashi Kuno (Eds.)*
7 Antarctic Research: The Matthew Fontaine Maury Memorial Symposium *H. Wexler, M. J. Rubin, and J. E. Caskey, Jr. (Eds.)*
8 Terrestrial Heat Flow *William H. K. Lee (Ed.)*
9 Gravity Anomalies: Unsurveyed Areas *Hyman Orlin (Ed.)*
10 The Earth Beneath the Continents: A Volume of Geophysical Studies in Honor of Merle A. Tuve *John S. Steinhart and T. Jefferson Smith (Eds.)*
11 Isotope Techniques in the Hydrologic Cycle *Glenn E. Stout (Ed.)*
12 The Crust and Upper Mantle of the Pacific Area *Leon Knopoff, Charles L. Drake, and Pembroke J. Hart (Eds.)*
13 The Earth's Crust and Upper Mantle *Pembroke J. Hart (Ed.)*
14 The Structure and Physical Properties of the Earth's Crust *John G. Heacock (Ed.)*
15 The Use of Artificial Satellites for Geodesy *Soren W. Henricksen, Armando Mancini, and Bernard H. Chovitz (Eds.)*
16 Flow and Fracture of Rocks *H. C. Heard, I. Y. Borg, N. L. Carter, and C. B. Raleigh (Eds.)*
17 Man-Made Lakes: Their Problems and Environmental Effects *William C. Ackermann, Gilbert F. White, and E. B. Worthington (Eds.)*
18 The Upper Atmosphere in Motion: A Selection of Papers With Annotation *C. O. Hines and Colleagues*
19 The Geophysics of the Pacific Ocean Basin and Its Margin: A Volume in Honor of George P. Woollard *George H. Sutton, Murli H. Manghnani, and Ralph Moberly (Eds.)*
20 The Earth's Crust: Its Nature and Physical Properties *John C. Heacock (Ed.)*
21 Quantitative Modeling of Magnetospheric Processes *W. P. Olson (Ed.)*
22 Derivation, Meaning, and Use of Geomagnetic Indices *P. N. Mayaud*
23 The Tectonic and Geologic Evolution of Southeast Asian Seas and Islands *Dennis E. Hayes (Ed.)*
24 Mechanical Behavior of Crustal Rocks: The Handin Volume *N. L. Carter, M. Friedman, J. M. Logan, and D. W. Stearns (Eds.)*
25 Physics of Auroral Arc Formation *S.-I. Akasofu and J. R. Kan (Eds.)*
26 Heterogeneous Atmospheric Chemistry *David R. Schryer (Ed.)*
27 The Tectonic and Geologic Evolution of Southeast Asian Seas and Islands: Part 2 *Dennis E. Hayes (Ed.)*
28 Magnetospheric Currents *Thomas A. Potemra (Ed.)*
29 Climate Processes and Climate Sensitivity (Maurice Ewing Volume 5) *James E. Hansen and Taro Takahashi (Eds.)*
30 Magnetic Reconnection in Space and Laboratory Plasmas *Edward W. Hones, Jr. (Ed.)*
31 Point Defects in Minerals (Mineral Physics Volume 1) *Robert N. Schock (Ed.)*
32 The Carbon Cycle and Atmospheric CO_2: Natural Variations Archean to Present *E. T. Sundquist and W. S. Broecker (Eds.)*
33 Greenland Ice Core: Geophysics, Geochemistry, and the Environment *C. C. Langway, Jr., H. Oeschger, and W. Dansgaard (Eds.)*
34 Collisionless Shocks in the Heliosphere: A Tutorial Review *Robert G. Stone and Bruce T. Tsurutani (Eds.)*
35 Collisionless Shocks in the Heliosphere: Reviews of Current Research *Bruce T. Tsurutani and Robert G. Stone (Eds.)*
36 Mineral and Rock Deformation: Laboratory Studies —The Paterson Volume *B. E. Hobbs and H. C. Heard (Eds.)*
37 Earthquake Source Mechanics (Maurice Ewing Volume 6) *Shamita Das, John Boatwright, and Christopher H. Scholz (Eds.)*
38 Ion Acceleration in the Magnetosphere and Ionosphere *Tom Chang (Ed.)*
39 High Pressure Research in Mineral Physics (Mineral Physics Volume 2) *Murli H. Manghnani and Yasuhiko Syono (Eds.)*
40 Gondwana Six: Structure, Tectonics, and Geophysics *Gary D. McKenzie (Ed.)*
41 Gondwana Six: Stratigraphy, Sedimentology, and Paleontology *Garry D. McKenzie (Ed.)*
42 Flow and Transport Through Unsaturated Fractured Rock *Daniel D. Evans and Thomas J. Nicholson (Eds.)*
43 Seamounts, Islands, and Atolls *Barbara H. Keating, Patricia Fryer, Rodey Batiza, and George W. Boehlert (Eds.)*

44 Modeling Magnetospheric Plasma *T. E. Moore and J. H. Waite, Jr. (Eds.)*

45 Perovskite: A Structure of Great Interest to Geophysics and Materials Science *Alexandra Navrotsky and Donald J. Weidner (Eds.)*

46 Structure and Dynamics of Earth's Deep Interior (IUGG Volume 1) *D. E. Smylie and Raymond Hide (Eds.)*

47 Hydrological Regimes and Their Subsurface Thermal Effects (IUGG Volume 2) *Alan E. Beck, Grant Garven, and Lajos Stegena (Eds.)*

48 Origin and Evolution of Sedimentary Basins and Their Energy and Mineral Resources (IUGG Volume 3) *Raymond A. Price (Ed.)*

49 Slow Deformation and Transmission of Stress in the Earth (IUGG Volume 4) *Steven C. Cohen and Petr Vaníček (Eds.)*

50 Deep Structure and Past Kinematics of Accreted Terranes (IUGG Volume 5) *John W. Hillhouse (Ed.)*

51 Properties and Processes of Earth's Lower Crust (IUGG Volume 6) *Robert F. Mereu, Stephan Mueller, and David M. Fountain (Eds.)*

52 Understanding Climate Change (IUGG Volume 7) *Andre L. Berger, Robert E. Dickinson, and J. Kidson (Eds.)*

53 Plasma Waves and Instabilities at Comets and in Magnetospheres *Bruce T. Tsurutani and Hiroshi Oya (Eds.)*

54 Solar System Plasma Physics *J. H. Waite, Jr., J. L. Burch, and R. L. Moore (Eds.)*

55 Aspects of Climate Variability in the Pacific and Western Americas *David H. Peterson (Ed.)*

56 The Brittle-Ductile Transition in Rocks *A. G. Duba, W. B. Durham, J. W. Handin, and H. F. Wang (Eds.)*

57 Evolution of Mid Ocean Ridges (IUGG Volume 8) *John M. Sinton (Ed.)*

58 Physics of Magnetic Flux Ropes *C. T. Russell, E. R. Priest, and L. C. Lee (Eds.)*

59 Variations in Earth Rotation (IUGG Volume 9) *Dennis D. McCarthy and Williams E. Carter (Eds.)*

60 Quo Vadimus Geophysics for the Next Generation (IUGG Volume 10) *George D. Garland and John R. Apel (Eds.)*

61 Cometary Plasma Processes *Alan D. Johnstone (Ed.)*

62 Modeling Magnetospheric Plasma Processes *Gordon R. Wilson (Ed.)*

63 Marine Particles: Analysis and Characterization *David C. Hurd and Derek W. Spencer (Eds.)*

64 Magnetospheric Substorms *Joseph R. Kan, Thomas A. Potemra, Susumu Kokubun, and Takesi Iijima (Eds.)*

65 Explosion Source Phenomenology *Steven R. Taylor, Howard J. Patton, and Paul G. Richards (Eds.)*

66 Venus and Mars: Atmospheres, Ionospheres, and Solar Wind Interactions *Janet G. Luhmann, Mariella Tatrallyay, and Robert O. Pepin (Eds.)*

67 High-Pressure Research: Application to Earth and Planetary Sciences (Mineral Physics Volume 3) *Yasuhiko Syono and Murli H. Manghnani (Eds.)*

68 Microwave Remote Sensing of Sea Ice *Frank Carsey, Roger Barry, Josefino Comiso, D. Andrew Rothrock, Robert Shuchman, W. Terry Tucker, Wilford Weeks, and Dale Winebrenner*

69 Sea Level Changes: Determination and Effects (IUGG Volume 11) *P. L. Woodworth, D. T. Pugh, J. G. DeRonde, R. G. Warrick, and J. Hannah*

70 Synthesis of Results from Scientific Drilling in the Indian Ocean *Robert A. Duncan, David K. Rea, Robert B. Kidd, Ulrich von Rad, and Jeffrey K. Weissel (Eds.)*

71 Mantle Flow and Melt Generation at Mid-Ocean Ridges *Jason Phipps Morgan, Donna K. Blackman, and John M. Sinton (Eds.)*

72 Dynamics of Earth's Deep Interior and Earth Rotation (IUGG Volume 12) *Jean-Louis Le Mouël, D.E. Smylie, and Thomas Herring (Eds.)*

73 Environmental Effects on Spacecraft Positioning and Trajectories (IUGG Volume 13) *A. Vallance Jones (Ed.)*

74 Evolution of the Earth and Planets (IUGG Volume 14) *E. Takahashi, Raymond Jeanloz, and David Rubie (Eds.)*

75 Interactions Between Global Climate Subsystems: The Legacy of Hann (IUGG Volume 15) *G. A. McBean and M. Hantel (Eds.)*

76 Relating Geophysical Structures and Processes: The Jeffreys Volume (IUGG Volume 16) *K. Aki and R. Dmowska (Eds.)*

77 The Mesozoic Pacific: Geology, Tectonics, and Volcanism—A Volume in Memory of Sy Schlanger *Malcolm S. Pringle, William W. Sager, William V. Sliter, and Seth Stein (Eds.)*

78 Climate Change in Continental Isotopic Records *P. K. Swart, K. C. Lohmann, J. McKenzie, and S. Savin (Eds.)*

79 The Tornado: Its Structure, Dynamics, Prediction, and Hazards *C. Church, D. Burgess, C. Doswell, R. Davies-Jones (Eds.)*

80 Auroral Plasma Dynamics *R. L. Lysak (Ed.)*

81 Solar Wind Sources of Magnetospheric Ultra-Low Frequency Waves *M. J. Engebretson, K. Takahashi, and M. Scholer (Eds.)*

82 Gravimetry and Space Techniques Applied to Geodynamics and Ocean Dynamics (IUGG Volume 17) *Bob E. Schutz, Allen Anderson, Claude Froidevaux, and Michael Parke (Eds.)*

83 Nonlinear Dynamics and Predictability of Geophysical Phenomena (IUGG Volume 18) *William I. Newman, Andrei Gabrielov, and Donald L. Turcotte (Eds.)*

84 Solar System Plasmas in Space and Time *J. Burch, J. H. Waite, Jr. (Eds.)*

85 **The Polar Oceans and Their Role in Shaping the Global Environment** *O. M. Johannessen, R. D. Muench, and J. E. Overland (Eds.)*

86 **Space Plasmas: Coupling Between Small and Medium Scale Processes** *Maha Ashour-Abdalla, Tom Chang, and Paul Dusenbery (Eds.)*

87 **The Upper Mesosphere and Lower Thermosphere: A Review of Experiment and Theory** *R. M. Johnson and T. L. Killeen (Eds.)*

88 **Active Margins and Marginal Basins of the Western Pacific** *Brian Taylor and James Natland (Eds.)*

89 **Natural and Anthropogenic Influences in Fluvial Geomorphology** *John E. Costa, Andrew J. Miller, Kenneth W. Potter, and Peter R. Wilcock (Eds.)*

90 **Physics of the Magnetopause** *Paul Song, B.U.Ö. Sonnerup, and M.F. Thomsen (Eds.)*

91 **Seafloor Hydrothermal Systems: Physical, Chemical, Biological, and Geological Interactions** *Susan E. Humphris, Robert A. Zierenberg, Lauren S. Mullineaux, and Richard E. Thomson (Eds.)*

92 **Mauna Loa Revealed: Structure, Composition, History, and Hazards** *J. M. Rhodes and John P. Lockwood (Eds.)*

93 **Cross-Scale Coupling in Space Plasmas** *James L. Horwitz, Nagendra Singh, and James L. Burch (Eds.)*

94 **Double-Diffusive Convection** *Alan Brandt and H.J.S. Fernando (Eds.)*

95 **Earth Processes: Reading the Isotopic Code** *Asish Basu and Stan Hart (Eds.)*

96 **Subduction Top to Bottom** *Gray E. Bebout, David Scholl, Stephen Kirby, and John Platt (Eds.)*

97 **Radiation Belts—Models and Standards** *J. F. Lemaire, D. Heynderickx, and D. N. Baker (Eds.)*

Maurice Ewing Volumes

1 **Island Arcs, Deep Sea Trenches, and Back-Arc Basins** *Manik Talwani and Walter C. Pitman III (Eds.)*

2 **Deep Drilling Results in the Atlantic Ocean: Ocean Crust** *Manik Talwani, Christopher G. Harrison, and Dennis E. Hayes (Eds.)*

3 **Deep Drilling Results in the Atlantic Ocean: Continental Margins and Paleoenvironment** *Manik Talwani, William Hay, and William B. F. Ryan (Eds.)*

4 **Earthquake Prediction—An International Review** *David W. Simpson and Paul G. Richards (Eds.)*

5 **Climate Processes and Climate Sensitivity** *James E. Hansen and Taro Takahashi (Eds.)*

6 **Earthquake Source Mechanics** *Shamita Das, John Boatwright, and Christopher H. Scholz (Eds.)*

IUGG Volumes

1 **Structure and Dynamics of Earth's Deep Interior** *D. E. Smylie and Raymond Hide (Eds.)*

2 **Hydrological Regimes and Their Subsurface Thermal Effects** *Alan E. Beck, Grant Garven, and Lajos Stegena (Eds.)*

3 **Origin and Evolution of Sedimentary Basins and Their Energy and Mineral Resources** *Raymond A. Price (Ed.)*

4 **Slow Deformation and Transmission of Stress in the Earth** *Steven C. Cohen and Petr Vaníček (Eds.)*

5 **Deep Structure and Past Kinematics of Accreted Terrances** *John W. Hillhouse (Ed.)*

6 **Properties and Processes of Earth's Lower Crust** *Robert F. Mereu, Stephan Mueller, and David M. Fountain (Eds.)*

7 **Understanding Climate Change** *Andre L. Berger, Robert E. Dickinson, and J. Kidson (Eds.)*

8 **Evolution of Mid Ocean Ridges** *John M. Sinton (Ed.)*

9 **Variations in Earth Rotation** *Dennis D. McCarthy and William E. Carter (Eds.)*

10 **Quo Vadimus Geophysics for the Next Generation** *George D. Garland and John R. Apel (Eds.)*

11 **Sea Level Changes: Determinations and Effects** *Philip L. Woodworth, David T. Pugh, John G. DeRonde, Richard G. Warrick, and John Hannah (Eds.)*

12 **Dynamics of Earth's Deep Interior and Earth Rotation** *Jean-Louis Le Mouël, D.E. Smylie, and Thomas Herring (Eds.)*

13 **Environmental Effects on Spacecraft Positioning and Trajectories** *A. Vallance Jones (Ed.)*

14 **Evolution of the Earth and Planets** *E. Takahashi, Raymond Jeanloz, and David Rubie (Eds.)*

15 **Interactions Between Global Climate Subsystems: The Legacy of Hann** *G. A. McBean and M. Hantel (Eds.)*

16 **Relating Geophysical Structures and Processes: The Jeffreys Volume** *K. Aki and R. Dmowska (Eds.)*

17 **Gravimetry and Space Techniques Applied to Geodynamics and Ocean Dynamics** *Bob E. Schutz, Allen Anderson, Claude Froidevaux, and Michael Parke (Eds.)*

18 **Nonlinear Dynamics and Predictability of Geophysical Phenomena** *William I. Newman, Andrei Gabrielov, and Donald L. Turcotte (Eds.)*

Mineral Physics Volumes

1 **Point Defects in Minerals** *Robert N. Schock (Ed.)*

2 **High Pressure Research in Mineral Physics** *Murli H. Manghnani and Yasuhiko Syona (Eds.)*

3 **High Pressure Research: Application to Earth and Planetary Sciences** *Yasuhiko Syono and Murli H. Manghnani (Eds.)*

Geophysical Monograph 98

Magnetic Storms

Bruce T. Tsurutani
Walter D. Gonzalez
Yohsuke Kamide
John K. Arballo

Editors

American Geophysical Union

Published under the aegis of the AGU Books Board

Cover: Red aurora observed in Rikubetsu, Japan, on October 21, 1989. Photograph by H. Tsuda.

Library of Congress Cataloging-in-Publication Data
Magnetic storms / Bruce T. Tsurutani ... [et al.].
       p. cm. -- (Geophysical Monograph ; 97)
   Includes bibliographical references.
   ISBN 0-87590-080-1 (alk. paper)
   1. Magnetic Storms. I. Tsurutani, Bruce T. II. Series.
QC835.M34 1997
538'.744--dc21   97-11445
                                                        CIP

ISBN 0-87590-080-1
ISSN 0065-8448

Copyright 1997 by the American Geophysical Union
2000 Florida Avenue, N.W.
Washington, DC 20009

Figures, tables, and short excerpts may be reprinted in scientific books and journals if the source is properly cited.

Authorization to photocopy items for internal or personal use, or the internal or personal use of specific clients, is granted by the American Geophysical Union for libraries and other users registered with the Copyright Clearance Center (CCC) Transactional Reporting Service, provided that the base fee of $1.50 per copy plus $0.35 per page is paid directly to CCC, 222 Rosewood Dr., Danvers, MA 01923. 0065-8448/97/$01.50+0.35.
   This consent does not extend to other kinds of copying, such as copying for creating new collective works or for resale. The reproduction of multiple copies and the use of full articles or the use of extracts, including figures and tables, for commercial purposes requires permission from AGU.

Printed in the United States of America.

CONTENTS

Preface
Bruce T. Tsurutani, Walter D. Gonzalez, Yohsuke Kamide, and John K. Arballo ix

Overview
Magnetic Storms: Current Understanding and Outstanding Questions
Y. Kamide, R. L. McPherron, W. D. Gonzalez, D. C. Hamilton, H. S. Hudson, J. A. Joselyn, S. W. Kahler, L. R. Lyons, H. Lundstedt, and E. Szuszczewicz 1

Solar Origins of Storms
Solar Coronal Dynamics and Flares as a Cause of Interplanetary Disturbances
Takeo Kosugi and Kazunari Shibata 21

The Solar Antecedents of Geomagnetic Storms
Hugh S. Hudson 35

Prominence Eruptions and Geoeffective Solar Wind Structures
James Chen 45

Heliospheric Observations of Solar Disturbances and Their Potential Role in the Origin of Geomagnetic Storms
Bernard V. Jackson 59

Interplanetary Origins of Storms
The Interplanetary Causes of Magnetic Storms: A Review
Bruce T. Tsurutani and Walter D. Gonzalez 77

Magnetic Clouds and the Quiet-Storm Effect at Earth
C. J. Farrugia, L. F. Burlaga, and R. P. Lepping 91

Storm Dynamics
The Role of Magnetosphere-Ionosphere Coupling in Magnetic Storm Dynamics
Ioannis A. Daglis 107

Magnetotail Dynamics During Storms
Dynamics of the Magnetotail During Magnetic Storms: Review of ISEE 3 and GEOTAIL Observations
Susumu Kokubun 117

Storm/Substorm Relationships
The Role of Substorms in the Generation of Magnetic Storms
R. L. McPherron 131

Physics of Magnetic Storms
Gordon Rostoker, Erena Friedrich, and Matthew Dobbs 149

Magnetospheric Processes/Computer Modeling
Modeling Convection Effects in Magnetic Storms
R. A. Wolf, J. W. Freeman, Jr., B. A. Hausman, R. W. Spiro, R. V. Hilmer, and R. L. Lambour 161

CONTENTS

Modeling of Ring Current Formation and Decay: A Review
Margaret W. Chen, Michael Schulz, and Larry R. Lyons 173

Modeling of the Contribution of Electromagnetic Ion Cyclotron (EMIC) Waves to Stormtime Ring Current Erosion
J. U. Kozyra, V. K. Jordanova, R. B. Horne, and R. M. Thorne 187

Ionospheric and Thermospheric Storm-Time Processes
How Does the Thermosphere and Ionosphere React to a Geomagnetic Storm?
T. J. Fuller-Rowell, M. V. Codrescu, R. G. Roble, and A. D. Richmond 203

Magnetic Storm Associated Perturbations of the Upper Atmosphere
Gerd W. Prölss 227

Storm Forecasting
AI Techniques in Geomagnetic Storm Forecasting
Henrik Lundstedt 243

Review of Techniques for Magnetic Storm Forecasting
Thomas R. Detman and Dimitris Vassiliadis 253

PREFACE

Magnetic storms is a topic that is central to Space Weather. In addition to theoretically understanding the flow of energy from the Sun through interplanetary space to the Earth's magnetotail, magnetosphere, and ionosphere, there are important practical implications for satellite and astronaut safety, ground electrical power systems, and commercial and military telecommunications. This volume covers the broad scope of phenomena related to geomagnetic storms, from their solar and coronal origins to interplanetary manifestations, magnetospheric effects, and ionospheric and atmospheric consequences. The purpose is to summarize our current understanding of the chain of processes that occur between the Sun and Earth.

Results from these studies show that energy and momentum do not flow only in one direction. There are also extremely important flows in the opposite direction. Another exciting topic is that oxygen ions dominate the energetics of the storm time ring-current and that ion injections are associated with substorms occurring during the main phases of magnetic storms. Such ions must have their ultimate origin in the Earth's ionosphere. However, it is still not well understood by what process thermal oxygen is extracted from the ionosphere, and what further (ionospheric and magnetospheric) energization processes the thermal oxygen must undertake. This should be a prime topic for further study.

How thermal oxygen ions are energized, ascend lines of force into the magnetosphere, and become trapped is not currently known. One suggestion is heating via ULF waves, forming ion conics, but that is only part of the problem. Particle pitch angle scattering and trapping onto low-latitude orbits (~90 degree pitch angles) is almost certainly accomplished by interaction with plasma waves, but by which modes and what instabilities?

A further complication is that these results tell us that one cannot simply assume that convection, compression, and radial diffusion of the preexisting plasma sheet and quiet-time ring-current gives the full story of the formation of the storm-time ring-current. Strong and continuous coupling to the ionosphere clearly must be occurring, as it appears that the percentage of oxygen increases with increasing storm intensity (Dst).

The L value of the source of oxygen is also not well understood at this time. If, as suggested in the recent literature, substorms occur at lower and lower magnetic latitudes as the storm main phase develops, then it is possible that the ionospheric oxygen ion injection occurs over a wide variety of L shells during the storm. Thus, during the peak of the storm when the substorms are occurring at the lowest L shells, these (substorm) plasma injections would experience the least energization from magnetospheric electric fields. If this scenario is correct, this would imply that the innermost portion of the ring-current would have a strong component of low-energy oxygen ions. This speculation can easily be tested using available satellite data.

The role of the substorm electric fields versus storm-time electric field has now been discussed once again. The "stochastic" substorm fields may not only be important for the radial diffusion of ring-current particles, but they may also play a fundamental role in the pre-acceleration of ionospheric ions that eventually form the ring-current.

The papers in this volume are based on the invited lectures at a Chapman Conference on Magnetic Storms, held at the Jet Propulsion Laboratory in Pasadena, California, February 12-16, 1996. Scientists representing 20 countries attended, and, to our knowledge, this was the first conference devoted solely to the topic of geomagnetic storms. It is therefore particularly fitting that the name Chapman be associated with it, since Sydney Chapman led much of the early work on magnetic storms and substorms.

We wish to dedicate this book to Baron Alexander von Humboldt. Although famous for many other discoveries (the Humboldt current is named after him), lesser known is his discovery of "Magnetisches Ungewitter," or magnetic storms. From May 1806 until June 1807 in Berlin, he and a colleague observed the local magnetic declination every half hour from midnight to morning. The observations were made using a microscope to identify accurately the pointing direction of a magnetic needle. On December 21, 1806, for six consecutive hours he observed strong magnetic deflections, and also noted the presence of correlated northern lights (aurora) overhead. When the aurora disappeared at dawn, the magnetic perturbations did as well. This was reported in *Annales der Physik* (29, 425, 1808).

Part of the work of compiling this volume was performed at the Jet Propulsion Laboratory, California Institute of Technology, Pasadena, California, under contract with the National Aeronautics and Space Administration. We wish to thank K.-H. Glassmeier, G. Musmann, and I. Mann for their assistance in translating the von Humbolt references used here. We also wish to acknowledge L. F. Burlaga, M. Dryer, C. J. Farrugia, and R. P. Lepping for their participation in discussions relating to the correct usage of the terms CMEs, interplanetary CMEs, driver gases, and magnetic clouds. Additionally, we want to thank the following individuals for their dedicated efforts in reviewing and commenting on one or more of the submitted papers:

S.-I. Akasofu	D. C. Hamilton	R. L. McPherron
M. W. Chen	C. M. Ho	G. W. Proelss
I. A. Daglis	J. C. Joselyn	V. A. Sergeev
C. J. Farrugia	S. W. Kahler	R. M. Thorne
B. G. Fejer	L. J. Lanzerotti	R. A. Wolf
D. E. Gary	R. P. Lepping	

Bruce T. Tsurutani
California Institute of Technology

Walter D. Gonzalez
Instituto Nacional de Pesquisas Espaciais

Yohsuke Kamide
Nagoya University

John K. Arballo
California Institute of Technology

Dedication

Alexander von Humboldt was one of the last renaissance men (see: his five volume book *Cosmos*). Throughout his long expedition of South America (see: *Voyage de Humboldt et Bonpland 1799-1804*), he made regular and precise measurements of the barometric pressure, temperature and humidity of the atmosphere, and the strength, dip and inclination of the Earth's magnetic field. He studied Andean volcanoes and related seismic activity. He was the first to study the geography, geology, and climatology of South America, leading to modem geophysics. With Bonpland, he collected 60,000 biological specimens, spanning 6,000 species, 3,000 of which were new to the scientific world. For over 30 years he held the world mountain climbing record of 19,286 feet (determined by a barometer; Mount Chimborazo outside Quito). Through his scientific analyses and reports, the terms "isodynamic," "isocline," "Jurassic," and "magnetic storm" have come into common usage. From a six month whirlwind tour of Russia sponsored by Tsar Nicolas I (April-October, 1829) covering 9,700 miles, he studied the mineral deposits of Russia, leading to the discovery of diamonds. Upon his successful return, Humboldt's efforts also led to a worldwide installation of meteorological and magnetic stations, in Russia and Alaska (1829-1835). Through Gauss and the Göttingen Association (Humboldt was the originator and leader of the "Magnetische Verein"), a network was constructed in western Europe. Through Humboldt's contacts with the President of the Royal Society, he stimulated the placing of permanent stations in the outer reaches of the British empire (Canada, Africa and Australia). Herschel, Sabine and Airy led this effort. During his long lifetime (he died at age 90 in Berlin), he also served as Chamberlain to Kings Frederick III and Frederick Wilhelm IV of Prussia. For an excellent overview of his life, we recommend *Humboldt and the Cosmos* by Douglas Botting (Prestel, New York, 1994).

Magnetic Storms: Current Understanding and Outstanding Questions

Y. Kamide[1], R.L. McPherron[2], W.D. Gonzalez[3], D.C. Hamilton[4], H.S. Hudson[5], J.A. Joselyn[6], S.W. Kahler[7], L.R. Lyons[8], H. Lundstedt[9], and E. Szuszczewicz[10]

There is much evidence in recent Yohkoh and Ulysses observations that the most intense magnetic storms are a manifestation of fast magnetic clouds, perhaps originating in Coronal Mass Ejections (CMEs). However, uncertainties exist as to what magnetic configuration is formed for CMEs and how it changes over time, how CMEs interact with the interplanetary medium, and how geoeffectiveness depends on the different size of ejections. The constituents of the ring current in the magnetosphere as a function of storm time are now observationally identified. In particular, the ionospheric component shows the largest increase at $L < 4$ during the main phase of magnetic storms, indicating that the frequent occurrence of intense substorms is important. There seems to be a consensus that, of the two processes playing essential roles in enhancing the storm-time ring current, the enhanced electric field driven by southward interplanetary magnetic fields dominates the effects of the induced electric field resulting from substorm expansion onsets. This creates a new controversy regarding the relative importance of the two processes, although they are not mutually exclusive. There is a need to evaluate observations and models in kinetic, chemical, and electrodynamic coupling between the ionospheric and thermospheric magnetic storms in a more quantitative manner. For example, global, not regional, observations of the electron density and the neutral wind are required in order to understand the ionospheric/thermospheric storms within the framework of the chain of processes from the Sun to the Earth. In view of the important effects of magnetic storms on a wide variety of human-societal systems, prediction schemes continue to be upgraded.

[1] Solar-Terrestrial Environment Laboratory, Nagoya University, Toyokawa, Japan.
[2] Institute of Geophysics and Planetary Physics, University of California at Los Angeles, Los Angeles, CA, USA.
[3] Instituto Nacional de Pesquisas Espacias, Sao Jose dos Campos, Sao Paulo, Brazil.
[4] Department of Physics, University of Maryland, College Park, MD, USA.
[5] Institute for Astronomy, University of Hawaii, Honolulu, HI, USA.
[6] Space Environment Center, National Oceanic and Atmosphere Administration, Boulder, CO, USA.
[7] Phillips Laboratory, Hanscom AFB, MA, USA.
[8] Aerospace Corporation, Los Angeles, CA, USA.
[9] Solar-Terrestrial Division, IRF-STL, Lund, Sweden.
[10] Science Applications International Corporation, McLean, VA, USA.

Magnetic Storms
Geophysical Monograph 98
Copyright 1997 by the American Geophysical Union

1. INTRODUCTION

This collaborative overview has a number of goals:
1. To summarize our current understanding of the chain of processes occurring from the Sun to Earth and their association with magnetic storms.
2. To address important findings and ideas in specific subject areas.
3. To identify outstanding questions on magnetic storms in terms of solar-interplanetary-magnetosphere-ionosphere-thermosphere-ground processes.
4. To recommend the direction for future investigations.
5. To stimulate further studies integrating all aspects of geomagnetic storms.

Each section that follows is the result of extensive discussions and attempts to cover physical processes occurring in more than one region in the solar-terrestrial environment.

2. SOLAR ORIGIN OF MAGNETIC STORMS

H. S. Hudson

Geomagnetic storms occur when the solar wind with unusual characteristics, namely, strong, persistent interplanetary fields oriented southward in the Earth's frame of reference, sweeps past the Earth. These characteristics have been associated with interplanetary stream structures and/or discrete eruptive events which originate at the Sun, but which are difficult to observe. The traditional means for viewing the corona consists of coronagraphs and radio observations. Both of these techniques have their limitations: the coronagraph data show us only the limb, not the disk center, and the current radioheliographic observations in the important meter-wave spectral regions have limited capabilities. Accordingly the Yohkoh soft X-ray data have been filling an important gap in our knowledge. The success of the SOHO X-ray imaging and spectroscopy will continue and extend this capability, as will the future observations from the GOES meteorological satellites.

The X-ray observations show us quite a few things related to mass ejections. Ejections appear to occur on a wide range of spatial and temporal scales, ranging from loops rising at a few km/sec out of active regions to ejecta at speeds exceeding 1,000 km/sec. The faster ejecta are associated with flares and with X-ray jets, a phenomenon discovered for the first time with the Yohkoh data. We would like in particular to identify the X-ray counterparts of the classical Coronal Mass Ejections (CMEs), as seen in a coronagraph. Until recently this has remained a bit unclear because of the limited amount of coronagraphic observational material available (at least prior to the SOHO launch). Theoretically, too, the Yohkoh observations have a strong bias towards higher temperatures, and some portion of the CME material arising in the cooler corona or lower atmosphere could in principle remain undetected.

In spite of these uncertainties it now seems clear that Yohkoh can observe phenomena extremely closely related to the CME phenomenon. There are two major channels: first, we observe a dimming (depletion) signature in many events. In this we find that a previously bright part of the corona disappears on a relatively short time scale, almost invariably at the time of flare-like arcade brightening. The streamer-associated event reported by *Hiei et al.* [1993], for example, shows a clear dimming signature in a large volume surrounding the re-forming streamer [*Hudson*, 1996]. The material dimming in this case, and in many others, is too diffuse to allow an easy determination of the flow speed.

The second and more direct X-ray signature is simply to observe the material moving outwards. We now have many examples of this, one of which is shown in Figure 1 [*Hudson* et al., 1996a]. This event was observed not far from disk center, as are some of the dimming events, which offers us the exciting possibility of actually seeing the mass motions of a geoeffective CME directly as it occurs. Whether this capability ever turns out to be really useful in a practical sense depends upon calibrating it with observations from the L1 Lagrangian point and, of course, from the Earth.

Why do these ejections take place? There is much evidence that the main interplanetary result of a coronal mass ejection has the form of a flux rope, or something quite similar; at the solar end we also seem to require a flux-rope geometry to support the H-alpha filament that is often observed to erupt in association with the ejection. It is therefore tempting to consider the problem purely in terms of MHD theory. On the other hand, solar flares - often observed in close association with the CME launch - are characterized by gross violations of ideal MHD. We do not know at present why the flux-rope configuration erupts unstably, nor how it is related to the accompanying flare (if any). Understanding this phenomenon theoretically must be one of the top priorities in solar physics today.

The top-level problems can be summarized briefly:

1. The soft X-ray observations show large-scale ejecta probably related to CMEs. Can we use these observations to establish the geometry of the coronal evolution, and thereby to develop a predictive theory?

2. The classical theoretical scenario for flare and CME formation involves large-scale magnetic reconnection. Is the reconnection physically important in the ejection, or is it a byproduct?

3. SOLAR-INTERPLANETARY COUPLING

S. W. Kahler

To predict the largest geomagnetic storms at the Earth, we would first need to observe the timing, positions, and configurations of the eruptions of CMEs. A high initial CME speed appears to be an important factor in the resulting geoeffectiveness of an interplanetary CME (ICME) [*Gosling et al.*, 1991]. However, in a geomagnetic storm the most important factor is the strength of the southward field component associated with the ICME [*Tsurutani et al.*, 1988]. Because the resulting interplanetary magnetic clouds are low-beta plasmas, it is important to observe the magnetic fields of ICMEs. However, it is a well-known limitation that those fields cannot be observed. Coronal fields are currently extrapolated from photospheric field measurements using a potential field and source surface above which the field is radial [*Zhao and Hoeksema*, 1994].

Although the intensities of strong fields can be inferred from microwave measurements, and the topology of strong fields can be inferred from soft X-ray images, such as those from the Yohkoh SXT, these measurements fall far short of revealing the large-scale structures of CME magnetic fields extending over tens of solar degrees.

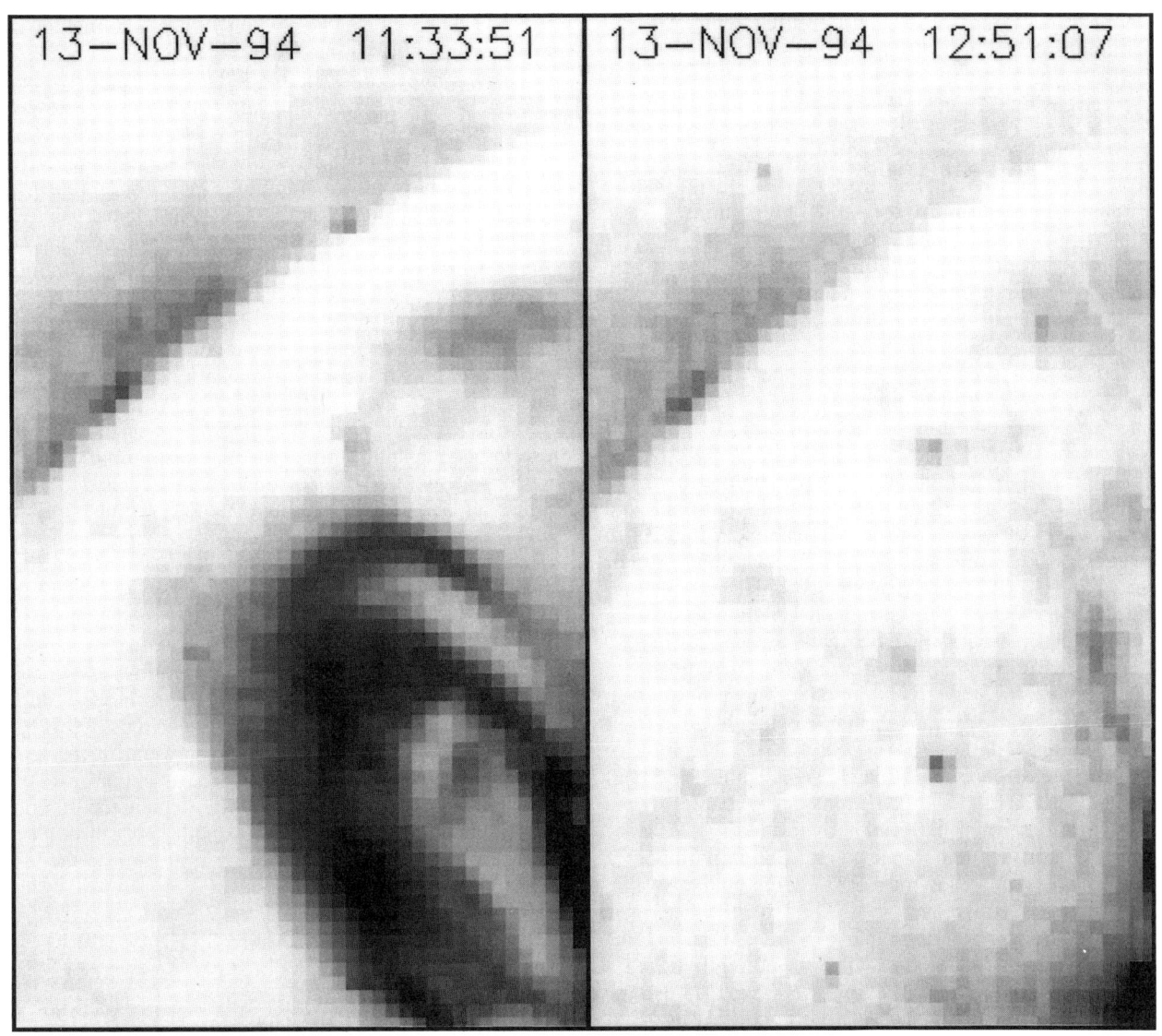

Fig. 1. Soft X-ray images of a moving coronal cloud near a C-class solar flare of Nov. 13, 1994. In the time span between the images, the cloud was observed to move outwards with a projected velocity of about 100 km/s, exactly contemporaneous with the flare brightening (to the SW of the cloud, not shown in these images). The mass was of the order of 4×10^{14} g. From *Hudson et al.* [1996a].

The best solar observations of CMEs are those of CME material made in white-light coronagraphs. These often show a three-part structure consisting of a bright leading shell surrounding a dark cavity and an embedded bright tongue. These features are identified with the bright dome, cavity and quiescent prominence, respectively, of the pre-eruptive coronal streamer [*Hundhausen*, 1988]. The coronagraph observations suffer from the limitation of being projected onto the plane of the sky, so that while the spatial extent of the CME source region is sought, even the mean solar longitude of that region is undetermined. In addition, the CMEs most important for geomagnetic storms are those arising from around disk center, and these are not usually detected in coronagraphs [*Hundhausen*, 1993]. However, several possible ways of detecting CME signatures on the disk were reported at this conference. *Hudson et al.* [1996c] have found events in which the Yohkoh X-ray images show a dimming which may reflect the mass loss due to CMEs. D. Gary reported that it may be possible to image either mass depletions or ejections in the radio range, although the background levels make these observations difficult. Optical observations of filament disappearances may also yield limited temporal and positional information of CMEs. In addition, post-flare loop systems, such as those seen in the Yohkoh observations on 14 April 1994 (Figure 2), can show the aftermath of CMEs leading to geomagnetic storms [*McAllister et al.*, 1996]. B. Jackson reviewed the several ways in which ICMEs are tracked through the interplanetary medium from remote observations. These include ground-based radio scintillation techniques, and for the fastest events, radio kilometric observations from space. White-light imaging of the interplanetary medium could also detect ICMEs directed toward the Earth [*Jackson et al.*, 1996]. If direct observations of ICMEs headed toward the Earth are considered sufficiently important, a spacecraft positioned at 60° to 90° from the Sun-Earth line could provide the necessary coronal observations [*Pizzo*, 1994].

The second requirement for storm prediction is to understand how the CME interacts with coronal and interplanetary structures. Most CMEs appear to arise in streamers, which consist of closed magnetic loops at their bases with overlying current sheets which extend into interplanetary space. The current sheets are imbedded in high-beta plasmas with slow solar wind speeds. There may be a single current sheet or multiple current sheets contained in a streamer envelope. Since most of the open magnetic fields at several solar radii in height arise from coronal holes [*Wang and Sheeley*, 1994], a temporary change in the coronal hole structure is inevitable for all but the smallest CMEs. The CME may act to excite Alfven waves in the coronal hole fields or to modify the high-speed flows from the coronal holes. The importance of these effects is an area of current study reported by W. D. Gonzalez and S. Bravo.

Even with a good understanding of the CME dynamics and interactions with the solar wind, we currently have no models of the structure of the quiet solar wind to use for calculating such interactions. There is the further problem of interplanetary wave fields, which may be important for storm effects [*Tsurutani et al.*, 1990], but cannot be predicted from a knowledge of coronal magnetic conditions.

A further problem is to deduce the three-dimensional structure of CMEs in the interplanetary medium. A few ICMEs appear to have been detected at high solar latitudes by Ulysses. One of these, observed on 27 February 1994 at a distance of 3.5 AU and a heliolatitude of S54°, may have provided a conduit for energetic particles from a solar event on 27 February [*Pick et al.*, 1995]. Thus even 7 days after formation, the ICME may still have been magnetically attached to the Sun near the active region. A geomagnetic storm beginning on 21 February suggests that the same ICME also propagated in the ecliptic plane [*Hudson et al.*, 1996b]. While a flux tube is a popular model for ICMEs, that event suggests a very broad latitudinal extent for the ICME, perhaps more like a pancake than a tube. The challenge then is to include this large angular extent in the models of ICMEs.

The reconnection of magnetic flux in CMEs is a basic element of the *Kopp and Pneuman* [1976] model for post-flare loops commonly seen in the Yohkoh X-ray observations. Coronal candidates for large-scale (> 10°) disconnection events are numerous [*McComas*, 1995], but interplanetary evidence for their existence, usually taken to be dropouts in the solar wind heat flux, is minimal at best [*Lin and Kahler*, 1992]. Helios observations of continued mass outflow for hours after the CME also argue against this possibility [*Webb et al.*, 1996]. Thus, the basic parameters and topology of magnetic reconnection in CMEs are unresolved and presents a problem for future study.

We have been concerned with the large-scale CMEs that lead to geomagnetic storms, but T. Kosugi reported at the conference that transient coronal X-ray events occur on a variety of size scales, ranging over jets, impulsive flares, long-decay flares, and giant arcades [*Hudson*, this issue]. While a plot of the event frequency versus event size might show a decreasing power-law distribution similar to that for solar flare radio and X-ray peak flux events, the white-light CME mass observations show a peak in the frequency distribution with a clear decrease toward low-mass events that is not an instrumental effect [*Hundhausen et al.*, 1994]. This suggests that most small ejections observed in X-rays in the lower corona are not present in the upper corona. Why the size scales of CMEs should preclude small mass injections remains to be explained. It would also be of interest to follow the small ejecta through the corona to determine whether they continue into the solar wind.

4. INTERPLANETARY-MAGNETOSPHERIC COUPLING

W. D. Gonzalez

Concerning the overall power input to the magnetosphere, as reviewed by V. Vasyliunas, one approximation to it would be to compute the magnetospheric dynamo power, electric field times the current, where the electric field is that governed by magnetopause reconnection and the current refers to the total tail current density. At the moment, there are models to compute this power input using only large-scale steady reconnection [e.g., *Gonzalez et al.*, 1994]. However, it is desirable to extend those models trying to incorporate localized and time-dependent reconnection features, such as those related to the so called "flux transfer events." Toward this goal, future magnetopause observations with improved time and space resolutions will be of great importance.

We have learned in this conference about the crucial importance of understanding better the interplanetary

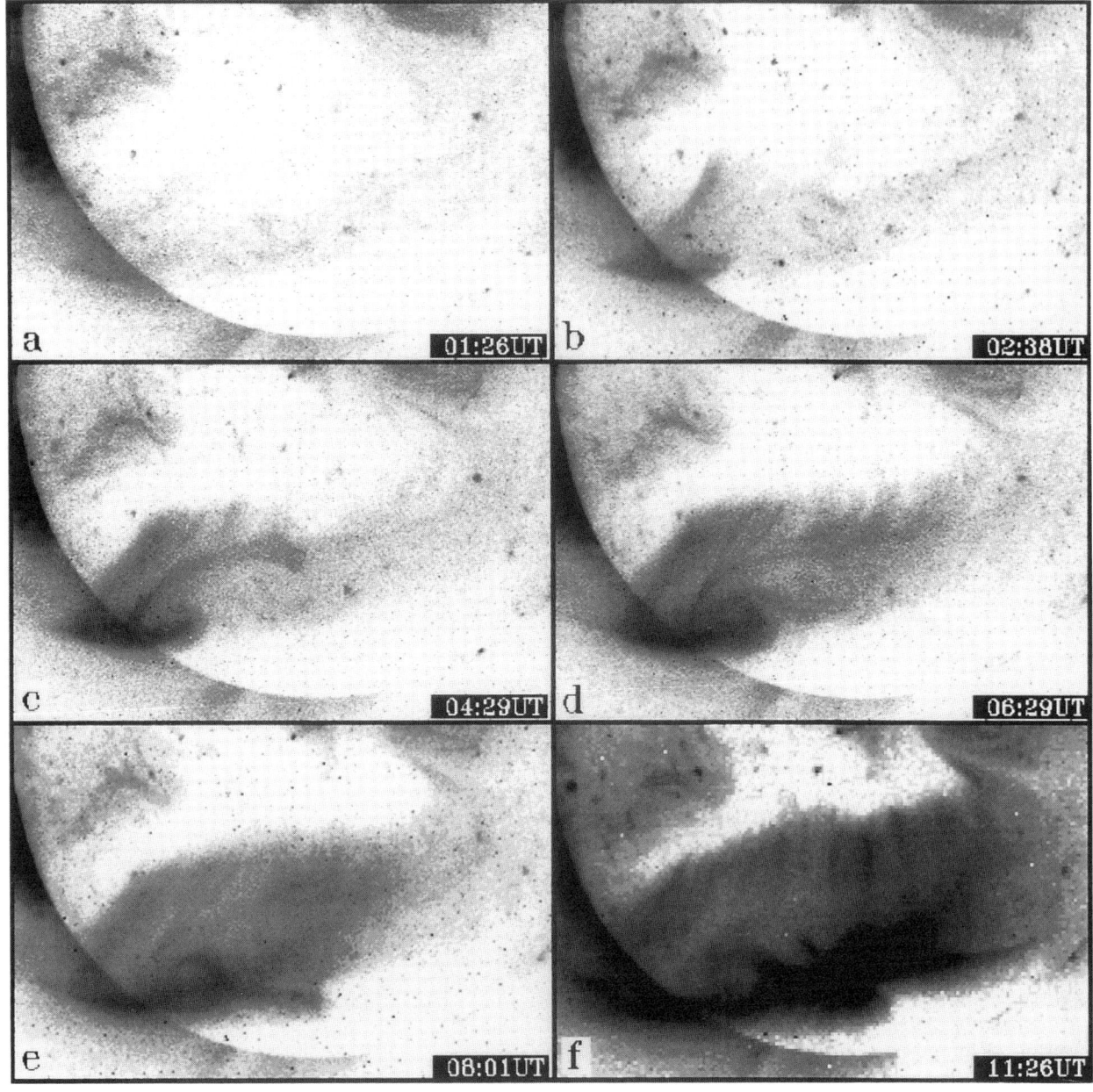

Fig. 2. Yohkoh SXT images showing the development of a large coronal arcade over the southern polar crown on April 14, 1994. A geomagnetic storm followed on April 16-17. From *McAllister et al.* [1996].

structures that involve large-amplitude and long-duration negative B_z fields, in order to be able to follow the development of intense geomagnetic storms. B. T. Tsurutani and W. Gonzalez and C. J. Farrugia reviewed our present knowledge of such structures, grouped in "sheath" and "driver gas" field types. The first type involves compressed (shocked) and draped fields, whereas the second one refers mainly to the ejected field itself such as in a magnetic cloud. Intense B_s (negative B_z) fields can be found in each of the two types of structures and when one finds them simultaneously in both regions the resultant magnetic storm tends to be of a larger intensity and usually composed of a two-stage main phase [see *Tsurutani et al.*, 1988, 1992; *Gonzalez et al.*, 1989]. However, in order to distinguish carefully the two B_s structures, it is important to realize the geometry of the given situation. The satellite at the inner Lagrangian region of the Sun-Earth system, such as WIND, can cross the shocked plasma and miss the driver gas when the latitudinal and/or longitudinal position of the solar ejecta is sufficiently far from the Sun-satellite axis. In such a case a B_s structure of the "sheath" field type can be monitored, but this does not mean that another B_s structure of the "driver gas" type did not exist. This is particularly important also when one is trying to find a solar source for such an ejecta. Nevertheless, one also expects that the Mach number of the observed shock, for crossings far from the center of the ejecta, should not be as high as it is typically found for central crossings. Therefore, if one finds a situation that the Mach number is fairly high (like 3) and a "driver gas" has not been observed, this could also mean that the monitored structure is not a regular ejecta (ICME). Among other alternative structures, it could well be a high speed stream emanating from non-recurrent coronal holes [e.g., *Gonzalez et al.*, 1996].

Ongoing computer simulation studies are helping us to understand more about these interplanetary structures. For instance, D. Odstrcil presented the results of a three-dimensional, time-dependent MHD model of the interaction between an interplanetary plasma cloud and the heliospheric current sheet, through which substantial B_s fields can be generated.

5. STORM DYNAMICS/RING CURRENT

D. C. Hamilton

The global magnetic fluctuations in a geomagnetic storm are caused by a major intensification of the ring current, by a factor of three to twenty or even more in great storms. The ring current is largely carried by ions in the 20-200 keV energy range. The major questions concerning the dynamics of the ring current during magnetic storms are the origin of the ring current ions, their energization and transport into the ring current region ($L = 2-7$), and their loss during the storm recovery phase. Progress in answering all these questions has been made in recent years on both the observational and modeling fronts. Many of these advances were reported at this conference.

At one time it was thought that a simple inward displacement of the pre-existing ion population, with its concomitant energization, might be sufficient to explain magnetic storms. However, composition measurements from the AMPTE/CCE and CRRES spacecraft in recent years have shown that storms involve major injections of new plasma into the ring current. That plasma can come from the solar wind or the ionosphere, but the ionospheric component shows the largest increase during storms, particularly for very large storms and during the maximum phase of the solar cycle. I. A. Daglis reported data from AMPTE/CCE and CRRES verifying the ionospheric feeding of O^+ ions to the ring current early in storms and the larger abundance of O^+ compared to H^+ in very large storms. D. C. Hamilton and M. E. Greenspan reported that the quiet ring current is more than 95% protons. In magnetic storms the O^+ fraction increased dramatically and in moderate storms (~100 nT) varied from 20% near solar minimum to 44% near solar maximum. M. Grande reported CRRES ring current measurements of tracer ions of solar wind origin that indicate the rapid entry of solar wind ions into the ring current. A population of iron ions in the ring current with mean charge state +10 was replaced within 90 minutes by a population with mean charge state of +20, presumably reflecting the changing composition of the incident solar wind.

A. Korth and R. Friedel presented synoptic particle data from the CRRES mission showing the spatial reconfiguration of the ring current during storms. The degree of inward displacement of the peak of the ring current energy density is closely related to the size of the storm. The peak moves in to $L = 3.5$ in moderate storms ($Dst \sim -120$ nT) and in to $L = 2.5$ in great storms ($Dst \sim -300$ nT). In great storms, the slot region of the radiation belts is filled, and in fact a new belt of energetic protons can be formed at low L. The March 1991 storm produced such a new belt whose formation was very successfully modeled by M. K. Hudson et al. J. R. Wygant reported electric field measurements from CRRES that indicated the dawn-dusk electric field can penetrate to $L = 2$ or less during large storms and that these electric fields can play an important role in the energization and inward displacement of the ring current particles.

In general the modeling of magnetic storms as reviewed by M. Chen is becoming more and more realistic and agreement with observations is improving. The main piece still missing is the mechanism by which the injection of ionospheric ions (O^+) into the ring current is enhanced in storms, particularly large storms. These ions are apparently injected into the outer magnetosphere/inner plasma sheet ($L > 7$) at modest energies (< 20 keV) rather than being directly injected into the ring current at low L-shells ($L = 2-7$). In a storm main phase this low energy population, consisting of a mixture of ionospheric and solar wind ions, is transported inward and energized to form the storm-time ring current. The enhanced and variable convection electric fields that produce this transport are now reasonably well-modeled, and it appears that we are not missing any major theoretical components. However, inclusion of all known processes in the models is difficult and more detailed comparisons with observations are still required.

In one such comparison, L. M. Kistler reported AMPTE/CCE measurements of the pitch angle distributions of H^+ and O^+ ions at $L = 3-6$ during the main phase of several storms. She found a large variety of distributions that on the whole were well-explained by a drift model containing charge exchange losses. She found that differences in drift paths due to the non-dipolar aspects of the magnetic field were important. Current comprehensive storm models include charge exchange and energy loss due to Coulomb collisions, but in general do not include losses due to scattering by plasma waves. In a paper by Kozyra et al. (presented by R. M. Thorne), it was suggested that there could be times during a storm when this mechanism plays an important role in the loss rate in some regions of the ring current. More study is required in this area.

The Dessler-Parker-Sckopke (D-P-S) relation predicts a direct proportionality between the kinetic energy content of the ring current and the perturbation magnetic field at the center of the Earth. The standard proxy for that perturbation field is the *Dst* index. W. Campbell, however, provocatively questioned whether the *Dst* index was a reliable indicator of ring current strength. He argued that other ionospheric and Earth currents present during magnetic storms contribute to *Dst*. When observational tests of the D-P-S relation are made, generally an estimate of the ring current energy content is made from a spacecraft observation, the perturbation field predicted by D-P-S to arise from that ring current is calculated, and that calculated field is compared to the observed *Dst*. This should produce a poor comparison if there are major contributions to *Dst* from currents other than the ring current [see *Carovillano and Siscoe*, 1973]. M. E. Greenspan and D. C. Hamilton presented a survey of ~70 magnetic storms from AMPTE/CCE in which the measured ring current energy (deduced from observations at one local time) was compared with the *Dst* index at the same universal time. Although scatter was large, on the average the measured ring current energy was only modestly less than required by the D-P-S relation. Another result from this large study was a local time variation that illustrated on a statistical basis the expected asymmetry of the ring current near storm maximum. Measurements of the ring current particle population (including those of Greenspan) have consistently found 50%-100% of the energy predicted from *Dst* and the D-P-S relation, at least near storm maximum. This appears to put the contributions of other currents to *Dst* in the minority category.

6. RING CURRENT FORMATION PROCESSES

L. R. Lyons

A fundamental aspect of the physics of magnetic storms is the injection of the trapped, ring-current particles that are responsible for the depression in *Dst*. Papers presented at the Chapman Conference contributed significantly to our understanding of this important topic and to the identification of critical outstanding questions. In this summary, the ring-current injection problem is divided into three parts: what particles cause the depression in *Dst*, what processes cause the injection of these particles, and what processes determine the strength of a particular storm?

It is known that ions in the energy range ~ 20 - 200 keV contribute most of the energy density to the storm-time ring current. Also, as shown by D. C. Hamilton in his presentation, it is known that the majority of the enhancement in ring-current energy during the main phase of storms occurs at $L < 4$ [e.g., *Smith and Hoffman*, 1973; *Lui et al.*, 1987; *Hamilton et al.*, 1988]. This implies that it is the injections of ~20 - 200 keV ions to $L < 4$ that needs to be explained in order to understand storm-time ring current formation. Consistent with this, particle injections at $L > 4$ are about the same in intensity during storms as during non-storm-time substorms, implying that it is the injections to $L < 4$ that are unique to storms [*Lyons and Williams*, 1975].

Two processes have been considered as being the fundamental cause of the main phase injection of ring current particles: the induced electric fields associated with the magnetic field dipolarizations that occur during the expansion phase of substorms and the enhanced convection electric fields that are driven by a strongly negative z-component, B_z, of the interplanetary magnetic field (IMF). The main phase of magnetic storms has been described as consisting of a sequence of intense substorms [e.g., *Chapman*, 1962; *Kamide*, 1979]. This suggests that the storm-time ring current is the result of a succession of the substorm particle injections [*Akasofu*, 1968] which are associated with dipolarizations of the magnetic field. On the other hand, as discussed at this meeting by B. T. Tsurutani, it has been found that there is a one-to-one relation between large ($Dst < -100$ nT) storms and large, negative IMF B_z's (< -10 nT) that last for at least three hours [*Gonzalez and Tsurutani*, 1987]. This relation suggests that enhanced convection is responsible for storms, since large, negative IMF B_z's lead to strongly enhanced magnetospheric convection.

The effectiveness of substorm-associated magnetic field dipolarizations in injecting ring current particles to the $L < 4$ region can be estimated with a very simple calculation. Assume a dipole magnetic field, and represent a dipolarization by a change in equatorial B_z given by $\Delta B_z = 50$ nT everywhere. Assuming the radial component of the induced electric field $E_r = 0$, we obtain an induced electric field in the equatorial plane that is in the azimuthal direction and is given by $E_\phi = -(LR_E/2)(\partial B_z/\partial t)$. (This result also holds if ΔB_z occurs over only a slice in MLT, provided $E_r = 0$).

This E_ϕ causes an inward radial displacement of particles given by $\Delta(L^{-3}) = 4.8 \times 10^4 \Delta B_z$, or $\Delta(L^{-3} = 2.4 \times 10^{-3})$ or $\Delta B_z = 50$ nT, which gives an inward displacement ΔL that decrease rapidly with decreasing L. For example, $\Delta L = 3.2$ for particles injected to synchronous orbit, but ΔL

= 0.2 for particles injected to $L = 4$. We thus find that the substorm dipolarizations can readily inject particles from much further out to near synchronous orbit, where substorm injections are often observed. However, the dipolarizations cannot displace particles significant distances below $L = 4$. This suggests that substorms are not fundamental to the injection of ring current particles to $L < 4$.

A number of presenters at the meeting addressed the impact of substorms on the ring current injection process, and they found a common result in agreement with the simple calculation given above. R. L. McPherron, G. Rostoker, E. Friedrich, and V. A. Sergeev compared *Dst* traces to times of well-defined substorm expansion onsets, and they all found no increase in the downward slope of *Dst* in association with the expansion phase. In addition, R. L. McPherron found that *Dst* starts down almost simultaneously with southward turnings of IMF, without the time delay necessary for a substorm growth phase. He also found no correlation between residuals in the *AL*-solar wind correlation and the *Dst*-solar wind correlation, in contradiction to what he expected if a process not directly driven by the solar wind electric field affected both *AL* and *Dst*. R. A. Wolf addressed the effects of substorm dipolarizations using his Magnetospheric Specification Model, and found that dipolarizations have little effect on the development of a symmetric ring current within the model.

Based on the independently obtained simple calculation, model results, and observational results, it can be concluded that a major new understanding has emerged at this meeting: Substorm expansions do not cause the particle injections responsible for the majority of the storm-time ring current!

If substorm expansions are not responsible for storm-time ring current injections, we must consider injection by enhanced convection as suggested by the association between the IMF and storms. The effects of enhanced convection are illustrated in Figure 3 (courtesy of M. W. Chen), which shows the equatorial separatrix between open and closed drift trajectories for various values of the cross-polar-cap potential drop $\Delta\Phi$. Plasma sheet particles only have access to regions beyond this separatrix, whereas particle trajectories encircle the Earth within the separatrix.

It can be seen that for quiescent conditions (represented by the $\Delta\Phi = 30$ kV separatrix in Figure 3) inward convection of particles from the plasma can only populate regions beyond about $L = 5$. However, unlike substorm dipolarizations, strongly enhanced convection (represented by the $\Delta\Phi = 150$ kV and 230 kV separatricies in Figure 3) can bring particles well within $L = 4$, where particles can be transferred to closed trajectories following a period of enhanced $\Delta\Phi$.

For enhanced convection to form the storm-time ring current, it is necessary that the enhanced electric fields penetrate to $L < 4$, and J. R. Wygant reported CRRES

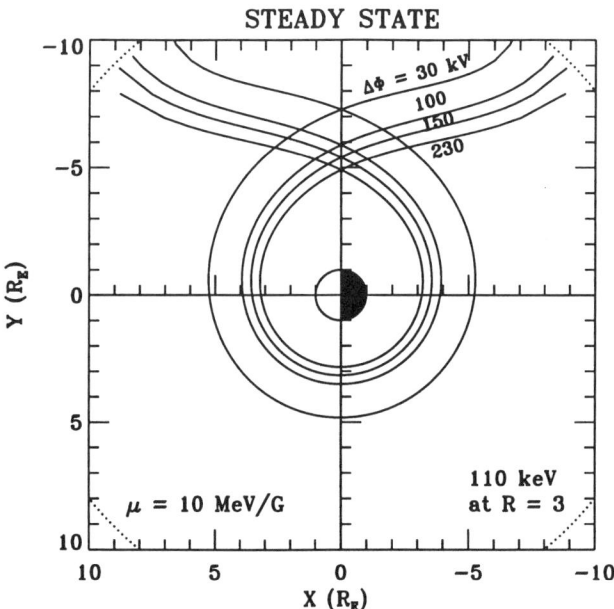

Fig. 3. Illustration of equatorial separatricies between open and closed drift trajectories for singly-charged ions having a magnetic moment $m = 10$ MeV/G for different values of the total cross-tail potential drop $\Delta\Phi$. The trajectories are traced in a model magnetic field consisting of a uniform southward field added to the geomagnetic dipole field. The electric field includes the effect of coronation, shielded quiescent convection, and unshielded storm-time convection. The total cross-tail potential drop $\Delta\Phi$ includes a quiescent contribution of 30 kV.

satellite observations showing that the required electric fields indeed do occur in the $L = 2 - 4$ region at times of ring current particle injection. M. W. Chen reported calculations of the distribution of ring current particles versus energy and L that can be expected from enhanced convection, and found that such convection can account for the storm-time ring current particles observed at $L < 4$ [*Chen et al.*, 1994]. It thus seems plausible that enhanced convection is responsible for the ring current particle injections.

In addition to convection, what processes determine the intensity of storms? Motion of particles across L-shells can be increased by fluctuations of an enhanced convection electric field, leading to random changes in the radial position of particles that resembles radial diffusion [*Lyons and Schulz*, 1989]. This suggests that an electric field that fluctuates during a storm main phase may lead to a larger particle injection than a steady electric field of the same average magnitude. A subsidiary question is whether having a pre-existing ring current leads to a larger |*Dst*| than staring from non-storm-time conditions. Plots of *Dst* during large storms often show evidence for two or more periods of significant ring current injection, suggesting that multi-pulsed injections may lead to an enhanced *Dst* (see also Section 4), and multiple ring current injections to $L < 4$ have been seen during very large (< -300 nT) storms [*Hamilton et al.*, 1988; *Roeder et al.*, 1996].

Two papers addressed this point at the conference. R. A. Wolf compared ring current formation in the Magnetospheric Specification Model for a storm with steady enhanced convection to a storm of the same total duration but having two 3-hr pulses of enhanced convection, the average magnitude of the electric fields over the time period of the enhancements being the same in both storms. He found that a larger ring current developed during the pulsed storm than during the storm with steady, enhanced convection.

M. W. Chen compared ring current formation in model storms having steady convection to that in model storms having fluctuating enhanced fields of the same average magnitude. She found that enhanced steady convection gives ring current injection over a three hour period, but that longer periods of enhanced steady convection do not give further ring current increases. On the other hand, fluctuations were found to have little effect for a 3-hr storm but give an increasing ring current with increasing time for longer storms. After a 12-hour period of enhanced fluctuating electric fields, the total energy of the ring current was nearly a factor of two larger than after a 3 hr storm. These new results of M. W. Chen and R. A. Wolf imply that the temporal structure of the enhanced electric fields during the main phase of a storm can significantly affect the magnitude of the storm, and they also suggest that a pre-existing ring current may enhance the magnitude of a storm.

M. W. Chen found that to fully account for the ring current energy enhancement during storms, it is necessary to have an enhancement in the source population in the plasma sheet that is convected inward to form the ring current [*Chen et al.*, 1994]. She also identified observational support [*Hamilton et al.*, 1988] for such an enhancement. This suggests that processes that contribute to the enhancement of the source ion population should affect the strength of storms, and a number of possible contributors to this enhancement were described during the conference.

V. M. Vasyliunas mentioned in a comment that enhanced energization by large, storm-time electric fields will occur as particles convect inwards within the plasma sheet, leading to an enhancement in the source population. In support of this idea, enhanced magnetic fields in the lobe during the main phase of storms were reported by S. Kokubun and by C. M. Ho. This implies that the plasma energy density with the tail plasma sheet is enhanced during the main phase. J. E. Borovsky mentioned that the densities and mean energies of solar wind ions are enhanced following interplanetary disturbances. This suggests that the solar wind plasma which feeds the plasma sheet can have enhanced pressures during storms, leading to enhanced plasma sheet pressures. W. Baumjohann reported that large magnetic dipolarization events occur at $r \sim 10 - 20\ R_E$ in the tail during storms. He suggested that these events may be related to bursty convection events but are of larger scale and that they may give enhanced energization and inward transport during storms.

The effects of storm-time substorms was mentioned by several presenters. I. A. Daglis reported that plasma sheet ions often have an unusually high concentration of oxygen during storms. This implies that the substorm-associated, ionospheric plasma injected into the plasma sheet may be enhanced during storms, and that this plasma can be an important contributor to the plasma that is injected into the ring current. If this were the case, it would imply that substorms can have a significant effect on the strength of storms by providing an enhanced source population, even if substorms do not cause the particle injections into $L < 4$. On the other hand, M. Grande noted in his presentation that enhancement of ionospheric plasma is a fundamental aspect of substorms, and may not be unique to storms.

G. Rostoker took another approach and suggested that substorms may facilitate the penetration of enhanced convection to lower L, and thus may increase the ring current injection by the convection electric field. Y. I. Feldstein addressed the issue of the location of substorms during storms, and noted that observations show that they can occur at unusually low latitudes. The effects of such low-latitude substorms, however, were not addressed during the conference.

While our opinions on ring current formation during storms may change in the future, what now seems to be true is that:

1. Ring-current particle injection to $L < 4$ is primarily responsible for the main-phase decrease in *Dst*.
2. The ring-current particle injection to $L < 4$ occurs via an enhanced convection electric field.
3. Fluctuations in convection can enhance the storm-time ring current.
4. Substorm expansion phase injections do not cause significant direct enhancements of the storm-time ring current, but may affect the source plasma that is injected in to form the ring current.
5. Processes that enhance the plasma-sheet "source" population enhance the storm-time ring current.

The main outstanding questions related to the ring current formation problem seem at this time to be

1. What is the physics of the formation of the large storm-time electric fields?
2. What is the relative importance of the various processes that may enhance the plasma-sheet source population for the ring current?
3. Does a pre-existing ring current give larger |*Dst*|?
4. What and where are the sources for ionospheric O^+ ions and their subsequent trapping and acceleration to ring current energies?
5. What are the characteristics and effects of the low-latitude substorms that occur during storms?

6. What are the contributions of wave-particle interactions, addressed in J. U. Kozyra's paper that was presented by R. M. Thorne, to ring current particle losses during and following a storm main phase?

7. STORM-SUBSTORM RELATIONSHIPS

R. L. McPherron

One of the most significant issues discussed at this conference was the relation of substorms to magnetic storms. Many researchers believe that the relation is intimate with substorms being the mechanism that injects particles into the ring current. In fact, the name "substorm" was coined by *Chapman* [1962] because he believed that each substorm was a miniature storm and that a magnetic storm was no more than a superposition of substorms. The discovery that substorms energize and inject particles at synchronous orbits [*DeForest and McIlwain*, 1971] led to the idea that substorms are the cause of storms. *Mauk* [1986] subsequently developed a theory to explain the energization/injection through the dipolarization of the tail field during the substorm expansion phase. This theory and others like it suggest that it is the induced electrical field created by the rapid change in field during the substorm expansion that is the cause of ring current injection. Some researchers go so far as to claim that the main phase is a substorm, hence the belief that substorms are the cause of magnetic storms.

There is considerable evidence to suggest that the idea that substorm expansions are responsible for ring current injection is wrong. A simple way to demonstrate this is to plot high-resolution *Dst* and *AL* indices on the same graph. It is found that the *Dst* index usually begins to decrease at the beginning of a substorm growth phase, not after the first expansion phase onset. This result has been demonstrated statistically in three ways. First, *Burton et al.* [1975] and *Fay et al.* [1986] found that the prediction filter that relates the rectified solar wind electric field to *Dst* injection is a Gaussian pulse centered at 20 minutes delay. This implies that injection follows a change in the dayside reconnection rate by only 20 minutes. This time constant is far too short compared to the average 55 minutes duration of a substorm growth phase to be related to a substorm expansion phase. Second, in a paper just accepted for publication, *Iyemori and Rao* [1996] show that a superposed epoch analysis of the *Dst* index relative to substorm onset does not appear to support this idea. They find that substorm onsets during the storm main phase appear to decrease the rate of decrease of *Dst*, and during the recovery phase actually increase *Dst* (less negative). Third, R. L. McPherron (this conference) demonstrated that more than 85% of the variance in hourly *Dst* can be accounted for by the solar wind alone. Ring current injection appears to be directly driven by the solar wind without the need for substorm expansions. In fact, it is easy to show that if the solar wind control of both *AL* and *Dst* is removed from the two time series there is virtually no correlation between the residuals.

These results all seem to indicate that the substorm expansion phase does not inject the particles responsible for the *Dst* perturbation. This raises an interesting question about the measure of the *Dst* index. *Rostoker et al.* [1997] point out that there are possible "errors" in the *Dst* index that might be systematically related to the expansion phase onset. If these effects are positive and sufficiently large, and of sufficient duration, they may bias the index so that expansion phase effects are hidden. This possibility suggests that the *Iyemori and Rao* [1996] result must be used cautiously until the importance of such errors is known. However, as we discuss next, there are theoretical reasons for believing that substorm expansion onset is not important in increasing the *Dst* magnitude.

There appears to be theoretical justification for the results summarized above. *Siscoe and Petschek* [1996] argue that the results of *Iyemori and Rao* [1996] are expected. They appeal to the generalized virial theorem which states that if no energy flows across the boundary of a volume of space, then

$$\frac{\Delta B}{B_0} = \frac{2K + M}{3 U_M}$$

where ΔB is the magnitude of the magnetic perturbation at the Earth's center $(-Dst)$, B_0 is the equatorial field of the Earth, K is the total thermal energy of particles in the plasma sheet and ring current, M is the magnetic energy of the magnetopause current, ring current, and tail current within the volume, and U_M is the energy in the Earth's dipole field above the Earth's surface. Evidently, if all the magnetic energy is converted to particle energy, the factor of 2 guarantees that *Dst* will decrease. However, if more than half the magnetic energy goes into the ionosphere as Joule heating and particle precipitation, then *Dst* should decrease as observed. According to *Weiss et al.* [1992], the observed partition ratio is about 3:1, that is three times as much energy dissipated as Joule heating and precipitation compared to the increase in ring current energy. Similar values of the partition ratio have been obtained in theoretical estimates and in numerical simulations by *Harel et al.* [1981]. Although this ratio is not precisely known, if it is larger than 2, the change in *Dst* will be positive at expansion onset. *Siscoe and Petschek* [1996] provide theoretical arguments based on a lumped circuit theory of magnetospheric currents that shows that this ratio must exceed 2. Thus, substorm expansions do not enhance the ring current, they deplete it! Only if the term substorm is used in the more general sense to include the growth phase can it be said that substorms cause storms.

What then is the cause of a magnetic storm? It appears likely that it is fluctuations in the global convection electric field as reviewed by M. W. Chen (this conference). *Harel et al.* [1981] used the Rice Convection Model to show it is possible to account for changes in the *Dst* index simply by drift of particles convected sunward from the tail and trapped into circular drift paths by fluctuations in the electric field. No collapse of the tail field is needed to explain the observed particle energies. Particles deposited by earlier substorms are further energized and radially diffused if on the night side, as new particles are brought in from the tail.

As persuasive as the above may seem there are some aspects of storm morphology that appear to require a substorm expansion phase. For example, more than half of the ring current energy is sometimes contained in ionospheric O^+ ions with energy of order 100 keV. The only known mechanisms for injecting these ions into the ring current are field-aligned potential drops above the discrete aurora of the expansion phase and perpendicular heating (that leads to ion conic distributions). On injection these particles have only a few keV of energy and some mechanisms like those of *Mauk and McIlwain* [1974] and of *Thorne and Horne* [1994] is required to energize them to 100 keV. Furthermore, quantitative modeling by *Chen et al.* [1994] requires large and rapid electric field fluctuations to achieve energies sufficient for the ring current.

These results suggest that our understanding of magnetic storms is incomplete. Clearly we do not yet understand the role of substorms in storm dynamics. Are the inductive electric fields of the expansion phase required for ring current injection and energization? How are ionospheric ions injected (during quiet convection or during the expansion phase), and how are they accelerated to the observed energies so quickly? Perhaps even more fundamental, is the proxy we use for the ring current, the *Dst* index, good enough to distinguish the various processes? The hourly index from four stations is obviously inadequate if injection occurs on a 20 minute time scale. The various corrections include the secular variation, the quiet day variation, the magnetopause current, and the rate of ring current decay. The tail current is assumed to be insignificant, while the partial ring current and the substorm current wedge are assumed to average to zero in the calculation of the *Dst* index. Can changes in one of these current systems account for the results of superposed epoch analysis discussed above? Although the *Dst* index seems to be highly predictable with linear models, there are still difficulties in trying to forecast the development of magnetic storms.

The primary difficulty is determining the functional dependence of the ring current decay parameter on the state of the magnetosphere. This parameter is generally taken as a constant or a simple function of the solar wind coupling function or the *Dst* index itself. Since it represents charge exchange and other processes such as wave-particle interactions, it probably depends on the magnetospheric state in a complex way. In obtaining the solar wind coupling functions to predict the ring current injection rate, high resolution measurements must be used since these functions are very nonlinear. However, there is no good model that allows one to propagate measurements from a monitor far upstream to the Earth with high reliability. On a longer term, there are no good techniques for remote sensing the properties of the solar wind before it arrives at an upstream monitor, let alone predict what structures on the Sun will create geoeffective structures in space.

8. IONOSPHERIC-THERMOSPHERIC STORMS

E. Szuszczewicz

The "ionospheric storm" has generally come to be known as the response of the ionosphere to energetic inputs from magnetospheric storm or substorm events triggered by the dynamics of the solar wind and the IMF at 1 AU. The ionospheric storm response most often studied has been the dramatic variation of the electron densities at the F2-region peak, with those variations often involving 100% enhancements or reductions in what has been defined as positive and negative storm phases, respectively. Underlying these responses is a complicated chain of physical processes that involve kinetic, chemical, and electrodynamic coupling between the ionosphere and thermosphere. This coupling takes place not just at altitudes in and around the F2 peak but at altitudes extending down through the E-region where conductivities can undergo equally dramatic variations and dynamo fields are driven by perturbed thermospheric winds interacting with the ambient plasma distributions. The temporal and spatial variations in the roles of all the forcing functions have often resulted in surprisingly different behaviors of the ionosphere at locations separated by only a few tens of degrees in latitude or longitude. In fact, in 1987 this apparent scientific dilemma was summarized by *Wrenn et al.* [1987] who noted that, "Fifty years since its discovery (i.e., the discovery of positive and negative storm characteristics) and several hundred published papers later, there is still no complete explanation of the effect."

Since the work of *Wrenn et al.* in 1987 there have been considerable advances in our understanding, with the advances primarily brought about by insights developed through an improved database and interpretations of large-scale numerical simulations of the ionospheric-thermospheric system under storm-time conditions [*Codrescu et al.*, 1992; *Proelss*, 1993; *Fuller-Rowell and Codrescu*, 1994; *Yeh et al.*, 1994]. In the papers presented at the conference there was general agreement that numerical simulations of ionospheric responses to storms have provided a better understanding of the dynamics of the ionospheric-thermospheric domain and have permitted the scientific community to identify the likely processes responsible for the observations. Lack of uniqueness in the numerical simulations and associated uncertainties in energetic inputs and in the relative and quantitative roles of contributing forces placed an emphasis on using the qualifier "likely" in model identification of the cause-effect processes, since model results are seldom in convincing quantitative agreement with the observations. It was also agreed that a true test of model integrity and our understanding of coupling terms and cause-effect relationships can only be gathered through the comparison of a global-scale simulation with a global-scale database. Only then would a truly self-consistent solution be rigorously enforced on the simulation and only then would an accurate quantitative understanding evolve with regard to the influences of storm onset time, intensity and the temporal and spatial distributions of ionospheric responses to storm-time conditions.

Future needs were well articulated at the conference and many of the contributions represented important beginnings to a more comprehensive and quantitative empirical and theoretical understanding of ionospheric storms. Those needs and relevant conference presentations are delineated below:

1. Experimental efforts need to transition from site-specific and regional studies to more global investigations. This will impose self-consistency on our understanding and on the output products of global-scale models. In this way, we can definitively and uniquely unfold UT/LT effects along with the consequences of storm onset times, intensities, and duration. The conference presentations of S. Y. Ma et al., E. Szuszczewicz et al., and U. J. Lindqwister et al. are pursuing this approach. Ma et al. and Szuszczewicz et al. are employing global networks of ionosondes which readily provide long-duration, around-the-clock observations of the heights and densities of the F-region peak, while Lindqwister et al. is providing complementary measurements of total electron content with a worldwide network of ground-based GPS receivers. Such long-term around-the-clock observations make it possible to globally monitor and quantitatively define the entire ionospheric storm process from onset through recovery.

2. Models should continue to be upgraded in an effort to accurately and self-consistently include all forcing functions; they need to be tested and validated against global-scale databases, and there should be concerted efforts involving model-model comparisons. Such efforts are underway in whole or in part:

(a) The paper of Emery et al. described modeling of ionospheric storm characteristics using the TIGCM (Thermosphere-Ionosphere General Circulation Model) with time-dependent empirical specification of high-latitude inputs from AMIE (Assimilative Mapping of Ionospheric Electrodynamics); and the model results for F-region densities, heights and meridional winds were tested against a broad distribution of observations covering the model-predicted parameters. Absent from the model were self-consistent representations of plasmaspheric fluxes, ionospheric electric fields, and thermospheric coupling to the mesosphere. Several improvements in the model are underway in the form of the TIME-GCM (Thermospheric-Ionospheric-Mesospheric Electrodynamics - General Circulation Model, R. G. Roble and A. D. Richmond) which include coupling to the mesosphere and a self-consistent methodology for the calculation of ionospheric electric fields.

(b) Supplementary modeling work was also reported by T. Fuller-Rowell and M. V. Codrescu with emphasis on the theoretical explanation of a number of aspects of apparent randomness in storm observations. While their model (CTIM, Coupled Thermosphere Ionosphere Model) is not as complete as the TIGCM with AMIE inputs nor as the TIME-GCM, it does provide an empirical high-latitude energy input specification through the NOAA-TIROS measurements and derived values for auroral input power. The model has also been extensively exercised in parameter studies of UT/LT and storm-time onset effects to help establish the likely consequences of their ultimate controls over the evolution of the ionospheric response.

3. In parallel with upgrades in modeling efforts there must be an expansion in the database which defines the prevailing conditions and cause-effect relationships. While (for example) ionosondes can provide a global specification of F-region heights and densities along with meridional winds at mid-latitudes, there is an ever-increasing need for empirical specification of electric fields, thermospheric winds at all latitudes, and thermospheric composition. In the case of deficiencies in electric field measurements, B. G. Fejer and L. Scherliess have worked to specify the characteristics of low-latitude disturbance electric fields during storm-time conditions using extensive incoherent-scatter radar observations of ionospheric drifts. Their results address the effects of short-lived penetration fields and longer-lasting dynamo fields. In the former case, they found that direct-penetration electric field patterns were in excellent agreement with the RCM (Rice Convection Model), improving overall understanding of the phenomena and establishing a baseline upon which to develop an overall model for global electric field distributions. There was little-to-no progress reported at the conference on thermospheric winds and composition.

4. There is the need to recognize that there is more to storm-time dynamics than F2-region characteristics. Admittedly, these characteristics are the easiest to measure (although the data set is still never complete), but unless measurements are expanded to include the E- and F1-regions, the full spectrum of kinetic and electrodynamic coupling terms in the ionospheric-thermospheric system

will never be determined. From this perspective it is important to recognize that the E- and F1-regions are the primary sources of dynamo fields and the principle domains supporting ionospheric currents. There has been little treatment of the storm-time characteristics of the E- and F1-regions in the literature to date, and the subject was not addressed in the conference.

5. For significant advances to be made, particularly in the emerging framework of "space weather and climatology," it is important that studies of ionospheric storms become more interdisciplinary and begin to focus on the development of a predictive methodology. Some of this was explicitly treated in the conference while other important interdisciplinary subjects were only treated implicitly with respect to their relevance to the understanding of ionospheric storms.

Considering ionospheric storm responses due to inputs from other solar-terrestrial domains, it is worthwhile to make reference to the work of T. Kosugi and K. Shibata and H. S. Hudson who describe the EUV and X-ray characteristics of dynamic solar events, some of which (e.g., fast magnetic clouds and their upstream sheath fields) are directly associated with the triggering of ionospheric storms. It is reasonable to expect that precursor ionospheric energy inputs, as those provided by bursts of solar radiation, can pre-condition the ionosphere and alter its early-phase response to subsequent storm-time enhancements in particles, winds, and electric fields. Thus, future ionospheric storm studies should track the varying solar electromagnetic spectrum before and during the dynamic evolution of the ionospheric-thermospheric system since bursty portions of the EUV and X-ray spectra can affect ionospheric plasma densities, conductivities and ultimately the intensity of dynamo-driven electric fields.

Independent of any pre-conditioning of the ionospheric/thermospheric system, the actual triggering of an ionospheric storm is traceable to the dynamics of the IMF and solar wind conditions at 1 AU. Those upstream conditions are traceable to dynamic solar events. Therefore, to truly understand the forces driving ionospheric storms and ultimately to predict their occurrences and intensities it is important to understand and predict the associated solar events and the interplanetary evolution of the attendant particle and field distributions. Work in progress presented at the conference includes the modeling of CMEs and prominences by J. Chen, with attention to the pre-eruptive structure and associated consequences in the interplanetary medium. Additional efforts include those of B. T. Tsurutani and W. D. Gonzalez who have reviewed and analyzed the large interplanetary database and established empirically-based prediction methodologies for the IMF B_z component at 1 AU associated with fast mass ejecta/driver gases resulting from dynamic solar events.

The next sequence in the chain of events from the Sun to the Earth involves the coupling of the interplanetary medium to the magnetosphere and the bi-directional coupling between the magnetosphere and the ionosphere. This is the final link that drives the ionospheric storm process through enhanced particle precipitation, cross polar cap potentials, and convection electric fields at high latitudes. Modeling work like that of the RCM and MSFM (Magnetospheric Specification and Forecast Model) by R. A. Wolf and colleagues needs to continue in order to properly identify cause-effect terms that transfer energy from the interplanetary medium through the magnetosphere, into the high-latitude ionosphere, and ultimately to the equatorial ionosphere through penetrating electric-field events.

Considering inverse processes (e.g., the influence of the ionosphere on the magnetosphere) it is important to note that advances in our understanding of several features of storm-time magnetospheric phenomena cannot proceed without including the role of the ionosphere. This includes the influences of temporally- and spatially-varying ionospheric conductivities, and associated influences on electric field and current coupling processes between the ionosphere and the magnetosphere. And there is the important role of the ionosphere as an oxygen-ion source for heavy particles in the near-Earth magnetotail during a storm (e.g., I. A. Daglis).

There is the need for an organizational infrastructure that can integrate the activities of the contributing disciplines and carry the interaction into a seamless first-principle understanding of the full chain of events from the Sun to the Earth, including the influence on humankind and its systems. An effort to spawn this structure and support it is embodied partly in the National Space Weather Initiative (as discussed by R. M. Robinson), with a "test" scenario reported in the collaborative storm study of D. J. Knipp.

9. ARTIFICIAL INTELLIGENCE AND STORM FORECASTING

H. Lundstedt

The variety of AI methods that exists today was shown by H. Lundstedt. Since each technique has weaknesses and strengths, integrating them would result in even more powerful methods, called Intelligent Hybrid Systems. A neural network can find the rules a fuzzy system needs. A genetic algorithm can optimize the parameters of a neural network. Trained neural networks can be understood by extracting symbolic concept descriptions from them. Many therefore foresee that the next break-through in technology is the combination of these AI methods and other state-of-the-art technologies into intelligent hybrid systems.

Two intelligent hybrid systems, the Rice Magnetospheric Specification and Forecast Model

(MSFM) and the Lund Space Weather Model were discussed. The MSFM uses both MHD models and neural network predictions to be able to specify the particle fluxes at arbitrary points in the magnetosphere. The Lund model predicts different indices and physical parameters and also tries to explain the effects of magnetic storms on power systems, satellites, communication, and climate.

At present predictions of geomagnetic activity 1-3 days ahead are inaccurate [*Joselyn*, 1995]. However, several promising methods are in progress. T. Detman showed in his invited talk how geomagnetic activity index A_p can be predicted by using solar magnetic field data, Wang-Sheeley's model and the Russell-McPherron principle. The Lund Space Weather Model will also be able to predict geomagnetic activity 1-3 days ahead from solar data. A module of that model, consisting of a MHD-model and a radial-basis function neural network, predicts solar wind velocity from solar magnetic data 1-3 days ahead [*Wintoft and Lundstedt*, 1996]. The predictions of geomagnetic storms, described by the *Dst* index have now reached a very high accuracy. Using a recurrent neural network and only solar wind data as input, about 85% of the observed *Dst* variance is predictable [*Wu and Lundstedt*, 1996]. This is shown in Figure 4. The correlation coefficient between the predicted and observed *Dst* is about 0.91. The root-mean square is 16 nT. Even two hours ahead predictions are very accurate; 82% of the observed *Dst* variance is predictable.

Predictions of substorms activity using neural networks were also discussed by Takalo et al. and Lundstedt. In their study *Gleisner et al.* [1997] found that with the solar wind variables n, V, B_y, B_z, as input to the network, 76% of the *AE* index variance is accounted for. Even more accurate predictions of *AE* are expected with the use of recurrent neural networks.

The most accurate predictions are today obtained using intelligent hybrid systems. Explaining these systems is therefore an exciting challenge. Very complex and general models can be developed with these hybrid systems, and the enormous flow of solar-terrestrial data can be assimilated and studied with these systems. Altogether it promises many interesting future results.

10. SYSTEM EFFECTS - WHAT DO USERS WANT?

J. A. Joselyn

As presented by W. H. Campbell, geomagnetic storms affect a wide variety of systems (e.g., radio communications, pipeline currents, and low-altitude satellite orbits). From a customer point of view, a significant unsolved problem is how to apply rapidly advancing scientific knowledge in the most beneficial way to increasingly vulnerable systems. The tasks include describing the level of observed geomagnetic activity in a way that is universally meaningful, predicting the behavior

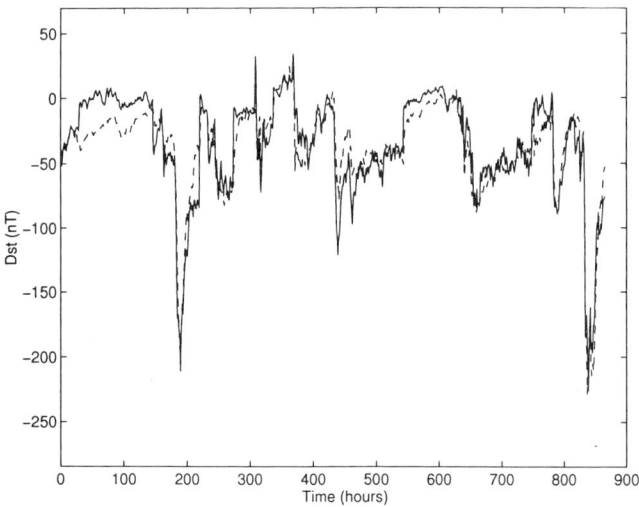

Fig. 4. Prediction of *Dst* one hour in advance from only solar wind data with use of an Elman recurrent neural network. The solid line represents the observed *Dst* and the dashed line the predicted *Dst*.

of that representation as far into the future and as accurately as possible, and then communicating that information effectively. This need has been formalized as the goal of the National Space Weather Program as expressed in the NSWP Strategic Plan (1995): "Timely, accurate, and reliable space environment observations, specifications, and forecasts within the next 10 years."

At the present time, agencies requiring real-time information about the space environment are served by one of ten Regional Warning Centers of the International Space Environment Services. These are located in Australia, Canada, China, the Czech Republic, France, India, Japan, Poland, Russia, and the USA. In the USA, these services of alerts and warnings, forecasts and summaries are provided by Space Weather Operations (SWO) in Boulder, Colorado, which is operated jointly by the National Oceanic and Atmospheric Administration (NOAA) Space Environment Center and the US Air Force 50th Weather Squadron. Subscribers to geomagnetic activity services receive information in the form of *K* and *A* indices, introduced by J. Bartels in 1932. An explanation of these indices is attached as an appendix for those readers unfamiliar with their formulation.

At NOAA's SWO, the choice was made more than 30 years ago to use the *A* index to define a storm and describe its severity, as shown in Table 1. These categories guarantee that a minor storm includes one or more *K*'s of 5 or greater; that a major storm includes one or more *K*'s of 6 or greater, and that a severe storm includes a *K* of 7 or greater.

Alerts (messages informing subscribers of the occurrence of the specific condition) are issued when the *K*

TABLE 1. A and K Indices and Corresponding Activity Categories

Category	A Index Range	Typical K Indices
Quiet	0-7	0, 1, 2
Unsettled	8-15	3
Active	16-29	4
Minor Storm	30-49	5
Major Storm	50-99	6
Severe Storm	100-400	7, 8, 9

index at Boulder is observed to be 4 or greater, or when a 24-hour "running" A index is observed to be 20 or greater. Warnings are issued if these conditions are expected at Fredericksburg, VA (the SEC "control" observatory) within 72 hours.

Who are the users of this information, and what do they need? The Space Environment Center endeavors to track and survey customers, and desires to improve services to meet their needs. The following information distills some of the information that has been gleaned from customer surveys and face-to-face meetings at user conferences and trade association meetings. There are more than 2000 entries in the SWO customer data base; these have been cast into 4 groups for the purpose of this discussion. We find that K and A indices as expressions of geomagnetic activity are not meeting the specific needs of certain users, but at present, alternative indices are not obvious. An important, unaddressed problem is the need for a better expression of geomagnetic activity that is both meaningful and predictable.

The first group includes utilities that must contend with induced currents in power lines and pipelines, and geophysical exploration companies; these constitute about 11 percent (by number) of SEC/SWO customers. Utilities have indicated that they would like descriptions of the rate of change of the magnetic field, the location and severity of ionospheric electrojet currents, and products that allow for geological features that can strongly modulate general, global conditions. Geophysical explorers are interested in the amount of variation expected in declination (compass swing) during storm conditions, and would like reliable forecasts of quiet conditions. K and A indices do not have adequate temporal and spatial resolution to meet these needs.

High Frequency (HF) and amateur radio operators, the aviation industry, and satellite communicators (TV and telephone links, GPS) constitute 44% of SEC customers. They need to know about unusual radio wave propagation conditions and periods of ionospheric scintillation that can compromise channel performance. Again, K and A indices are not the best proxies for these conditions, although users have learned to extrapolate from the information given to understand whether any problems they may be having are environmental.

There are only a few control centers concerned with satellite health (6% of the data base), but they represent a large national economic investment. These operations centers are worried about incidents of surface and deep dielectric charging, magnetopause crossings (if the geomagnetic field is being used for orbital alignment), and atmospheric frictional drag on low-altitude satellites. For this customer set, the SEC has recently added an alert condition in addition to the K and A warnings of geomagnetic storms. This new alert reports observations at geosynchronous orbit of energetic (greater than 2 MeV) electron fluxes exceeding a threshold of 10^3 electrons/cm²-s-sr. However, we do not produce a quantitative forecast of these "killer electrons"; research is needed to better understand the conditions that lead to high levels of energetic electrons.

Thirty-eight percent of our consumers of geomagnetic storm alerts and warnings are the media and general public, who want to see auroras when they occur at lower latitudes; and homing pigeon racers who have learned that storms are correlated with high losses of birds in flight. Again, these diverse customers must take the warnings that are issued, and extrapolate to meet their needs. Education is a major component of service to many of these persons.

What can be done? In response to user surveys and conferences and in coordination with the National Space Weather Program, there are plans in agencies in the US and worldwide to integrate and interpret data from various sensors (e.g., high-resolution ground-based magnetometers; satellite measurement of particles precipitating into the polar caps) and adapt them for operational use. New displays and generic descriptors are being suggested as visualization techniques (and hardware capabilities) improve. Models of global and regional magnetospheric and ionospheric parameters (e.g., particle fluxes and Total Electron Content) and algorithms using solar wind data are being tested for accuracy and "hardened" for real-time use. Better communication technologies are being explored, as the Internet spreads into common use. Choosing the most beneficial ideas, and transitioning them into operational products is not a trivial task.

Appendix: Explanation of K and A indices

The K index is a quasi-logarithmic number between 0 and 9 that is assigned at the end of specified 3-hour periods (00-03, 03-06, etc.), by measuring the maximum deviation (in nanoteslas) of the observed field beyond expected quiet field conditions, for each of the three magnetic field vector components. The largest of the maxima is converted to a K index by using a look-up table appropriate for that particular observing site. This process standardizes the data by correcting for expected geophysical biases between observing sites. At individual stations, to combine the eight daily K indices into one number representative of overall activity for the whole day, each K is converted to an "a" index which linearizes the quasi-logarithmic K index; then the eight "a"s are arithmetically averaged to yield a daily "A" index. *Menvielle and Berthelier* [1991, 1992] describe the details of constructing the K and A indices and the observatory networks that comprise global indices.

11. SUMMARY

Y. Kamide

Geomagnetic storms are multi-faceted phenomena that originate at the Sun and occur in the solar wind, the magnetosphere, the ionosphere, and in the thermosphere. What causes storms; how is the ring current formed and intensified during magnetic storms; what are storm effects in the ionosphere; how do changes in the magnetosphere-ionosphere system affect the Earth's upper atmosphere; and how can we predict the storm occurrence and intensity? These are some of the general questions relating to magnetic storms. It is important to note that the challenge facing modern space physics lies historically in the study of geomagnetic storms. In the mid-1800s, extraordinary fluctuations in the Earth's magnetic field (such as atmospheric storms) were called geomagnetic storms. Information on the solar wind and the IMF gathered by spacecraft over the last 30 years, however, has modified the concept of the magnetic storm.

Magnetic storms must be understood as a chain of processes from the Sun to the Earth. Interdisciplinary scientists and students recently met at the Jet Propulsion Laboratory, California Institute of Technology, Pasadena, California to discuss the many different aspects of magnetic storms and to test their scenarios on the cause-and-effect relationship of a variety of observations made during magnetic storms. They addressed not only the general issues suggested above but also specific questions, including how to identify individual substorm expansions during the main phase of magnetic storms, what makes the decay "constant" of the ring current variable, and whether there is a process priming the magnetosphere-ionosphere system before the sharp increase in the ring current?

Intriguing discussions using new data sets from, for example the Yohkoh and Ulysses satellites, focused on how CMEs are created and how they interact with interplanetary structures. Considerable uncertainties exist, however, regarding their magnetic configuration which plays a key role in revealing the physics for different sizes of ICMEs and such changes as dimming of coronal features. One of the major problems is that it is difficult if not impossible at present to visualize the three-dimensional structure of CMEs and the corresponding magnetic field lines extending into interplanetary space. This is also important in accounting for relevant interplanetary structures, i.e., the so-called "geoeffectiveness," which lead to the development of magnetic storms.

Many questions continue with regard to the formation and population/composition of the ring current during magnetic storms, as well as the loss mechanism of ring current particles during the storm recovery phase, that sometimes takes only several hours, but occasionally several weeks. The relative importance of the solar wind and the ionosphere as the origin of ring current ions in the 20-200 keV energy range seems to have been identified observationally. It depends on the storm phases. The ionospheric component has been shown to be most abundant during the main phase of intense magnetic storms. This implies that ionospheric ions are accelerated upward along parallel electric fields or by perpendicular heating (involving plasma waves) leading to ion conic distributions, both of which are associated with substorm expansion onsets. Both sources would appear to have field-aligned distributions near the equatorial plane. No observations, however, have clearly shown at what L shell these ions are injected, although observations of field-aligned ionospheric ions indicate that injection occurs over a broad range of L. These ions are trapped through an efficient magnetospheric process, becoming the main contributors to the storm-time ring current at $L < 4$. Therefore, the successive occurrence of intense substorms appears to be a necessary condition for the main phase of magnetic storms. This in fact supports the classical notion that a magnetic storm consists of continual substorm activity.

On the other hand, an enhanced solar wind electric field of sufficient duration directly drives magnetospheric convection, which, in turn, enhances the ring current. In the course of this process of the extraction of power from the solar wind flow required to build up the storm-time ring current, substorms are often triggered. In this scenario, the action of substorm activity is merely a byproduct of the overall energy process in the magnetosphere-polar ionosphere coupling. Although the ring current as expressed by Dst can be reproduced mainly by changes in the large-scale electric field driven by southward IMF, the role of substorms is an issue suitable for further study.

Ionosphere-thermosphere coupling is another important aspect of geomagnetic storms. Even ionospheric composition can change during major magnetic storms in ways yet to be described properly by models of kinetic, chemical, and electromagnetic interactions between the ionosphere and the thermosphere. How does the heat source at high latitudes induce a global neutral wind system which in turn propagates from high to low latitudes even into the opposite hemisphere? It is extremely important that we try to understand global maps of the electron density and the neutral wind as well as the ionospheric layers moving up and down in the framework of the chain of processes from the Sun to the Earth. For example, the large-scale convection electric field in the magnetosphere not only controls the boundary condition for the neutral wind, but can also vary drastically as a result of the thermospheric neutral wind.

In view of the crucial effects of geomagnetic storms on such human-societal systems as radio communications and satellite drag, there is a strong need for prediction schemes continually to be upgraded. It is essential to integrate our observation and modeling efforts in the Space Weather project.

One important aspect of magnetic storms that was not discussed explicitly during the conference is how to monitor their intensity. The intensity of a magnetic storm is commonly defined by the minimum Dst value during the magnetic storm, which is equivalent to the maximum Dst magnitude at the main phase. It may be natural to assume that when a major event is observed near the Earth, something major is also occurring throughout the chain of processes from the Sun to the Earth. Is this assumption really correct in magnetic storms? Is a large disturbance in the interplanetary structure really necessary and sufficient to generate a major magnetic storm? Is it physically meaningful to rely on the minimum Dst value to define the storm intensity?

During the panel discussion, it was pointed out that the increase in the ring current during the largest magnetic storms in terms of the Dst decrease often goes through two steps at the main phase. That is, well before the Dst decrease has fully recovered to its pre-storm level, a succeeding decrease tends to occur. If this is truly the case, the largest magnetic storms consist of two medium-size storms. Our future efforts should then be directed toward looking for a signature for a two-stage structure in the southward IMF, not one large southward IMF turning. In other words, what is perhaps needed in generating an intense magnetic storm are two separate regions of southward IMF in an interplanetary structure. One of the key factors in generating intense magnetic storms is perhaps a slow decay rate of the ring current.

Acknowledgments. The work of D. C. Hamilton was supported by NASA grants NAG5-2800 and NAG5-716. L. R. Lyons is grateful to M. W. Chen for her collaborations on ring current studies and for valuable discussions concerning this summary. He is also grateful to M. Schulz for numerous stimulating interactions concerning storm-time dynamics over the past twenty-five years. The work of L. R. Lyons was supported by NASA grant NAGW-3968, NSF grant ATM-9522288, and the Aerospace Sponsored Research Program. The work of R. L. McPherron was initiated while he was a visiting professor at the Solar-Terrestrial Environment Laboratory in Japan. He thanks Professors S. Kokubun and Y. Kamide, and the laboratory for their support. The work was completed at UCLA with partial support from grants by the National Science Foundation ATM-95-02124 and the National Aeronautics and Space Administration NAG5-1167. H. Lundstedt is grateful to the conference covenors, W. D. Gonzalez, Y. Kamide, and B. T. Tsurutani, and Session Chair R. L. McPherron for inviting him to give a talk at the Chapman Conference. Y. Kamide would like to thank all the panel members for their extensive discussions and analysis during the finalization process of this joint paper. The work of Y. Kamide at the Solar-Terrestrial Environment Laboratory was supported in part by the Ministry of Education, Science, Culture and Sports (Monbusho) under a Grant-in-Aid for Scientific Research, Category B.

REFERENCES

Akasofu, S.-I., *Polar and Magnetospheric Substorms*, p. 5, D. Reidel, Norwell, Mass., 1968.

Burton, R. K., R. L. McPherron, and C. T. Russell, An empirical relationship between interplanetary conditions and Dst, *J. Geophys. Res.*, 80, 4204, 1975.

Carovillano, R. L., and G. L. Siscoe, Energy and momentum theories in magnetospheric processes, *Rev. Geophys.*, 11, 289, 1973.

Chapman, S, Earth storms: Retrospect and prospect, *J. Phys. Soc. Japan*, 6, Suppl. A-I, 17, 1962.

Chen, M. W., M. Schulz, and L. R. Lyons, Simulations of phase space distributions of stormtime proton ring current, *J. Geophys. Res.*, 99, 5745, 1994.

Codrescu, M. V., R. G. Roble, and J. M. Forbes, Interactive ionospheric modeling: A comparison between TIGCM and ionosonde data, *J. Geophys. Res.* 97, 8591, 1992.

DeForest, S. E., and C. E. McIlwain, Plasma clouds in the magnetosphere, *J. Geophys. Res.*, 76, 3587, 1971.

Fay, R. A., C. R. Garrity, R. L. McPherron, and L. F. Bargatze, Prediction filters for the Dst index and the polar cap potential, in *Solar Wind-Magnetosphere Coupling*, edited by Y. Kamide and J. A. Slavin, p. 111, Terra Scientific Pub. Co., Tokyo, 1986.

Fuller-Rowell, T., and M. V. Codrescu, Response of the thermosphere and ionosphere to geomagnetic storms, *J. Geophys. Res.*, 99, 3893, 1994.

Gleisner, H., H. Lundstedt, and P. Wintoft, The response of auroral electrojets to the solar wind modeled with neural networks, *J. Geophys. Res.*, 102, in press, 1997.

Gonzalez, W. D., and B. T. Tsurutani, Criteria of interplanetary parameters causing intense magnetic storms ($Dst < -100$ nT), *Planet. Space Sci.*, 35, 1101, 1987.

Gonzalez, W. D., B. T. Tsurutani, P. S. McIntosh, and A. L. Clua de Gonzalez, Coronal hole-active region-current sheet (CHARCS) association with intense interplanetary and geomagnetic activity, *Geophys. Res. Lett.*, 23, 2577, 1996.

Gonzalez, W. D., B. T. Tsurutani, A. L. C. Gonzalez, E. J. Smith, F. Tang, and S.-I. Akasofu, Solar wind-magnetosphere coupling during intense magnetic storms (1978-1979), *J. Geophys. Res.*, 94, 8835, 1989.

Gonzalez, W. D., J. A. Joselyn, Y. Kamide, H. W. Krohel, G. Rostoker, B. T. Tsurutani, and V. M. Vasyliunas, What is a geomagnetic storm?, *J. Geophys Res.*, 99, 5771, 1994.

Gosling, J. T., D. J. McComas, J. L. Phillips, and S. J. Bame, Geomagnetic activity associated with Earth passage of interplanetary shock disturbances and coronal mass ejections, *J. Geophys. Res.*, 96, 7831, 1991.

Hamilton, D. C., G. Gloeckler, F. M. Ipavich, W. Studemann, B. Wilken, and G. Kremser, Ring current development during the great geomagnetic storm of February 1986, *J. Geophys. Res.*, *93*, 14343, 1988.

Harel, M., R. A. Wolf, R. W. Spiro, P. H. Reiff, C.-K. Chen, W. J. Burke, F. J. Rich, and M. Smiddy, Quantitative simulation of a magnetospheric substorm, 2. Comparison with observations, *J. Geophys. Res.*, *86*, 2242, 1981.

Hiei, E., A. Hundhausen, and D. Sime, Reformation of a coronal helmet streamer by magnetic reconnection after a coronal mass ejection, *Geophys. Res. Lett.*, *20*, 2785, 1993.

Hudson, H. S., Yohkoh observations of coronal mass ejections, in *Magnetodynamic Phenomena in the Solar Atmosphere - Prototypes of Stellar Magnetic Activity*, edited by Y. Uchida, T. Kosugi and H. S. Hudson, p. 89, Kluwer Pub., Dordrecht, 1996.

Hudson, H. S., L. W. Acton, and S. F. Freeland, A long-duration solar flare with mass ejection and global consequences, *Ap. J.*, *470*, 629, 1996a.

Hudson, H. S.. L. W. Acton, D. Alexander, K. L. Harvey, S. W. Kahler, H. Kurokawa, and J. R. Lemen, The solar origins of two high-latitude interplanetary disturbances, in *Solar Wind Eight*, edited by D. Winterhalter, J. T. Gosling, S. R. Habbal, W. S. Kurth, and M. Neugebauer, p. 84, AIP Press, Woodbury, N.Y., 1996b.

Hudson, H. S., L. W. Acton, D. Alexander, S. L. Freeland, J. R. Lemen, and K. L. Harvey, Yohkoh/SXT soft X-ray observations of sudden mass loss from the solar corona, in *Solar Wind Eight*, edited by D. Winterhalter, J. T. Gosling, S. R. Habbal, W. S. Kurth, and M. Neugebauer, p. 88, AIP Press, Woodbury, N.Y., 1996c.

Hundhausen, A. J., The origin and propagation of coronal mass ejections, in *Solar Wind Six*, edited by V. Pizzo, D. G. Sime, and T. E. Holzer, p. 181, NCAR TN-306, *Vol. 1*, Boulder, Colorado, 1988.

Hundhausen, A. J., Sizes and locations of coronal mass ejections: SMM observations from 1980 and 1984-1989, *J. Geophys. Res.*, *98*, 13177, 1993.

Hundhausen, A. J., A. L. Stanger, and S. A. Serbicki, Mass and energy contents of coronal mass ejections: SMM results from 1980 and 1984-1988, *Proc. of the Third SOHO Workshop*, ESA SP-373, 409, 1994.

Iyemori, T. and D. R. K. Rao, Decay of the *Dst* field of geomagnetic disturbances after substorm onset and its implication to substorm relation, *Ann. Geophys.*, *14*, 608, 1996.

Jackson, B. V., A. Buffington, P. L. Hick, S. W. Kahler, R. C. Altrock, R. E. Gold, and D. F. Webb, The solar mass ejection imager, in *Solar Wind Eight*, edited by D. Winterhalter, J. T. Gosling, S. R. Habbal, W. S. Kurth, and M. Neugebauer, p. 536, AIP Press, Woodbury, N.Y., 1996.

Joselyn, J. A., Geomagnetic activity forecasting: The state of the art, *Rev. Geophys.*, *33*, 383, 1995.

Kamide, Y., Relationship between substorms and storms, in *Dynamics of the Magnetosphere*, ed. by S.-I. Akasofu, p. 425, D. Reidel, Dordrecht, Holland, 1979.

Kopp, R. A., and G. W. Pneuman, Magnetic reconnection in the corona and the loop prominence phenomena, *Solar Phys.*, *50*, 85, 1976.

Lin, R. P., and S. W. Kahler, Interplanetary magnetic field connection to the Sun during electron heat flux dropouts in the solar wind, *J. Geophys. Res.*, *97*, 8203, 1992.

Lui, A. T. Y., R. W. McEntire, and S. M. Krimigis, Evolution of the ring current during two geomagnetic storms, *J. Geophys. Res.*, *92*, 7459, 1987.

Lyons, L. R., and M. Schulz, Access of energetic particles to storm time ring current through enhanced radial diffusion, *J. Geophys. Res.*, *94*, 5491, 1989.

Lyons, L. R., and D. J. Williams, The storm and post-storm evolution of energetic (35-560 keV) radiation belt electrons, *J. Geophys. Res.*, *80*, 3985, 1975.

Mauk, B. H., Quantitative modeling of the "convection surge" mechanism of ion acceleration, *J. Geophys. Res.*, *91*, 13423, 1986.

Mauk, B. H., and C. E. McIlwain, Correlation of Kp with the substorm-injected plasma boundary, *J. Geophys. Res.*, *79*, 3193, 1974.

McAllister, A. H., M. Dryer, P. McIntosh, H. Singer, and L. Weiss, A large polar crown CME and a "problem" geomagnetic storm: April 14-23, 1994, *J. Geophys. Res.*, *101*, 13497, 1996.

McComas, D. J., Tongues, bottles, and disconnected loops: The opening and closing of the interplanetary magnetic field, *Rev. Geophys., Suppl.*, *33*, 603, 1995.

Menvielle, M., and A. Berthelier, The K-derived planetary indices: Description and availability, *Rev. Geophys.*, *29*, 415, 1991, and, *Rev. Geophys.*, *30*, 91, 1992.

The National Space Weather Program, The Strategic Plan; A Report Prepared by the Working Group for the National Space Weather Program of the Committee for Space Environment Forecasting of the Office of the Federal Coordinator for Meteorological Service and Supporting Research; FCM-P30-1995, Washington, D. C., August 1995.

Pick, M., L. J. Lanzerotti, A. Buttighoffer, E. T. Sarris, T. P. Armstrong, G. M. Simnett, E. C. Roelof, and A. Kerdraon, The propagation of sub-MeV solar electrons to helio latitudes above 50°, *Geophys. Res. Lett.*, *22*, 3373, 1995.

Pizzo, V. J., The findings of the SPINS science workshop, *SEL Workshop Report*, Space Environment Laboratory, Boulder, CO, 1994.

Proelss, G. W., Common origin of positive ionospheric storms at middle latitudes and the geomagnetic activity effect at low latitudes, *J. Geophys. Res. 98*, 5981, 1993.

Roeder, J. L., J. F. Fennell, M. W. Chen, M. Schulz, M. Grande, and S. Livi, CRRES observations of the composition of the ring-current ion population, *Adv. Space Res.*, *17*, 17, 1996.

Rostoker, G., W. Baumjohann, W. D. Gonzalez, Y. Kamide, S. Kokubun, R. L. McPherron, and B. T. Tsurutani, Comment on "Decay of the *Dst* field of geomagnetic disturbance after substorm onset and its implication to storm-substorm relation" by Iyemori and Rao, *Ann. Geophys.*, *15*, in press, 1997.

Siscoe, G. L., and H. E. Petschek, Why substorms might not be sub-storms, *Geophys. Res. Lett.*, submitted, 1996.

Smith, P. H., and R. A. Hoffman, Ring current particle distributions during the magnetic storms of December 16-18, 1971, *J. Geophys. Res.*, *78*, 4731, 1973.

Thorne, R. M., and R. B. Horne, Energy transfer between energetic ring current H^+ and O^+ by electromagnetic ion cyclotron waves, *J. Geophys. Res.*, *99*, 17275, 1994.

Tsurutani, B. T., W. D. Gonzalez, F. Tang, S.-I. Akasofu, and E. J. Smith, Origin of interplanetary southward magnetic fields responsible for major magnetic storms near solar maximum (1978-1979), *J. Geophys. Res.*, *93*, 8519, 1988.

Tsurutani, B. T., T. Gould, B. E. Goldstein, W. D. Gonzalez, and M. Sugiura, Interplanetary Alfven waves and auroral (substorm) activity: IMP 8, *J. Geophys. Res.*, *95*, 2241, 1990.

Tsurutani, B. T., W. D. Gonzalez, F. Tang, and Y. T. Lee, Great magnetic storms, *Geophys. Res. Lett.*, *19*, 73, 1992.

Wang, Y.-M., and N. R. Sheeley, Jr., Global evolution of interplanetary sector structure, coronal holes, and solar wind streams during 1976-1993: Stackplot displays based on solar magnetic observations, *J. Geophys. Res.*, *99*, 6597, 1994.

Webb, D. F., R. A. Howard, and B. V. Jackson, Comparison of CME masses and kinetic energies near the Sun and in the inner heliosphere, in *Solar Wind Eight*, edited by D. Winterhalter, J. T. Gosling, S. R. Habbal, W. S. Kurth, and M. Neugebauer, p. 540, AIP Press, Woodbury, N.Y., 1996.

Weiss, L. A., P. H. Reiff, J. J. Moses, R. A. Heelis, and B. D. Moore, Energy dissipation in substorms, *Proc. of the Internat. Conf. on Substorms (ICS-1)*, ESA SP-335, 23, 1992.

Wintoft, P., and H. Lundstedt, Space weather modeling with intelligent hybrid systems: Predicting the solar wind velocity, *Adv. Space Res.*, in press, 1996.

Wrenn, G. L., A. S. Rodgers, and H. Smith, Geomagnetic storms in the Antarctic region, *J. Atm. Terr. Phys.* *49*, 901, 1987.

Wu, J.-G. and H. Lundstedt, Prediction of geomagnetic storms from solar wind data using Elman recurrent neural networks, *Geophys. Res. Lett.*, *23*, 319, 1996.

Yeh, K. C., S. Y. Ma, and K. H. Lin, Global ionospheric effects of the October 1988 geomagnetic storm, *J. Geophys. Res.*, *99*, 6201, 1994.

Zhao, X., and J. T. Hoeksema, A coronal magnetic field model with horizontal volume and sheet currents, *Solar Phys.*, *151*, 91, 1994.

W. D. Gonzalez, Instituto Nacional de Pesquisas Espacias, Postal 515, 12200 Sao Jose dos Campos, Sao Paulo, Brazil

D. C. Hamilton, Department of Physics, University of Maryland, College Park, MD 20742, USA

H. S. Hudson, Institute for Astronomy, University of Hawaii, Honolulu, HI 96822, USA

J. A. Joselyn, Space Environment Center, National Oceanic and Atmosphere Administration, Boulder, CO 80303, USA

S. W. Kahler, Phillips Laboratory, Hanscom AFB, MA 01731-3010, USA

Y. Kamide, Solar-Terrestrial Environment Laboratory, Nagoya University, Toyokawa 442, Japan

L. R. Lyons, Aerospace Corporation, Box 92957, MC-260, Los Angeles, CA 90009-2957, USA

H. Lundstedt, Solar-Terrestrial Division, IRF-STL, Box 43, S-221 00 Lund, Sweden

R. L. McPherron, Institute of Geophysics and Planetary Physics, University of California at Los Angeles, Los Angeles, CA 90024, USA

E. Szuszczewicz, Science Applications International Corporation, McLean, VA 22102, USA

Solar Coronal Dynamics and Flares as a Cause of Interplanetary Disturbances

Takeo Kosugi[1] and Kazunari Shibata

National Astronomical Observatory, Mitaka, Tokyo 181, Japan

Flares and coronal mass ejectios (CMEs) are the most energetic and eruptive among various types of solar coronal magnetic activity, and as such they might be responsible for major geomagnetic storms. It has not yet been fully understood, however, how these two major types of solar activity are interrelated. This article is aimed to clarify this point, first, by summarizing observations over the past several decades of various types of eruptive phenomena, and then, by presenting recent observations from the *Yohkoh* satellite. The *Yohkoh* observations show that the solar corona is intermittently expanding and restructuring itself through magnetic reconnection, that flares occur in association with rapid expansion and restructuring of the surrounding active-region corona, and further that both flares (irrespective of whether they are of impulsive or long-duration type) and large-scale arcade formations (believed to be intimately related to CMEs) show features such as "loop-with-a-cusp" structure and "plasmoid/filament ejection" in common. We conclude that flares, CMEs, and possibly some other related phenomena can be interpreted as manifesting different aspects of a common process. Several fundamental subprocesses are involved in this process, in which plasmoid ejection and magnetic reconnection play dominant roles. A schematic model is presented for understanding flares and CMEs in a unified way.

1. INTRODUCTION

The Sun is, for astronomers, the only star that can be observed with moderate, though far from sufficient, spatial resolution to investigate fundamental processes involved in stellar magnetic activity. A variety of magnetic activities has been known to occur in the atmosphere of the Sun, from relatively steady ones such as coronal heating and solar wind acceleration to violently dynamical ones such as flares and coronal mass ejections (CMEs).

The solar flare [e.g., *Svestka*, 1976; *Zirin*, 1988] is an explosive phenomenon that usually occurs in a single active region around a group of sunspots and lasts for a relatively short period of time (minutes to hours). Though relatively compact in size (typically $\sim 10^{19}$ cm^2 in area), it is the most intense and energetic among various types of solar activity. Flares are most readily detected as sudden brightenings in chromospheric lines (such as in Hα), but such chro-

[1]Also at Nobeyama Radio Observatory of NAOJ, Minami-maki, Minami-saku, Nagano 384-13, Japan

mospheric brightenings are not the flare proper. The primary energy release of flares takes place in the corona, releasing as much as 10^{32} erg in the case of the largest flare, energizing electrons and ions up to ~MeV and ~GeV/n, respectively, and heating plasmas up to temperatures exceeding 3×10^7 K. A large amount of energy flows down from the corona in the form of particle flux as well as heat flux, resulting in chromospheric (and sometimes even photospheric) brighteings as a byproduct. The released energy originates, without doubt, from coronal magnetic fields, as no other sources can supply such a large amount of energy rapidly enough. A major problem to be answered here is, thus, in what configuration and dynamics of magnetic fields, and how, the energy is stored relatively stably and then abruptly and explosively released as flares.

It has long been known that flares are accompanied by ejections of material and/or shock waves in the corona with increasing probabilities of association for larger flares. These flare ejecta and/or shocks might be interpreted as resulting from a pressure pulse caused by flare energy deposition into the dense chromosphere. Alternatively their existence may be interpreted as inherently magnetic in nature. In any case, these flare ejecta and/or shocks, at least some of them, may propagate outwards and give rise to interplanetary disturbances, and as such they have attracted the attention of geomagnetic storm researchers.

Such a simple view as connecting solar flares directly to geomagnetic storms has been challenged since the 1970's when so-called coronal transients or coronal mass ejections (CMEs) were found (and then almost continuously monitored until 1989) with white-light coronagraphs onboard *OSO-7*, *Skylab*, *P78-1*, and *SMM* [e.g., *MacQueen*, 1980; *Hildner et al.*, 1989]. CMEs are gigantic upward mass motions observed in the outer corona and typically characterized by a three-component structure, i.e., a balloon-shaped leading edge or expanding shock front, followed by a cavity or void, and an embedded filament material. Altogether these three components expand at velocities from a few tens to several hundreds km s^{-1} (rarely exceeding 1,000 km s^{-1}), and convey, say, up to 10^{32} erg of energy and 10^{16} g of mass. The energy may be comparable to that of the largest flare. As CMEs are an erupting phenomenon expanding outwards from the upper corona into interplanetary space, it is not surprising if they are statistically more geoeffective than flares; in fact geomagnetic storms are known to be better associated with CMEs (especially those with larger velocities and masses) than flares [*Gosling et al.*, 1991].

Then, how are CMEs related to flares and what causes CMEs to erupt? A CME cannot be the simple aftermath of a flare explosion for several reasons. First, CMEs are occasionally not associated with any chromospheric flares, or more rigorously not associated with any Hα events that are conventionally classified as flares. Even though CMEs are often flare-associated, the three-component structure seems to be incompatible with a simple, pressure-pulse explosion model. In fact, CMEs have usually a much wider extension in latitude than any single active region in which flares occur, and seem to start erupting, at least in some well-studied cases, well prior to the corresponding flare occurrence [*Harrison et al.*, 1985; *Harrison*, 1986]. Thus we are faced with a controversy on the so-called "solar flare myth" [*Gosling*, 1993, 1995; *Hudson et al.*, 1995], in which it is debated whether the statement that solar flares do not but CMEs do play a central role in producing major interplanetary disturbaces could be totally acceptable or not. Even apart from this controversy, there is no doubt that we need to clarify how these two phenomena are interrelated.

With these problems in mind, this article reviews eruptive phenomena in the solar corona which may potentially be a cause of interplanetary disturbances. During the past several decades, many observations of eruptive phenomena have been made in visible light and in radio, as well as in X-rays, regardless of whether they are associated with flares or not. We first summarize these observations in a retrospective way in order to avoid oversimplifying the problem, i.e., flare *versus* CME. It is pointed out that flares produce several types of ejecta that may reach interplanetary space. New observations by the *Yohkoh* satellite have recently revealed a clear relation of prominence (or filament) eruption to flare occurrence. This and other related observations by *Yohkoh* suggest that basically a single mechanism, magnetic reconnection, provides the key to understand a variety of these dynamical phenomena. We will discuss this unified view briefly. (No discussion is made about the heliospheric response to ejecta from the low corona and interplanetary causes of magnetic storms. For these topics, see, e.g., *Farrugia et al.* [1997] and *Tsurutani and Gonzalez* [1997] in these proceedings. Also we do not discuss the formation of shocks nor sweeping up of materials from the solar wind.)

TABLE 1. Key Words for Understanding Eruptive Phenomena in the Corona

before ~1950
    (with Hα spectrographs/coronagraphs)
        – Flares as a chromospheric phenomenon ($T\sim10^4$ K)
        – Prominence Eruption, Flare Sprays, and Surges
        – Disparition Brusque (filament disappearance)
~1950's
    (with radio instruments)
        – Type III Bursts: electron beams propagating outwards
        – Type II Bursts: shock waves propagating outwards
        – Type IV Bursts: moving plasmoids or expanding bottles
        – Type I Storms: long-lasting activity with particle acceleration
        – Microwave Bursts: particle acceleration in flares
~1960's
    (with X-ray, EUV, and microwave instruments)
        – Impulsive vs Gradual Phases of a flare
        – Particle Acceleration vs Heating in a flare
since ~1970
    (with radio imagers)
        – Three Subclasses of Moving Type IV Bursts
            (advancing fronts, expanding arches, and plasmoids)
    (with soft X-ray imagers)
        – Flares as a coronal phenomenon ($T\sim10^7$ K)
        – Eruptive Flares   vs   Confined Flares
            (two-ribbon flares)   (compact, simple-loop flares)
    (with space-borne, white-light coronagraphs)
        – Coronal Mass Ejections (CMEs)
            (composing of leading edge, cavity, and prominence material)
    (with hard X-ray imagers)
        – Coronal sources
        – Footpoint vs Loop-top sources
            (composing of leading edge, cavity, and prominence material)
    (with hard X-ray imagers)
        – Coronal sources
        – Footpoint vs Loop-top sources

2. ERUPTIVE PHENOMENA IN THE SOLAR CORONA – A RETROSPECTIVE VIEW

As in other fields in astronomy, our understanding of the solar corona has been rapidly extended or sometimes even revised with the advent of new observational techniques. The development of our knowledge on dynamical phenomena seen in the solar corona is briefly summarized in Table 1 from this viewpoint.

2.1. *Hα Flares and Ejecta*

Before radio telescopes began to be widely used in the 1950's, solar activity was monitored only with optical telescopes operating mainly in Hα. Since the Hα line emission is radiated or absorbed by low-temperature ($T\sim10^4$ K) plasmas, these telescopes are the most suitable to observe chromospheric features and phenomena.

Flares observed in Hα are basically brightenings in the chromosphere. In the case of large-sized flares one can see two bright ribbons, the separation of which gradually increases with time. Later in the decay phase a system of loops connecting the two ribbons forms or becomes observable in Hα ("post-flare loops") in the corona; this is interpreted as reflecting a cooling process of hot flare loops originally heated up to ~10^7 K and containing large amounts of material injected into the corona. Indeed material can be seen falling along each leg of a post-flare loop in Hα.

Prominences and dark filaments [e.g., *Tandberg-Hanssen*, 1974; *Priest*, 1989] are physically similar to each other, i.e., cool ($T\sim10^4$ K) chromospheric material elevated up into the hot ($T\sim 10^6$ K) corona; the former are emission features seen outside the solar limb with dark sky as background while the latter are absorption features seen inside the solar disk against the bright background. They look like a twisted or sheared bundle of ropes overlying along the magnetic field reversal line (neutral line). They appear either in an active region (active-region filament) or outside active regions (quiet-region filament). Quiet-region filaments are huge structures, sometimes longer than one solar radius, and are quite stable (quiescent filament) for more than a week. Abruptly, however, they may erupt partially or as a whole into the upper corona at velocities typically of several tens km s^{-1}, reaching a great height before eventually escaping from detection (prominence eruption or *disparition brusque*; Hα filaments on the disk may also suddenly disappear ("wink") due to the Doppler effect). It is to be noted that prominence eruptions occur in association with or without signatures of chromospheric flares.

The most dynamical Hα phenomenon is the flare spray [e.g., *Bruzek*, 1974]. It is the flare-associated ejection of an active-region filament, whose velocity sometimes reaches as high as 1,000 km s^{-1}, larger than the gravitational escape velocity from the Sun. The kinetic energy and mass involved in a spray may exceed 10^{31} erg and 10^{15} g, respectively. We will discuss the timing of spray onset later.

Surges are the jet-like ejection of Hα material, seen either in emission or in absorption [e.g., *Bruzek*, 1974]. The velocity is usually about one hundred km s^{-1}, smaller than the escape velocity and the ejected material eventually returns back downwards. They are associated with small flare-like Hα brightenings, which are not necessarily classified as flares because of their small brightening areas.

2.2. *Radio Manifestation of Flare Ejecta*

The advent of new types of observations, first in radio with multifrequency spectrometers and interferometers in the early 1950's, and then in X-rays with detectors onboard satellites in the early 1960's, has revealed various new aspects of a flare explosion, in which a variety of coronal dynamical processes are involved. Each type of coronal dynamical process preferentially appears in a specific phase, as is schematically shown in Figure 1.

The precursor phase of a flare is an interval lasting for a few minutes to several tens of minutes, when

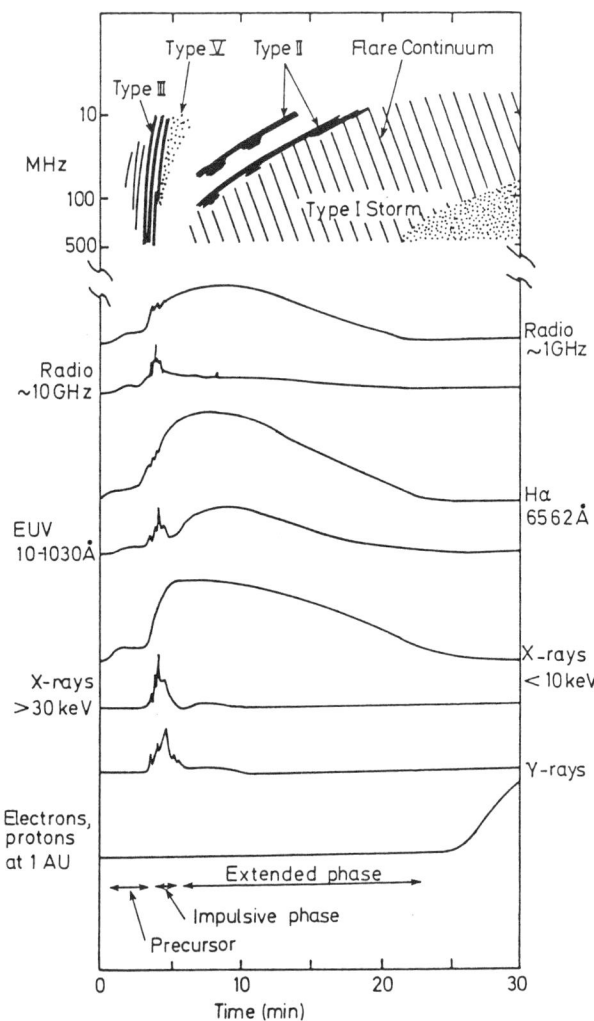

Fig. 1. Schematic time history of a typical solar flare as observed in various electromagnetic radiation and particles (after *Dulk et al.* [1985]).

weak enhancements of microwave, EUV, and soft X-ray intensity levels are recorded together with weak Hα brightenings. In addition to these weak enhancements which may be interpreted as resulting from preflare heating, several forms of filament activation, such as untwisting as well as eruptive motions, have been reported to take place prior to the impulsive phase by a few tens of minutes [e.g., *Bruzek*, 1974]. Also meter-wave type III bursts are sometimes detected, though they are usually weak in comparison with type III bursts seen in the impulsive phase.

Then a major fraction of the total flare energy is released rather violently in a short period of less than several minutes typically. The energy release rate sometimes reaches 10^{30} erg s^{-1}. This phase is characterized by various signatures of impulsive particle acceleration, i.e., impulsively varying total fluxes such as seen in hard X-rays (bremsstrahlung emission from energetic electrons), microwaves (gyrosynchrotron emission from energetic electrons), and γ-rays (nuclear reaction lines such as the positron annihilation line, nuclear de-excitation lines, and so on). Hence this phase is called the impulsive phase. A small fraction of impulsively accelerated electrons escapes upwards from the acceleration site and forms electron beams, which are detected as meter-wave type III bursts. Coronal plasmas are rapidly heated to much more than $\sim 3 \times 10^7$ K as a direct result of the primary energy release. Also a larger amount of $\sim 10^7$-K plasma is created as a byproduct of thermalization of nonthermal energy in the dense chromosphere; cool chromospheric material is rapidly heated and "evaporates" into the corona ("chromospheric evaporation"). As a consequence, strong upward as well as turbulent motions are detected, which is most remarkable during the rising part of the impulsive phase because no stationary component predominates. (Note that this phase is also called the flash phase based upon a sudden spectral broadening, say more than 10 Å, of the Hα linewidth in flare "kernels", or the explosive phase based upon an explosive increase in area of Hα ribbons.)

The impulsive phase is followed by the so-called gradual or extended phase, which may be a mixture of prolonged, though usually weakened, particle acceleration (seen mainly in microwaves, hard X-rays, and γ-rays), gradual heating (seen mainly in soft X-rays and Hα), and the aftermath of the impulsive phase. The most dramatic events in this phase are the type II and moving type IV bursts seen in meter-wave radio observations. Type II bursts are believed to be the radio manifestation of shock waves initiated in the impulsive phase at a lower height, becoming observable at a certain plasma-frequency level in the corona, and then propagating upwards, while moving type IV bursts represent upward-moving material in the corona. Figures 2 and 3 show trajectories of type II/IV burst pairs in space as well as in height-time plots for two representative cases observed with the Culgoora Radioheliograph operating at 80 MHz. In the first case (Figure 2; 1968 October 23-24 event [*Kai*, 1970]), a moving type IV burst source, depicted by thick lines, is widely extended, moves in the height-time plot at ~1,400 km s^{-1} on the line extrapolated from the preceding type II burst, and thus is interpreted as "advancing shock front", which emanates from the lower corona in the impulsive phase. On the other hand, in the latter case (Figure 3; 1969 March 1 event [*Riddle*, 1970]), two isolated moving type IV sources ("plasmoids") are observed to move upwards together at ~270 km s^{-1}. An interesting feature seen in Figure 3 is the height-time relationship comparing a type II burst, an Hα flare spray, and the "plasmoid"-type moving type IV sources; the Hα flare spray starts erupting almost simultaneously with the type II burst onset, which is suggestive of the start of the spray eruption in the impulsive phase, while the moving type IV sources seem to appear behind the (extrapolated) height of the spray. A third, but the most frequently detected, subclass of moving type IV bursts is the "expanding arch", which may be a radio manifestation of flare-associated expanding loops (not shown). For more details of these radio ejecta, see *Smerd and Dulk* [1971] and *Stewart* [1985].

One remark should be placed here on the classical type IV burst definition. In Figure 1 we see "flare continuum" in the extended phase after the passage of a harmonic pair of type II bursts. This is not any subclass of the moving type IV bursts discussed above but a stationary type IV burst; moving type IV bursts are usually too weak to be observable in dynamic spectral observations without imaging. Rather the flare continuum is a manifestation of the so-called second-phase acceleration or trapping of energetic electrons in the corona [*Robinson*, 1985]. The flare continuum is followed by a "storm continuum", lasting in some cases for more than a day after the flare occurrence, which may be manifestation of a long-lasting acceleration process high in the corona [*Kai, Melrose, and Suzuki*, 1985].

2.3. *Coronal Mass Ejections (CMEs) and Eruptive Flares*

White-light coronagraph observations from space, first made by *OSO-7* and *Skylab*, enabled us to find coronal mass ejections (CMEs) and have greatly contributed to extend our understanding of eruptive phenomena in the corona. Since many reviews are now available on CMEs [e.g., *MacQueen*, 1980; *Dryer*, 1982; *Hildner et al.*, 1989] as well as on their relation to flares [e.g., *Kahler*, 1992], and since the most essential points have been given in the Introduction of this article, only a few words are added here.

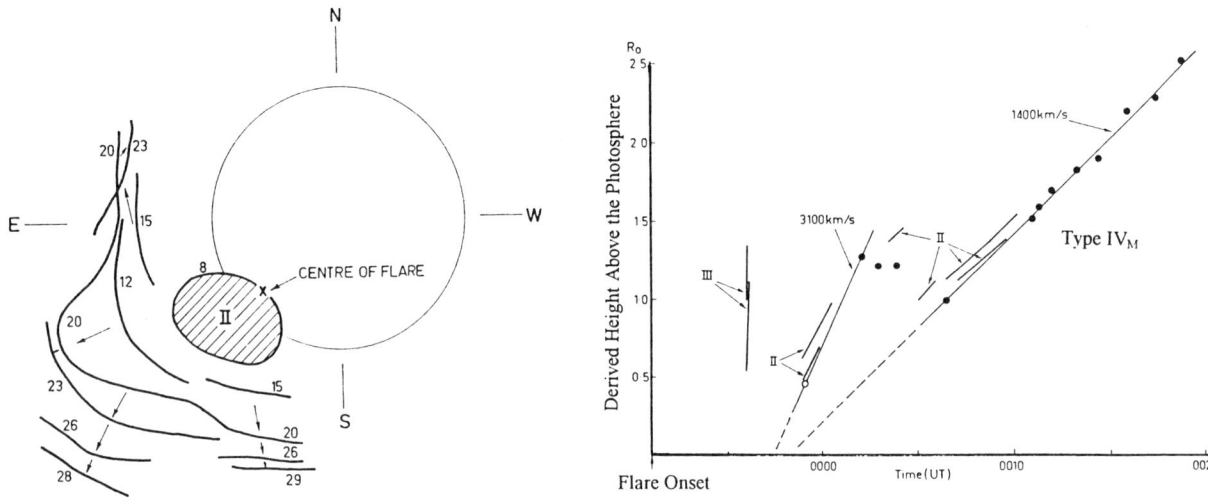

Fig. 2. The 1968 October 23–24 event observed with the Culgoora Radioheliograph at 80 MHz (after *Kai* [1970]). (left) The type II burst location is shown by hatched area, and the evolution of the "advancing-front" moving type IV burst by thick lines with numbers denoting the time in minutes from the start of the flare. (right) Height–time plot for type III, type II, and moving type IV bursts.

Since the early stage of CME study it has been known that CMEs are full of variety in shape, in velocity and mass, and in their association with other solar phenomena such as flares, prominence eruptions, type II/IV radio bursts, and so on. Some CMEs are associated with flares, some others with prominence eruptions, and the remainder with both or neither. No simple one-to-one correspondence has been found in the CME–flare or CME–prominence eruption relationship, which may imply that they are interrelated in a complicated manner. Basic questions such as what CMEs are and what causes CMEs to erupt are still to be answered. Related to the first question, i.e., what CMEs are, it is to be pointed out that coronal mass ejections can be detected not only in white light but also in radio [e.g., *Gopalswamy and Kundu*, 1995] and in X-rays [e.g., *Hudson*, 1997]. Thus multi-wavelengths observations is of crucial importance to comprehensively unveil the origin of CMEs.

In 1973–1974 long-term soft X-ray imaging observations were conducted for the first time by *Skylab*. High-temperature ($T \sim 10^7$ K) flares, or the flares proper, were clearly imaged. Based upon the images, a classification of flares into two distinct classes was advocated, i.e., into "eruptive flares" and "simple-loop flares" [e.g., *Priest*, 1981; *Svestka et al.*, 1992]. The former class is characterized by relatively large soft X-ray loops connecting the two underlying Hα ribbons, an increase in size and height of the loops with time, and a good association with meter-wave type II/IV bursts. Often flares of this class are long lasting (long-duration events; LDEs). On the other hand, the latter class shows a compact, simple-loop brightening in soft X-rays with relatively short duration, and an impulsively varying time history in hard X-rays as well as in microwaves, so that flares of this class are sometimes called either compact or confined flares, and also impulsive flares when temporal behavior is concerned. Flares of this class are occasionally believed not to be associated with any eruptive signatures, which we believe incorrect because many cases have been known in which compact, impulsive flares are associated with Hα flare sprays as well as meter-wave type II and/or type IV bursts. Thus it has not been clear whether the two classes are really distinct in the basic physics involved, a question left to be answered by *Yohkoh*.

3. INTERMITTENT EXPANSION AND RESTRUCTURING OF THE CORONA – NEW OBSERVATIONS FROM *YOHKOH*

Since the *Skylab* era, the solar corona seen in soft X-rays has been known to be highly inhomogeneous, consisting of three portions, i.e., active regions, loops connecting active regions (or "quiet corona"), and coronal holes, with tiny X-ray bright points appear-

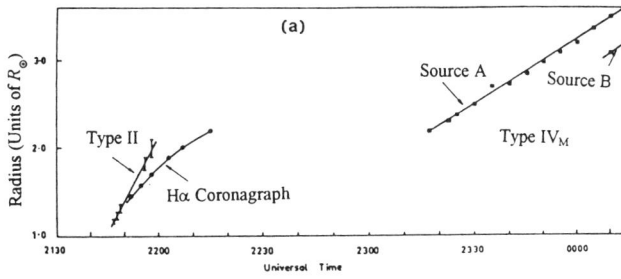

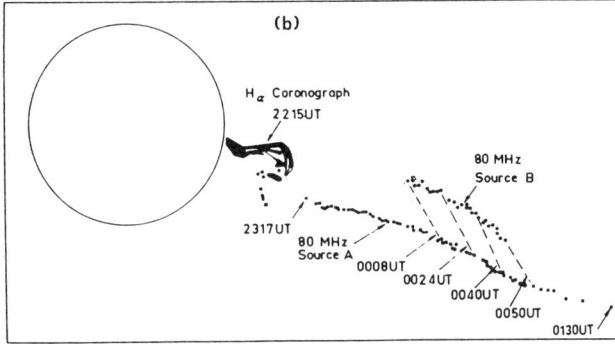

Fig. 3. The 1969 March 1 event observed with the Culgoora Radioheliograph at 80 MHz (after *Riddle* [1970]). (a) Temporal relationship between the type II burst, the Hα flare spray, and the "plasmoid" moving type IV burst sources. (b) Spatial comparison between the Hα flare spray and the moving type IV burst sources.

ing sporadically everywhere. The *Yohkoh* Soft X-ray Telescope (SXT) [*Tsuneta et al.*, 1991] has recently revealed, thanks to its higher spatial resolution (~2.5 arcsec pixels), much lower X-ray scattering in its telescope optics, and higher time cadence of image acquisition with better time coverage, that the corona is full of marvelous complexity and dynamical behavior.

Active regions are filled with many long, thin loops, which are believed to trace magnetic flux tubes or bundles of field lines. Much longer but similarly thin loops connect two active regions whose separation is sometimes greater than the solar radius. Some of the loops are twisted or S-shaped; others show even a kink-like shape [*Acton et al.*, 1992]. It is astonishing to see a loop whose diameter is much smaller than its length but has an almost constant cross-sectional area along the loop [*Klimchuk et al.*, 1992]. This cannot be interpreted as a current-free magnetic loop; instead it suggests current flowing along the loop. Relatively frequently but intermittently, we see loops expand outwards, but only rarely to shrink. It is

noteworthy mentioning that, although no precise estimate has been made, the rate of mass loss due to this kind of "magnetic expansion" may not be negligibly small in comparison with the steady solar wind flow [*Uchida et al.*, 1992].

One of the most spectacular findings by *Yohkoh* SXT is that the X-ray corona drastically changes its morphology or topology through "magnetic reconnection", evidence for which, and whose significance for our understanding of flares and CMEs, we will discuss below.

3.1. *Large-Scale Arcade Formation*

A large-scale restructuring of coronal magnetic fields often takes place in association with a *disparition brusque* (or disappearance of a quiet-region filament) in the form of "propagating arcade formation", in which a closed-loop arcade is formed progressively from one end to the other in an open-field area. The spatial and temporal scales of the event sometimes exceed $l \sim 10^6$ km and $\tau \sim 10^5$ s, respectively.

One of the largest-sized, but relatively faint, examples was reported by *Tsuneta et al.* [1992a]. It occurred near the north pole when a polar crown filament disappeared on 1991 November 12 (see Figure 1 of their paper). The arcade covered almost half the northern hemisphere, with the arcade length and width of 1.5 $R_\odot$ and 0.5 $R_\odot$, respectively. The arcade formation began at ~00:00 UT, propagated westwards ($v = 20$–40 km s^{-1}), and reached the west limb at ~12:00 UT, when the process was still in progress as revealed by increasing height and footpoint separation ($v = 2$–4 km s^{-1}). The arcade above the west limb showed a cusp shape, which, together with the increasing height and footpoint separation, may be evidence for magnetic reconnection, i.e., merging of anti-parallel magnetic fields above the cusp region.

The 1991 November 12 event mentioned above is an example of the most simplistic cases. Other events may be slightly more complicated. Nonetheless the basics of the process mentioned above may apply to most of the cases; for examples, see *Watanabe et al.* [1992], *McAllister et al.* [1992], *Kano* [1994], *Hanaoka et al.* [1994], and *Hiei et al.* [1993].

When we see such an arcade formation in soft X-rays, the Hα filament associated with it has already disappeared, or more rigorously already erupted into the corona to more than 100,000 km above the photosphere. For a beautiful example, see *Hanaoka et al.* [1994]. Thus it is phenomenologically quite clear that the eruption of filament cannot be a result of the arcade formation; rather they may be two aspects of a single process.

Arcade formations, or more rigorously large-scale restructurings of coronal magnetic fields that reveal themselves as propagating arcade formations in soft X-rays, may give rise to interplanetary disturbances. In fact, a circumpolar arcade formation on 1994 April 14 (Figure 4; *McAllister et al.* [1996]) was accompanied by a major geomagnetic storm two days later.

3.2. Long-Duration, Cusp-Shaped Flares

Soft X-ray arcade formations discussed above take place outside active regions and are not so intense as flares. In fact, most of them are too weak to be recognized in the *GOES* total intensity plot unless we have imaging observations. Even with images, however, what we actually observe cannot be necessarily classified without any ambiguity into the two categories. Long-duration events (LDEs), defined as such flares lasting for more than one hour in the *GOES* total intensity plot, may correspond to intermediate cases, with typical spatial and temporal scales being $l \sim 10^5$ km and $\tau \sim 10^4$ s, respectively.

When an LDE occurs near the solar limb (e.g., Figure 5; *Tsuneta et al.* [1992b]), we see at least in the late phase of the event a soft X-ray bright loop with a cusp above its apex. This "loop-with-a-cusp" structure increases in height and footpoint separation with time. Hence, the structure and evolution are similar to those found in the arcade formation discussed in the previous subsection. In addition, thanks to its higher intensity we can derive temperatures of individual portions in the structure; we find that the cusp area, especially its outer boundary, has higher temperatures than inside the bright loop. These observational facts are all suggestive, again, that we observe the magnetic reconnection process going on [*Tsuneta*, 1993]. It is to be noted here that the magnetic field configuration suggested by the "loop-with-a-cusp" structure is just homologous to a classical model for two-ribbon flares now known as the Carmichael / Sturrock / Hirayama / Kopp and Pneuman (CSHKP) model [*Svestka and Cliver.*, 1992].

Another feature to be remarked in the 1992 February 21 event (the flare shown in Figure 5) is that expansion and restructuring of the active-region coronal magnetic fields are intermittently repeated from ten hours prior to the flare (see Figures 2 and 3 of *Tsuneta et al.* [1992b]), resulting in formation of a

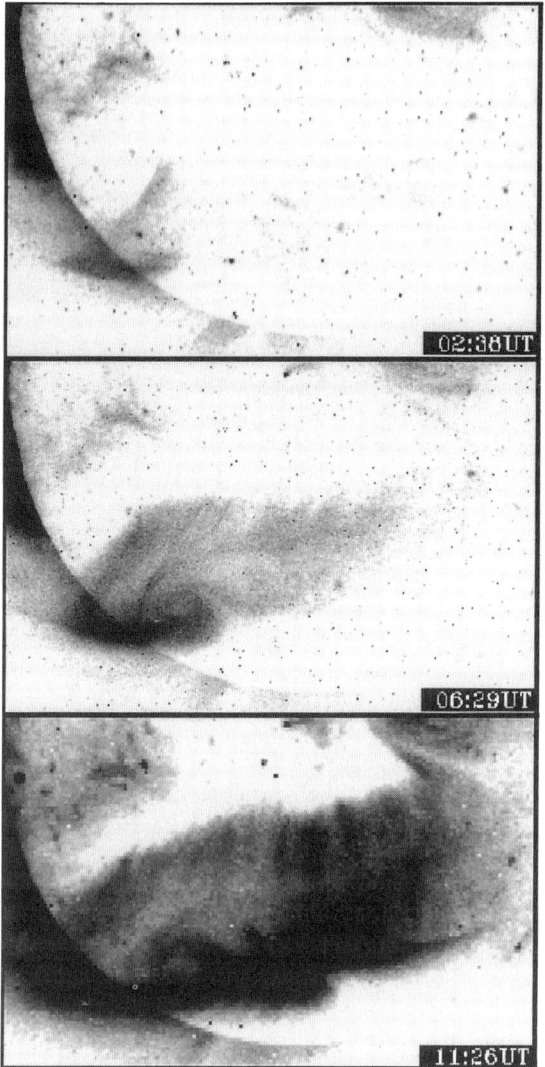

Fig. 4. Circumpolar arcade formation event on 1994 April 14 observed with *Yohkoh* SXT (courtesy A. McAllister).

helmet-streamer-like, bright arch several hours prior to the flare. This suggests that a current sheet is formed as a result of global MHD instability. Note that the expansion and restructuring are in progress in a much wider volume than the bright core portion of the flare.

3.3. Compact Flares with Erupting Signatures

Though both LDEs and large-scale arcade formations usually show a clear "loop-with-a-cusp" structure, no such structure had been found for impulsive flares; rather the apparent shape of impulsive flares

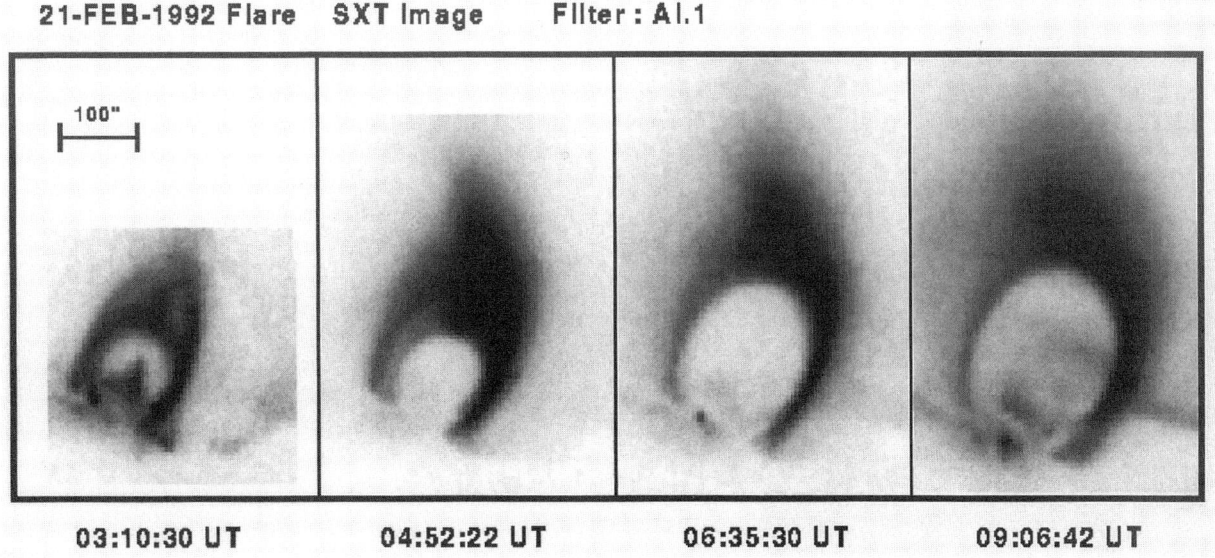

Fig. 5. LDE flare on 1992 February 21 observed with *Yohkoh* SXT (courtesy S. Tsuneta).

seen in soft X-rays is just a simple loop. The apparent lack of cusp structure in impulsive flares had been thought to be evidence against magnetic reconnection models such as the CSHKP model. Thus, some theoreticians have considered loop-flare models in which the primary energy release is assumed to take place inside the simple loop [*Spicer*, 1977; *Uchida and Shibata*, 1988]. The situation is, however, a little bit complicated; impulsive flares may not be necessarily confined in a simple loop, as discussed at the end of Section 2.

When a simple loop is seen in soft X-rays, impulsive flares brighten in hard X-rays at the two footpoints of the loop. This double footpoint structure was first recognized for a small number of impulsive flares in the first hard X-ray images taken by *SMM* and *Hinotori* in the early 1980s [*Hoyng et al.*, 1981; *Duijveman et al.*, 1982] and has finally been confirmed as a general trend by Sakao and colleagues [*Sakao*, 1994; *Sakao et al.*, 1992, 1994a,b] using the *Yohkoh* Hard X-ray Telescope (HXT) [*Kosugi et al.*, 1991]. They also confirmed that the two sources brighten almost simultaneously, with time lags less than a few tenths of a second. This is interpreted as due to precipitation of electrons that are accelerated to high energies near the apex of the loop; for a more detailed discussion, see *Kosugi* [1996]. Thus the remaining question is whether the acceleration takes place inside the loop or somewhere else.

An interesting finding has been made by Masuda and colleagues [*Masuda*, 1994a,b; *Masuda et al.*, 1994, 1995] in this context. They have found, in several impulsive flares occurring near the solar limb, an isolated, blob-like source ("loop-top impulsive source") at an altitude of more than 10^4 km above the photosphere. This source looks isolated from the double footpoint sources in the X-ray energy range above ~30 keV. Surprisingly, at least in some cases, the source is located well above the apex of the corresponding soft X-ray loop (Figure 6); this source may be better named "above-the-loop-top source". Although this hard X-ray source is weak in comparison with the double footpoint sources by about an order of magnitude, it varies its intensity almost similarly to the footpoint sources, i.e., impulsively, as far as the effective temporal resolution of several to ten seconds (limited by poor photon statistics) is concerned. It shows a relatively hard spectrum; if the emission is assumed to be of thermal origin, we get the temperature $T \sim 2 \times 10^8$ K and the emission measure $EM \sim 10^{44}$–10^{45} cm^{-3}.

The existence of an impulsive hard X-ray source above the soft X-ray flaring loop is of crucial importance. Without doubt, some energetic process, including particle acceleration, is in progress outside the bright soft X-ray loop. Moreover, soft and hard X-ray images combined together remind us of the "loop-with-a-cusp" structure seen in LDEs and arcade for-

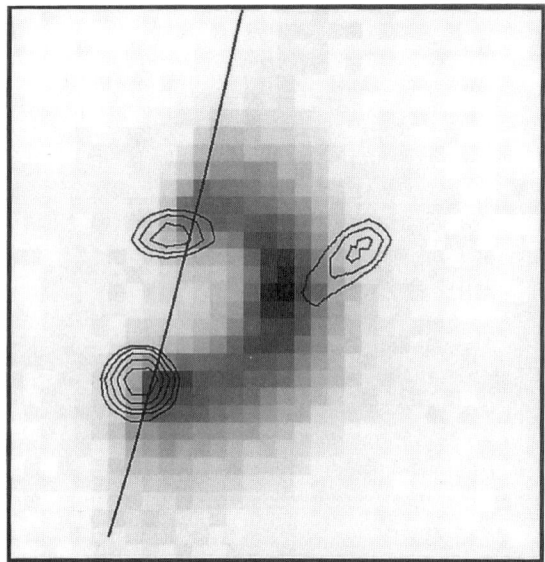

Fig. 6. "Above-the-loop-top" hard X-ray source seen over the limb (1992 January 13 flare). A hard X-ray image in the HXT M2-band (32.7–52.7 keV; in contours at 18, 25, 35, 50, and 71 % of the peak brightness) is overlaid upon the corresponding soft X-ray image through the SXT Be filter (in grey scale). The west solar limb is denoted by solid line. The field of view covers 78×78 arcsec, with the solar north to the top, west to the right (courtesy S. Masuda).

mations, which was interpreted in Sections 3.1 and 3.2 as evidence for magnetic reconnection. In this context it is reasonable to hypothesize that the "above-the-loop-top" hard X-ray source is evidence for similar, but more violent, magnetic reconnection taking place in a smaller spatial scale in impulsive flares.

Now we have arrived at a point where we claim, contrary to the widely-accepted classification scheme of solar flares, that impulsive flares do not differ much from LDEs or further from large-scale arcade formations in their basic physics. A few consequences of this hypothesis need be checked, however. The presence of a vertical current sheet may be easily understood if a rising plasmoid stretches anti-parallel magnetic fields upwards; see, e.g., *Svestka and Cliver* [1992]. Also the X-type reconnection scheme requires, in addition to the downward reconnection outflow, an upward outflow which may collide with the plasmoid. In fact, *Shibata et al.* [1995] have found faint upward ejecta seen in soft X-rays in association with impulsive flares for all the events analyzed by *Masuda* [1994a]. Thus it is evident that some eruptive process is really involved even in impulsive flares. Where are electrons accelerated? An interesting new technique has been used in an attempt to answer this question by *Aschwanden et al.* [1996a,b,c]. They have examined energy-dependent peak delays of hard X-rays measured with the highly sensitive BATSE detector onboard *CGRO* and interpreted them in terms of time-of-flight distances of electrons along the path from the acceleration site to the thick-target hard X-ray emission site, i.e., the double footpoint sources, and compared the distances with *Yohkoh* HXT and SXT images. They have concluded that the acceleration site is above the apex of the soft X-ray loop and coincides with the "above-the-loop-top" hard X-ray source (or even above), which is consistent with the idea that electrons are accelerated where the reconnection outflow collides with the underlying magnetic field structure and forms a shock. A cartoon of our reconnection "flare" model is shown in Figure 7, which we believe to be applicable not only to impulsive flares but also to LDEs and further to large-scale arcade formations.

3.4. Soft X-ray Jets and Microflares

An apparently different type of magnetic reconnection can be seen as "X-ray jets" [*Shibata et al.*, 1992, 1994a,b; *Shimojo et al.*, 1996]. An X-ray jet has a typical length 5×10^3–4×10^5 km, velocity 30–300 km s^{-1}, and kinetic energy 10^{25}–10^{28} erg. Many jets occur in emerging flux regions, are accompanied by tiny soft X-ray brightening loops and sometimes by Hα surges. They are classified into two morphological types, either of which can be interpreted as resulting from magnetic reconnection due to collisions of an emerging flux with the overlying magnetic structure [*Shibata et al.*, 1994b].

Thanks to the high sensitivity of *Yohkoh* SXT, many tiny transient brightenings not counted as flares have been observed in active regions and extensively studied by *Shimizu et al.* [1992, 1994]. They are most plausibly the soft X-ray counterparts of hard X-ray microflares [*Lin et al.*, 1984]. In fact, their occurrence frequency as a function of their peak intensity follows a power-law distribution, i.e., $dN/dI \propto I^{-\alpha}$ with $\alpha \cong 1.5$–1.6, which is similar to what was found for flares [e.g., *Dennis*, 1985]. It has not yet been clear observationally, however, whether the morphology of these microflares is the same as shown in Fig-

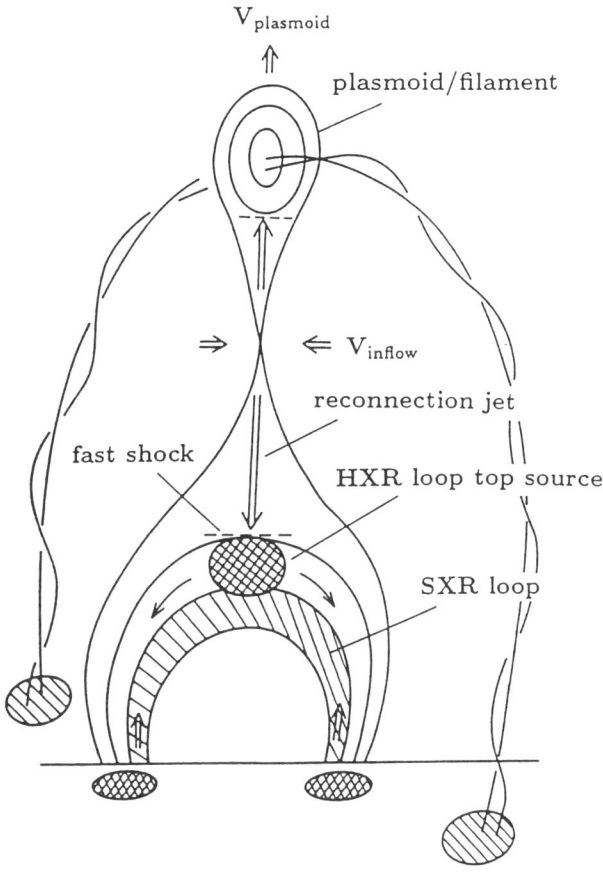

Fig. 7. Plasmoid-driven reconnection model which are applicable to both impulsive and LDE flares (after *Shibata et al.* [1995]).

ure 7, but theoretically we have no reason to adopt any other models than magnetic reconnection.

4. SUMMARY AND CONCLUSIONS

In the preceding sections, we have discussed in parallel two topics that are apparently independent of each other.

At first, mainly in Section 2, much effort has been devoted to establishing that flares are really accompanied by several types of ejecta, some of which may reach interplanetary space. Flare-associated ejecta may be divided into two categories. One includes Hα flare sprays, shock waves (revealing themselves as meter-wave type II bursts and "advancing-front" moving type IV bursts), and plasmoid ejections (seen as "plasmoid" moving type IV bursts), all of which seem to be launched in the impulsive phase of flares.

The other category includes prominence eruptions (or *disparition brusques*) and CMEs which, at least for many cases, are initiated in the precursor phase or well prior to the impulsive phase. Also to be recalled is the "expanding arch" form of moving type IV burst, which may be a radio signature of intermittent expansion of loops that is now frequently observed in soft X-rays by *Yohkoh* SXT. Thus the corona is full of material/shock wave ejections from the precursor phase, through the impulsive phase, and perhaps into the gradual phase of flares.

The other topic discussed mainly in Section 3 is the similarity or common features seen among various types of dynamical processes seen in X-rays. The key observational features here are the "plasmoid/filament ejection" and the "loop-with-a-cusp structure", which are seen in common from large-scale arcade formations, through LDEs, to impulsive flares. We have advocated our new view that most, though not necessarily all, of the eruptive phenomena can be understood in a unified way as resulting from a single, common process, whose essence is schematically shown in a cartoon of Figure 7.

We believe that the statements given in the preceding two paragraphs can be united if we take the following assumptions into account:

1. The violence of energy release or time scale of explosion of "generalized flares" is determined mainly by the size scale, the intensity of magnetic field, and perhaps other factors in the volume which are ready to destabilize. The total energy released is determined by the same parameters as well. Here we use the term "generalized flare" to denote any type of explosive phenomenon from large-scale arcade formations to flares (hopefully further to tiny microflares). Also this term may be used to denote the individual phases of a flare.

2. "Flares" are universally associated with plasmoid ejections, irrespective of whether they are seen in a specific observation channel, and also irrespective of the velocity of ejections.

Scaling laws, related to item 1) above have been discussed by *Shibata* [1996] based upon the reconnection model. Here we only summarize the time scales and energetics of "generalized flares" thus derived by *Shibata* [1996] in Table 2. It is remarkable and encouraging that we find similar time scales for the wide variety of "flares" if we adopt values normalized by the Alfvèn transit time. Also to be remarked is that total energy values estimated for impulsive flares, LDEs, and large-scale arcade formations have almost

TABLE 2. Comparison of Various Types of "Flares"

type	size (10^4 km)	τ (s)[a]	τ_A (s)[b]	τ/τ_A	energy (erg)
microflares	0.5–4	60–600	5	12–120	10^{26}–10^{29}
impulsive flares	1–10	60–3×10^3	10	6–300	10^{29}–10^{32}
LDE flares	10–40	3×10^3–10^5	90	30–1000	10^{30}–10^{32}
arcade formations	30–100	10^4–2×10^5	400	25–500	10^{29}–10^{32}

[a]Typical time scale of the phenomenon.
[b]Alfvèn transit time, i.e., size divided by the Alfvèn velocity.

the same maximum values. This simple fact may explain why interplanetary disturbances, as well as geomagnetic storms, are given rise to by any of the three classes of "generalized flares".

Acknowledgements. We owe much to our colleagues in the *Yohkoh* team. The Ministry of Education, Science, Sports and Culture of Japan supported TK's attendance at this Chapman Conference. The *Yohkoh* satellite is a Japanese national project with international collaboration with U.S.A. and U.K., with its operation having been continuously supported by ISAS, NASA, and SERC.

REFERENCES

Acton, L. W., S. Tsuneta, Y. Ogawara *et al.*, The Yohkoh mission for high-energy solar physics, *Science*, *258*, 618, 1992.

Aschwanden, M. J., H. S. Hudson, T. Kosugi, and R. C. Schwartz, Electron time-of-flight measurements during the Masuda flare 1992 January 13, *Astrophys. J.*, *464*, 985, 1996a.

Aschwanden, M. J., M. J. Wills, H. S. Hudson *et al.*, Electron time-of-flight distances and flare loop geometries compared from *CGRO* and *Yohkoh* observations, *Astrophys. J.*, *468*, 398, 1996b.

Aschwanden, M. J., T. Kosugi, H. S. Hudson *et al.*, The scaling law between electron time-of-flight distances and loop lengths in solar flares, *Astrophys. J.*, *470*, 1198, 1996c.

Bruzek, A., Optical evidence for plasma ejections and waves in the solar corona, in *Coronal Disturbances*, IAU Symp. No. 57, edited by G. Newkirk, Jr., pp.323–332, D. Reidel, Dordrecht, 1974.

Dennis, B. R., Solar hard X-ray bursts, *Solar Phys.*, *100*, 465, 1985.

Dryer, M., Coronal transient phenomena, *Space Sci. Rev.*, *33*, 233, 1982.

Duijveman, A., P. Hoyng, and M. E. Machado, X-ray imaging of three flares during the impulsive phase, *Solar Phys.*, *81*, 137, 1982.

Dulk, G. A., D. J. McLean, and G. J. Nelson, Solar flares, in *Solar Radiophysics*, edited by D. J. McLean and N. R. Labrum, chapter 4, Cambridge Univ. Press, Cambridge, 1985.

Farrugia, C. J., L. F. Burlaga, and R. P. Lepping, Magnetic clouds and the quiet-storm effect at Earth, in these proceedings, 1997.

Gopalswamy, N. and M. R. Kundu, Surprises in the radio signatures of CMEs, in *Coronal Magnetic Energy Releases*, edited by A. O. Benz and A. Krüger, pp.223–232, Springer-Verlag, Berlin, 1995.

Gosling, J. T., The solar flare myth, *J. Geophys. Res.*, *98*, 18,937, 1993.

Gosling, J. T., Reply, *J. Geophys. Res.*, *100*, 3479, 1995.

Gosling, J. T., D. J. McComas, J. L. Phillips, and S. J. Bame, Geomagnetic activity associated with Earth passage of interplanetary shock disturbances and coronal mass ejections, *J. Geophys. Res.*, *96*, 7831, 1991.

Hanaoka, Y., H. Kurokawa, S. Enome *et al.*, Simultaneous observations of a prominence eruption followed by a coronal arcade formation in radio, soft X-rays, and Hα, *Publ. Astr. Soc. Japan*, *46*, 205, 1994.

Harrison, R. A., Solar coronal mass ejections and flares, *Astron. Astrophys.*, *162*, 283, 1986.

Harrison, R. A., P. W. Waggett, R. D. Bentley *et al.*, The X-ray signature of solar coronal mass ejections, *Solar Phys.*, *97*, 387, 1985.

Hiei, E., A. Hundhausen, and D. Sime, Reformation of a coronal helmet streamer by magnetic reconnection after a coronal mass ejection, *Geophys. Res. Lett.*, *20*, 2785, 1993.

Hildner, E., J. Bassi, J. L. Bougeret *et al.*, Coronal mass ejections and coronal structures, in *Energetic Phenomena on the Sun*, edited by M. R. Kundu *et al.*, chapter 6, Kluwer, Dordrecht, 1989.

Hoyng, P., M. E. Machado, A. Duijveman *et al.*, Hard X-ray imaging of two flares in active region 2372, *Astrophys. J. Lett.*, *244*, L153, 1981.

Hudson, H. S., The solar antecedents of magnetic storms, in these proceedings, 1997.

Hudson, H. S., B. Haisch, and K. T. Strong, Comments on "the solar flare myth" by J. T. Gosling, *J. Geophys. Res.*, *100*, 3473, 1995.

Kahler, S. W., Solar flares and coronal mass ejections, *Ann. Rev. Astron. Astrophys.*, *30*, 113, 1992.

Kai, K., Expanding arch structure of a solar radio outburst, *Solar Phys.*, *11*, 310, 1970.

Kai, K., D. B. Melrose, and S. Suzuki, Storms, in *Solar Radiophysics*, edited by D. J. McLean and N. R. Labrum, chapter 16, Cambridge Univ. Press, Cambridge, 1985.

Kano, R., The time evolution of X-ray structure during filament eruption, in *X-ray Solar Physics from Yohkoh*, edited by Y. Uchida *et al.*, pp.273–276, Universal Academy Press, Tokyo, 1994.

Klimchuk, J. A., J. R. Lemen, U. Feldman *et al.*, Thickness variations along coronal loops observed by the soft X-ray telescope on *Yohkoh*, *Publ. Astr. Soc. Japan*, *44*, L181, 1992.

Kosugi, T., Solar flare energy release and particle acceleration as revealed by *Yohkoh* HXT, in *High Energy Solar Physics*, edited by R. Ramaty *et al.*, pp.267–274, AIP Press, New York, 1996.

Kosugi, T., K. Makishima, T. Murakami *et al.*, The hard X-ray telescope (HXT) for the SOLAR-A mission, *Solar Phys.*, *136*, 17, 1991.

Lin, R. P., R. A. Schwartz, S. R. Kane *et al.*, Solar hard X-ray microflares, *Astrophys. J.*, *283*, 421, 1984.

MacQueen, R. M., Coronal transients: a summary, *Phil. Trans. R. Soc. London, Ser. A*, *58*, 363, 1980.

Masuda, S., Hard X-ray sources and the primary energy release site in solar flares, Ph.D. thesis, University of Tokyo, 1994a.

Masuda, S., Vertical structure of hard X-ray sources in solar flares, in *Proc. of Kofu Symposium*, edited by S. Enome and T. Hirayama, NRO Report No.*360*, pp. 209–212, 1994b.

Masuda, S., T. Kosugi, H. Hara *et al.*, A loop-top hard X-ray source in a compact solar flare as evidence for magnetic reconnection, *Nature*, *371*, 495, 1994.

Masuda, S., T. Kosugi, H. Hara *et al.*, Hard X-ray sources and the primary energy-release site in solar flares, *Publ. Astr. Soc. Japan*, *47*, 677, 1995.

McAllister, A., Y. Uchida, S. Tsuneta *et al.*, The structure of the coronal soft X-ray source associated with the dark filament disappearance of 1991 September 28 using the *Yohkoh* soft X-ray telescope, *Publ. Astr. Soc. Japan*, *44*, L205, 1992.

McAllister, A. H., M. Dryer, P. McIntosh, and H. Singer, A large polar crown coronal mass ejection and a "problem" geomagnetic storm: April 14–23, 1994, *J. Geophys. Res.*, *101*, 13,943, 1996.

Priest, E. R. (ed.), *Solar Flare Magnetohydrodynamics*, Gordon and Breach, New York, 1981.

Priest, E. R. (ed.), *Dynamics and Structure of Quiescent Solar Prominences*, Kluwer, Dordrecht, 1989.

Riddle, A. C., 80 MHz observations of a moving type IV solar burst, March 1, 1969, *Solar Phys.*, *13*, 448, 1970.

Robinson, R. D., Flare continuum, in *Solar Radiophysics*, edited by D. J. McLean and N. R. Labrum, chapter 15, Cambridge Univ. Press, Cambridge, 1985.

Sakao, T., Characteristics of solar flare hard X-ray sources as revealed with the hard X-ray telescope aboard the *Yohkoh* satellite, Ph.D. thesis, University of Tokyo, 1994.

Sakao, T., T. Kosugi, S. Masuda *et al.*, Hard X-ray imaging observations by *Yohkoh* of the 1991 November 15 solar flare, *Publ. Astr. Soc. Japan*, *44*, L83, 1992.

Sakao, T., T. Kosugi, S. Masuda *et al.*, Particle acceleration in the 15 November, 1991 solar flare observed with HXT, in *X-ray Solar Physics from Yohkoh*, edited by Y. Uchida *et al.*, pp. 91–94, Universal Academy Press, Tokyo, 1994a.

Sakao, T., T. Kosugi, S. Masuda *et al.*, Hard X-ray imaging observations of footpoint sources in impulsive flares, in *Proc. of Kofu Symposium*, edited by S. Enome and T. Hirayama, NRO Report No.*360*, pp. 169–172, 1994b.

Shibata, K., New observational facts about solar flares from *Yohkoh* studies – evidence of magnetic reconnection and a unified model of flares, *Adv. Sp. Res.*, *17*, Nos. 4/5, 9, 1996.

Shibata, K., Y. Ishido, L. W. Acton *et al.*, Observations of X-ray jets with the *Yohkoh* soft X-ray telescope, *Publ. Astr. Soc. Japan*, *44*, L173, 1992.

Shibata, K., N. Nitta, K. T. Strong *et al.*, A gigantic coronal jet ejected from a compact active region in a coronal hole, *Astrophys. J. Lett.*, *431*, L51, 1994a.

Shibata, K., N. Nitta, R. Matsumoto *et al.*, Two types of interactions between emerging flux and coronal magnetic field, in *X-ray Solar Physics from Yohkoh*, edited by Y. Uchida *et al.*, pp.29–32, Universal Academy Press, Tokyo, 1994b.

Shibata, K., S. Masuda, M. Shimojo *et al.*, Hot plasma ejections associated with compact-loop solar flares, *Astrophys. J.*, *451*, L83–L85, 1995.

Shimizu, T., S. Tsuneta, L. W. Acton *et al.*, Transient brightenings in active regions observed by the soft X-ray telescope on *Yohkoh*, *Publ. Astr. Soc. Japan*, *44*, L147, 1992.

Shimizu, T., S. Tsuneta, L. W. Acton *et al.*, Morphology of active region transient brightenings with the *Yohkoh* soft X-ray telescope, *Astrophys. J.*, *422*, 906, 1994.

Shimojo, M., S. Hashimoto, K. Shibata *et al.*, Statistical

study of solar X-ray jets observed with the *Yohkoh* soft X-ray telescope, *Publ. Astr. Soc. Japan*, *48*, 123, 1996.

Smerd, S. F. and G. A. Dulk, 80 MHz radioheliograph evidence on moving type IV bursts and coronal magnetic fields, in *Solar Magnetic Fields*, IAU Symp. No.*43*, edited by R. Howard, pp.616–641, D. Reidel, Dordrecht, 1971.

Spicer, D. S., An unstable arch model of a solar flare, *Solar Phys.*, *53*, 305, 1977.

Stewart, R. T., Moving type IV bursts, in *Solar Radiophysics*, edited by D. J. McLean and N. R. Labrum, chapter 14, Cambridge Univ. Press, Cambridge, 1985.

Svestka, Z., *Solar Flares*, D. Reidel, Dordrecht, 1976.

Svestka, Z. and E. W. Cliver, History and basic characteristics of eruptive flares, in *Eruptive Solar Flares*, IAU Colloq. No.*133*, edited by Z. Svestka *et al.*, pp.1–11, Springer-Verlag, Berlin, 1992.

Svestka, Z., R. V. Jackson, and M. E. Machado (eds.), *Eruptive Solar Flares*, IAU Colloq. No.*133*, Springer-Verlag, Berlin, 1992.

Tandberg-Hanssen, E., *Solar Prominences*, D. Reidel, Dordrecht, 1974.

Tsuneta, S., Solar flares as an ongoing magnetic reconnection process, in *The Magnetic and Velocity Fields of Solar Active Regions*, IAU Colloq. No.*141*, edited by H. Zirin *et al.*, pp.239–248, ASP, San Franscisco, 1993.

Tsuneta, S., L. Acton, M. Bruner *et al.*, The soft X-ray telescope for the SOLAR-A mission, *Solar Phys.*, *136*, 37, 1991.

Tsuneta, S., H. Hara, T. Shimizu *et al.*, Global restructuring of the coronal magnetic fields observed with the *Yohkoh* Soft X-ray Telescope, *Publ. Astr. Soc. Japan*, *44*, L63, 1992a.

Tsuneta, S., T. Takahashi, L. W. Acton *et al.*, Observation of a solar flare at the limb with the *Yohkoh* Soft X-ray Telescope, *Publ. Astr. Soc. Japan*, *44*, L211, 1992b.

Tsurutani, B. T. and W. D. Gonzalez, The interplanetary causes of magnetic storms: a review, in these proceedings.

Uchida, Y. and K. Shibata, A magnetodynamic mechanism for the heating of emerging magnetic flux tubes and loop flares, *Solar Phys.*, *116*, 291, 1988.

Uchida, Y., A. McAllister, K. T. Strong *et al.*, Continual expansion of the active-region corona observed by the *Yohkoh* soft X-ray telescope, *Publ. Astr. Soc. Japan*, *44*, L155, 1992.

Watanabe, T., Y. Kozuka, M. Ohyama *et al.*, Coronal / interplanetary disturbances associated with disappearing solar filaments, *Publ. Astr. Soc. Japan*, *44*, L193, 1992.

Zirin, H., *Astrophysics of the Sun*, Cambridge Univ. Press, Cambridge, 1988.

Takeo Kosugi and Kazunari Shibata, National Astronomical Observatory, 2-21-1 Ohsawa, Mitaka, Tokyo 181, Japan.

Solar Antecedents of Geomagnetic Storms

HUGH S. HUDSON

Solar Physics Research Corp.. ISAS, 3-1-1 Yoshinodai, Sagamihara-shi, Kantatawa-ken, Japan 229

Geomagnetic storms occur when solar wind with unusual characteristics touches the Earth's magnetosphere, but their ultimate origins lie in the structure and dynamics of the solar atmosphere. We observe this atmosphere above the solar limb with coronagraphs, and generally across the whole visible hemisphere at radio and XUV wavelengths. The soft X-ray telescope on *Yohkoh* is giving us new, comprehensive views of the hot solar corona, starting from September 1991, and still newer data are now arriving from SOHO. These new data reveal solar mass loss of several types, including large-scale effects probably related to coronal mass ejections. The X-ray images often show a feature (brightening, ejection, or other disturbance) associated with such an event. The disturbances (sometimes visible only as a coronal depletion or "dimming") include solar flares and large-scale loop systems outside the active latitudes. In addition there are other, previously unobserved forms of coronal mass loss, including the gradual ejection of loops from active regions and the impulsive formation of X-ray jets. For events probably associated with CMEs, the X-ray imaging data tend to show a close temporal relationship between the mass ejection and any accompanying brightening.

1. INTRODUCTION

Terrestrial magnetic storms start at the Sun, a fact discovered long ago from the recognition of the solar rotation period in storm statistics. A study of the surface of the Sun and its lower atmosphere therefore offers us the earliest possible notification that a terrestrial disturbance will happen, if only we can decode the signatures and understand their physics properly. The best-known solar antecedent is the coronal mass ejection [see *Hundhausen*, 1997] for a recent review), the dominant direct cause of a geomagnetic storm [eg *Kahler*, 1992; *Gosling*, 1991]. There also exists a connection between slowly-varying solar features and certain geomagnetic storms, the so-called "M-region" relationship known for many decades.

Since *Skylab* we have known of the close relationship between coronal mass ejections and the formation of large arcades of magnetic loops in the solar corona. These loops emit soft X-rays, and the hot plasma trapped in them supplies the energy to the chromosphere that results in Hα "ribbons" and other manifestations [*Harvey et al.*, 1996]. Our new X-ray data come from the *Yohkoh* soft X-ray telescope, which observes these events much more comprehensively than *Skylab* could. The SOHO coronal instruments successfully began observations at about the time of this conference and are now providing similarly improved observations relative to those of the earlier coronagraphs.

This article will focus on X-ray observations of the discrete events (transients), rather than on the "corotating" or M-region structures. At the time of writing there is

Magnetic Storms
Geophysical Monograph 98
Copyright 1997 by the American Geophysical Union

almost no literature describing the *Yohkoh* SXT or SOHO views of the solar antecedents of storms in this second (non-transient) category. *Crooker and Cliver* [1992] note that the M-region identification with coronal holes may need revision in terms of "streamer/hole ensembles", so this is an interesting area for future work. This paper will also not deal with the storm-related work on filament eruptions (for recent *Yohkoh* observations, see [*Hanaoka et al.*, 1994; *Kano*, 1994]), the role of helicity [*Martin et al.* 1994; *Bothmer and Schwenn,* 1994; *Rust,* 1994; {*Rust and Kumar*, 1994], nor the interplanetary developments [*e.g. Weiss et al.*, 1996].

2. BACKGROUND: THE CORONA AND THE SOLAR WIND

We can image the solar corona in white light during eclipses and via coronagraphs, but only above the solar limb. We can also image the corona in front of the solar disk by going to long or short wavelengths, either because the lower temperature of the photosphere makes it appear dark (X-rays) or because the height of optical depth unity rises into the corona directly (the "photosphere" defined in this way reaches 1 A.U. at about 30 kHz). Soft X-radiation is the natural emission channel for the hot corona, but there are coronal emission lines of highly-ionized atoms even in the visible spectrum, and their successful identification historically provided the evidence necessary to establish the high temperature of the corona. The complicated structure we see in the lower solar atmosphere in an X-ray image is mainly defined by small-scale magnetic fields. These simplify with height and eventually map into the solar-wind flow, which has a predominantly bidirectional magnetic structure dominated by the flow itself. Most of the magnetic field lines advected out into the heliosphere map back onto the solar photosphere in coronal holes, identified as the sources of the high-speed flow of the solar wind, and perhaps into smaller areas elsewhere. The origin of the slow component of the solar wind is less well known, but it probably originates in or near the closed-field regions of the solar atmosphere at lower latitudes [*e.g. Withbroe et al.*, 1991; *Gosling*, 1996].

In addition to the quasi-steady coronal structure, there are transients of many kinds. These usually take the form of additional ejections of magnetic field in discrete events. The mass entrained in this field may then also become a part of the outward flow of the solar wind. Specifically, coronal mass ejections (CMEs) occur several times a day during sunspot maximum and have a strong solar-cycle modulation in phase with the sunspot number [*Webb and Howard*, 1994; *cf. Hundhausen*, 1993]. The defining observations of CMEs are those of white-light coronagraphs, but there are strong relationships with certain signatures obtained with *in situ* observations made by interplanetary spacecraft [*e.g. Gosling*, 1990]. The soft X-ray observations provide a fundamentally different mode for the observation of coronal mass ejections, and the *Yohkoh* observations represent the first comprehensive data set of this type. *Yohkoh* began in September, 1991, and continues through the time of writing of this paper (Figure 1). We need to "calibrate" the observed phenomena against traditional coronagraph observations in white light, but few comparisons have yet been made either with the SOHO space-based coronagraphs or with the Mauna Loa K-coronameter.

3. THE *Yohkoh* SOFT X-RAY OBSERVATIONS

Yohkoh SXT responds to soft X-rays from lines and continuum near 1 keV (10 Å). Such X-rays represent the main emission domain of plasmas at coronal temperatures, and the *Yohkoh* passbands lie at the hot end of this domain. [*Tsuneta et al.* 1991] give a full description of the instrument, which uses a grazing-incidence mirror, a set of broad-band filters sensitive in the range 0.3 - 3 keV, and a CCD sensor with 1024 x 1024 pixels 2.45" square [see *Acton et al.*, 1992], for an overview of the data). The telemetry capacity of *Yohkoh* allows the transmission of about 20 whole-Sun images, with 2 x 2-pixel summations (about 5" pixels), per 97-minute orbital period of the spacecraft. *Yohkoh* SXT provides a global view of the corona within its field of view, which is a square 0.70^0 across. The spacecraft pointing almost always keeps the full solar disk in view, except during special operations. A flare mode normally triggers at about the C2 level, 2×10^{-3} ergs (cm^2 sec)$^{-1}$ in the 1-8 Å band. This results in the loss of full-Sun imaging for an extended interval, in exchange for more telemetry devoted to high-resolution observations of the flare itself with a maximum field of view 10 arc min square.

There are several substantial interpretative differences between soft X-ray and white-light observations. Most of our knowledge of CME origins comes from the white-light data, so it is important to understand the differences in the data being reported here. The X-rays show the entire hemisphere, offer a better view near the Sun, and provide some information on temperatures. In contrast to the relatively direct interpretation of white-light brightness and polarization, the X-ray brightness unfortunately is a complicated function of density and temperature. For *Skylab* observations, for example, [*Kahler* 1976] finds the signal to depend approximately as the square of the electron pressure. It is generally a still stronger function of temperature for the SXT observations because of the bias towards shorter wavelengths[1]. This strong dependence means that X-ray coronal features, which reflect any temperature non-uniformity, will have more contrast than white-light features.

---

[1]For SXT, the response S through its thinnest filter varies with temperature approximately as $d(\ln S)/d(\ln T) = 2-6$ at typical coronal temperatures.

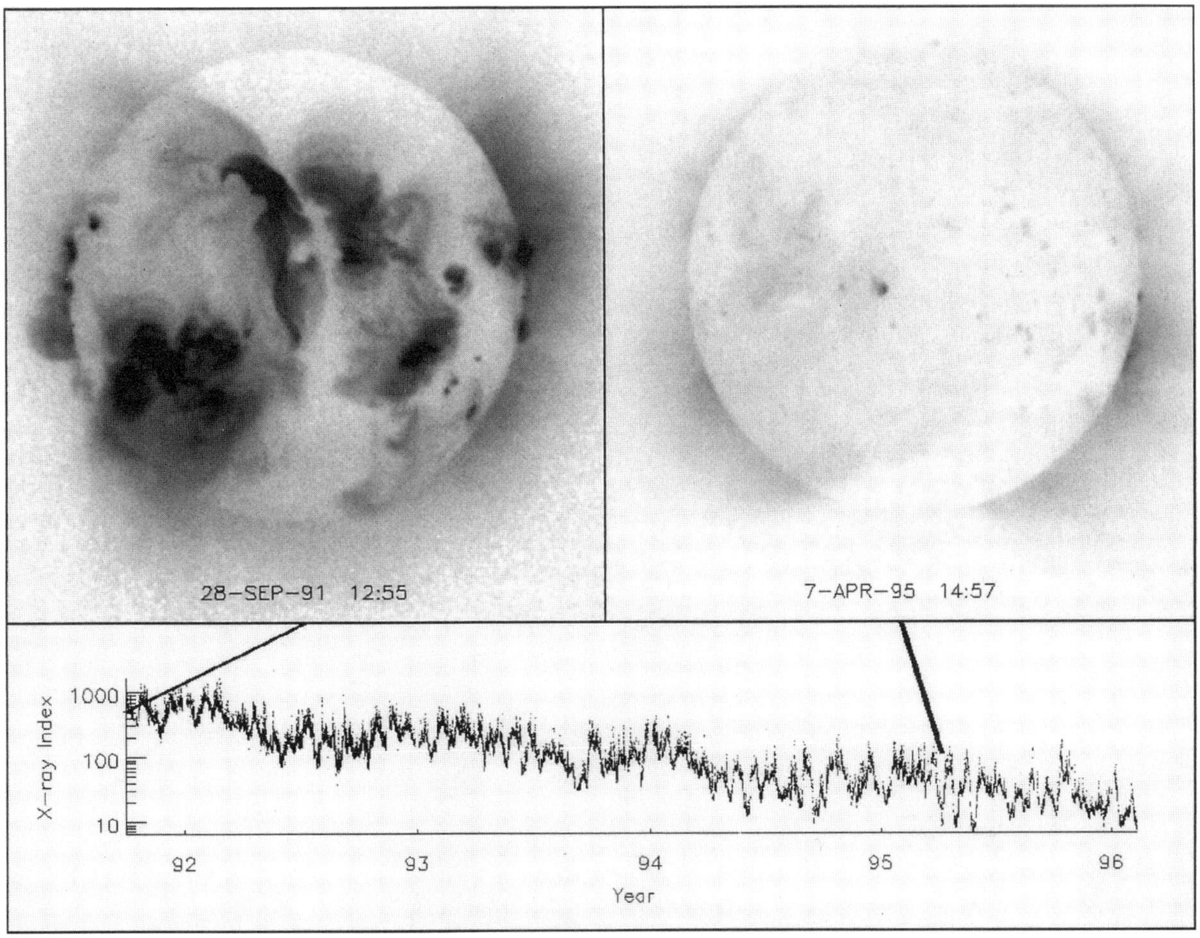

Figure 1. The solar-cycle dependence of coronal brightness, illustrated in two soft X-ray images from *Yohkoh* and in the time series of total fluxes measured in such images. These data come from the standard *Yohkoh* full-Sun movie, and therefore do not include major flares.

Because of this strong dependence, and because of the presence of scattered light even in low-scattering optics such as those of *Yohkoh* SXT, faint or cool features may be obscured by bright or hot ones. On the other hand, there is no need for the "plane of the sky" approximation in soft X-rays, except in the sense that the corona is optically thin in X-radiation.

4. SXT OBSERVATION OF SOLAR MASS LOSS

At the time of launch, the expectation may have been that the X-ray images would reveal coronal mass ejections more or less resembling those observed with a coronagraph. This turned out not to be the case: whereas some CMEs probably are easily observed in soft X-rays, they seem different; at the same time *Yohkoh* detects different kinds of mass-loss events that had not been predicted. Approximately in order of size, the *Yohkoh* data show (Table 1) the following things.

TABLE 1. X-ray Ejection Phenomena

Active-region loops	*Uchida et al.,* [1992]
Filament eruptions	e.g *Kahler,* [1992]; *Hanaoka et al.,* [1994]
X-ray jets	*Shibata et al.,* [1992]; *Strong et al.,* [1992]
Fast flare ejecta ($>10^3$ km s^{-1})	(no literature)
Slow flare ejecta ($<10^3$ km s^{-1})	*Shibata et al.,* [1995]; *Hudson et al.,* [1996c]
CME analogs	*Klimchuk et al.,* [1994]; *Watanabe et al.,* [1994]
Streamer events	*Hiei et al.,* [1993]; *Hudson,* [1996]

4.1. Soft X-ray Jets

The common occurrence of X-ray jets, invariably associated with flaring loops at their feet, provides one possible channel of mass ejection. *Yohkoh* SXT has made the first observations of such jets [*Shibata et al.*, 1992; *Strong et al.*, 1992; *Shimojo et al.*, 1996] because of its greatly improved time coverage. The jets have recently been successfully associated with Type III bursts [*Aurass et al.*, 1994; *Kundu et al.*, 1995; *Raulin et al.*, 1996], in spite of their different time scales. An X-ray jet consists of a collimated stream of plasma suddenly flowing outwards; the Type III burst comes from non-thermal electrons streaming out in this plasma column. The U-bursts (similar to Type III bursts, but on apparently closed field lines) may have similar soft X-ray counterparts [*Pick et al.*, 1994].

This identification therefore means that - in some cases at least - the outward flow of the jet may continue into the outer corona and contribute to the solar wind. Electrons from Type III bursts have long been detected directly in the solar wind directly [*e.g. Lin* 1985].

4.2. Flare Ejecta

A large fraction of solar flares produce ejecta in varying forms, ranging from the slow (<100 km s^{-1}) outward motion of compact blobs to faster (>1000 km s^{-1}) motions. [*Shibata et al.* 1995] found ejecta in all 8 of a list of strong, impulsive limb flares observed by *Yohkoh*. We distinguish these from jets (above) but do not know yet whether this distinction is physically justified. Certainly, many flare ejecta are not jet-like from the point of view of collimation and velocity. One of the present puzzles in the analysis of the soft X-ray data is the weakness or nonexistence of ordinary soft X-ray jets in flares, in spite of the known good correlation between type III bursts and flares [*e.g. Kundu* 1965]; in other words, the ratio of jet brightness to flare brightness appears to be much smaller for ordinary flares in active regions than in the quiet Sun.

4.3. Filament Eruptions

The behavior of Hα filaments provided the early best guide to the occurrence of a solar eruptive event (*disparition brusque*), and *Skylab* data clearly showed the X-ray coronal structures of the filament channels that represent the structure in which such eruptive events appear to arise. The new X-ray data show many beautiful examples of large arcades of this "global restructuring" type [*Tsuneta et al.*, 1992b; *Watanabe et al.*, 1992; *Hanaoka et al.*, 1994; *Khan et al.*, 1994; *Harvey et al.*, 1996]. The sense of chirality of the filament, if observed, appears to be related to that of the magnetic cloud resulting from the eruption [*Bothmer and Schwenn*, 1994; *Rust*, 1994]; this offers encouragement that the coronal configuration can eventually be linked to the "geoeffectiveness" of the event, since this depends upon the field orientation and flow properties.

4.4. Expanding Active-region Loops

One of the first discoveries in the new X-ray data was the tendency for some active-region loops to expand at intermediate speeds (10-50 km s^{-1}), rather than remain static [*Uchida et al.*, 1992]. This observation suggests that a magnetically-driven outward flow from active regions may contribute to the slow component of the solar wind. Such a mechanism also may be of interest more generally for stellar mass loss.

4.5. CME-like Eejecta at the Limb

Klimchuk et al. [1994], in an early study of the *Yohkoh* data based upon the standard movie images, found expanding features at the limb with roughly the same parameters (size, speed) as those seen in coronagraphs, except with somewhat lower projected velocities. Recently [*Gopalswamy et al.* 1996] have described a slow ejection at the limb that appears to incorporate the three elements of a "classic" CME - a front, a cavity, and an embedded filament. In this case the filament was well-observed with the Nobeyama 17 GHz radioheliograph, and the front clearly was a slowly-rising magnetic structure. The mass of this event was estimated as 1.2 x 10^{14} g, on the low side for a true CME but not very different from the mass found by [*Hudson et al.*, 1996c] for a flare-associated *Yohkoh* SXT mass-ejection event. It has not yet been possible to do a thorough calibration of the *Yohkoh* data set against coronagraph data, but there are many common events with the Mark III K-coronameter observations of the Mauna Loa Solar Observatory (A. Hundhausen, personal communication 1995).

5. THE "DIMMING" SIGNATURE

We have recently noticed a relatively obvious and natural effect in the *Yohkoh* SXT data: at the times expected for CME launches, *i.e.* near the beginnings of LDE flares or large arcade events, a large volume of the soft X-ray corona may suddenly become a factor of 2-3 dimmer. This effect was probably first described in the *Skylab* data [*Rust and Hildner*, 1978]. The *Skylab* observations, however, were distinctly limited in sampling and photometry, and were not well optimized for detecting such effects. The *Yohkoh* SXT data also have sampling limitations; with more frequent images the coverage would have been better and motions could have been measured more easily. The new results will be described in more detail below.

In some of the cases studied thus far, the dimming appears to be amorphous and unstructured. It results in a decrease of the coronal surface brightness of order 50% directly above the accompanying brightening. In other cases a structured mass flows outward from the region that dims. The dimming or outward mass flow apparently can either be sharply localized above the brightening, or it can be widespread in the vicinity of the brightening. [*Hudson et al.*, 1996c] point out that the radiative cooling time for such space and time scales greatly exceeds the time scale of the dimming, consistent with an interpretation in terms of material ejection.

The launching of a CME should in principle result in a decrease of coronal brightness, without new material supplied simultaneously from below. White-light decreases have been called "coronal depletions" [*Hansen et al.*, 1974], and a probably related X-ray signature "transient coronal holes" [*Rust*, 1983]. We observe similar effects in the *Yohkoh* data in a variety of forms, and use the term "dimming" for neutrality of interpretation. Table 2 lists some of the events studied thus far.

TABLE 2. X-ray Coronal Dimming Events

9-Nov-91	X1.1 flare	[*Watari et al.*, 1996]
24-Jan-92	Re-formation	[*Hiei et al.*, 1993]
		[*Hudson*, 1996]
21-Feb-92	M3.2 flare	[*Tsuneta et al.*, 1992b]
		[*Tsuneta*, 1996]
17-Apr-92	Arcade on limb	[*Lemen et al.*, 1996]
28-Aug-92	C1.5 flare	
16-Jan-93	Transient CH	[*Hundhausen and Hiei*, private comm. 1996]
24-Feb-93	Arcade in south	[*Lemen et al.*, 1996]
21-Mar-93	Arcade in south	[*Lemen et al.*, 1996]
9-Jul-93	Behind limb	[*Gosling et al.*, 1994]
		[*Lemen et al.*, 1996]
20-Feb-94	M4.0 flare	[*Hudson et al.*, 1996a]
27-Feb-94	M2.8 flare	[*Hudson et al.*, 1996b]
14-Apr-94	Arcade on disk (dimming?)	[*Alexander et al.*, 1994] [*McAllister et al.*, 1996]
25-Oct-94	C4.7 flare	[*Manoharan et al.*, 1966]
13-Nov-94	C1.2 flare	[*Hudson et al.*, 1996c]

The limb flare of 21 February 1992, studied previously by [*Tsuneta et al.*, 992a], shows a sudden dimming directly above the prominent cusp seen in this event [*Tsuneta*, 1996]. The dimming occurred just at the onset of flare brightening, *i.e.* the two effects appear to have been simultaneous. The movie representation of the data gives a distinct impression of outward motion. Another similar example occurred 28 August 1992, except in this case it was possible to see large-scale filamentary structures expanding out of the void region.

Large-scale clouds adjacent to the sites of solar flares have been observed to disappear during the flare brightening. In the best-studied event, 13 November 1994, the coronal cloud moves outward in a direction consistent with local vertical and a (constant) projected velocity consistent with the range expected from a CME [*Hudson et al.*, 1996c]. Another good example occurred in the event of 27 February 1994 [*Hudson et al.*, 1996b]. Many large-scale ejective events [*Manoharan et al.*, 1996] show a two-lobed structure (two sets of loops expanding in roughly opposite directions) strongly suggesting non-vertical motions.

The first *Yohkoh* SXT event directly associated with coronagraph observations consisted of a streamer disruption followed by a re-formation [*Hiei et al.*, 1993]. This event turns out also to provide an excellent example of coronal dimming [*Hudson*, 1996], which allows a relatively precise determination of the time of a CME launch. The onset of dimming preceded the onset of arcade brightening, although this conclusion is ambiguous because part of the event clearly occurred beyond the limb. In other large arcade events, the standard *Yohkoh* movie clearly shows the dimming to occur on an extremely large scale (on the order of 1 R_0. The event of 24 February 1993, for example, appears to unmask a large region of the south polar coronal hole (Figure 2).

We can measure the masses of the dimming regions. For the ejected cloud of the 13 November 1994 event, we find a lower limit of 4×10^{14}g [*Hudson et al.*, 1996c]. Only a lower limit is possible because of confusion with the brighter parts of the flare. Other difficulties in mass estimation include the lack of complete knowledge of the differential emission-measure distribution, and of the conceptual problem of estimating the replenished mass. This latter difficulty results from the continuous nature of the outward flow during the interval of flare brightening, and from the contrast problem (the difficulty of detecting faint features near bright ones).

For the 24 January 1992 large-scale event [*Hiei et al.* 1993], a rough estimate is about 10^{15}g [*Hudson*, 1996]. These estimates are within errors of the typical range of CME masses, although this means very little given the broad range of this property. A more explicit comparison will be interesting but will require simultaneous coronagraph and X-ray data.

6. TIMING AND CAUSALITY

The relative timing of the X-ray dimming signature and the X-ray brightening could point towards the direction of causality. Is there a major time lag (an hour or more) between the material ejection and the brightening, as could appear to be the case from coronagraph observations [*e.g. Hundhausen*, 1997]? Figure 2 shows an example of the relative timing for a well-observed large arcade event, 24

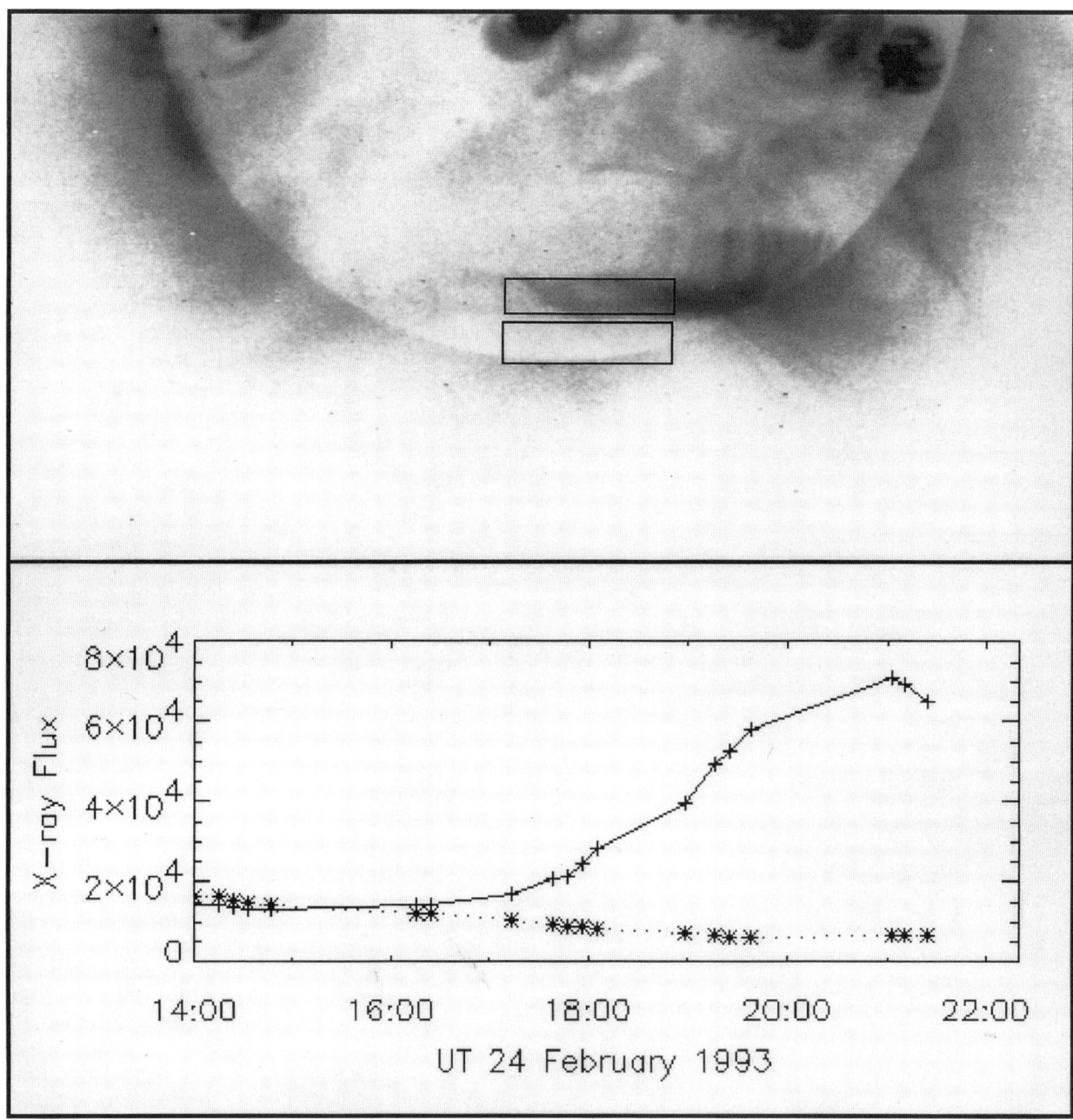

Figure 2. Example of mass loss associated with a large arcade event, that reported by [*Lemen et al.*, 1996]. The light curves at right show the total soft X-ray fluxes observed in the two boxes shown on the image at the left. The arcade event itself, as seen in the image, appeared to have a highly twisted ridge and later developed a streamer-like structure above the southwest limb.

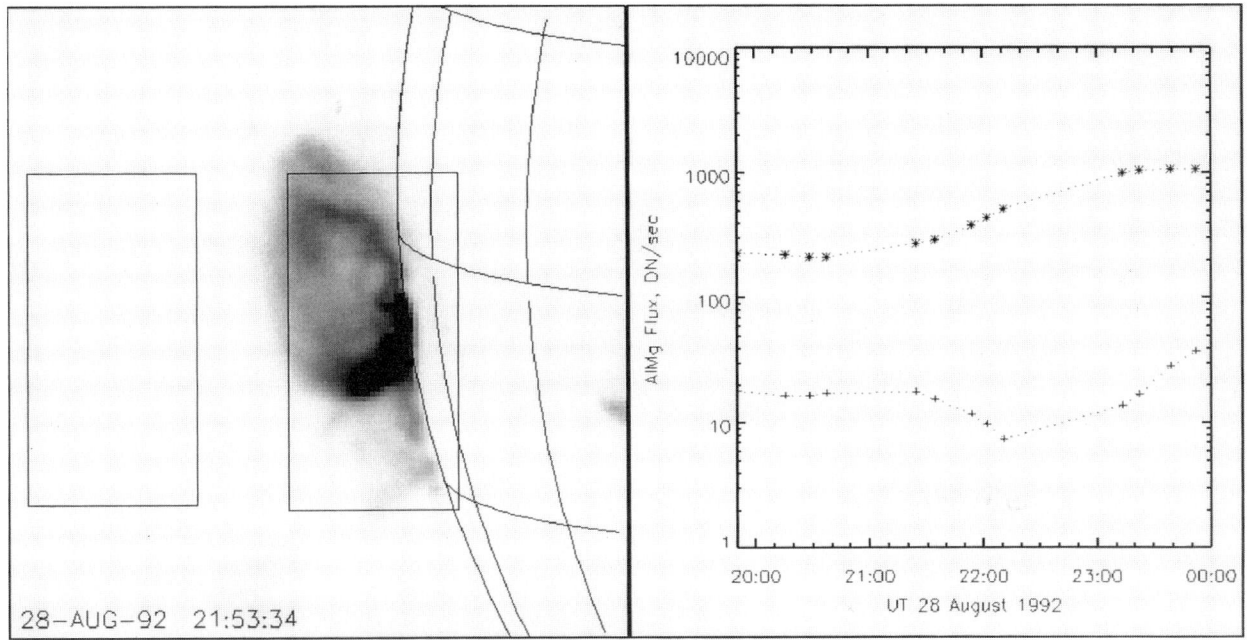

Figure 3. Example of mass loss associated with an LDE flare, 28 August 1992, in the same format as the arcade event of Figure 2. The flare arcade here appears to develop on an axis perpendicular to the line of sight, *i.e.* approximately north-south. In the lower light curve, the initial decrease is the "dimming" from the area above the flare; its subsequent increase is due to the development of cusp structures that grow to high altitudes late in the event.

February 1993 [*Lemen et al.*, 1996]. There is a gradual dimming over the south coronal hole throughout the time of brightening, which occurs at or before 17:14:10 UT in the (twisted?) spine of the arcade event. The data sampling of this event is relatively good and there appears to be no obscuration of any part of the event by the solar limb.

Figure 3 shows similar time series for the flare of 28 August 1992. Here the dimming appears to be centered over the site of the arcade structure, rather than to one side as in the event of Figure 2. This may be a perspective effect, or it may reflect a real difference between the flare-associated and arcade-associated dimming events.

In both events represented here, the dimming appears to proceed almost simultaneously with the brightening. In these examples and several others, there seems to be no well-defined time lag, and hence no particular reason to assign a direction of causality between the launch of coronal mass outward (our interpretation of the dimming) and the flare (our interpretation of the brightening). This is consistent with recent discussions of the optical coronagraph data as well [*Feynman and Hundhausen*, 1994; *Harrison*, 1995], in the sense that they are inconclusive on this point.

The 28 August 1992 event is especially interesting from the point of view of the dynamics of the expansion. As shown in the difference images of Figure 4, there appears to be a network of filamentary structures embedded in the expanding material above the flare. This opens the exciting possibility of direct measurement of the flow field. Already from the difference images themselves, one can see that the flow is everywhere outwards and shows no trace of the inward flow that might be expected if large-scale magnetic reconnection were the source of the flare energy.

7. CONCLUSIONS

The new data provide much greater sensitivity to solar disturbances, to the extent that the Dodson-Prince "problem storms" with no clear antecedents in solar flares [*Dodson and Hedeman*, 1964] may no longer represent problems. As [*McAllister et al.* 1992] point out in the context of the 14 April 1994 event, the solar antecedents of geomagnetic storms need not be intense flare events in active regions. The arcade resulting from such a disturbance may have a much larger spatial extent, develop more slowly, and have a lower peak temperature than an ordinary flare, and in the past it may have escaped detection by all techniques except for the *disparition brusque* or 1083 nm imaging observations, which have had very low time resolution. Essentially all of the *Ulysses* events at high heliographic latitude during the south polar passage could plausibly be identified with solar antecedents [*Lemen et al.*, 1996; *Weiss et al.*, 1996].

The *Yohkoh* SXT data show several forms of mass ejection from the solar corona, including new types for which a geomagnetic link, if any, remains to be established.

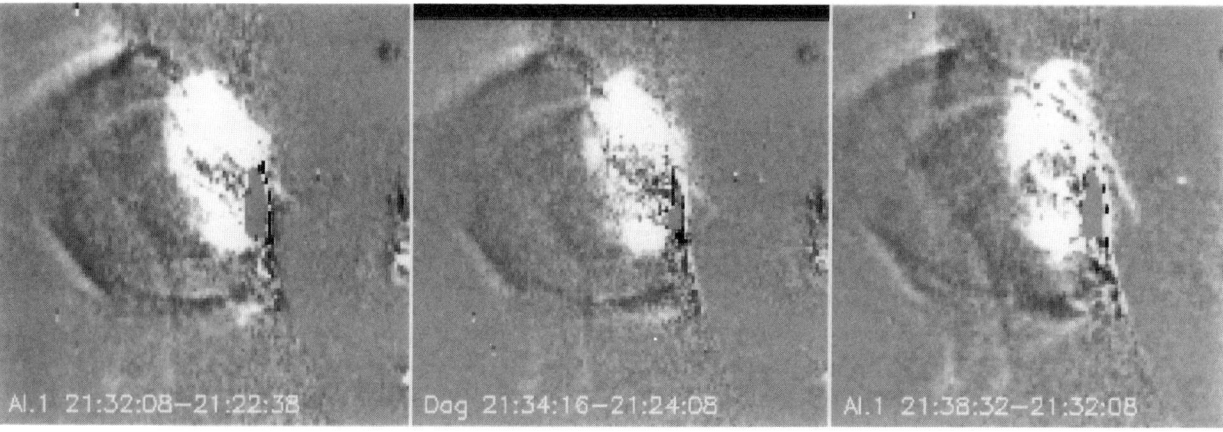

Figure 4. Difference images showing the expansion of a filamentary loop structure in the dimming episode of the flare of 28 August 1992. These are shown pairwise, comparing exposures in the same filter, and the intervals overlap as indicated. Each frame shows bright (late) outside dark (early) features, showing that the outward motion is general. Note an elongated, twisted element almost parallel to the solar limb.

As expected from the coronagraph data, events probably associated with CMEs often show clearly measurable dimmings of the X-ray corona near the site of the flare or arcade brightening. These data have clarified the question of timing differences between the ejection and brightening. As pointed out, the *Yohkoh* data have better sensitivity and sampling than the white-light coronagraph data, and view the entire visible hemisphere. With these observational advances it now appears that there is usually no delay, and that the two processes occur more or less simultaneously. This seems quite consistent with the traditional reconnection picture in any case, and points (as do the nonthermal effects such as particle acceleration) to the rise phase of flare soft X-radiation as the most interesting interval for physical understanding.

In some of the cases that have been studied, the *Yohkoh* SXT observations show details of the origin of the ejected material. For example, the 13 November 1994 cloud event and others have the appearance of a large structure with approximately a single twist; the structure appears to be anchored at one end in a flaring active region. The 21 February 1992 event and others, on the contrary, appear to dim symmetrically above the arcade development. Finally some of the large-scale arcade events appear to show large-scale dimming both at and remote from the arcade location. Are these signatures really different? Are they consistent with the currently-favored theoretical picture of the unstable outward motion of a large-scale flux rope or "sheared-core" structure [*Hirayama* 1974; *Moore and LaBonte*, 1979; *Gosling*, 1990; *Low*, 1994]?

Further study of these and related events, especially in conjunction with the SOHO coronagraph observations, has two obvious objectives. First, we do not understand the physics of the event onset nor its development. Much of the theoretical work is in the form of cartoons and frustratingly simple numerical models, neither of which is capable of predicting the observed effects quantitatively. Second, since we have found that we can see the mass ejections leave the Sun, at least in the best cases, it may be possible in the future to predict the speed and magnetic-field orientation of the interplanetary magnetic cloud that results from the ejection. This would be of great interest in the "space weather" community interested in the prediction of magnetic storms.

Acknowledgements. NASA supported this work under contract NAS 8-37334. *Yohkoh* is a mission of the Institute of Space and Astronautical Sciences (Japan), with participation from the U. S. and U. K. I thank D. Webb for bringing the 28 August 1992 event to my attention; A. Hundhausen and E. Hiei for the 16 January 1993 event; R. Moore for sharing several sheared core" flares showing coronal dimming; and L. W. Acton for preparing Figure 1.

REFERENCES

Acton, L. W., S. Tsuneta, Y. Ogawara, R. Bentley, M. Bruner, R. Canfield, L. Culhane, G. Doschek, E. Hiei, and T. Hirayama, The *Yohkoh* mission for high-energy solar physics, *Science 258,* 618-621, 1992.

Alexander, D., H. S. Hudson, G. Slater, A. McAllister, and K. Harvey, The large scale coronal eruptive event of April 14, 1994, in *Solar Dynamic Phenomena and Solar Wind Consequences: Proceedings of the Third SOHO Workshop*, ESA SP-373, pp. 187-190, 1994.

Aurass, H., K.-L. Klein, P. C. H. Martens, First detection of correlated electron beams and plasma jets in radio and soft X-ray data, *Solar Phys.*, *155,* 203-206, 1994.

Bothmer, V., and R. Schwenn, Eruptive prominences as sources of magnetic clouds in the solar wind, *Space Sci. Revs.*, *70,* 215-220, 1994.

Canfield, R. C., H. S. Hudson, K. D. Leka, D. L., Mickey, T. R. Metcalf, J. Wuelser, L. W. Acton, K. T. Strong, T. Kosugi, and T. Sakao, The X Flare of 1991 November 15: Coordinated Mees/*Yohkoh* Observations, *Publ. Ast. Soc. Japan, 44*, L111-L115, 1992.

Crooker, N. U., and E. W. Cliver, Postmodern view of M-regions, *J. Geophys. Res., 99*, 23,383-23,390, 1994.

Dodson, H. W., and E. R. Hedeman, Problems of differentiation of flares with respect to geophysical effects, *Planet. Space Sci. 12(5)*, 329-341, 1964.

Feynman, J., and A. J. Hundhausen, Coronal mass ejections and major solar flares: The great active center of March 1989, *J. Geophys. Res., 99*, 8451-8464, 1994.

Gopalswamy, N., Y. Hanaoka, M. R. Kundu, S. Enome, J. R. Lemen, M. Akioka, and A. Lara, *Astrophys. J.*, to be published, 1996.

Gosling, J., Coronal mass ejections and magnetic flux ropes in interplanetary space, in *Physics of Magnetic Flux Ropes*, edited by C. T. Russell, E. R. Priest, and L. C. Lee, pp. 343-364, AGU, Washington D.C., 1990.

Gosling, J. T., D. J. McComas, J. L. Phillips, and S. J. Bame, Geomagnetic activity associated with Earth passage of interplanetary shock disturbances and coronal mass ejections, *J. Geophys. Res., 96*, 7831-7839, 1991.

Gosling, J. T., Corotating and transient solar wind flows in three dimensions, *Annu. Rev. Astron. Astrophys., 24*, 35-73, 1996.

Gosling, J. T., D. J. McComas, J. L. Phillips, L. A. Weiss, V. J. Pizzo, B. E. Goldstein, and R. J. Forsyth, A new class of forward-reverse shock pairs in the solar wind, *Geophys. Res. Lett. 21*, 2271-2274, 1994.

Hanaoka, Y., H. Kurokawa, S. Enome, H. Nakajima, K. Shibasaki, M. Nishio, T. Takano, C. Torii, H. Sekiguchi, and S. Kawashima, Simultaneous observations of a prominence eruption followed by a coronal arcade formation in radio, soft X-rays, and Hα, *Publ. Astr. Soc. Japan, 46*, 205-216, 1994.

Hansen, R. T., C. G. Garcia, S. F. Hansen, and E. Yasukawa, Abrupt depletions of the inner corona, *Publ. Astr. Soc. Pacific, 86*, 500-515, 1974.

Harrison, R., The nature of solar flares associated with coronal mass ejection, *Astr. Astrophys., 304*, 585-595, 1995.

Harvey, K., A. McAllister, H. Hudson, D. Alexander, J. R. Lemen, and H. P. Jones, Comparison and relation of HeI 1083 nm two-ribbon flares and large-scale coronal arcades observed by *Yohkoh*, in *Solar Drivers of Interplanetary and Terrestrial Disturbances*, ASP Conference Proceedings, to be published 1996.

Hiei, E., A., Hundhausen, and D. Sime, Reformation of a coronal helmet streamer by magnetic reconnection after a coronal mass ejection, *Geophys. Res. Lett., 20*, 2785-2788, 1993.

Hirayama, T., Theoretical model of flares and prominences, *Solar Phys., 34*, 323-338, 1974.

Hudson, H. S., *Yohkoh* observations of coronal mass ejections, in *Magnetodynamic Phenomena in the Solar Atmosphere: Prototypes of Stellar Activity*, edited by Y. Uchida, T. Kosugi, and H. S. Hudson, Kluwer, Dordrecht, p. 89, 1996.

Hudson, H. S., L. W. Acton, D. Alexander, K. L. Harvey, S. W. Kahler, H. Kurokawa, and J. R. Lemen, The solar origins of two high-latitude interplanetary disturbances, *Solar Wind 8*, to be published, 1996a.

Hudson, H. S., L. W. Acton, D. Alexander, S. L. Freeland, J. R. Lemen, K. L. Harvey, *Yohkoh* SXT observations of sudden mass loss from the solar corona, *Solar Wind 8*, to be published, 1996b.

Hudson, H. S., L. W. Acton, and S. L. Freeland, A long-duration solar flare with mass ejection and global consequences, *Astrophys. J., 470*, 629, 1996c.

Hundhausen, A. J., Sizes and locations of coronal mass ejections - SMM observations from 1980 and 1984-1989, *Geophys. Res. Lett., 98*, 13,177-13,200, 1993.

Hundhausen, A. J., in *Cosmic Winds and the Heliosphere*, edited by J. R. Jokipii, C. P. Sonett, and M. S. Gianpapa, U. Arizona, Tucson, to be published 1997.

Kahler, S. W., Determination of the energy or pressure of a solar X-ray structure using X-ray filtergrams from a single filter, *Solar Phys., 48*, 255, 1978.

Kahler, S. W., Solar flares and coronal mass ejections, *Annu. Rev. Astron. Astrophys., 30*, 113-141, 1992.

Kano, R., The time evolution of X-ray structure during filament eruption, in *X-ray Solar physics from Yohkoh*, edited by Y. Uchida, T. Watanabe, K. Shibata, and H. Hudson, pp. 273-276, Universal Academy Press, Tokyo, 1994.

Khan, J. I., Y. Uchida, A. H. McAllister, and Ta. Watanabe, *Yohkoh* soft X-ray observations related to a prominence eruption and arcade flare on 7 May 1992, in *X-ray Solar physics from Yohkoh*, edited by Y. Uchida, T. Wata\-nabe, K. Shibata, and H. Hudson, pp. 201-204, Universal Academy Press, Tokyo, 1994.

Klimchuk, J. A., L. W. Acton, K. L. Harvey, H. S. Hudson, K. L. Kluge, D. G. Sime, K. T. Strong, and Ta. Watanabe, Coronal eruptions observed by *Yohkoh*, in *X-ray Solar Physics from Yohkoh*, edited by Y. Uchida, T. Watanabe, K. Shibata, and H. Hudson, pp. 181-186, Universal Academy Press, Tokyo, 1994.

Kundu, M., *Solar Radio Astronomy*, Interscience, New York, 1965.

Kundu, M. R., J.-P. Raulin, N. Nitta, H. S. Hudson, M. Shimojo, K. Shibata, and A. Raoult, Detection of nonthermal radio emission from coronal X-ray jets, *Ap. J., 447*, L135-L139, 1995.

Lemen, J. R., L. W. Acton, D. A. Alexander, A. B. Galvin, K. Harvey, T. Hoeksema, X. Zhao, and H. S. Hudson, Solar identification of solar-wind disturbances observed at *Ulysses*, *Solar Wind 8*, to be published, 1996.

Lin, R. P., Energetic solar electrons in the interplanetary medium, *Solar phys., 100*, 537-561, 1985.

Low, B. C., Magnetohydrodynamic processes in the solar corona: Flares, coronal mass ejections, and magnetic helicity, *Phys. Plasmas, 1*, 1684-1690, 1994.

Manoharan, P. K., L. van Driel-Gesztelyi, M. Pick, and P. Demoulin, 1996, Evidence for large-scale magnetic reconnection obtained from radio and X-ray measurements, *Ap. J. (Lett.)*, L73-L76, 1996.

Martin, S. F., R. Bilimoria, and P. W. Tracadas, Magnetic field configuration basic to filament channels and filaments, in *Solar Surface Magnetism*, edited by R.J. Rutten and C.J. Schrijver, pp 303-338, NATO ASI Series C 433, 1994.

McAllister, A. H., M. Dryer, P. McIntosh, H. Singer, and L. Weiss, A large polar crown CME and a "problem" geomagnetic storm: April 14-23, 1994, *J. Geophys. Res., 101*, 13,497-13,515, 1996.

Moore, R. L., and B. LaBonte, The filament eruption in the 3B flare of July 29, 1973 - Onset and magnetic field configuration, in *Proc. Symposium on Solar and Interplanetary Dynamics*, Reidel, Dordrecht, pp. 207-210, 1979.

Pick, M., A. Raoult, G. Trottet, N. Vilmer, K. Strong, and A. Magalhaes, Energetic electrons and magnetic field structures in the corona, *Proceedings of Kofu Symposium*, edited by S. Enome and T. Hirayama, pp. 263-266, NRO Report No. 360, 1994.

Raulin, J.-P, M. R. Kundu, H. S. Hudson, N. Nitta, and A. Raoult, Metric type III bursts associated with soft X-ray jets, *Astr. Astrophys., 306*, 299-307, 1996.

Rust, D. M., Coronal disturbances and their terrestrial effects, *Space Science Reviews, 34*, 21-36, 1983.

Rust, D. M., 1994, Spawning and shedding helical magnetic fields in the solar atmosphere, *Solar Phys., 21*, 241-244, 1994.

Rust, D. M., and E. Hildner, Expansion of an X-ray coronal arch into the outer corona, *Solar Phys., 48*, 381-387, 1978.

Rust, D. M., and A. Kumar, Helical magnetic fields in filaments, *Solar Phys., 155*, 69, 1994.

Shibata, K., Y. Ishido, L. W. Acton, K. T. Strong, T. Hirayama, Y. Uchida, A. H. McAllister, R. Matsumoto, S. Tsuneta, T. Shimizu, H. Hara, T. Sakurai, K. Ichimoto, Y. Nishino, and Y. Ogawara, Observations of X-ray jets with the *Yohkoh* soft X-ray telescope, *Publ. Astr. Soc. Japan, 44*, L173 -L180, 1992.

Shibata, K., S. Masuda, M. Shimojo, H. Hara, T. Yokoyama, S. Tsuneta, T. Kosugi, and Y. Ogawara, Hot plasma ejections associated with compact-loop solar flares, *Ap. J., 451*, L83-L85, 1995.

Shimojo M., Hashimoto, S., Shibata, K., Hirayama, T., Hudson, H. S., and Acton, L. W., Statistical study of solar X-ray jets observed with the *Yohkoh* soft X-ray telescope, *Publ. Astr. Soc. Japan, 48*, 123-136, 1996.

Strong, K. T., K. L. Harvey, T. Hirayama, N. Nitta, T. Shimizu, T., and S. Tsuneta, Observations of the variability of coronal bright points by the soft X-ray telescope on *Yohkoh*, *Publ. Astr. Soc. Japan, 44*, L161-L166, 1992.

Tsuneta, S., Structure and dynamics of magnetic reconnection in a solar flare, *Ap. J., 456*, 840-849, 1996.

Tsuneta, S., Acton, L, Bruner, M., Lemen, J., Brown, W., Caravalho, R., Catura, R., Freeland, S., Jurcevich, B. Morrison, M., Ogawara, Y., Hirayama, T., and Owens, J., *Solar Phys., 136*, 37, 1991.

Tsuneta, S., T. Takahashi, L. W. Acton, M. E. Bruner, K. L. Harvey, and Y. Ogawara, Observation of a solar flare at the limb with the *Yohkoh* Soft X-ray Telescope, *Publ. Astr. Soc. Japan, 44*, L63-L67, 1992a.

Tsuneta, S., H. Hara, T. Shimizu, L. W. Acton, K. T. Strong, H. S. Hudson, and Y. Ogawara, Global restructuring of the coronal magnetic fields observed with the *Yohkoh* Soft X-ray Telescope, *Publ. Astr. Soc. Japan , 44*, L211-L215, 1992b.

Uchida, Y., A. McAllister, K. T. Strong, Y. Ogawara, T. Shimizu, R. Matsumoto, and H. S. Hudson, *Publ. Ast. Soc. Japan,44*, L155-L158, 1992.

Watanabe, T., Y. Kozuka, M. Ohyama, M. Kojima, K. Yamaguchi, S. Watari, S. Tsuneta, J. A. Joselyn, K. Harvey, and L. W. Acton, Coronal/interplanetary disturbances associated with disappearing solar filaments, *Publ. Astr. Soc. Japan, 44*, L193-L197, 1992.

Watanabe, Ta., M. Kojima, Y. Kozuka, S. Tsuneta, J. R. Lemen, H. S. Hudson, J. A. Joselyn, and J. A. Klimchuk, Interplanetary consequences of transient coronal events, in *X-Ray Solar Physics from Yohkoh,* edited by Y. Uchida, T. Watanabe, K. Shibata, and H. S. Hudson, pp. 207-210, Universal Academy Press, Tokyo, 1994.

Watari, S., Z. Smith, H. A. Garcia, T. Detman, and M. Dryer, Coronal change at the south-west limb observed by *Yohkoh* on 9 November 1991 and the subsequent interplanetary shock at Pioneer-Venus-Orbiter, *Solar Phys.*, to be published, 1996.

Webb, D. F., and R. A. Howard, The solar cycle variation of coronal mass ejections and the solar wind mass flux, *J. Geophys. Res., 99*, 4201-4220, 1994.

Weiss, L. A., J. T. Gosling, A. McAllister, A. J. Hundhausen, J. T. Burkepile, J. L. Phillips, K. T. Strong, and R. J. Forsyth, A comparison of interplanetary coronal mass ejections at Ulysses with *Yohkoh* soft X-ray coronal events, *J. Geophys. Res., 101*, 13,497-13,515, 1996.

Withbroe, G., W. C. Feldman, and H. S. Ahluwahlia, The solar wind and its coronal origins, in *Solar Interior and Atmosphere*, edited by A. N. Cox, W. C. Livingston, and M. S. Matthews, pp. 1087-1106, U. Arizona Press, Tucson, 1991.

Prominence Eruptions and Geoeffective Solar Wind Structures

James Chen

Plasma Physics Division, Naval Research Laboratory, Washington, D.C.

It has been well established that severe geomagnetic disturbances are caused by solar wind streams that impose long periods of strong southward interplanetary magnetic field (IMF) on the Earth's magnetosphere. Such "geoeffective" solar wind structures are closely associated with eruptions at the Sun. The relationships between solar activity and geoeffective solar wind structures have been a major research issue for several decades and have been based primarily on empirical associations. In this paper, we describe a recent theoretical effort to understand the physical mechanisms responsible for eruption and propagation of prominences as well as the relationships between eruptive prominences, coronal mass ejections, and their heliospheric consequences. We also discuss empirical and theoretical studies of magnetospheric response to the arrival of geoeffective solar wind structures and a technique to identify such structures using real-time solar wind measurements.

1. INTRODUCTION

Since the first observations of solar flares [*Carrington*, 1860; *Hodgson*, 1860], the influences of solar eruptions on the geomagnetic environment have attracted considerable attention [e.g., *Hale*, 1931]. In the early works, the coupling between the Sun and the Earth was thought to be primarily due to the solar particles impinging on the Earth [*Chapman*, 1950]. In the modern paradigm [*Dungey*, 1960], the solar wind magnetic field plays a central role: a southward interplanetary magnetic field (IMF) causes magnetic reconnection at the dayside magnetopause, rapidly enhancing injection of magnetic and particle energy into the magnetosphere and modifying the large-scale current systems. When spacecraft measurements of solar wind properties became available, it was empirically established that prolonged periods of strong southward IMF ($B_z < 0$) indeed lead to geomagnetic storms [e.g., *Rostoker and Fälthammar*, 1967; *Hirshberg and Colburn*, 1969; *Russell et al.*, 1974; *Gonzalez and Tsurutani*,

1987]. A fraction of the solar wind energy ultimately precipitates into the ionosphere, producing aurorae and other disturbances in the near-Earth electric and magnetic fields.

It has long been realized that there are two distinct categories of magnetic storms, those that recur with the solar rotation period of 27 days [*Maunder*, 1905] and those that are nonrecurrent. Recurrent storms tend to be moderate, and their frequency is anticorrelated with sunspot numbers [*Greaves and Newton*, 1929]. In contrast, large storms tend to be nonrecurrent and occur near solar maximum. The immediate cause of magnetic storms, i.e., long periods of strong southward IMF, is the same for both categories, but the solar origins are typically different. The recurrent form of storms is associated with the Earth crossing of magnetic sectors corresponding to the (polar) open-field regions (coronal holes) [*Neupert and Pizzo*, 1974; *Sheeley et al.*, 1976]. The sector pattern rotates with the Sun with the 27-day period. Coronal holes tend to be localized to high heliolatitudes during solar maximum but tend to extend toward the ecliptic plane, more often affecting the Earth during solar minimum. This accounts for the anticorrelation between recurrent storms and sunspot numbers. A brief review of the solar origins of recurrent storms can be found in *Crooker and Cliver* [1994]. For nonrecurrent storms, association with solar flares was

Magnetic Storms
Geophysical Monograph 98
This paper is not subject to U.S. copyright.
Published in 1997 by the American Geophysical Union

found [e.g., *Hale*, 1931; *Newton*, 1943], leading to the view that solar flares produced the geoeffective solar wind disturbances (e.g., shocks) responsible for nonrecurrent geomagnetic storms. However, with the discovery of coronal mass ejections (CMEs) [*Tousey*, 1973; *Brueckner*, 1974; *MacQueen et al.*, 1974; *Gosling et al.*, 1974], it gradually became clear that the disturbed solar wind streams associated with the occurrence of CMEs provide a better correlation with large nonrecurrent magnetic storms [*Gosling et al.*, 1991]. In addition, *Joselyn and McIntosh* [1981] found that nonrecurrent storms tend to be correlated with eruptive prominences. Indeed, CMEs are more closely associated with prominence eruptions than flares [*Gosling et al.*, 1974; *Munro et al.*, 1979; *Sheeley et al.*, 1983; *Webb and Hundhausen*, 1987]. Even in the recurrent class of storms, CMEs propagating in the streamer belt [*Crooker and Cliver*, 1994] and shock-compressed southward IMF [*Tsurutani et al.*, 1995] may be responsible for large storms near solar minimum. As a result, the earlier notion that solar flares cause such storms has been called into question [*Kahler*, 1992; *Gosling*, 1993].

In terms of the solar wind structures that cause geomagnetic storms, interplanetary magnetic clouds [*Burlaga et al.*, 1981], a clearly identifiable class of solar wind structures describable as magnetic flux ropes with twisted field lines, have been associated with geomagnetic storms [*Wilson*, 1987; *Zhang and Burlaga*, 1988]. Magnetic clouds constitute 1/3 (perhaps more) of all interplanetary CMEs [*Gosling*, 1990]. Perhaps 1/2 or more of large geomagnetic storms are correlated with magnetic clouds [*Burlaga et al.*, 1987; *Tsurutani et al.*, 1988], and a large fraction of magnetic clouds impinging on the Earth cause storms [*Wilson*, 1987]. This is because magnetic clouds can impose long periods of southward IMF on the magnetosphere. More generally, interplanetary CMEs and CMEs preceded by shocks are correlated with large storms [*Gosling et al.*, 1991]. Magnetic clouds in turn are strongly associated with eruptive filaments [*Wilson and Hildner*, 1986; *Rust*, 1994; *Bothmer and Schwenn*, 1994].

In the last several decades, much has been learned about the likely sources of geoeffective solar wind streams. A number of good reviews have been written on CMEs [see, for example, *Howard et al.*, 1985; *Hundhausen*, 1987, 1996; *Kahler*, 1987], on prominences [e.g., *Tandberg-Hanssen*, 1974; *Priest*, 1989], and on magnetic storms [e.g., *Gonzalez et al.*, 1994]. However, issues concerning the causal relationships between different types of solar activity (e.g., solar flares versus CMEs) and terrestrial consequences are being debated [*Kahler*, 1992; *Gosling*, 1993, 1995; *Hudson et al.*, 1995]. Neither the possible magnetic geometries of the pre-eruptive coronal structures nor the structural relationships between the eruptive solar structures and their (presumed) heliospheric counterparts have been fully understood. A critically important element of the Sun-Earth plasma coupling, the propagation of the initial solar structure to 1 AU, has been difficult to determine observationally. Theoretically, fully three-dimensional structures and their dynamics over large distances are difficult to analyze.

It should be clear to the reader by now that eruptive prominences/CMEs and geomagnetic storms are intimately linked but that the actual physical processes of solar eruption and subsequent evolution of the ejecta have eluded satisfactory understanding. In the present paper, we will summarize a recent theoretical effort to model the physical mechanisms responsible for solar eruptions and the subsequent propagation of the resulting ejecta. In section 2, the focus will be on the physics of eruptive prominences and how they propagate through the ambient medium. We then examine how eruptive prominences and CMEs can be related to 1 AU solar wind structures (section 3) and how the magnetosphere responds to the passage of such structures (section 4).

2. MODELS OF ERUPTIVE PROMINENCES

2.1. *Initial Structures and Eruption*

Prominences, or filaments in the disk view, have been studied for a number of decades [e.g., *Tandberg-Hanssen*, 1974; *Priest*, 1989]. Eruptive prominences often exhibit helical structures [e.g., *Schmahl and Hildner*, 1977], suggestive of underlying magnetic fields with twisted field lines, such as magnetic flux ropes [*Kuperus and Raadu*, 1974] and arcades [*Kippenhahn and Schlüter*, 1957]. A significant recent understanding is that eruptive prominences and CMEs often occur together in a three-part structure: a bright outer rim (the classic CME loop seen in "white light" coronagraphs) located above a low-density (i.e., dark) cavity which contains the prominence [*Michels et al.*, 1980; *Fisher et al.*, 1981; *Illing and Hundhausen*, 1985, 1986]. The observational and morphological properties of the initial solar structures and the immediate aftermath of eruption have been well documented [e.g., *Illing and Hundhausen*, 1985, 1986] and extensively reviewed [*Hundhausen*, 1996].

In some early CME models, the bright "loop" seen in the plane of the sky was taken to be a current loop [*Mouschovias and Poland*, 1978; *Anzer*, 1978; *Van Tend*, 1979]. The current loop rises (expands) because of the Lorentz self-force $\mathbf{J} \times \mathbf{B}$, where $\mathbf{B}$ is the field due to the current $\mathbf{J}$ of the flux rope. However, later observational analyses appeared to indicate that the looplike CMEs may be the two-dimensional (2-D) projection of shell-like structures with significant extent along the line of sight [e.g., *Howard et al.*, 1982; *Webb*, 1988; *Hundhausen*, 1996]. One configuration that can lead to the appearance of a shell is a magnetic arcade viewed along its axis [*Kippenhahn and Schlüter*, 1957; *Pneuman*, 1980]. Under quasi-static shearing of photospheric footpoints of the magnetic field, arcades grow, and the stored magnetic energy may be released suddenly as a result of loss of equilibrium or instability [*Low*, 1977; *Birn and Schindler*, 1981]. In this scenario, plasmoids or flux ropes are formed as a result of magnetic reconnection [*Mikić et al.*, 1988; *Priest and Forbes*, 1990; *Forbes*, 1990; *Linker et al.*, 1990; *Gosling*, 1990; *Forbes and Isenberg*, 1991; *Wu et al.*, 1991; *Finn et al.*, 1992; *Mikić and Linker*, 1994]. The resulting flux rope rises by the unbalanced Lorentz force and plasma pressure determined by the ambient medium.

In another scenario, eruption of magnetic arcades is thought to be triggered and driven by impulsive pressure pulses produced by flares [*Dryer*, 1982; *Wu et al.*, 1983; *Steinolfson*, 1985] and by waves produced by such eruptions [*Sakai and Nishikawa*, 1983]. However, substantial issues have been raised regarding the possible causal role of flares as the driver of CMEs. Timing considerations of flare-associated CMEs [e.g., *Hundhausen*, 1996] have shown that CMEs are frequently launched before the onset of flares. Prominence eruption has also been postulated as a possible driving mechanism of CMEs [*Pneuman*, 1980; *Hu*, 1983]. However, eruptive prominences, when they occur with CMEs, tend to lag behind and have speeds slower than the bright CMEs (the outer rims) [e.g., *Illing and Hundhausen*, 1986]. Thus, it is difficult to think of prominence eruptions as the drivers of CMEs. It was suggested that prominence eruption is a consequence of the CME [*Hundhausen*, 1987]. Alternatively, based on a theoretical model, *Chen* [1996] proposed that the observed bright rim-cavity-prominence features are three parts of one magnetically organized flux rope, so that prominence eruption is neither the cause nor the result of CME eruption. In this model, the flux rope erupts in response to injection of magnetic flux.

Models of prominences and CMEs have also been proposed using the flux rope topology. One class of models envisions that the dense cold ($\lesssim 10^4$ K) prominence is contained inside a flux rope [*Kuperus and Raadu*, 1974; *Démoulin and Forbes*, 1992; *Chen and Garren*, 1993; *Low and Hundhausen*, 1995; *Kumar and Rust*, 1996; *Chen*, 1996], corresponding to the so-called "inverse-polarity" configuration (in contrast to the "normal" polarity configuration [*Kippenhahn and Schlüter*, 1957]). The azimuthal field (transverse to the axis of a filament) may thread the chromosphere, exhibiting possible signatures consistent with observed chromospheric fibril patterns [*Foukal*, 1971; *Martin et al.*, 1992]. In this class of models, it is the cavity-prominence structure that corresponds to the flux rope and not the bright CME loop, in contrast to the assumption of *Mouschovias and Poland* [1978] and *Anzer* [1978]. The models of *Kuperus and Raadu* [1974], *Démoulin and Forbes* [1992], and *Low and Hundhausen* [1995] describe 2-D equilibrium configurations, while *Chen and Garren* [1993] and *Chen* [1996] investigated the dynamics of initial equilibrium flux ropes based on a class of three dimensional (3-D) equilibria identified previously [*Xue and Chen*, 1983; *Chen*, 1989]. The calculation of *Kumar and Rust* [1996] does not include the initial equilibrium but is otherwise similar to that of *Chen and Garren* [1993]. In these calculations, the cavity and the prominence were treated as one average structure. However, it has been found that it is necessary to treat the hot cavity plasma and the cold prominence plasma separately in order to obtain solutions consistent with the eruptive behavior near the Sun *and* the heliospheric properties at 1 AU and beyond [*Chen*, 1996]. In a somewhat different scenario, *Wu et al.* [1995] postulated that a helmet streamer is an equilibrium arcade and simulated initiation of eruption by forcing a cylindrical flux rope into the arcade through the photosphere. Thus, the end products of most models are flux ropes: some start with flux ropes and others produce them from initial arcades. In the following, we focus on the flux rope as a possible initial structure and describe a physical model of the eruption and propagation of prominences.

2.2. *Forces on a Flux Rope*

Figure 1 illustrates a current loop of major radius R and minor radius a. The center of mass of the apex is at height Z from the photosphere with the footpoints separated by $2s_0$. It carries a toroidal (loop-aligned) and poloidal (locally azimuthal) currents with densities J_t and J_p, respectively. The magnetic field has toroidal, B_t, and poloidal, B_p, components, so that field lines are twisted. The electric current vanishes outside the minor radius as does the B_t component, but the poloidal magnetic component B_p extends outside the current loop. There may also be an ambient magnetic field B_s due to a separate current system. In this paper, we take B_s to be perpendicular to the plane of the loop. We will generally use the term "flux rope" to specify the overall magnetic structure consisting of the "current loop" and its magnetic field. In most contexts, however, the distinction is not important. We adopt the hypothesis that such a structure exists and that the current loop is rooted below the photosphere [*Chen*, 1989]. The precise subphotospheric current structure will not enter the analysis.

The model structure depicted in Figure 1 is idealized but is fully three-dimensional. The basic dynamics of flux ropes have been explored in detail using an integrated magnetohydrodynamic (MHD) formulation in which physical quantities averaged over toroidal sections of the flux rope are evolved [*Chen*, 1989]. In the remainder of section 2 and in section 3,

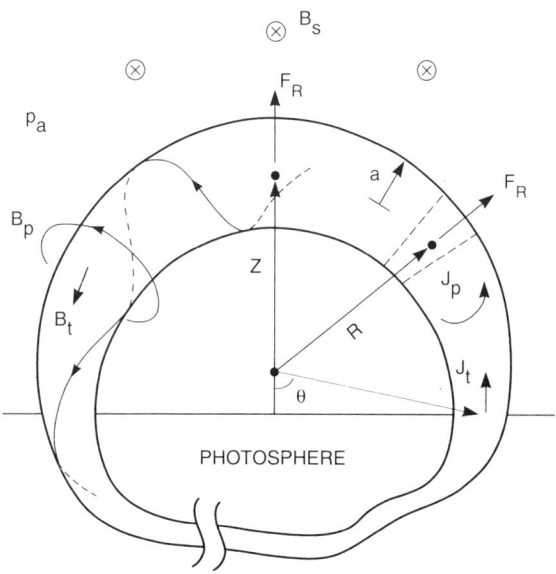

Figure 1. Geometry of toroidal flux rope. The current loop of major radius R and minor radius a is embedded in its own poloidal field (B_p). The toroidal current J_t, poloidal current J_p, toroidal field B_t, and poloidal field B_p are indicated. The flux rope consists of the current loop and the poloidal field. The cold prominence material is assumed to be contained inside the flux rope. The overlying ambient field B_s is indicated. (From *Chen* [1996].)

prominence eruption and propagation are discussed following the recent work of *Chen* [1996]. The basic forces are given using standard MHD,

$$\mathbf{f} = (1/c)\mathbf{J} \times \mathbf{B} - \nabla p + \rho \nabla \phi, \quad (1)$$

where $\mathbf{J} = (c/4\pi) \nabla \times \mathbf{B}$, ρ is the mass density, p is the pressure, and ϕ is the gravitational potential. By toroidal, we mean that the flux rope can be approximated as a section of a torus characterized by major radius R and minor radius a. The major radial force integrated over a toroidal section of unit length is given by

$$M\frac{d^2 Z}{dt^2} = \frac{I_t^2}{c^2 R}\left[\ln\left(\frac{8R}{a}\right) + \frac{1}{2}\beta_p - \frac{1}{2}\frac{\overline{B_t}^2}{B_{pa}^2}\right.$$
$$\left. + 2\left(\frac{R}{a}\right)\frac{B_s}{B_{pa}} - 1 + \frac{\xi_i}{2}\right] + F_g + F_d, \quad (2)$$

where $\beta_p = 8\pi(\overline{p} - p_a)/B_{pa}^2$, $\overline{p}$ is the average pressure inside the loop, p_a the ambient coronal pressure, $\overline{B_t}$ the average toroidal field inside the loop, $B_{pa} \equiv B_p(a) = 2I_t/ca$, and $\xi_i \equiv 2 \int r B_p^2(r) dr/(a^2 B_{pa}^2)$ is the internal inductance. Here, r is the minor radial coordinate ($0 \le r \le a$), I_t is the total toroidal current $I_t \equiv 2\pi \int J_t(r) r dr$, and $M = \pi a^2 \overline{n}_T m_i$ is the total mass per unit length of the loop. The terms F_g and F_d are the gravity and drag terms, respectively.

The minor radius $a(t)$ of the apex satisfies

$$M\frac{d^2 a}{dt^2} = \frac{I_t^2}{c^2 a}\left(\frac{\overline{B_t}^2}{B_{pa}^2} - 1 + \beta_p\right). \quad (3)$$

This equation simply states that $\overline{B_t}$ and $\overline{p}$ tend to increase a while B_{pa} and p_a tend to decrease a. A drag term is also included in the actual calculations. These equations are supplemented by constraints on physical quantities such as the poloidal and magnetic fluxes and a thermodynamic (polytropic) expansion law with $\gamma > 1$.

Prominences and their associated cavities can remain nearly stationary and stable for days and weeks. One must then conclude that the forces acting on them are nearly balanced on this time scale. Thus, we must show that it is possible to balance the major radial and minor radial forces. Demanding $d^2 Z/d^2 t = 0$ and $d^2 a/dt^2 = 0$, we obtain

$$B_{pa} = (R_0/a_0)\Lambda_0^{-1}\left\{-B_{s0} + \left[B_{s0}^2 - \left(\frac{a_0}{R_0}\right)^2 \Lambda_0\right.\right.$$
$$\left.\left. \times \left[8\pi(\overline{p} - p_a) + 4\pi m_i R_0 g_0 (n_a - \overline{n}_c - \overline{n}_p)\right]\right]^{1/2}\right\}, \quad (4)$$

where $\Lambda_0 \equiv \ln(8R_0/a_0) - 1.5 + \xi_i/2$, $g_0 = g(Z_0)$, and $B_{s0} = B_s(Z_0)$. All other quantities also refer to the initial values. The minor radial condition, $(\overline{B_t}^2/B_{pa}^2 - 1 + \beta_p) = 0$, has been used to eliminate $\overline{B_t}^2/B_{pa}^2$. The sign convention is such that $B_{s0} < 0$ provides downward $I_t B_{s0}$. This equilibrium condition has been obtained previously and will not be discussed in detail here. The point is that force balance is possible in flux ropes embedded in ambient plasmas with free boundaries. This constraint admits a wide variety of quasi-stationary flux ropes under coronal conditions.

2.3. Initial Flux Rope

Consider now what kind of equilibrium flux rope structures in the solar corona are dictated by (4). Let the apex height be $Z_0 = 1.5 \times 10^5$ km at center of mass, the major radius $R_0 = 3.75 \times 10^5$ km, the minor radius $a_0 = 7.5 \times 10^4$ km, and the footpoint separation $2s_0 = 6 \times 10^5$ km. These dimensions are chosen to be consistent with quiescent prominences [e.g., *Tandberg-Hanssen*, 1974]. The ambient coronal temperature is taken to be $T_a = 2 \times 10^6$ K. We assume that the low-density cavity plasma is at $\overline{T}_0 = T_a$. The ambient density at $Z = Z_0$ is taken to be $n_a = 3 \times 10^8$ cm^{-3}. The plasma density inside the flux rope is averaged over the flux rope so that the spatial distribution of the prominence gas is not treated. We use $\overline{n}_c = (1/4)n_a = 7.5 \times 10^7$ cm^{-3} for the cavity plasma. These values correspond to $p_a = 0.16$ dyn/cm^2 and $\overline{p}_c = 0.04$ dyn/cm^2. The average density of the cold prominence component is $\overline{n}_p$. The parameter $\chi_0 \equiv \overline{n}_p/\overline{n}_c$ proves to be important. If we take $\chi_0 = 30$ for this example, the total mass of the flux rope is $M_T = 4.75 \times 10^{16}$ g. This corresponds to a rather massive prominence. Finally, we choose the ambient magnetic field to be $B_{s0} = -0.5$ G. Equation (4) then gives the poloidal magnetic $B_{pa} = 4.4$ G, so that $\beta_p \simeq -0.16$. By setting $d^2a/dt^2 = 0$ in (3), we find $\overline{B}_t = 4.8$ G. The magnetic field on the flux rope axis, which is entirely toroidal, is $\sim 3\overline{B}_t$ [Chen, 1996] so that $B_t(r=0) \simeq 15$ G. That is, the magnetic field increases with height inside the lower half of the flux rope. These magnetic field values are consistent with those estimated in observed quiescent prominences [Rust, 1967; Leroy et al., 1983]. It is satisfying that the flux rope dimensions chosen to be consistent with quiescent prominences indeed lead to magnetic field values via (4) in agreement with the observationally estimated values.

Given a flux rope carrying toroidal and poloidal currents, J_t and J_p, respectively, one can define the magnetic fluxes using B_p and B_t. In particular, the poloidal flux due to B_p is

$$\Phi_p = cLI_t, \qquad (5)$$

where I_t is the toroidal current and L is the effective inductance of the flux rope. The inductance L is defined in *Chen and Garren* [1993] and will not be repeated here. For the purpose of the discussion, we simply mention that the dominant scaling dependence is $L(t) \propto R(t)$, so that $I_t \propto \Phi_p/R$. This scaling has also been used by *Kumar and Rust* [1996] in deriving a number of scaling relations. In this integrated formalism, equations (5) embodies the three-dimensionality of the system: the flux rope has a finite length ($\sim \pi R$), thus finite inductance L, and a finite poloidal flux Φ_p. The magnetic energy in the poloidal field is given by $U_p = (1/2)\Phi_p^2/c^2L$. For this loop, the initial poloidal flux is $\Phi_0 = 5.4 \times 10^{21}$ maxwell (Mx) corresponding to the initial poloidal magnetic field energy $U_{p0} = 4.4 \times 10^{31}$ erg. The magnetic energy in the toroidal component is $U_{t0} = 1.1 \times 10^{31}$ erg. The toroidal magnetic flux is simply given by $\Phi_t = \pi \overline{B}_t a^2$, which is $\Phi_t = 4.1 \times 10^{21}$ Mx.

We have shown equation (4) also to illustrate why it is important to treat the cold prominence component ($\overline{n}_p \neq 0$) separately from the tenuous cavity component $\overline{n}_c$: $\overline{n}_p \gg \overline{n}_c$ primarily affects the gravity term while $\overline{n}_c$ mainly affects the pressure, $\overline{p} = 2\overline{n}_c k\overline{T}$. One consequence is that if $\overline{n}_p = 0$, B_{pa} is limited to relatively small values because the value inside the square brackets must be positive. As a result, the amount of the current and hence the strength of the field that can be supported by the initial flux rope is weaker than for the $\overline{n}_p > 0$ case for which this restriction does not exist. In the previous flux rope models of prominences [*Chen and Garren*, 1993; *Kumar and Rust*, 1996], the cavity and prominence are treated as one average structure. This proves to be inadequate to model the full range of flux rope structures observed at 1 AU and beyond. We will return to this point in section 3.

2.4. A Driven Eruption Mechanism

Rewriting (5) as $I_t(t) = \Phi_p(t)/cI_t(t)$, we see that the multiplicative factor in (2) can be written as

$$\frac{I_t^2}{c^2R} = \frac{\Phi_p^2}{c^4L^2R} \propto \frac{\Phi_p^2(t)}{R^3(t)}. \qquad (6)$$

Thus, increasing Φ_p can act as a driver. Suppose that we prescribe a flux injection profile $d\Phi_p/dt$. Physically, this corresponds to injection of the poloidal flux of the flux rope. Conceptually, this is equivalent to prescribing the footpoint shear except that it is on the fast dynamical timescale (tens of minutes), as opposed to the slow shearing over days and weeks. Figure 2 shows a profile of $d\Phi_p/dt$. The poloidal flux $\Phi_p(t)$ normalized to the initial value Φ_0 is also shown. The functional form is chosen merely to mimic a packet of poloidal magnetic flux $\Delta\Phi_p$ (i.e., energy) with different rise and falloff timescales. This flux injection profile corresponds to the injection of $\Delta\Phi_p = 2.7 \times 10^{22}$ Mx and $\Delta U_p = 2.1 \times 10^{32}$ erg. Note that $d\Phi_p/dt \to 0$ for $t > 400$ min. Thus Φ_p is conserved thereafter. The toroidal flux Φ_t is taken to be conserved throughout the process. We want to examine the dynamical consequences of this process.

Figure 3 shows the flux rope dynamics for 100 hours: the apex distance from the photosphere (Figure 3a), the apex speed (Figure 3b), and the minor radius $a(t)$ and its expansion speed $w(t) \equiv da/dt$ (Figure 3c). The results are obtained by integrating equations (2) subject to a number of physical constraints such as the magnetic flux conservation after the flux injection is finished ($t \gtrsim 400$ min) and equation of state ($\gamma > 1$). (See *Chen* [1996] for detail.) We see that most of the acceleration occurs near the Sun, in agreement with the finding of *MacQueen and Fisher* [1982]. A detailed examination shows that beyond 2–3 $R_\odot$ above the photosphere, the apex experiences only gradual acceleration, and $V(t)$ and $w(t)$ both become nearly constant after $t \gtrsim 5$ hr. The nearly constant V is determined by the momentum coupling between the flux rope and the ambient solar wind, which is included through the drag term. The solar wind speed is smoothly increased from zero to V_{sw} according to empirical and theoretical solar wind models. Different values of V_{sw} have been

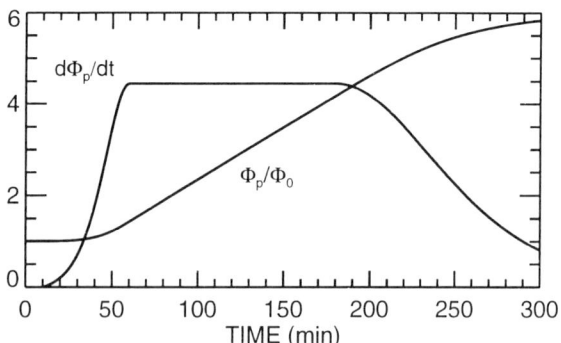

Figure 2. Flux injection profile $d\Phi_p/dt$ in 1.5×10^{17} Mx/s and $\Phi_p(t)/\Phi_0$, where $\Phi_0 = 5.4 \times 10^{21}$ Mx.

used (400 km/s and 600 km/s), but here we will use $V_{sw} = 400$ km/s, the nominal slow solar wind speed.

The apex reaches 1 AU (1.5×10^8 km) at $t = 73.6$ hr (3 days), where the speed is $V \simeq 570$ km/s and the minor radius is $a = 1.7 \times 10^7$ km. Thus, the current loop has a thickness $2a \simeq 0.23$ AU. The minor radial expansion speed is $w \simeq 60$ km/s. The magnetic field is $B_{pa} = 8.3$ nT and $\overline{B_t} = 9.7$ nT. The flux rope density is $\overline{n} \simeq 2.5$ cm^{-3} ($n_a = 5$ cm^{-3}), having evolved from the initial cavity plasma. The temperature is $\overline{T} \simeq 9 \times 10^4$ K, and the solar wind temperature is comparable at $T_a \simeq 9.5 \times 10^4$ K. These results are entirely consistent with observed magnetic clouds [e.g., *Lepping et al.*, 1990]. The reference flux rope represents a fast cloud with a relatively strong magnetic field embedded in a slow solar wind stream ($V_{sw} = 400$ km/s), similar to the one detected on 14–15 January 1988 [e.g., *Farrugia et al.*, 1993].

We have taken the identical initial system and reduced the magnetic flux injected. The dashed lines in Figure 3b show the velocity profiles for two such examples. (The height-time profiles are nearly the same as the solid line in Figure 3a and are not shown.) These have been chosen to show that the nearly constant apex speed can be either greater or less than the ambient V_{sw}. At $t = 100$ hr, the upper dashed line is $V = 411$ km/s while the lower dashed line is $V = 385$ km/s. Observationally, *Lepping et al.* [1990] found that magnetic clouds near 1 AU are often slightly faster than the ambient solar wind but can also be slower.

We note that the integrated MHD formalism does not concern itself with the detailed distribution of mass or forces inside the flux rope. The minor cross-section is assumed to remain circular, which is a good first approximation even under strong acceleration because of the fast Alfven speed in the flux rope. The momentum coupling is included using a drag term, F_d in equation (2). However, flux ropes are deformable, with the exact deformation dependent on the ambient magnetic field. These effects are not included in the integrated MHD calculations. Some idealized 2-D MHD simulations have now been carried out to address the interaction between propagating flux ropes and the ambient magnetized plasma [*Cargill et al.*, 1994, 1995, 1996; *Vandas et al.*, 1995]. We mention that the effective drag coefficient that can be inferred from these simulations is typically in the range of 1–4 [*Cargill et al.*, 1995, 1996]. (In the results shown here, a constant drag coefficient of 3 was used.) However, the flux rope interaction with the ambient plasma can be complex, and further simulations, especially in three dimensions, are needed. Note that the toroidal forces that are key to the flux rope dynamics are purely 3D effects and do not occur in straight cylindrical configurations. In the next section, we discuss the evolutionary consequences of this model flux rope system.

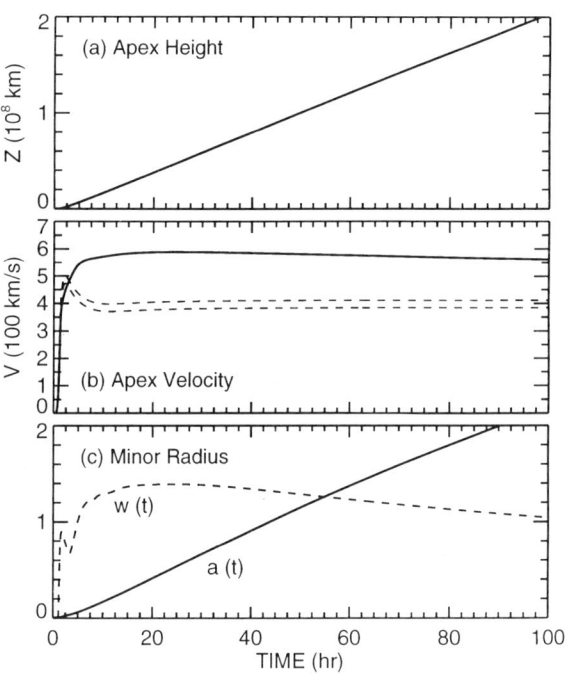

Figure 3. Long-time propagation of flux rope. (a) Apex height. (b) Apex velocity. (c) Minor radial dynamics: a (10^7 km) and w (in units of 50 km/s). Dashed lines: two examples with faster flux injection profiles, with roughly 1.5 times as much energy as that given by $d\Phi_p/dt$ shown in Figure 2.

It is of interest to consider the energetics of the flux rope expansion. The Lorentz force component $J_t B_p$, which leads to the logarithmic contribution in equation (2), is the primary driving force. This does work against the tension force, $J_p B_t$. As a result, magnetic energy is lost from the poloidal component, and a fraction of this energy goes into the magnetic energy in the toroidal component. The majority of the released magnetic energy, however, is released via the drag term at the rate of $F_d V$. A significant amount is also lost to the gravitational potential of the rising prominence material. The combined cavity-prominence structure is typically heavier than the ambient corona, and buoyant force [*Low*, 1981] is not likely to play a significant role in prominence eruptions and CMEs. It can also be shown that only a small amount of thermal energy ($\sim 10^{30}$ erg) is needed to keep the initially hot cavity plasma from cooling too rapidly. In contrast, *Kumar and Rust* [1996] assume that the magnetic energy released by the expanding flux rope (the same as above) is converted into the thermal energy of the plasma inside the flux rope. This is the main substantive difference between the model of *Chen and Garren* [1993] and that of *Kumar and Rust* [1996]. In the latter paper, the dynamically released magnetic energy is required to be partitioned into kinetic energy, gravitational energy, and thermal energy. The cold prominence gas is heated as the flux rope expands. Anomalous resistivity is implicitly assumed, but no physical mechanisms are included in the model equations. In this conception, it is assumed that helicity conservation [*Taylor*, 1974] determines the amount and time scale of magnetic energy conversion into thermal energy. However, Taylor's hypothesis does not provide time scales, nor does it specify the form of the energy. Had other forms of energy been allowed, e.g., drag dissipation, then the final energy partitioning would have been different. A more detailed discussion on magnetic helicity is given in *Chen* [1996].

We conclude this section with a brief comment on the triggering of prominence eruption. The main paradigm of prominence eruption based on initial closed-field structures is that the magnetic field footpoints rooted in the photosphere undergo shear, stressing the field in the corona and erupting after suffering loss of equilibrium arising from bifurcation of equilibrium solutions [*Low*, 1977] or formation of current sheets leading to reconnection and formation of plasmoids/flux ropes [e.g., *Mikić et al.*, 1988; *Forbes*, 1990; *Mikić and Linker*, 1994]. In this paradigm, the magnetic energy is slowly "stored" in the corona and suddenly "released." In contrast, the present scenario allows injection of magnetic flux into an existing flux rope to cause eruption. In this paradigm, which has been regarded as controversial [e.g., *Webb et al.*, 1994], a significant fraction of the magnetic energy released in eruption is injected into the corona immediately prior to and during the eruption. Thus, the flux rope is driven out of equilibrium, and the subsequent expansion is merely the "relaxation" of a nonequilibrium structure. Equation (6) provides the mathematical representation of this driven mechanism. The characteristic time scales below the photosphere are probably long while coronal relaxation occurs on much faster Alfven time scales. It is this fast relaxation process that we call eruption. This may result if the subphotospheric current system has gone out of equilibrium and must redistribute the magnetic energy while relaxing to a new equilibrium. Note that the recently proposed process of "helicity charging" [*Rust and Kumar*, 1994] is equivalent to the injection of poloidal flux.

3. HELIOSPHERIC CONSEQUENCES OF FLUX ROPE ERUPTIONS

The flux rope expands in a model interplanetary medium whose outward speed increases from zero to the asymptotic constant value V_{sw} at about 30 $R_\odot$. This solar wind profile is prescribed based on available solar wind observations and models. It is found that the speed of the apex approaches a nearly constant speed V, which is $\sim$570 km/s in the above example, while the minor radial extent increases according to equation (3). This flux rope reaches 1 AU in about 74 hr, with the minor radial size of $2a \simeq 0.23$ AU. The magnetic field is given by $B_{pa} = 8.3$ nT and $\overline{B_t} = 9.7$ nT. The minor radial expansion speed is $w \simeq 58$ km/s. The aspect ratio is $R/a \simeq 4.4$, which is not significantly different from the initial value of $R_0/a_0 = 5$. The particle density inside the flux rope is $\overline{n} \simeq 3$ cm^{-3}. These values are quite consistent with the macroscopic properties of magnetic clouds observed near 1 AU [*Burlaga et al.*, 1981; *Burlaga et al.*, 1990; *Lepping et al.*, 1990]. Thus, the model dynamics clearly indicate that magnetic clouds are the heliospheric counterpart of the initial cavity: the magnetic field and the hot tenuous plasma. There is significant variability in the particle densities observed inside clouds, which may result from the variability in the initial cavity plasma density or in the amount of prominence material that may be ionized and carried along as the cavity expands outward. The calculation described above shows that the tenuous cavity plasma is sufficient to provide magnetic cloud densities in the range of 2–5 cm^{-3} at 1 AU. This conclusion provides a specific theoretical and structural understanding of the well-known associations between prominence eruptions and observation of magnetic clouds in the solar wind [*Wilson and Hildner*, 1986; *Rust*, 1994; *Bothmer and Schwenn*, 1994], between CMEs and magnetic clouds [*Klein and Burlaga*, 1982; *Wilson and Hildner*, 1984; *Gosling*, 1990], and between prominence eruptions and geomagnetic storms [*Joselyn and McIntosh*, 1982; *Wright and McNamara*, 1983]. Our suggestion that a magnetic cloud is the 1 AU counterpart of the cavity in the initial cavity-prominence system is in agreement with that of *Tsurutani and Gonzalez* [1995] who argued that the low-β magnetic clouds must originate from the low-density, high-field (i.e., low-β) cavity. Note, however, the prominence gas is also a low-β structure because of the low temperatures. Thus, the low-β per se is not a unique signature of the cavity.

The magnetic field values given above are obtained from (2), (3), and (5). At any time t, the values of B_{pa} and $\overline{B_t}$ can be used to relate the macroscopic results to the magnetic field profile of the flux rope. A model field inside and immediately outside the current loop is [*Chen*, 1996]

$$B_p(r) = \begin{cases} 3B_{pa}\left(\frac{r}{a}\right)\left(1 - \frac{r^2}{a^2} + \frac{r^4}{3a^4}\right), & r \leq a \\ B_{pa}\frac{a}{r}, & r > a \end{cases} \quad (7a)$$

$$B_t(r) = \begin{cases} 3\overline{B_t}\left(1 - 2\frac{r^2}{a^2} + \frac{r^4}{a^4}\right), & r \leq a \\ 0, & r > a \end{cases} \quad (7b)$$

where $B_{pa} = B_p(a)$ and $\overline{B_t}$ is the average of $B_t(r)$ across the minor radius, with the relation $B_t(0) = 3\overline{B_t}$. For this profile, the maximum of $B_p(r)$ occurs at $r/a \simeq 0.66$ with $|B_p|_{\max} \simeq 1.24 B_{pa}$. The internal inductance is $\xi_i \simeq 1.2$, which is the value chosen for the reference flux rope. We have required that J_p and B_p vanish smoothly at $r = 0$ and that J_t and J_p vanish smoothly at $r = a$. The last condition also means that B_t vanishes smoothly at $r = a$. The above profile is constrained by the global loop dynamics through a, B_{pa}, and $\overline{B_t}$ computed using equations (2), (3), and (5). Thus, the quantities B_{pa}, $\overline{B_t}$, and a are functions of time, although no induction electric field is considered. In this respect, we find that the minor radius expands in near equilibrium even during the initial eruption with strong acceleration. In general, the loop field is embedded in some ambient magnetic field.

The profile (7) and Lundquist solution [*Burlaga*, 1988] have no remarkable differences. One distinction is that the latter has the fixed ratio $B_t(r=0)/|B_p|_{\max} \simeq 1.7$, whereas (7) yields $B_t(r=0)/|B_p|_{\max} \simeq 2.4\overline{B_t}/B_{pa}$ which is computed from the dynamical equations. Neither (7) nor the Lundquist solution is unique or necessarily in equilibrium for a given set of B_t, B_{pa}, a, p_a, and $\overline{p}$. (This is not a serious problem because $|\beta_p| \ll 1$ up to 1–2 AU.)

The magnetic field profiles inside magnetic clouds have been modeled using straight cylindrical flux ropes [*Burlaga*, 1988; *Suess*, 1988; *Cargill et al.*, 1994; *Farrugia et al.*, 1995]. In particular, *Burlaga* [1988] postulated that magnetic clouds are in a minimum energy state and can be modeled using the cylindrical constant-α force-free solution, the so-called Lundquist solution. This is based on the argument that the global helicity is conserved [*Taylor*, 1974] during the evolution of flux ropes. In contrast, equation (7) postulates that the observed structure is determined with no regard to the global helicity and that the field profile is characterized by one dominant scale length a in the absence of complicating effects such as interaction with the ambient solar wind. The underlying concept is that there is nothing to prevent magnetic surfaces from spreading themselves to minimize any spatial gradients and that the local profile of, for example, the apex is not strongly influenced by the magnetic properties near the footpoints which must be included in the global helicity. To the extent that the constant-α conjecture implies smoothing out local variations, the force-free Lundquist solution is a special case of single-scale-length solutions discussed here.

In describing the field (7) as seen by an observer at, say, 1 AU, we will assume that the cylinder axis lies in the ecliptic plane and is perpendicular to the Sun-Earth line. The z-direction is normal to the ecliptic and points toward north in the Earth frame with the cylinder axis in the y (East-West) direction. Figure 4 shows the magnetic field (7) with the expansion of the loop taken into account. The parameters (B_t, B_{pa}, a) have been obtained as functions of time from the reference flux rope as it expands past 1 AU. Figure 4a shows B_z as seen by the observer. The magnitude of the field, which has the maximum value of nearly 30 nT, is given in Figure 4b. Figure 4c shows the field rotation angle $\theta \equiv \sin^{-1}(B_z/B)$ where $B = (B_z^2 + B_y^2)^{1/2}$. The toroidal component (B_y) peaks near where $B_z = 0$. It is straightforward to generalize this exercise to different angles of approach. The results should resemble those of *Burlaga* [1988]. In these panels, the vertical dashed line in the middle indicates where $B_p = 0$, i.e., $B_z = 0$, which is the center of the flux rope. The two dashed lines on the sides mark leading and trailing edges of the current loop embedded inside the flux rope. These dashed lines correspond to the boundaries of magnetic clouds as defined by *Burlaga* [1988]. Recall that $B_p \sim r^{-1}$ outside the current loop, which is shown in Figure 4a. In displaying the field profile, we have included $B_y \simeq 6$ nT, the average ambient field at 1 AU. We have made no attempt to match the fields smoothly at the current loop boundary.

Recall that the presence of $\chi_0 = n_p/\overline{n}_c \neq 0$ can significantly affect the initial structure (section 2). In fact, if $n_p = 0$ is used, the resulting flux ropes at 1 AU tend to be on the small side ($2a \simeq 0.15$ AU) with relative weak fields ($|B| \lesssim 15$ nT) and high density ($\overline{n} \sim 10$–15 cm^{-3}) [*Chen and Garren*, 1993]. While these values are not unreasonable, Chen and Garren were not able to obtain larger magnetic clouds and stronger fields with lower density starting with a reasonable initial structure and eruption speeds. Inclusion of $n_p \neq 0$ removes this inadequacy: a wide range of initial values $5 \lesssim \chi_0 \lesssim 30$ can yield flux ropes corresponding to the range of observed magnetic clouds. This has two significant implications: (1) although the model has several parameters, it is not necessarily possible to produce desired results simply by adjusting the parameters if the initial structure does not have the necessary physical ingredients, and (2) the success

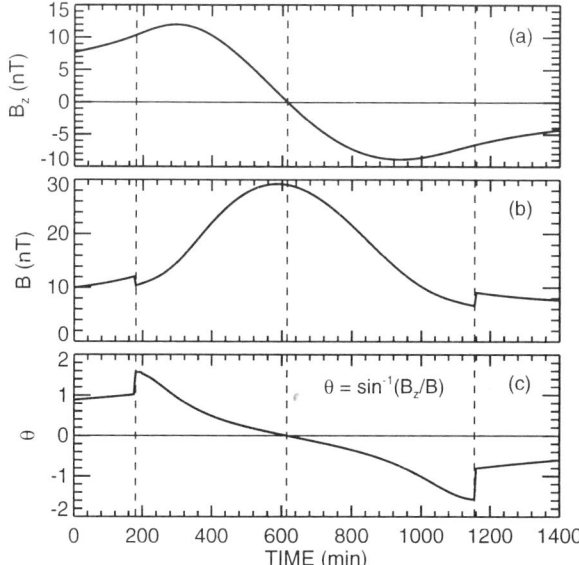

Figure 4. A magnetic profile for the reference flux rope at 1 AU, showing characteristics of an expanding magnetic cloud.

of any proposed prominence eruption/CME model cannot be determined by the initial eruption alone. The long-time consequences, e.g., at 1 AU and beyond, must also be consistent with observations.

4. TERRESTRIAL CONSEQUENCES

We now consider what happens to the magnetosphere-ionosphere system if a magnetic cloud impinges on the Earth. Empirically, it is well-known that long durations of strong southward IMF lead to large geomagnetic storms. Because of the long periods of strong southward IMF embedded in magnetic clouds, their geoeffectiveness has been well known [*Burlaga et al.*, 1981; *Klein and Burlaga*, 1982; *Wilson*, 1987; *Burlaga et al.*, 1987; *Zhang and Burlaga*, 1987; *Tsurutani et al.*, 1988]. In studies of large storm events, *Tsurutani et al.* [1988] and *Tsurutani et al.* [1992] have examined the global indices AE and D_{st}, in relation to the solar wind input. Their results indicate that perhaps 50% or more of large storms are caused by solar wind drivers such as magnetic clouds and structures with flux rope geometries. *Cumnock et al.* [1992], *Farrugia et al.* [1993], and *Knipp et al.* [1993] studied the ionospheric consequences of a large magnetic cloud (14 January 1988) interacting with the magnetosphere. *Cumnock et al.* [1992] investigated the cross-polar convection electric field based on the DMSP F-8 satellite, and *Knipp et al.* [1993] provided a detailed description of the ionospheric convection pattern using the Assimilated Mapping of Ionospheric Electrodynamics (AMIE) procedure. *Knipp et al.* [1993] found that the geomagnetic disturbance level as measured by ionospheric convection increases as B_y becomes comparable to B_z even while B_z is northward. The solar wind energy input indicated by the Akasofu ϵ parameter [*Perreault and Akasofu*, 1978] is consistent with this finding [*Farrugia et al.*, 1993]. (Note that D_{st} may not provide a good measure of disturbance in all aspects of geomagnetic activity [e.g., *Lanzerotti*, 1992]; we use D_{st} only as a traditional index.)

Recently, MHD simulation techniques have matured to the point that it has become possible to model long-duration storm events. In the first long-duration 3D simulation of magnetic storms, the interaction of magnetic clouds with the magnetosphere has been numerically investigated [*Chen et al.*, 1995; *Fedder et al.*, 1996]. In the former, a model magnetic cloud given by equation (7) was used as input to a global solar wind-magnetosphere simulation model [*Fedder and Lyon*, 1987]. Synthetic AE and D_{st} indices were calculated. Their behaviors in relation to the different phases of the model cloud were found to be in qualitative agreement with those of the actual AE and D_{st} during Earth passages of magnetic clouds [*Tsurutani et al.*, 1988]. An important finding is that the square of the cross-polar potential provides a good predictor of the energy deposition into the ionosphere due to Joule heating during storm conditions. Figure 5a shows the input B_y and B_z. These components are obtained from (7) assuming that the axis of the flux rope is in the ecliptic plane and perpendicular to the Earth-Sun line. Random noise has been superimposed. The vertical dashed line indicates when $B_z = 0$. Figure 5b shows the square of the cross-polar potential, Φ_p^2, obtained by the MHD model (thick line). The thin line is proportional to the ϵ parameter, measuring the solar wind energy input into the magnetosphere [*Perreault and Akasofu*, 1978]. The short arrows indicate the boundaries of the current loop.

In *Fedder et al.* [1996], the well-documented magnetic cloud of 14 January 1988 was used as input. In this MHD simulation, the solar EUV flux (parameterized by $F_{10.7}$) and the tilt of the Earth's magnetic axis were included, both of which affect the ionization levels of the polar ionospheres and therefore the energy precipitation into the ionosphere. The computed ionospheric cross-polar convection was compared with the available satellite measurements (DMSP F-8) [*Cumnock et al.*, 1992] and AMIE results [*Knipp et al.*, 1993]. The simulation results were found to be in reasonable agreement, until the simulated auroral oval moved to the low latitude simulation boundary at 58°, at which point the simulation results started to diverge from the observed values. The computed profile of Φ_p^2 is related to the input B_z profile in the same way as that shown in Figure 5. The Fedder-Lyon MHD model has also been used to simulate the onset of a substorm using the measured solar wind input during the polar passage of

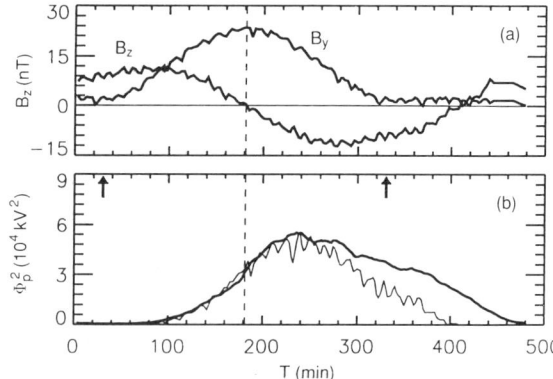

Figure 5. An MHD simulation of magnetic cloud-magnetosphere interaction. (From *Chen et al.* [1995].) (a) Input magnetic field. (b) The square of the computed cross-polar potential, Φ_p^2 (solid curve). The arrows indicate the edges of the current loop.

the VIKING satellite [*Fedder et al.*, 1995]. The simulated spatial and temporal distributions of energy precipitating into the ionosphere are in reasonable quantitative agreement with the images obtained by VIKING.

Although it is not the main topic of this review paper, we briefly touch on a practical byproduct of our increased understanding of the solar wind causes of geomagnetic storms: identification and prediction of geoeffective solar wind structures. This is important because severe geomagnetic disturbances can also have wide-ranging adverse consequences on advanced technological systems [e.g., *Lanzerotti*, 1979; *Allen et al.*, 1989; *Kappenman and Albertson*, 1990]. (A review of current forecasting techniques has been given by *Joselyn* [1995].) Note that solar wind streams with extended periods of strong southward IMF, streams that drive large geomagnetic storms, are magnetically well-organized (and prominent) structures. This means that one part of such a structure may contain information regarding the entire structure. *Chen et al.* [1996] exploited this realization to develop a pattern recognition technique to predict the approximate duration and severity of an impending storm. The basic idea is to monitor the solar wind magnetic field, speed, and density at, say, the Lagrange point L_1, and use these measured quantities to estimate the magnetic field structure in the solar wind stream that has yet to arrive. This method takes advantage of the fact that solar wind streams with long durations of unipolar (southward or northward) B_z component typically exhibit slow rotations of magnetic field. At any time, the rate of rotation inferred from the quantities measured up to that point is used to estimate the duration (τ) and maximum value of B_z (B_{zm}) of the segment of the stream before B_z reverses sign. The probability that a solar wind stream being encountered will cause a magnetic storm of certain severity is computed using Bayes theorem. This technique is intended to accurately predict the occurrence of large storms.

In Figure 6, we illustrate the application of this technique to the magnetic cloud observed on January 13–14, 1988, which produced a large storm characterized $D_{st} \simeq -150$ nT. In Figure 6a the observed $B_z(t)$ time series is shown (solid lines). (Note the similarity between the model B_z profile shown in Figure 4a and the observed profile shown in Figure 6a.) Using the magnetic field measured up to the present time, the profile of B_z field is estimated forward in time for the solar wind that has yet to come. One such estimate made at $t \simeq 10$ hr (indicated by a diamond) is shown by a dashed curve. This predicted profile remains nearly unchanged throughout the duration of this cloud. We see that the predicted B_z profile closely approximates the actual profile. That is, in this case, the approximate B_z profile of the solar wind stream is predicted for a period of $\sim$30 hours. Figure 6a shows the variation of the D_{st} index during this period, showing occurrence of a large storm of $D_{st} < -150$ nT coincident with the long $B_z < 0$ period. Figure 6c shows the profile of Φ_p^2 based on the estimated B_z profile shown in Figure 6a (dashed curve). This profile is obtained by scaling the Φ_p^2 shown in Figure 5b. That is, this is the Φ_p^2 that is predicted at $t \simeq 10$ hr (diamond in panel a).

This estimation process has the rudiments of an actual, real-time prediction of space weather conditions with forecasting time in excess of several hours. This example, of course, is a favorable case: had B_z field been southward ($B_z < 0$) first, then the forecasting time would have been shorter. However, this is still considerably greater than solar wind transit time from L_1 to the magnetopause, which is slightly less than one hour, assuming the wind speed of 400 km/s. This time can be much shorter because solar wind structures causing large storms, CMEs and magnetic clouds, often stream toward the Earth with speeds considerably in excess of the slow wind speed of 400 km/s. The January 1988 cloud is such an example, impinging on the Earth at an average speed greater than 600 km/s.

5. DISCUSSION

The early research work on the Sun-Earth coupling struggled with the correct empirical associations of solar origins and different types of geomagnetic disturbances such as recurrent and nonrecurrent storms. With the measurements of solar wind particles and magnetic fields becoming available, the focus shifted to the relationship between specific solar wind features and the corresponding terrestrial response. An obvious question was what specific solar events produced the geoeffective solar wind structures. After the discovery

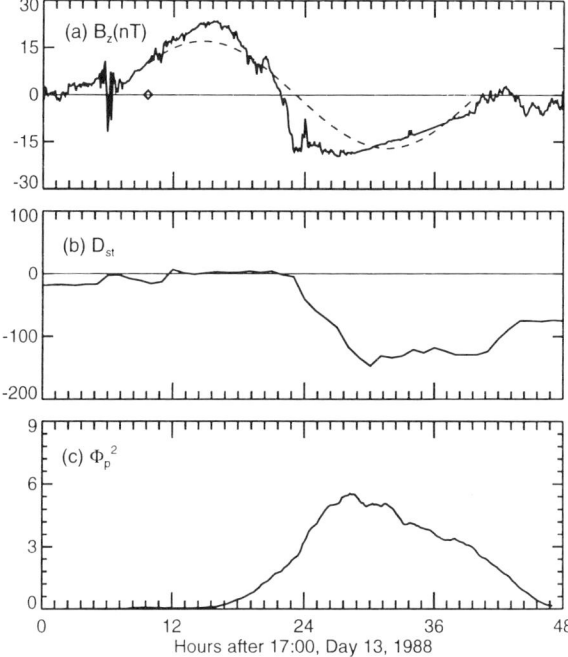

Figure 6. Identification and prediction of geoeffective solar wind streams. Application to the 13–14 January 1988 magnetic cloud. (a) The B_z measured by IMP-8. At $t = 10$ hr, the B_z profile is estimated forward in time based on B_y and B_z measured up to that point. The dashed curve is the estimated B_z profile. (b) The D_{st} index. (c) The cross-polar potential squared plotted in relative units, obtained by scaling (in time) the profile shown in Figure 5b to the predicted B_z profile given in Figure 6a (dashed curve). The Joule heating profile should be similar to Φ_p^2 [*Chen et al.*, 1995]. The order of magnitude of the Joule heating is 10^{10} W.

of CMEs and coronal holes in the early 1970's, both the recurrent and nonrecurrent storms now have well-established empirical causal associations with solar events: recurrent storms with Earth passages of sector boundaries and nonrecurrent storms with long-duration southward IMF structures attributable to CMEs and prominence eruptions. There are weak links in this chain of cause-effect associations. For example, CMEs and magnetic clouds observed at 1 AU and beyond may be the counterpart of the classic white light CMEs, or they may be the counterpart of the prominence or prominence cavity magnetic field. This precise correspondence is at the heart of the plasma connection between the Sun and the Earth. However, there are significant gaps in the empirical and theoretical understanding of solar eruptions and subsequent propagation of the resulting ejecta. At the present time, observations of solar eruptions are based on imaging of solar events, while beyond a certain distance from the Sun, observations of presumed ejecta rely primarily on *in situ* measurements of magnetic field and charged particles.

To date, no solar structures have been continuously observed from their points of origin at the Sun to the points of observation in the heliosphere. Theoretical understanding has also been difficult to obtain because the dynamics of 3-D magnetic structures over long periods of time must be calculated. This paper has summarized a recent effort representing the first step in quantifying the process of prominence eruption and propagation. The emphasis of this approach is to treat the entire process.

Note that the property required of magnetic field, $\nabla \cdot \mathbf{B} = 0$, means that any model of plasma structures must account for the global topology. Allowable magnetic structures include topological tori which have electric currents distributed throughout their spatial volumes and multipole fields whose source electric currents are primarily distributed in localized source regions (e.g., the Sun). Thus, flux ropes are an important "unit" of magnetic field in solar and astrophysical plasmas. The energy is dissipated along the way, primarily in the form of plasma energy outside the flux rope. The solar wind energy injected into the magnetosphere during geomagnetic storms represents a small fraction of the outwardly propagating magnetic energy of flux ropes.

Our understanding of the relationships between solar activity and large geomagnetic storms has progressed to the point where quantitative model calculations of solar eruption and subsequent evolution of geoeffective solar wind magnetic structures have begun to yield quantitatively promising results. The response of the magnetosphere-ionosphere system to the Earth passage of solar wind streams with long duration and strong southward IMF has been simulated whose results are in qualitative as well as quantitative agreement with satellite measurements. Although theoretical debates regarding the detailed physical mechanisms will undoubtedly continue, it has now become possible to discuss the plasma physics of Sun-Earth connection in specific and quantitative terms. We can look forward to a period of rapidly improving theoretical as well as observational predictive capabilities.

Acknowledgments. This work was supported by the Office of Naval Research. The IMP-8 data used in Figure 6 is courtesy of NASA/GSFC IMP-8 team and UCLA Space Physics Data Center.

REFERENCES

Allen, J., H. Sauer, L. Frank, and P. Reiff, Effects of the March 1989 Solar Activity, *EOS*, *70*, 1479, 1989.

Anzer, U., Can coronal loop transient be driven magnetically?, *Solar Phys.*, *57*, 111, 1978.

Birn, J., and K. Schindler, Two-ribbon flares: magnetostatic equilibria, in *Solar Flare Magnetohydrodynamics*, edited by E. R. Priest, p. 337, Gordon and Breach, New York, NY, 1981.

Bothmer, V., and R. Schwenn, Eruptive prominences as sources of magnetic clouds in the solar wind, *Space Sci. Rev.*, *70*, 215, 1994.

Brueckner, G. E., The behaviour of the outer solar corona ($3R_\odot$ to $10R_\odot$) during a large solar flare observed from OSO-7 in white light, in *Coronal Disturbances*, edited by G. Newkirk, p. 333, IAU, 1974.

Burlaga, L. F., E. Sittler, F. Mariani, and R. Schwenn, Magnetic loop behind an interplanetary shock: Voyager, Helios, and IMP 8 observations, *J. Geophys. Res.*, *86*, 6673, 1981.

Burlaga, L. F., Magnetic clouds and force-free fields with constant alpha, *J. Geophys. Res.*, *93*, 7217, 1988.

Burlaga, L. F., W. Behannon, and L. W. Klein, Compound streams, magnetic clouds, and major geomagnetic storms, *J. Geophys. Res.*, *92*, 5725, 1987.

Burlaga, L. F., R. P. Lepping, and J. A. Jones, Global configuration of a magnetic cloud, in *Physics of Magnetic Flux Ropes*, Geophys. Monogr. Ser., vol. 58, edited by C. T. Russell, E. R. Priest, and L. C. Lee, p. 373, AGU, Washington, DC, 1990.

Cargill, P. J., J. Chen, D. S. Spicer, and S. T. Zalesak, The deformation of flux tubes in the solar wind with applications to the structure of magnetic clouds and CMEs, in *Proceeding of the Third SOHO Workshop on Solar Dynamic Phenomena and Solar Wind Consequences*, ESA Spec. Publ. SP-373, p. 291, 1994.

Cargill, P. J., J. Chen, D. S. Spicer, and S. T. Zalesak, Geometry of interplanetary magnetic clouds, *Geophys. Res. Lett.*, *22*, 647, 1995.

Cargill, P. J., J. Chen, D. S. Spicer, and S. T. Zalesak, Magnetohydrodynamic simulations of the motion of magnetic flux tubes through a magnetized plasma, *J. Geophys. Res.*, *101*, 4855, 1996.

Carrington, R. C., Description of a singular appearance seen on the Sun on September 1, 1859, *Mon. Not. R. Astron. Soc.*, *20*, 13, 1860.

Chapman, S., Corpuscular influences upon the upper atmosphere, *J. Geophys. Res.*, *55*, 361, 1950.

Chen, J., Effects of toroidal forces in current loops embedded in a background plasma, *Astrophys. J.*, *338*, 453, 1989.

Chen, J., Theory of prominence eruption and propagation: Interplanetary consequences, *J. Geophys. Res.*, *101*, 1996.

Chen, J., and D. A. Garren, Interplanetary magnetic clouds: Topology and driving mechanism, *Geophys. Res. Lett.*, *20*, 2319, 1993.

Chen, J., S. Slinker, J. A. Fedder, and J. G. Lyon, Simulation of geomagnetic storms during the passage of magnetic clouds, *Geophys. Res. Lett.*, *22*, 1749, 1995.

Chen, J., P. J. Cargill, and P. J. Palmadesso, Real-time identification and prediction of geoeffective solar wind structures, *Geophys. Res. Lett.*, *23*, 625, 1996.

Crooker, N. U., and E. W. Cliver, Postmodern view of M-regions, *J. Geophys. Res.*, *99*, 23383, 1994.

Cumnock, J. A., R. A. Heelis, and M. R. Hairston, Response of the ionospheric convection pattern to a rotation of the interplanetary magnetic field on January 14, 1988, *J. Geophys. Res.*, *97*, 19460, 1992.

Démoulin, P., and T. G. Forbes, Weighted current sheets supported in normal and inverse configurations: A model for prominence observations, *Astrophys. J.*, *387*, 394, 1992.

Dryer, M., Coronal transient phenomena, *Space Sci. Rev.*, *33*, 233, 1982.

Dungey, J. W., Interplanetary magnetic field and the auroral zones, *Phys. Rev. Lett.*, *6*, 47, 1961.

Farrugia, C. J., M. P. Freeman, L. F. Burlaga, R. P. Lepping, and K. Takanashi, The Earth's magnetosphere under continued forcing: Substorm activity during the passage of an interplanetary magnetic cloud, *J. Geophys. Res.*, *98*, 7657, 1993.

Farrugia, C. J., V. A. Osherovich, and L. F. Burlaga, Magnetic flux rope versus the spheromak as models for interplanetary magnetic clouds, *J. Geophys. Res.*, *100*, 12,293, 1995.

Fedder, J. A., and J. G. Lyon, The solar wind-magnetosphere-ionosphere current-voltage relationship, *Geophys. Res. Lett.*, *14*, 880, 1987.

Fedder, J. A., S. P. Slinker, J. G. Lyon, and R. D. Elphinstone, Global numerical simulation of the growth phase and the expansion onset for a substorm observed by Viking, *J. Geophys. Res.*, *100*, 19,083, 1995.

Fedder, J. A., S. P. Slinker, J. Chen, J. G. Lyon, J. A. Cumnock, and M. R. Hairston, Global numerical simulation of the January 14–15, 1988 geomagnetic storm: Cross-polar convection, simulation and observations, *J. Geophys. Res.*, submitted, 1996.

Finn, J. M., P. N. Guzdar, and J. Chen, Fast plasmoid formation in double arcades, *Astrophys. J.*, *393*, 800, 1992.

Fisher, R., C. J. Garcia, and P. Seagraves, On the coronal transient-eruptive prominence of 1980 August 5, *Astrophys. J.*, *246*, L 161, 1981.

Forbes, T. G., Numerical simulation of a catastrophe model for coronal mass ejections, *J. Geophys. Res.*, *95*, 11,919, 1990.

Forbes, T. G., and P. A. Isenberg, A catastrophe mechanism for coronal mass ejections, *Astrophys. J.*, *373*, 294, 1991.

Foukal, P., Morphological relationships in the chromospheric H_α fine structure, *Solar Phys.*, *19*, 59, 1971.

Gonzalez, W. D., and B. T. Tsurutani, Criteria of interplanetary parameters causing intense magnetic storm ($D_{st} < -100$ nT), *Planet. Space Sci.*, *35*, 1101, 1987.

Gonzalez, W. D., J. A. Joselyn, Y. Kamide, H. W. Kroehl, G. Rostoker, B. T. Tsurutani, and V. M. Vasyliunas, What is a geomagnetic storm?, *J. Geophys. Res.*, *99*, 5771, 1994.

Gosling, J. T., Large-scale inhomogeneities in the solar wind of solar origin, *Rev. Geophys.*, *13*, 1053, 1975.

Gosling, J. T., Coronal mass ejections and magnetic flux ropes in interplanetary space, in *Physics of Magnetic Flux Ropes*, Geophys. Monogr. Ser., vol. 58, ed. by C.T. Russell, E.R. Priest, and L.C. Lee, p. 343, AGU, Washington, DC, 1990.

Gosling, J. T., The solar flare myth, *J. Geophys. Res.*, *98*, 18,937, 1993.

Gosling, J. T., Reply, *J. Geophys. Res.*, *100*, 3479, 1995.

Gosling, J. T., E. Hildner, R. M. MacQueen, R. H. Munro, A. I. Poland, and C. L. Ross, Mass ejections from the sun: A view from Skylab, *J. Geophys. Res.*, *79*, 4581, 1974.

Gosling, J. T., J. McComas, J. L. Phillips, and S. J. Bame, Geomagnetic activity associated with Earth passage of interplanetary shock disturbances and coronal mass ejections, *J. Geophys. Res.*, *96*, 7831, 1991.

Greaves, W. M. H., and H. W. Newton, On the recurrence of magnetic storms, *Mon. Not. R. Astron. Soc.*, *89*, 641, 1929.

Hale, G. E., The spectrohelioscope and its work, Part III. Solar eruptions and their apparent terrestrial effects, *Astrophys. J.*, *73*, 379, 1931.

Hirshberg, J., and D. S. Colburn, Interplanetary field and geomagnetic variations–A unified view, *Planet. Space Sci.*, *17*, 1183, 1969.

Hodgson, R., On a curious appearance seen in the Sun, *Mon. Not. R. Astron. Soc.*, *20*, 16, 1860.

Howard, R. A., D. J. Michels, N. R. Sheeley, and M. J. Koomen, The observation of a coronal transient directed at earth, *Astrophys. J.*, *263*, L101, 1982.

Howard, R. A., N. R. Sheeley, M. J. Koomen, and D. J. Michels, Coronal mass ejections: 1979–1981, *J. Geophys. Res.*, *90*, 8173, 1985.

Hu, W.-R., The dynamic process of a coronal transient associated with an eruptive prominence. I. Basic mechanism, *Astrophys. Space Sci.*, *92*, 373, 1983.

Hudson, H., B. Haisch, and K. T. Strong, Comment on 'The solar flare myth' by J. T. Gosling, *J. Geophys. Res.*, *100*, 3473, 1995.

Hundhausen, A. J., The origin and propagation of coronal mass ejections, in *Proceedings of the Sixth International Solar Wind Conference*, edited by V. J. Pizzo, T. Holzer, and D. G. Sime, p. 181, 1987.

Hundhausen, A. J., Coronal mass ejections: A summary of SMM observations from 1980 and 1984–1989, in *The Many Faces of the Sun*, edited by K. Strong, J. Saba, and B. Haisch, Springer-Verlag, 1996.

Illing, R. M. E., and A. J. Hundhausen, Observation of a coronal transient from 1.2 to 6 solar radii, *J. Geophys. Res.*, *90*, 275, 1985.

Illing, R. M. E., and A. J. Hundhausen, Disruption of a coronal streamer by an eruptive prominence and coronal mass ejection, *J. Geophys. Res.*, *91*, 10,951, 1986.

Joselyn, J. A., Geomagnetic activity forecasting: The state of the art, *Rev. Geophys.*, *33*, 383, 1995.

Joselyn, J. A., and P. S. McIntosh, Disappearing solar filaments: A useful predictor of geomagnetic activity, *J. Geophys. Res.*, *86*, 4555, 1981.

Kappenman, J. G., and V. D. Albertson, Bracing for the geomagnetic storms, *IEEE Spectrum*, March, 1990.

Kahler, S., Coronal mass ejections, *Rev. Geophys.*, *25*, 663, 1987.

Kahler, S. W., Solar flares and coronal mass ejections, *Annu. Rev. Astron. Astrophys.*, *30*, 113, 1992.

Kahler, S. W. and D. V. Reames, Probing the magnetic topologies of magnetic clouds by means of solar energetic particles, *J. Geophys. Res.*, *96*, 9419, 1991.

Kippenhahn, R., and R. Schlüter, *Z. Astrophys.*, *43*, 36, 1957.

Klein, L. W., L. F. Burlaga, Interplanetary magnetic clouds at 1 AU, *J. Geophys. Res.*, *87*, 613, 1982.

Knipp, D. J., et al., Ionospheric convection response to slow, strong variations in a northward interplanetary magnetic field: A case study for January 14, 1988, *J. Geophys. Res.*, *98*, 19273, 1993.

Kumar, A., and D. M. Rust, Interplanetary magnetic clouds, helicity conservation, and current-core flux ropes, *J. Geophys. Res.*, *101*, 15,667, 1996.

Kuperus, M., and M. A. Raadu, The support of prominences formed in neutral sheets, *Astro. Astrophys.*, *31*, 189, 1974.

Lanzerotti, L. J. (Ed.), Impacts of ionospheric/magnetospheric processes on terrestrial science and technology, in *Solar System Plasma Physics*, vol. III, edited by L. J. Lanzerotti, C. F. Kennel, and E. N. Parker, p. 317, North-Holland, New York, 1979.

Lanzerotti, L. J., Comments on 'Great magnetic storms' by Tsurutani et al., *Geophys. Res. Lett.*, *19*, 1991, 1992.

Lepping, R. P., J. A. Jones, and L. F. Burlaga, Magnetic field structure of interplanetary magnetic clouds at 1 AU, *J. Geophys. Res.*, *95*, 11,957, 1990.

Leroy, J. L., V. Bommier, and S. Sahal-Brechot, The magnetic field in the prominences of the polar crown, *Solar Physics*, *83*, 135, 1983.

Linker, J. A., G. Van Hoven, and D. D. Schnack, MHD simulations of coronal mass ejections: Importance of the driving mechanism, *J. Geophys. Res.*, *95*, 4229, 1990.

Low, B. C., Evolving force-free magnetic fields. I. The development of the preflare stage, *Astrophys. J.*, *212*, 234, 1977.

Low, B. C., Eruptive solar magnetic fields, *Astrophys. J.*, *251*, 352, 1981.

Low, B. C., and J. R. Hundhausen, Magnetostatic structures of the solar corona. II. The magnetic topology of quiescent prominences, *Astrophys. J.*, *443*, 818, 1995.

MacQueen, R. M., and R. R. Fisher, The kinematics of solar inner coronal transients, *Astrophys. J.*, *254*, 335, 1982.

MacQueen, R. M., J. A. Eddy, J. T. Gosling, E. Hildner, R. H. Runro, G. A. Newkirk, Jr., A. I. Poland, and C. L. Ross, The outer corona as observed from Skylab: Preliminary results, *Astrophys. J. Lett.*, *187*, L85, 1974.

Martin, S. F., W. H. Marquette, and R. Bilimoria, The solar cycle pattern in the direction of the magnetic field along the long axes of polar filaments, in *The Solar Cycle, ASP Conference Series*, vol. 27, edited by K. L. Harvey, p. 53, 1992.

Maunder, E. W., Magnetic disturbances, 1882 to 1903, as recorded at the Royal Observatory, Greenwich, and their association with sunspots, *Mon. Not. R. Astron. Soc.*, *65*, 2, 1905.

Michels, D. J., R. A. Howard, M. J. Koomen, and N. R. Sheeley, Jr., The solar mass ejection of 8 May 1979, in *Solar and Interplanetary Dynamics*, edited by M. Dryer and E. Tandberg-Hanssen, pp. 387–391, 1980.

Mikić, Z., D. C. Barnes, and D. D. Schnack, Dynamical evolution of a solar coronal magnetic field arcade, *Astrophys. J.*, *328*, 830, 1988.

Mikić, Z., and J. A. Linker, Disruption of coronal magnetic field arcades, *Astrophys. J.*, *430*, 898, 1994.

Mouschovias, T. C., and A. Poland, Expansion and broadening of coronal loop transients: a theoretical explanation, *Astrophys. J.*, *220*, 675, 1978.

Munro. R. J., J. T. Gosling, E. Hildner, R. M. MacQueen, A. I. Poland, and C. L. Ross, The association of coronal mass ejection transients with other forms of solar activity, *Solar Phys.*, *61*, 201, 1979.

Neupert, W. M., and V. Pizzo, Solar coronal holes as sources of recurrent geomagnetic disturbances, *J. Geophys. Res.*, *79*, 3701, 1974.

Newton, H. W., Solar flares and magnetic storms, *Mon. Not. R. Astron. Soc.*, *103*, 244, 1943.

Perreault, P. and S.-I. Akasofu, A study of geomagnetic storms, *Geophys. J. R. Astron. Soc.*, *54*, 547, 1978.

Pneuman, G. W., Eruptive prominences and coronal transients, *Solar Phys.*, *65*, 369, 1980.

Priest, E. R. (editor), *Dynamics and Structure of Quiescent Solar Prominences*, Kluwer Academic Publishers, Dordrecht, 1989.

Priest, E. R., and T. G. Forbes, Magnetic field evolution during prominence eruptions and two-ribbon flares, *Solar Phys.*, *126*, 319, 1990.

Rostoker, G., and C.-G. Fälthammar, Relationship between changes in the interplanetary magnetic field and variations in the magnetic field at the Earth's surface, *J. Geophys. Res.*, *72*, 5853, 1967.

Russell, C. T., R. L. McPherron, and R. K. Burton, On the cause of geomagnetic storms, *J. Geophys. Res.*, *79*, 1105, 1974.

Rust, D. M., Magnetic fields in quiescent solar prominences. I. Observations, *Astrophys. J.*, *150*, 313, 1967.

Rust, D. M., Spawning and shedding helical magnetic fields in the solar atmosphere, *Geophys. Res. Lett.*, *21*, 241, 1994.

Rust, D. M., and A. Kumar, Helicity charging and eruption of magnetic flux from the Sun, in *Proceeding of the Third SOHO Workshop on Solar Dynamic Phenomena and Solar Wind Consequences*, ESA Spec. Publ. SP-373, p. 39, 1994.

Sakai, J., and K.-I. Nishikawa, A model of 'disparition brusque' as an instability driven by MHD waves, *Solar Phys.*, *88*, 241, 1983.

Schmahl, E., and E. Hildner, Coronal mass-ejections-kinematics of the 19 December 1973 event, *Solar Phys.*, *55*, 473, 1977.

Sheeley, N. R., Jr., J. W. Harvey, and W. C. Feldman, Coronal holes, solar wind streams, and recurrent geomagnetic disturbances: 1973–1976, *Solar Phys.*, *49*, 271, 1976.

Sheeley, N. R., Jr., R. A. Howard, M. J. Koomen, and D. J. Michels, Associations between coronal mass ejections and soft X-ray events, *Astrophys. J.*, *272*, 349, 1983.

Steinolfson, R. S., Theories of shock formation in the solar atmosphere, in *Collisionless shocks in the heliosphere: Reviews of current research*, Geophys. Res. Monogr. Ser., vol. 35, edited by B. T. Tsurutani and R. G. Stone, p. 1, AGU, Washington, 1985.

Suess, S. T. Magnetic clouds and the pinch effect, *J. Geophys. Res.*, *93*, 5437, 1988.

Tandberg-Hanssen, E., *Solar Prominences*, D. Reidel, Dordrecht, Holland, 1974.

Taylor, J. B., Relaxation of toroidal plasma and generation of reverse magnetic fields, *Phys. Rev. Lett.*, *33*, 1139, 1974.

Tousey, R., The solar corona, *Space Res.*, *13*, 713, 1973.

Tsurutani, B. T., W. D. Gonzalez, F. Tang, S. I. Akasofu, and E. Smith, Origin of Interplanetary southward magnetic fields responsible for major magnetic storms near solar maximum (1978–1979), *J. Geophys. Res.*, *93*, 8519, 1988.

Tsurutani, B. T., W. D. Gonzalez, F. Tang, T. T. Lee, Great magnetic storms, *Geophys. Res. Lett.*, *19*, 73, 1992.

Tsurutani, B. T., W. D. Gonzalez, A. L. Gonzalez, F. Tang, J. K. Arballo, and M. Okada, Interplanetary origin of geomagnetic activity in the declining phase of the solar cycle, *J. Geophys. Res.*, *100*, 21,717, 1995.

Tsurutani, B. T., and W. D. Gonzalez, The future of geomagnetic storm predictions: Implications from recent solar and interplanetary observations, *J. Atmos. Terr. Phys.*, *57*, 1369, 1995.

Van Tend, W., The onset of coronal transients, *Solar Phys.*, *61*, 89, 1979.

Vandas, M., S. Fischer, M. Dryer, Z. Smith, and T. Detman, Simulation of magnetic cloud propagation in the inner heliosphere in two-dimensions 1. A loop perpendicular to the ecliptic plane, *J. Geophys. Res.*, *100*, 12,285, 1995.

Webb, D. F., Erupting prominences and the geometry of coronal mass ejections, *J. Geophys. Res.*, *93*, 1749, 1988.

Webb, D. F., T. G. Forbes, H. Aurass, J. Chen, P. Martens, B. Rompolt, V. Rusin, and S. F. Martin, Material Ejection, *Solar Phys.*, *153*, 73, 1994.

Webb, D. F., and A. J. Hundhausen, Activity associated with the solar origin of coronal mass ejections, *Solar Phys.*, *108*, 383, 1987.

Wilson, R. M., Geomagnetic response to magnetic clouds, *Planet. Space Sci.*, *35*, 329, 1987.

Wilson, R. M. and E. Hildner, Are interplanetary magnetic clouds manifestations of coronal transients at 1 AU?, *Solar Phys.*, *91*, 169, 1984.

Wilson, R. M., and E. Hildner, On the association of magnetic clouds with disappearing filaments, *J. Geophys. Res.*, *91*, 5867, 1986.

Wright, C. S. and L. F. McNamara, The relationships between disappearing solar filaments, coronal mass ejections, and geomagnetic activity, *Solar Phys.*, *87*, 401, 1983.

Wu, S. T., et al., Magnetohydrodynamic simulation of the coronal transient associated with the solar limb flare of 1980, June 29, 18:21 UT, *Solar Phys.*, *85*, 351, 1983.

Wu, S. T., M. T. Song, P. C. Martens, and M. Dryer, Shear-induced instability and arch filament eruption: A magnetohydrodynamic (MHD) numerical simulation, *Solar Phys.*, *134*, 353, 1991.

Wu, S. T., W. P. Guo, and J. F. Wang, Dynamical evolution of a coronal streamer-bubble system, I. A self-consistent planar magnetohydrodynamic simulation, *Solar Phys.*, *157*, 325, 1995.

Xue, M. L., and J. Chen, MHD equilibrium and stability properties of a bipolar current loop, *Solar Phys.*, *84*, 119, 1983.

Zhang, G. and L. F. Burlaga, Magnetic clouds, geomagnetic disturbances, and cosmic ray decreases, *J. Geophys. Res.*, *93*, 2511, 1988.

J. Chen, Code 6790, Naval Research Laboratory, Washington, DC 20375.

Heliospheric Observations of Solar Disturbances and Their Potential Role in the Origin of Geomagnetic Storms

Bernard V. Jackson

Center for Astrophysics and Space Sciences, University of California, San Diego

Ground-based interplanetary scintillation observations began heliospheric studies in the early 1960's using remote-sensing techniques. These were followed by coronagraph and Helios photometer white-light observations, and kilometric radio observations. The first solar wind *in situ* observations were measured by the Mariner 2 spacecraft three decades ago. These observations show heliospheric features which corotate with the Sun, as well as those which only occur as single events. For an observer at Earth both types of features can manifest themselves as abrupt time variations in the heliospheric plasma magnetic field, density, and velocity. On the corotating side, recent studies have shown that active regions are associated with sustained outflows of plasma at 1 AU in the form of a dense, slow-speed solar wind component which can interact with the less dense coronal hole regions of the solar wind. Recent work shows that the heliospheric manifestations of coronal mass ejections provide a significant portion of the solar wind mass (>15%) and energy outflow at solar maximum. Both of these types of features have implications in terms of their potential for magnetospheric interaction. We are now poised to take advantage of several technological advances currently underway which will undoubtedly play a key role in complete heliospheric plasma characterization in terms of fundamental interplanetary medium parameters from the Sun outward.

1. INTRODUCTION

From the perspective of a point in space such as Earth, solar disturbances can be either those which do not corotate and only pass Earth once, and corotating structures that have the potential to pass Earth on successive rotations. Both discrete events such as solar flares, and coronal mass ejections (CMEs) and corotating features (*e.g.* streamers) can be observed to be present in the remotely-sensed record near the solar surface. Observation of them can be used to judge persistence relative to the length of a solar rotation. Spacecraft *in situ* observations, which rely on the convection of material past them in the solar wind to determine structure, have difficulty determining the degree to which each structure persists. The *in situ* measurements must either repeat on successive solar rotations, or else be associated with persistent solar features in order that they be labeled corotating. Both the remotely-sensed and *in situ* record give evidence for non-radial expansion, evolution and solar wind interaction which can significantly modify the solar wind as it expands outward. This makes a straightforward extrapolation of solar surface features to 1 AU difficult.

The corona and solar wind are significantly modified by dramatic perturbations in the form of discrete ejections of mass and energy from the Sun (CMEs). These disturbances may extend outward in the form of plasma and magnetic field to the magnetosphere of Earth. The record does not clearly show whether portions of these disturbances remain rooted on the solar surface as they pass 1 AU. In the past, remote-sensing observations of the origins of these disturbances on the Sun have been restricted to coronal emission-line observations and the meter-wave and shorter radio wavelengths, but since the late 1960's there has been an addition of powerful tools for observation. These tools now include a wide variety of techniques from interplanetary scintillation (IPS) observations, soft X-ray observations of the corona near the solar surface, white-light observations of coronal and heliospheric features, kilometric radio data and spacecraft *in situ* observations. Unfortunately, the techniques most commonly used are each sensitive to a specific type of observation and more sensitive or more limited in one portion of the interplanetary medium than another. In addition, none of the remote sensing techniques has complete spatial or temporal coverage. Observations from spacecraft made *in situ* have the disadvantage that they do not observe the whole. Thus the basic physics, as well as the spatial and temporal evolution of heliospheric features, can be misinterpreted as they evolve away from the Sun.

Although this review is primarily limited to remote-sensing observations of solar wind features present in the inner heliosphere, in Section 2 of this paper I give a brief description of these features as measured and inferred from *in situ* observations and modeling. There have been a variety of techniques developed to observe these features remotely. I detail the most frequently used of these techniques, and I give examples of the data for each in Section 3. Section 4 shows the results of recent measurements and samples of the techniques developed to map these observations into three dimensions with spatial continuity. Concluding remarks are found in Section 5.

2. *IN SITU* OBSERVATIONS AND MODELING

In situ observations form the basis for ideas about the type of plasma impinging on the magnetosphere. As spacecraft first ventured above the magnetosphere [*Ness et al.*, 1964], observations of the solar wind from deep space probes and Earth orbiting spacecraft [*Neugebauer and Snyder*, 1966] mapped a variety of different solar wind features as postulated earlier by *Parker* [1958]. Some of these features could be observed to repeat with a cadence of a solar rotation while other features manifest themselves only once with no evidence of repetition. The most significant of the features which appeared to repeat with regularity were so-called sector boundaries, the locations of the boundary of outward to inward directed heliospheric magnetic field [*Wilcox and Ness*, 1965]. This regular magnetic field pattern manifests itself with a region of higher solar wind density followed (and preceded) by a rarefaction associated with regions of generally higher-speed solar wind. These dense solar wind regions which corotate are often thought of as regions where significant solar wind interaction has taken place to modify the structure as it moves outward from the Sun [*Hundhausen*, 1972]. At large solar distances (>1.5 AU) these interaction regions (or CIRs) are bounded by recurrent shocks [*Smith and Wolfe*, 1976]. At 1 AU, the corotating low density regions following these are found to be associated with regions of enhanced Alfven wave activity responsible for continuous substorms at Earth [*Tsurutani et al.*, 1995]. Superposed on this recurrent structure are significant non-recurrent events such as shocks which are recognized by their generally higher plasma density, magnetic fields and temperatures. When these solar wind features intersect the Earth, the magnetosphere responds through a coupling process involving magnetic field and solar wind pressure [*e.g., Tsurutani et al.*, 1988, 1995].

2.1. *Corotating Structures*

The dense regions that persist from solar rotation to solar rotation and their imbedded magnetic field structures, are usually depicted as extending in space from near the Sun to far beyond Earth in a nearly continuous fashion [*e.g., Dessler*, 1967; *Hundhausen*, 1972]. Figure 1 depicts these features in schematic form. Analysis of these features is usually organized around *in situ* magnetic field information, since the magnetic field observations are often those observed to repeat with the most regularity.

That dense corotating regions extend through space from near the solar surface to over 1 AU can be demonstrated by numerous types of measurements. Helios spacecraft *in situ* measurements as close as 0.3 AU distance from the Sun in the ecliptic show the existence of dense corotating structures [*Rosenbauer et al.*, 1977; *Schwenn*, 1990]. Superposed epoch analysis averages of the region near the location of magnetic field reversal are shown in Figure 2. These claimed *in situ* streamer (solar corotating dense features, next section) observations at 1 AU [*e.g., Borrini et al.*, 1981; *Gosling et al.*, 1981] give maximum densities

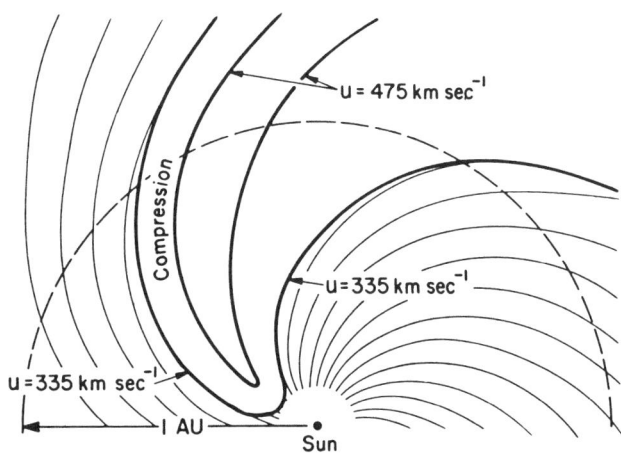

Figure 1. Schematic depiction of a corotating feature of enhanced slow solar wind (from *Hundhausen* [1972]).

which average slightly more than a factor of three times the ambient.

2.2. *Non-recurrent Disturbances/CMEs/ICMEs*

The *in situ* record also shows features which do not repeat from one rotation to the next. These often give rise to the largest geomagnetic storms at Earth [*Tsurutani et al.*, 1992]. When CMEs were first observed in coronagraph observations (next section) they became the likely solar manifestations of non-recurring solar wind features. Because of the oftentimes high speeds of some CMEs, they can be associated with shock waves moving outward in the interplanetary medium. However, because the observations require large extrapolations of material that evolves over time, it is not certain which portions of the brightening observed in coronagraphs are measured at

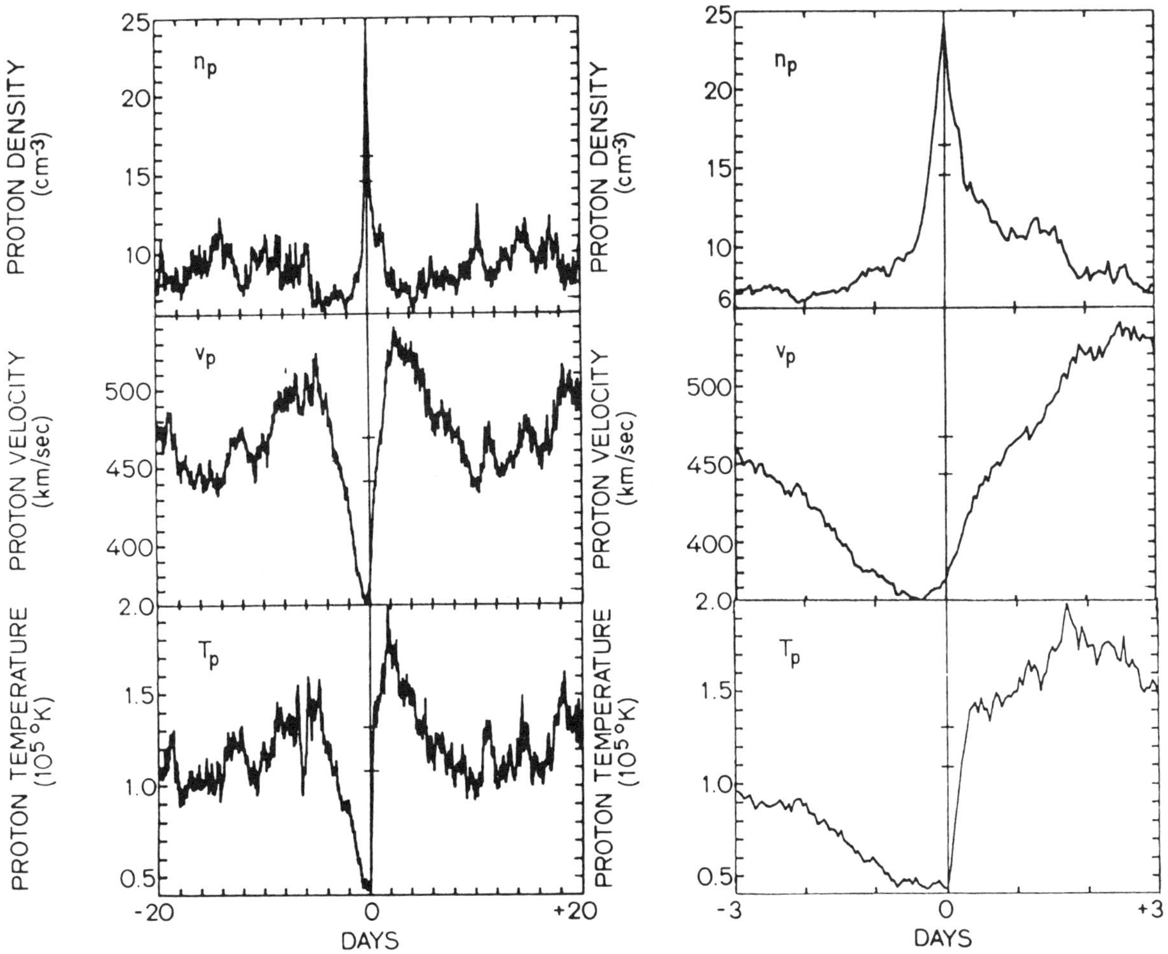

Figure 2. Composite averages of 74 corotating regions observed *in situ*. The left and right plots differ only in the number of days included in this analysis of superposed magnetic field reversal positions (from *Borrini et al.* [1981]).

1 AU. Some CMEs observed by coronagraphs have been traced outward until they arrive and are observed by spacecraft *in situ* where they show varying responses [*e.g., Burlaga et al.*, 1982; *Sheeley et al.*, 1985]. Figure 3 is a artist's depiction of these structures moving outward from the Sun. Figure 4 is an example of the *in situ* measurements of a CME that has been observed at 1 AU. The portion of the event directly behind the shock is in part the sheath of ambient solar wind that has been compressed and swept up by the event. As in Figure 4 the interplanetary CME manifestation (hereafter termed an ICME) is nearly always accompanied throughout by a significant increase in the magnetic field strength (above the ambient) that is smoothly-varying (with the absence of waves and discontinuities). The portion of the *in situ* plasma with an organized magnetic field that has a significant rotational component in Figure 4 is called a magnetic cloud [*Klein and Burlaga*, 1982]. The rotation is thought to imply that a large coherent current structure is imbedded in the convected plasma. In other instances intense, smooth fields without a significant rotational component can also signal the presence of ICME driver gas [*Tsurutani et al.*, 1988].

The heat flux in the solar wind is carried almost entirely by supra-thermal halo electrons with energies larger than about 100 eV. Usually this heat flux is uni-directional, flowing outward from the hot solar corona along the interplanetary magnetic field and connected with the hot corona at only one end. Occasionally, discrete events

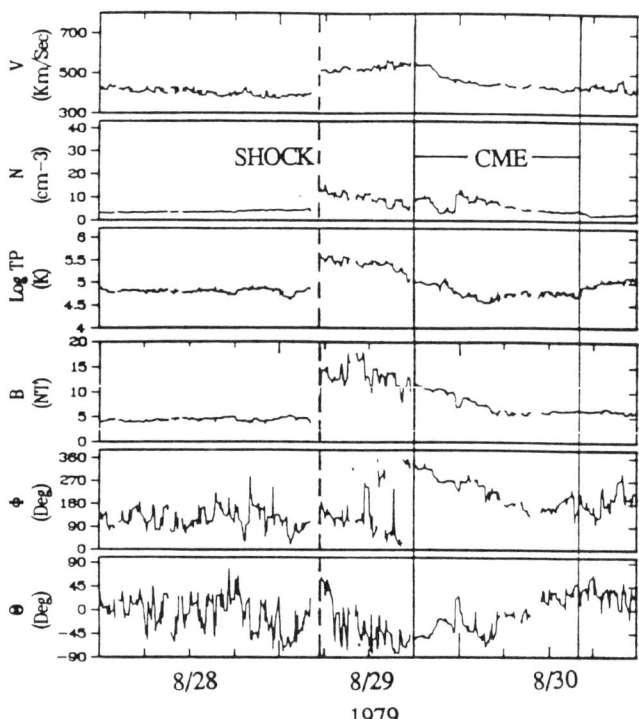

Figure 4. Solar wind plasma and field measurements encompassing a shock wave disturbance driven by a fast ICME observed *in situ* (from *Gosling* [1992]).

lasting for many hours are observed when the field-aligned electron flux becomes distinctly bi-directional [*Montgomery et al.*, 1974; *Pilipp et al.*, 1987; *Gosling et al.*, 1987]. If counterstreaming electron flux events can only occur on closed field lines either still rooted with both footpoints in the hot corona, or (in the case of a detached plasmoid) with a trapped counterstreaming electron flux in it, these events can be used as unique tracers of the passage of ICMEs [*Gosling*, 1992]. However, as argued by *Kahler and Reames* [1991], these *in situ* observations alone may not be sufficient to uniquely define the passage of a ICME at 1 AU.

As mass ejections traverse the heliosphere, they evolve with time so that *in situ* at 1 AU they often become difficult to distinguish from other structures (or enhanced mass in corotating features) present in the interplanetary medium (for a recent review see *Gosling* [1990]). A unique signature of ICMEs in *in situ* data will never be completely available simply because the CME driver gas may not fully intersect the spacecraft even though its effects (such as a shock) surrounding the CME may. In addition, it is not clear to what extent structures which primarily corotate with the Sun but evolve slowly differ from these discrete ejections of mass.

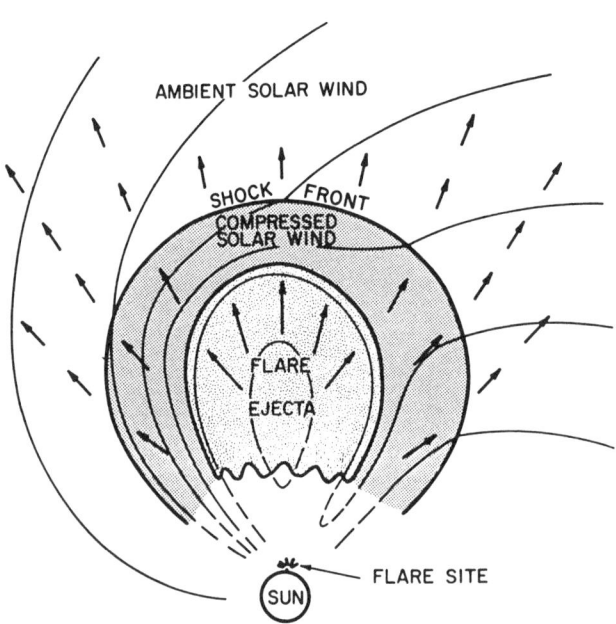

Figure 3. Schematic depiction of a CME moving outward in the heliosphere (from *Hundhausen* [1972]).

Other suggestions of ICME *in situ* characteristics at 1 AU that are important in how they interact with the magnetosphere are frequently discussed in the literature. The enhanced velocities of some ICMEs are thought to cause compression of the ambient medium [*Hundhausen*, 1972] as well as a shock ahead of the ICME. This outward motion may cause the ambient interplanetary magnetic field to drape around the ICME [*Gosling and McComas*, 1987], or an enhancement of the magnetic field from shock compression for ICMEs which are faster than the ambient (as in Figure 4). These effects can enhance the already southward component of the magnetic field associated with the ICME responsible for geomagnetic coupling during magnetic storms.

3. HELIOSPHERIC REMOTE SENSING TECHNIQUES

At times of a solar eclipse, coronal streamers are observed as enhancements of brightness that extend outward from near the solar surface. The reason for these stationary dense structures near the solar surface is essentially a sheet current system "pinch effect" as described for instance by *Pneuman and Kopp* [1971]. The regions of low solar surface brightness (coronal holes) have been known from the time of Skylab spacecraft observations [*e.g., Zirker*, 1977] to be locations from which high speed low density solar wind emanates [*Munro and Jackson*, 1977]. The solar wind which expands outward non-radially from polar coronal hole regions, especially at times of solar minimum, often provides the bulk of the solar wind flow. In terms of *in situ* observations, these regions are areas of magnetic field directed with the dominant polarity of that region on the solar surface [*Hundhausen*, 1977]. Coronal holes at 1 AU are generally observed following the passage of a reversal of the interplanetary magnetic field, and are thought to cause continuous recurrent substorms (last section). Coronal holes are generally observed to evolve slowly compared with solar rotation. Because denser regions dominate heliospheric observations from ecliptic-based measurements, remote-sensing studies of coronal holes are still in their infancy.

Solar flares observed by ground-based observatories and their associated radio signatures [*Wild and McCready*, 1950] showed the locations of these events near the solar surface long before spacecraft measurements of them provided coverage at other than optical or metric radio wavelengths. The largest solar flares are generally observed to be associated with expulsion of mass from the Sun. As ICMEs, these expulsions travel outward at speeds of from a few hundreds to a thousand km s^{-1}, some of which arrive at Earth several days following the event.

The enhancement of mass, entrained magnetic field, and sometimes enhanced velocity associated with these events plus their enhancement through interaction with the ambient solar wind, can potentially cause Earth's most energetic magnetic storms. By their general nature as enhancements of the background, these heliospheric features can be observed more easily than others, and they are thus frequently studied with instrumentation having sufficient temporal cadence to do so.

3.1. *Coronal Emission-line and X-ray techniques*

Imaging observations of coronal emission lines available from ground-based observatories and spaceborne solar X-ray observations [*Krieger et al.*, 1973] are limited to regions very near the solar surface. This is because the brightnesses of coronal features from these analyses are a function of the density squared and temperature, and coronal density near the solar surface (especially) diminishes rapidly with height above the Sun. Although these observations are not heliospheric, they none-the-less are commonly used to extrapolate features outward from the solar surface, and thus deserve mention here. In the case of spaceborne measurements, coronal observations in emission line or X-ray observations are available against the solar disk, giving some hope of mapping earthward-directed expulsion of mass away from the solar surface (CMEs?) associated with solar flare brightening observed in these same instruments.

An example of these data in the form of an image of the Sun in X-rays from the Soft X-ray Telescope (SXT) on board the *Yohkoh* spacecraft is shown in Figure 5a. Figure 5b gives a solar map constructed for the same day using solar limb scans (coronagraph techniques) from many days of ground-based instrumentation available from the Sacramento Peak Observatory. The figures show a striking example of a coronal hole which extends from north to south across the solar disk. As the Sun rotates, the solar wind emanating from this coronal hole is certain to impinge on the Earth's magnetosphere.

Using these coronal imaging techniques, observations of dynamic solar phenomena (solar flares and ejections of mass observed as direct motions and depletion of existing coronal material) can also be extrapolated to Earth's vicinity. That their extrapolation is fraught with difficulty is evidenced by recent discussion in the literature [*e.g., Gosling*, 1993; *Hudson et al.*, 1995] where the association of ICMEs and flares have been questioned primarily from groups measuring solar wind plasma *in situ* near Earth. Recent observations, especially those from the SXT which relate to this, are reviewed elsewhere in this volume [*Hudson*, 1997], and will not be repeated here.

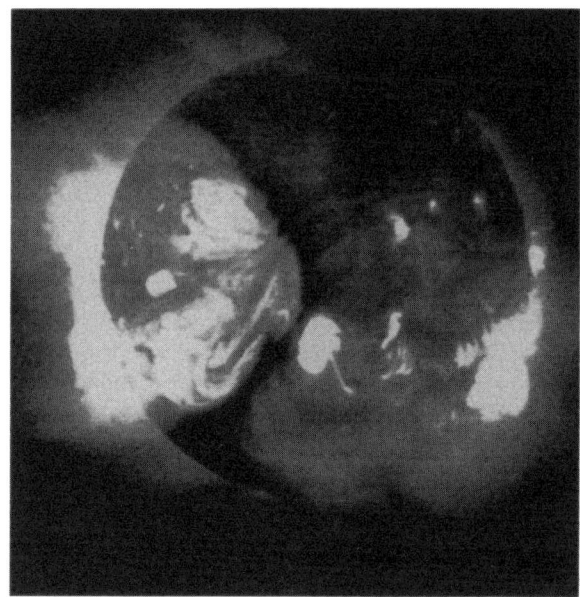

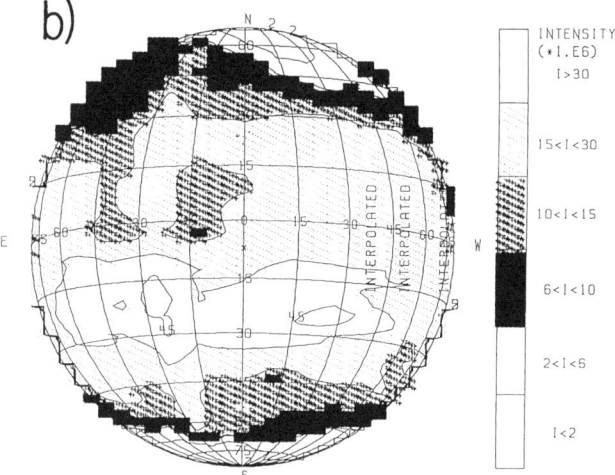

Figure 5. (a) An image of the Sun in soft X-rays showing a coronal hole extending from the northern to southern hemisphere. (b) Data construction from coronal emission line observations obtained at the solar limb.

3.2. Intensity Interplanetary Scintillation

Intensity IPS has been used since 1962 to observe the interplanetary medium [*Hewish et al.*, 1964]. The technique relies on measuring the rapidly fluctuating intensity level from point-like radio sources. Whether a single-site array is used or a system of multiple ones where the scintillation pattern is cross-correlated to measure velocities [*Coles and Kaufman*, 1978], the technique relies on the presence of small-scale (~200 km) density inhomogeneities in the interplanetary medium that move across the line-of-sight to the source. Heliospheric features are indicated in these data by either spatial or temporal changes of the scintillation level [*Houminer and Hewish*, 1974], changes of the IPS-determined speed [*Watanabe and Kakinuma*, 1984], or both [*Jackson*, 1984]. An example of the analysis from the Cambridge group using 1979 data is shown in Figure 6. Many different IPS disturbances are observed in the data each day. Over a time interval of a few days, it is possible to determine an approximate size and shape of the structures which move past Earth and to classify them as in *Gapper et al.* [1982].

These features, presumably modified significantly by their passage through the interplanetary medium on the way to 1 AU, are generally classified into structures that are either corotating or cut-off from the Sun. Some of these intensely scintillating features are remnants of mass ejections as discussed in *Rickett* [1975] or even *Hewish and Bravo* [1986]. However, *Hewish and Bravo* [1986] claim that most of the features they observe are best-associated with corotating solar wind features. Often, *in situ* measurements [*e.g., Behannon et al.*, 1991] can help to elucidate the differences between these two forms of disturbances.

Most of the experience with mass ejections comes from coronagraph observations in white light from disturbances close to the solar limb. Unfortunately, although intensity IPS observations measure disturbances in the solar wind, they do not have an interpretation as mass unless the proportionality between scintillation level enhancement (related to small-scale ~200 km density variations in the solar wind) and solar wind density is known. This proportionality has been determined in several ways with different results [*Tappin*, 1987; *Zwickl*, 1988; *Jackson et*

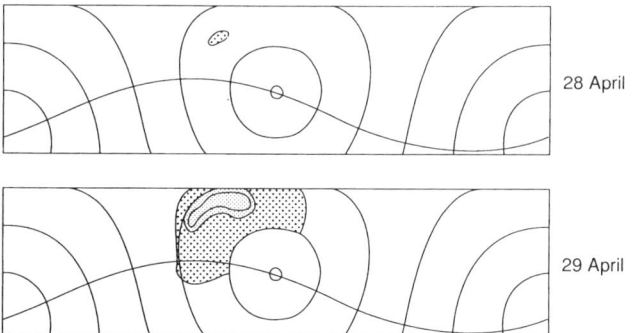

Figure 6. IPS heliospheric images as mapped from daily observations from the Cambridge array in April 1979. Sky maps in right ascension and declination show enhanced IPS levels (stippled) relative to average values (from *Jackson* [1992]).

al., 1997], that could possibly be due to the different structures measured. The technique maps a disturbance well at the Cambridge array frequencies (81.5 MHz) only beyond approximately 50° elongation (>0.75 AU) from the Sun, and thus distant from measurements of CMEs which are usually measured near the solar surface in the plane of the sky. Not only is the extent of an IPS disturbance difficult to discern given the signal to noise present in the observations, but at this solar distance most mass ejections have evolved significantly [Sheeley et al., 1985]. IPS simulation studies by Pizzo and Sime [1991] and others have shown that it is often possible to confuse a corotating structure with one that moves radially outward from the Sun. Add to this the observations discussed earlier including those by Woo and Schwenn [1992], which imply that the small-scale density variations within shocks are proportionally much larger relative to density than elsewhere, and the relation of intensity IPS measurements to different solar wind features becomes very difficult.

3.3. Kilometric Remote Sensing

Metric radio waves have regularly been used to monitor solar radio bursts in the solar corona since the late 1940's [Wild and McCready, 1950; Wild, 1970]. When kilometric receivers were flown on spacecraft above the ionosphere, observations of them became truly heliospheric [Steinberg, 1980]. Remote sensing observations of type II (from shock waves) and type III (from relativistic electrons) radio bursts indicate that they are greatly influenced by dense structures in the interplanetary medium near the Sun [Kundu et al., 1983; Smart et al., 1985; Steinberg et al., 1985]. The modeling of type III burst positions clearly indicates that the relativistic electrons which excite them propagate along the Archimedean spiral [Bougeret et al., 1984] and perhaps travel in dense regions of the interplanetary medium.

Shock waves, thought to be highlighted by the type II metric radiation, can be observed remotely to move outward in the lower corona in association with some flares and CMEs. Metric type II radiation is observed with CMEs having the highest speeds [Gosling et al., 1976] indicating an associated shock process. However, the association ends there; most metric type II radio bursts do not extend to the kilometric range even though CMEs move outward unabated into the interplanetary medium. Even so, both type II and type III kilometric radio bursts can appear as the interplanetary counterpart of their lower metric cousins. Figure 7 is the presentation of the dynamic spectrum primarily of type II radiation as observed from the Ulysses spacecraft [Stone et al., 1983] in 1991. While some direction and size information about the kilometric burst (and thus the shock) extent exists from modulation of the signal from a rotating spacecraft [i.e.. see Steinberg et al., 1985], scattering of radio radiation by the interplanetary medium restricts the usefulness of these observations. True kilometric imaging will not be possible until radio arrays consistent in size with several wavelengths of the radio signal are operated above the Earth's ionosphere, and even then scattering will remain a significant problem.

3.4. Thomson Scattering

Thomson-scattering analyses of sunlight from coronal plasma electrons have the advantage that they relate directly to a fundamental physical quantity (density) from an optically thin medium. Only when sunlight is reduced from the solar disk well-enough so that these electrons can

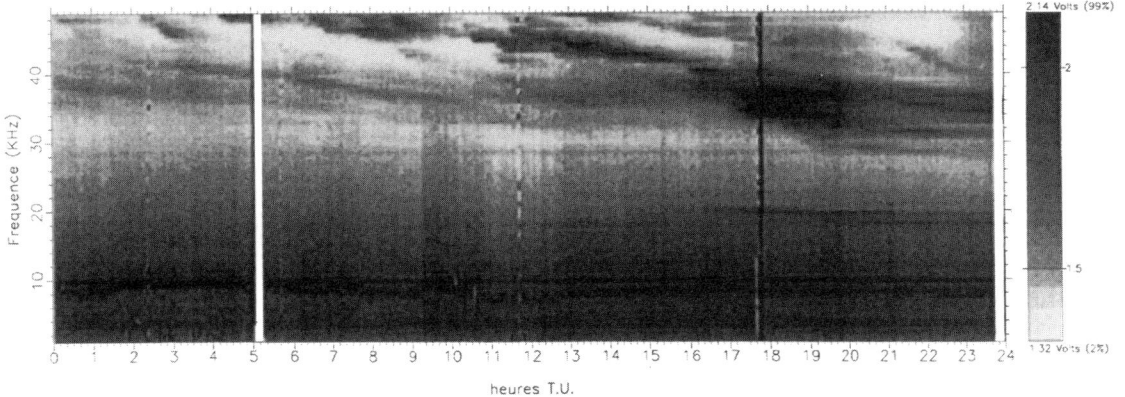

Figure 7. Dynamic spectrum of a type II kilometric radio burst on 21 January 1991 as observed from the Ulysses spacecraft.

be observed, can coronal features be discerned. A solar eclipse allows observation of structures which extend outward into the corona above the solar disk. Computer enhancement techniques from eclipse photographs [*e.g., Chapman*, 1979] have carried observations of coronal streamers as far as 13 R_S from the Sun (Figure 8). Streamers were some of the first structures observed by Earth-orbiting coronagraphs [*MacQueen et al.*, 1974; *Koomen et al.*, 1975] out to distances as large as 10 R_S. In coronagraph studies of coronal streamers within a few solar radii of the solar surface, *Poland* [1978] shows that a streamer, although changing in mass by as much as 50% at successive solar limb passages, can be present for several solar rotations. Thus, if extended outward, coronal streamers are persistent enough to be observed as density enhancements in the interplanetary medium.

Well over 2000 individual CME events have now been observed with spaceborne [*Howard et al.*, 1985a; *Hundhausen et al.*, 1994] and ground-based [*Fisher et al.*, 1981] coronagraphs. Figure 9a gives an example of these images as observed with the Solwind coronagraph (in operation from May 1979 - September 1985) and Figure 9b gives an example of these images from the SMM coronagraph (in operation from May 1980 - November 1989 with a three-year down time beginning late 1980). While these observations are spectacular, they are limited to the region near the Sun and generally to events only near the plane of the sky (*not* earthward directed). There

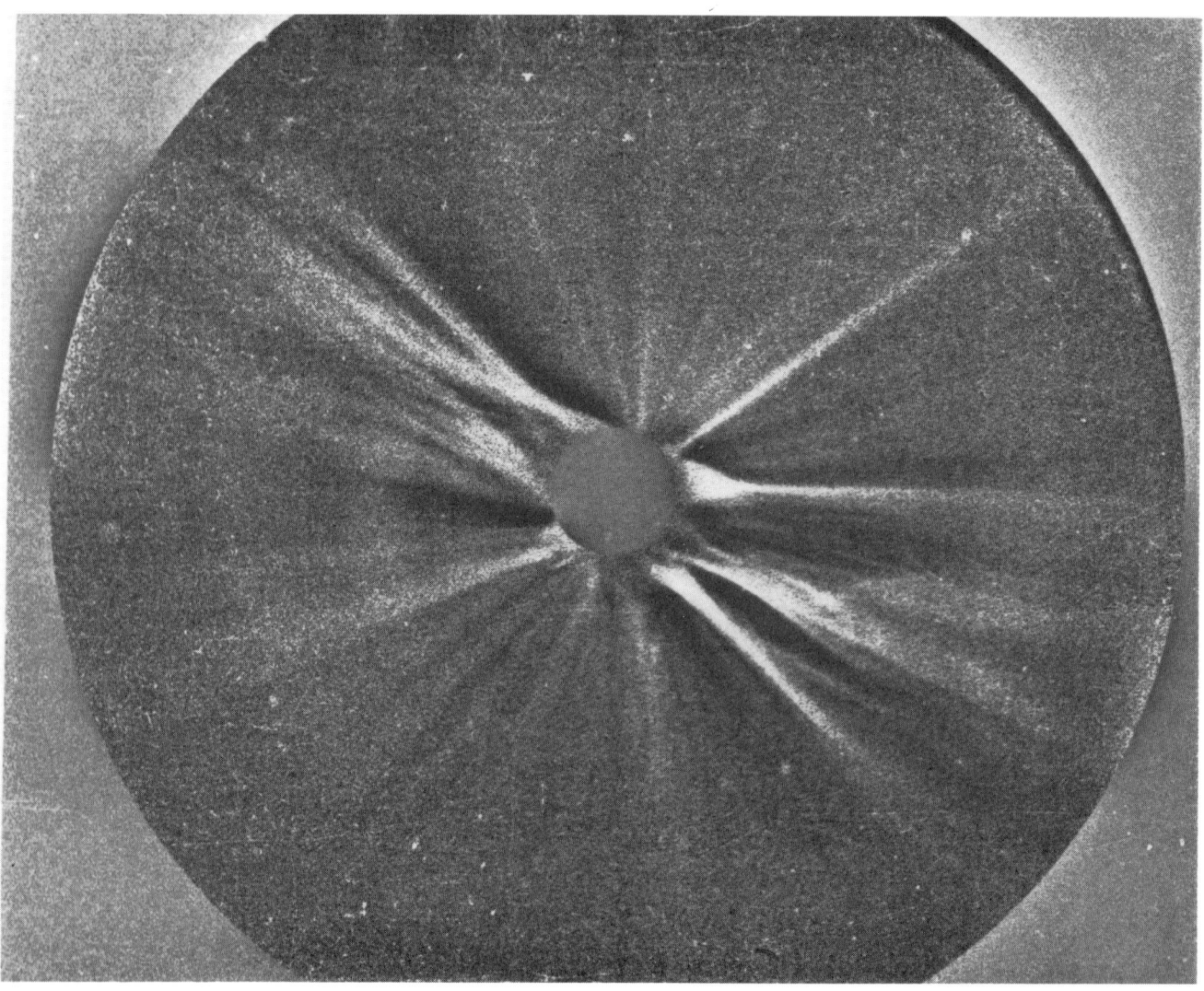

Figure 8. 30 June 1973 Eclipse observations (from *Chapman* [1979]).

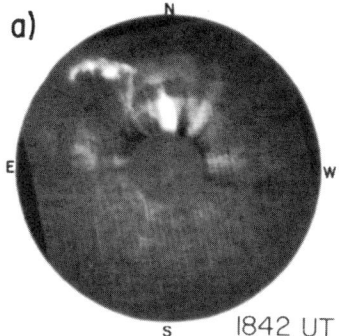

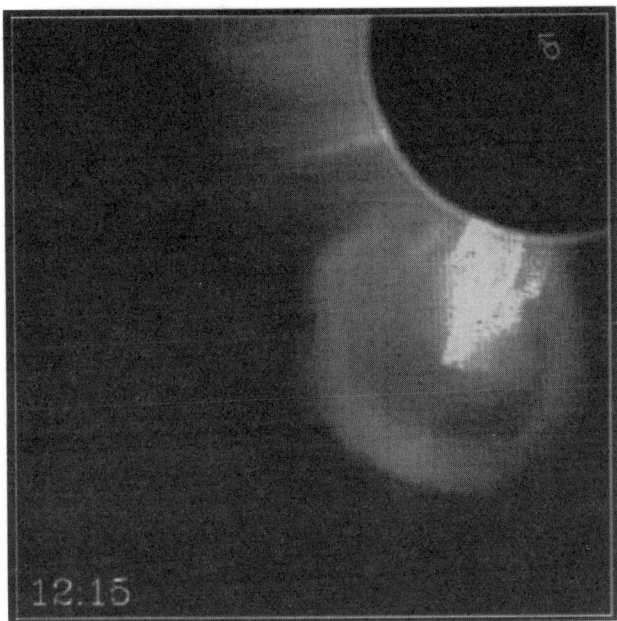

Figure 9. Coronagraph images. (*a*) The 24 May 1979 CME observed by the Solwind coronagraph obtained by subtracting a pre-CME image (at 1215 UT) from the image at 1842 UT. The outer field of view extends to ~8 R_S. (*b*) The 18 August 1980 CME observed by the SMM coronagraph. The outer field of view extends to ~6.5 R_S.

are some exceptions to this in the Thomson-scattering measurements, however.

The Helios spacecraft, the first of which was launched into heliocentric orbit in 1974, had on board three sensitive zodiacal-light photometers for the study of the zodiacal-light distribution [*Leinert et al.*, 1975]. These photometers swept the celestial sphere at 16°, 31° and 90° ecliptic latitude to obtain data fixed with respect to the solar direction, with a sample interval of about five hours. The two spacecraft were placed in heliocentric orbits with perihelia of about 0.3 AU.

The time histories of brightness in the different photometer sectors are used to provide information about the geometry of elongated structures in these data. Figure 10 shows an example. The brightness increases generally last for several days in an individual photometer sector. The displacement in time from east to west from one sector to another in one photometer represents the solar rotation of the structure. The extent of the structure in elongation (angular distance from the Sun) is shown primarily by its presence in the different photometers. The data allow heliographic latitudes and longitudes to be modeled for each elongated structure observed. In addition, these data also allow the determination of outward material velocity as well as excess densities of each structure [*Jackson*, 1991].

The Helios photometer data were first identified as a valid source of information for mapping mass ejections by *Richter et al.* [1982]. Since then, the Helios photometers have been used to image the interplanetary medium from 20 R_S out to 1 AU [*Jackson*, 1985b; *Jackson and Leinert*, 1985] and, in particular, disturbances in it [*Webb and Jackson*, 1990]. To date over 160 such events have been imaged by the Helios 1 and 2 spacecraft [*Jackson et al.*, 1994]. For several ICMEs, the motion of portions of the ejecta can be traced along the Sun-Earth line until it reaches Earth. One such event measured by the Solwind coronagraph as a "halo" CME on 27 November 1979 and observed in Helios 2 observations to move outward as an ICME, can be observed several days later *in situ* at Earth [*Jackson*, 1985a]. A schematic of the location of the Helios spacecraft and the photometer viewing angles is given Figure 11a for this ICME. Figure 11b indicates the outward motion of this ICME along the Sun-Earth line.

Figure 12 shows an example of the type of information available from the Helios spacecraft photometers in the form of contour images. The masses and mass ejection shapes obtained from these observations indicate that the material of a mass ejection observed in the lower corona moves coherently outward into the interplanetary medium. Mass estimates of coronal mass ejections observed by Helios are generally three times those determined by the Solwind and SMM coronagraphs for the same events [*Webb et al.*, 1996]. This difference is interpreted as due primarily to the inability of a coronagraph to observe the complete mass ejection over its entire height at any given instant in the low corona, and to some extent the accumulation of mass through interaction with the interplanetary medium (the sheath) for the faster ICMEs.

4. RECENT RESULTS

Several recent new advances in data analysis and our understanding of heliospheric processes are highlighted in the following subsections.

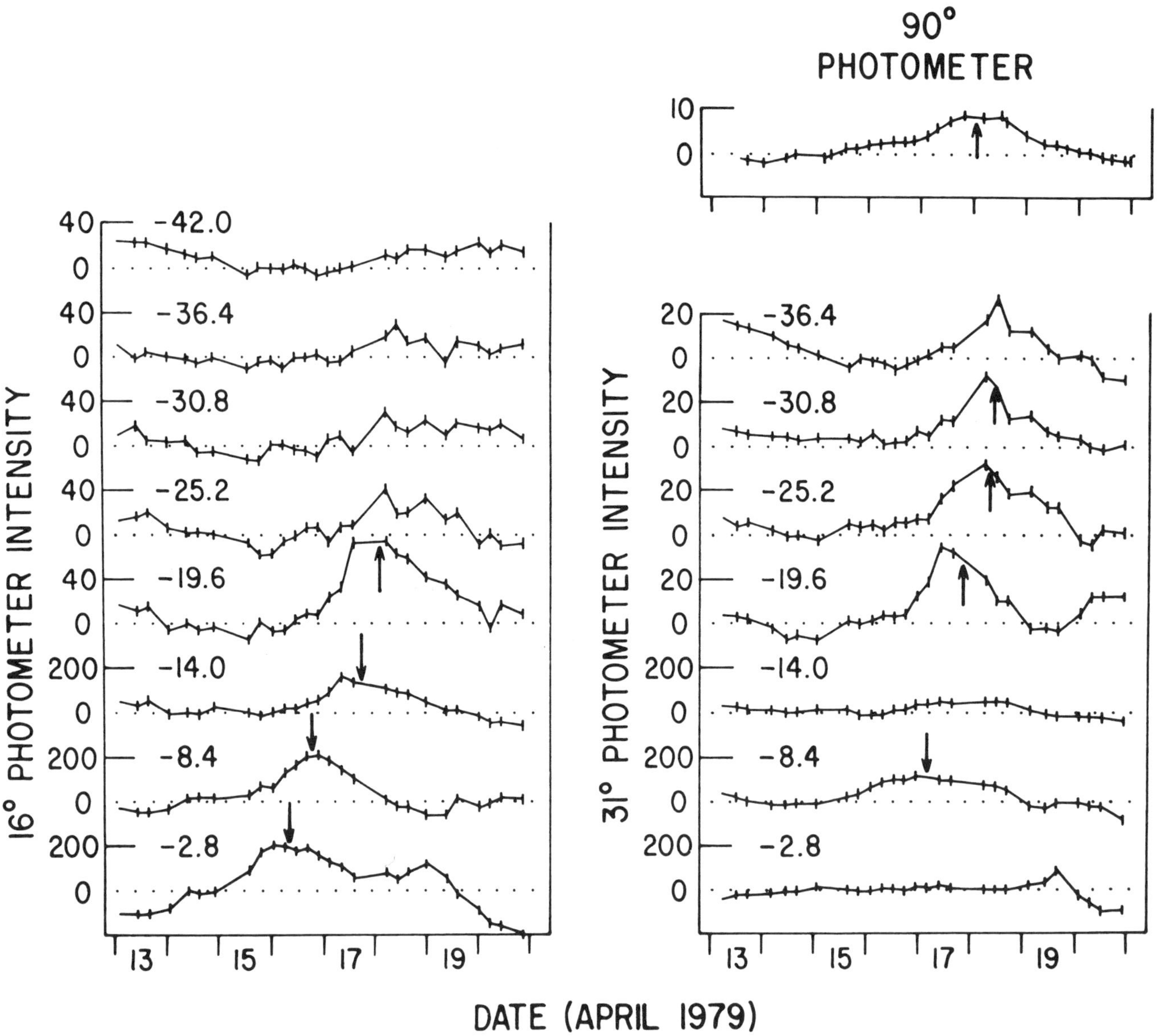

68 HELIOSPHERIC OBSERVATIONS

Figure 10. Time series plots from April 13 to 20, 1979, for the Helios B 16°, 31° and 90° photometers. The vertical scale from zero intensity is given on the plot ordinate in S10 units. The ecliptic longitude of the center of the photometer sectors relative to the sun are labeled on each graph. Faint vertical tic marks give times of the data measurements. Vertical arrows mark estimates of the structure centroid in individual photometer sectors (from *Jackson* [1991]).

4.1. *Slow Solar Wind from Active Regions*

A comparison of *Yohkoh* and Sacramento Peak Fe XIV data in the form of Carrington plots on a rotation by rotation basis shows that bright active regions look spatially similar in two dimensions to surface-projected IPS enhancements observed from 0.5 to 1.0 AU [*Hick et al.*, 1995; *Hick and Jackson*, 1996]. Calibrated in terms of density using IMP spacecraft data at 1 AU, the measurements indicate that these solar wind regions are very dense. The same regions observed in IPS velocities show up as some of the slowest areas observed on Carrington maps. This indicates that solar active regions are major contributors of slow solar wind (probably as observed in the form of outward-expanding active region loops in *Yohkoh* images [*Uchida et al.*, 1992]), and that these regions add significant mass to the interplanetary medium. These regions (and not the heliospheric current

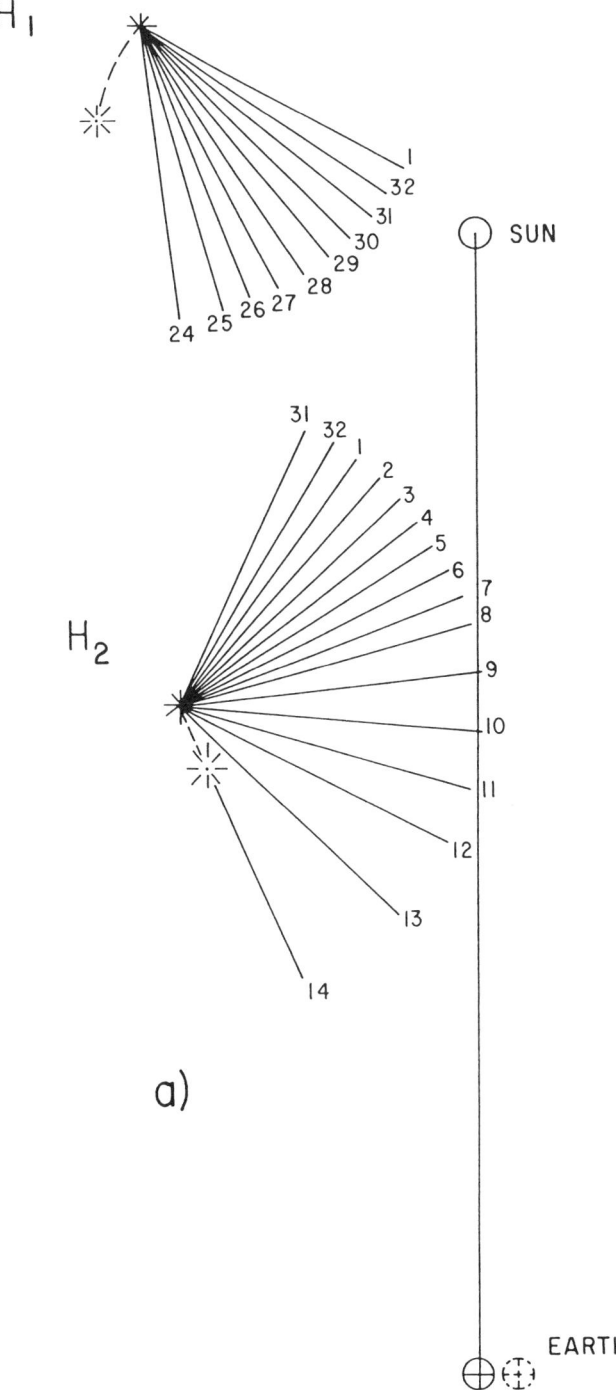
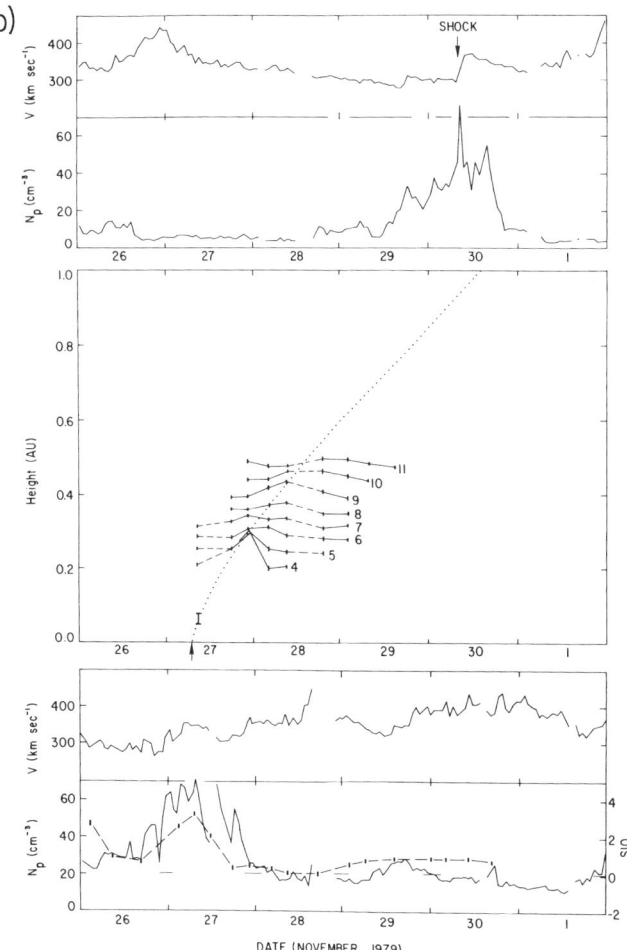

sheet mapped from the solar surface [*Hoeksema et al.*, 1983]) dominate the equatorial plasma densities of the quiet solar wind. This distinction would be unimportant except that present-day understanding is that corotating dense solar wind features are organized around the large scale heliospheric magnetic field reversal which divides the Sun into north and south magnetic polarities.

To help verify these observations, the perspective views available from solar rotation and outward solar wind motion in IPS observations have been used tomographically to deconvolve the heliosphere in three dimensions [*Jackson et al.*, 1997]. This analysis technique gives a more sharply defined picture of the

Figure 11. (*a*) Schematic of the Helios 1 and 2 locations and photometer view angles at the time of the 27 November, 1979 CME/ICME. (*b*) Motion of the 27 November ICME along the Sun-Earth line as shown in the height-time plot from the solar surface to 1 AU in the middle panel. The arrow depicts the time of the flare and the data point near the solar surface gives the time of the Solwind observations. Helios 2 photometer sector brightness measurements are given as traces based at the height of the intersection of the line of sight photometer sector projections on the ecliptic plane with the Sun-Earth line (see Figure 11a). Velocity and density at Earth and at the Helios 2 spacecraft are shown in the upper and lower panels, respectively (from *Jackson* [1985a]).

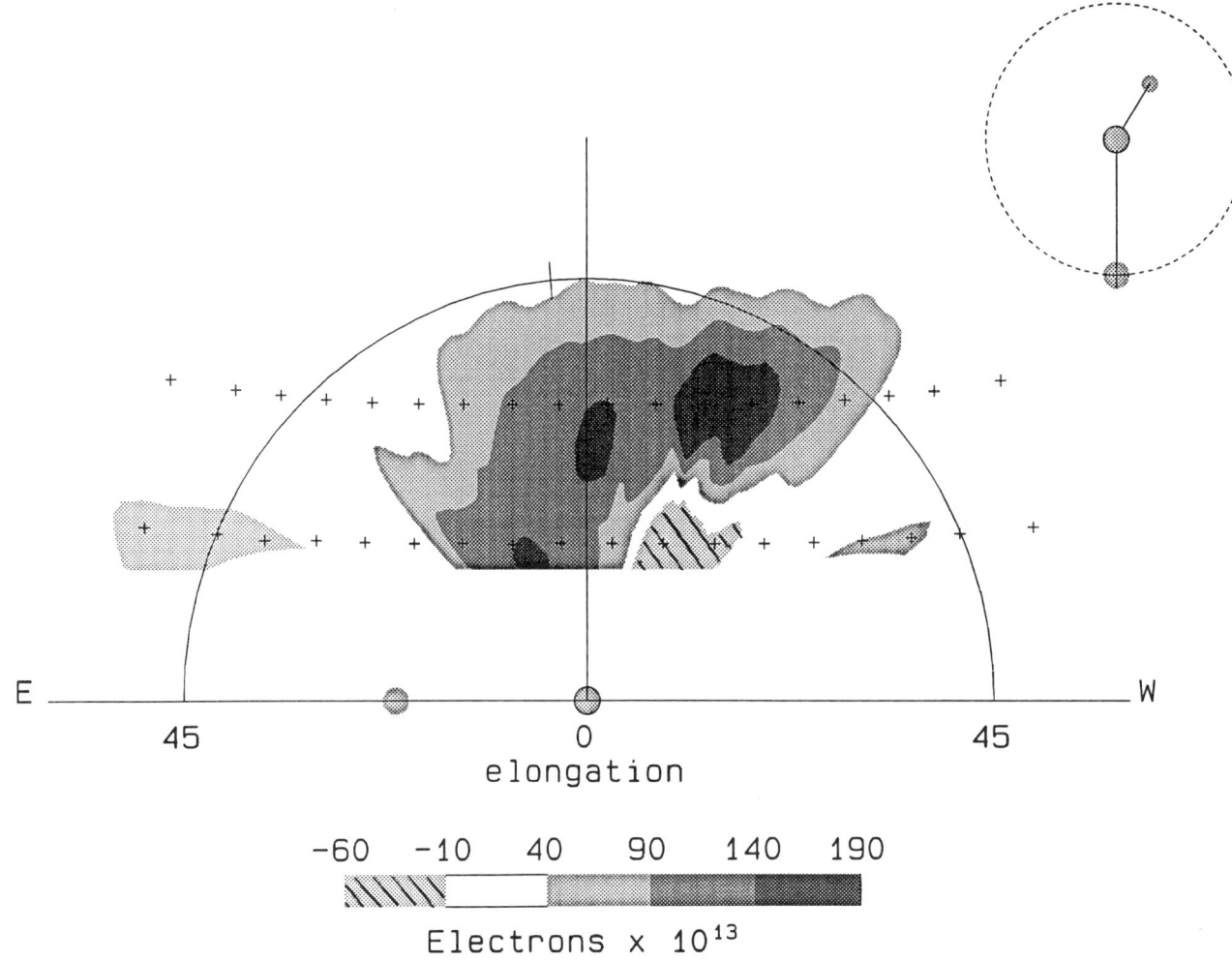

Figure 12. Electron excess number columnar density gray-scale plot of the May 24, 1979 mass ejection based on observations from the photometers on board Helios 2 at 0921 UT May 25. The lowest contour level is -6×10^{14} electrons cm^{-2} and are incremented in steps of 5×10^{14} electrons cm^{-2}. The origin of the plot represents the direction towards the Sun (as seen from Helios 2). A half-circle with the Sun in the center represents all directions with a constant solar elongation angle of 45°. The horizontal axis represents directions along the ecliptic. The elongation of the Earth as viewed from Helios 2 is indicated on the horizontal axis and shown as a diagram to the upper right of the figure (from *Jackson and Hick* [1994]).

spatial configurations of these features than the older "point-P" analysis method employed by *Hick et al.* [1995] with essentially the same result as before. A sample of these tomographic analyses are displayed in the form of Carrington maps in Figure 13.

4.2. CME three-dimensional geometry

The true three-dimensional geometry of mass ejections has long been a question for researchers. Much of the information of recent years suggests that a CME fills a large three-dimensional volume, but that CMEs may not be configured like either a simple arch or a bubble as once thought. The importance of the CME shape goes beyond a simple accounting of its structure and its cross section movement and evolution through the heliosphere. The CME shape is derived from the processes (as yet poorly known) which drive the CME outward from the Sun, and modelers need to know the organization of the mass within the ejection to determine how the energy needed to eject the material was distributed.

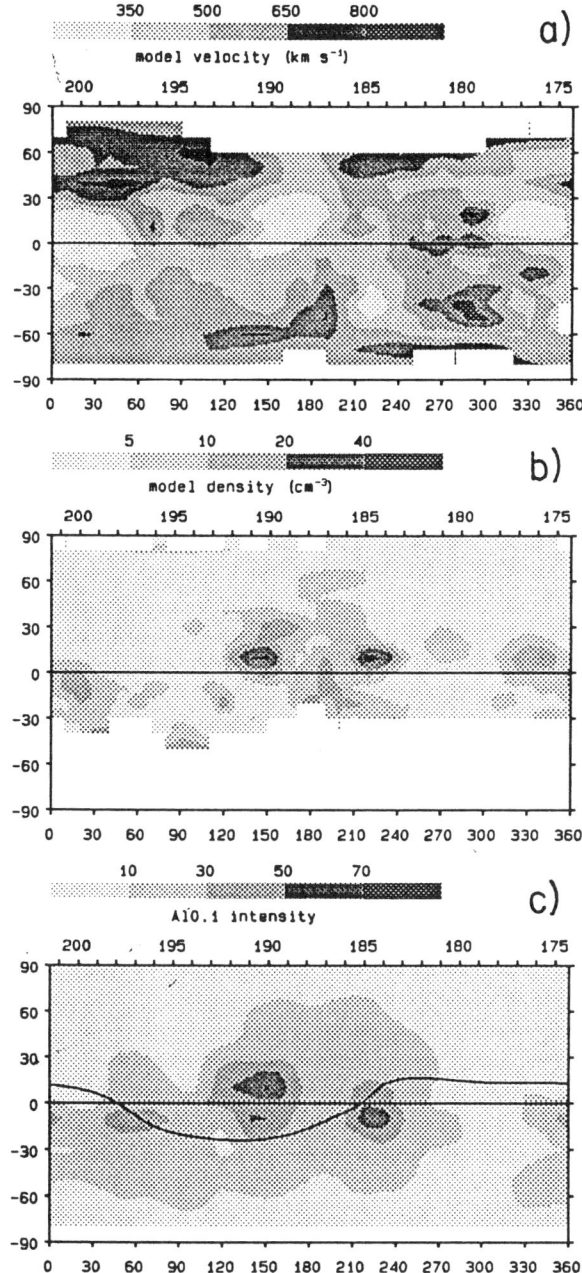

Figure 13. Solar Surface Carrington map displays for rotation 1884. (a) Deconvolved velocity projected to the solar surface. (b) Deconvolved density projected to the solar surface. (c) Yohkoh Solar X-ray Telescope Carrington map derived from central meridian solar images and degraded to 10° by 10° resolution. The potential field derived neutral line is overplotted on the map (from Jackson et al. [1997]).

Evidence about the large extents of CMEs/ICMEs is inferred from their positions observed *in situ* and the spatial locations of them near the Sun [*e.g., Sheeley et al.*, 1985] sensed either by coronagraphs or other techniques. That CMEs are extensive coronal features can be demonstrated by coronagraph data analyses [*e.g., Crifo et al.*, 1983; *MacQueen*, 1993]. However, the unique stereoscopic aspect of the Helios data can be fully utilized to produce three-dimensional models for specific events. For an early analysis see *Jackson et al.* [1985]. Two CMEs observed arising from the Sun on 7 May 1979 and 24 May 1979 were ideal in this respect because they were well-observed by both Solwind and as ICMEs by the Helios 2 photometers when the Helios spacecraft was widely separated from the Earth in solar longitude. From a computer program developed at UCSD, interplanetary and coronal densities in three dimensions have been reconstructed to least squares fit the brightness observed in both the Solwind or Helios views in an iterative sense. This CME/ICME tomographic reconstruction is shown for the 7 May and 24 May 1979 events in Figure 14a and 14b. The evidence from the tomography shows that ICMEs are extensive features that are varied in shape.

4.3. CME masses, energies and solar cycle dependence

The clear association of CME numbers with solar activity [*Hildner et al.*, 1976; *Howard et al.*, 1985b; *Webb and Howard*, 1994] implies that CMEs are manifest by the same processes that form strong solar magnetic fields. CMEs measured by coronagraphs show a number versus mass spectrum that indicates that the smallest CMEs are limited in number [*Jackson and Howard*, 1993]. This small mass limit does not appear to be associated with the observational limits of the various instruments used to observe them. This spectrum does not vary much with solar cycle. The total mass present in the solar wind in the form of CMEs observed by coronagraphs is difficult to estimate precisely by counting individual events over time, but current values place this at approximately 15% at solar maximum [*Jackson and Howard*, 1993; *Webb and Howard*, 1994]. These observations are strengthened by the Helios photometer observations.

Even at solar maximum, CMEs are directed outward more in the ecliptic than above the solar poles [*Howard et al.*, 1985a], and thus have more potential to hit Earth than the total numbers of them would indicate. Although the significance of the shape of their energy spectra has yet to be worked out, CMEs are at the energetic end of the scale in relationship to, *i.e.*, energies of solar flare brightening, and are among the most energetic solar phenomena as depicted in Figure 15. While there can be flare brightening associated with CMEs, this correspondence is by no means one-to-one. Occasionally, for instance, no

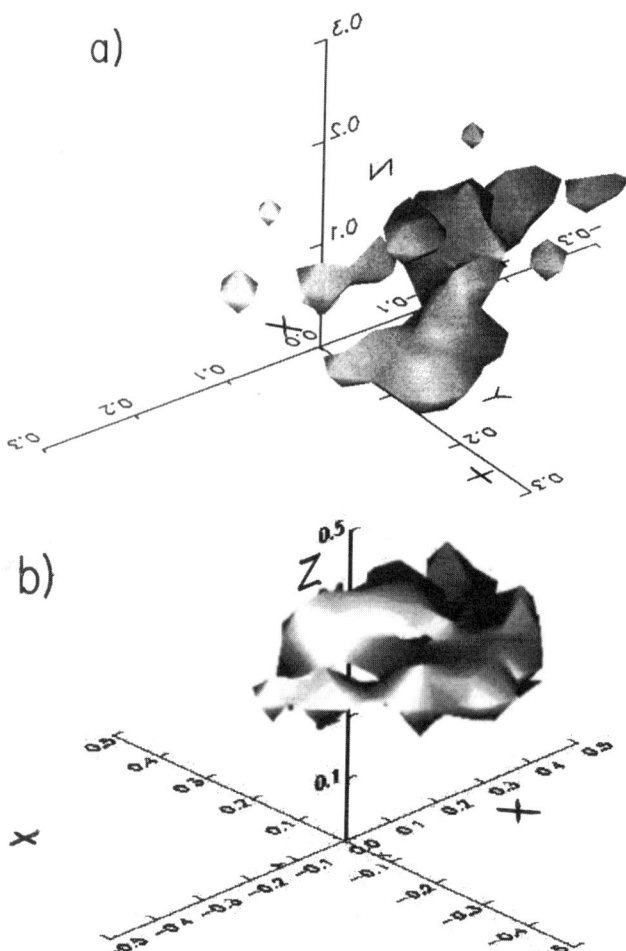

Figure 14. Three-dimensional reconstruction of two CMEs/ICMEs. The direction of the solar north pole is +Z; +X is in the heliographic longitude direction towards the Earth. The small plus on the images marks the ecliptic position of the Helios 2 spacecraft. (a) The 7 May 1979 ICME shown as a surface contoured at 200 e$^-$ cm^{-3} at 1440 UT 8 May 1979. The axes extend 0.3 AU outward from the Sun (from *Jackson and Froehling* [1995]). (b) The 24 May 1979 ICME shown as a surface contoured at 50 e$^-$ cm^{-3} at 0922 UT 25 May 1979. Axes lengths are 0.5 AU (from *Jackson and Hick* [1994]).

flare brightening is associated with a CME of comparable energy output even though observations should be adequate to show this correspondence. Thus, it is no wonder that observations of CMEs and their transit through the heliosphere towards Earth in the form of ICMEs is a primary objective of those who study effects upon Earth's magnetosphere.

5. CONCLUSIONS

From the perspective of the Earth, disturbances can be either those that repeat at the solar rotation rate or those which present themselves only one time. Since magnetic field reversals are observed to repeat with high regularity at Earth, and because heliospheric magnetic field changes can cause geomagnetic storms, it is natural to link corotating disturbances with other long-lasting solar features such as corotating density enhancements and coronal holes. These other solar features may not be as stable [*e.g., Jackson*, 1991] or as consistently associated with the large-scale heliospheric magnetic field reversal as previously thought, and this has lead to considerable confusion when using data averaged to determine the properties of corotating features.

Near the solar surface, observations of the corona show that a significant fraction of the energy involved in discrete events is manifest in the form of coronal motions

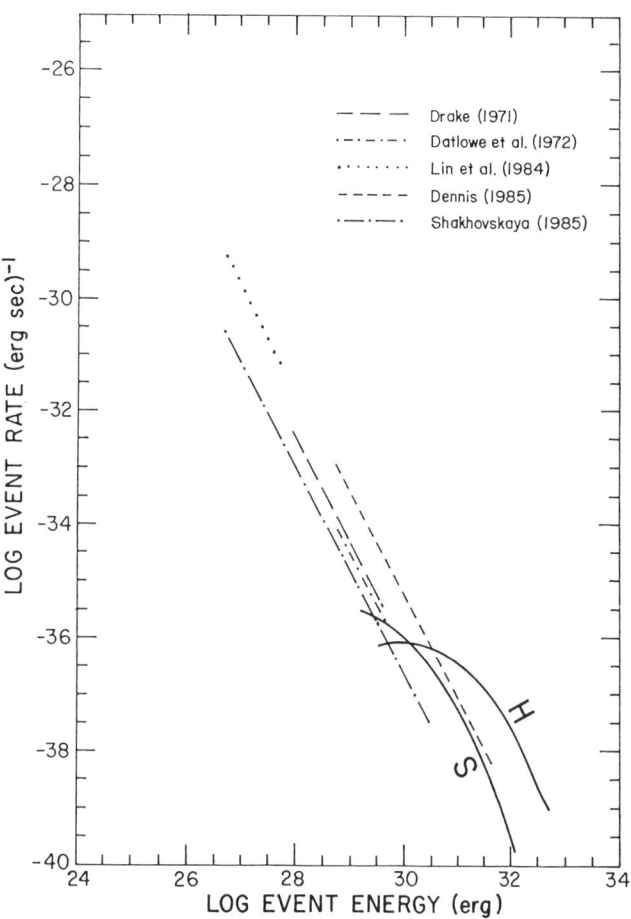

Figure 15. Comparison of solar flare total X-ray energy distributions from different observers derived by measurement of X-ray peak flux (from *Hudson* [1991]) with solar maximum CME number versus kinetic energy Solwind "S" data (from *Howard et. al.* [1985a]), and Helios "H" CME/ICME data.

(CMEs). As a CME moves outward into the interplanetary medium, the ICME can be observed remotely by a variety of techniques. The shapes and speeds of different structures, including those which corotate, evolve somewhat as they move outward through the heliosphere, and since none of the techniques that trace the events overlap well temporally or spatially, there are usually ambiguities in the observations. Thus, it becomes very advantageous to fill in with as much information as possible to describe these disturbances in the most complete fashion.

As more events are observed, general rules to describe their behavior will become well-accepted and their physics understood. In addition, new and better observations such as data from the Ulysses spacecraft which measure *in situ* heliospheric plasma out of the ecliptic will certainly help in the description, and thus advance our understanding of the physics involved. At some future date a set of instruments such as a combination of the LASCO coronagraphs [*Brueckner et al.*, 1995] which are now operating on SOHO, and the Solar Mass Ejection Imager [*Jackson et al.*, 1991; *Jackson et al.*, 1996] will be in operation simultaneously. Such a combination of instruments can remotely view the whole heliosphere in white light from extremely near the Sun to 1 AU, and will allow continuous mapping of solar observations until these same features can be measured *in situ* near Earth.

As our physical models of the processes involved become more accurate, so will our ability to forecast their effects on the magnetosphere. Some observations such as plasma density will be more refined than others. These, in conjunction with modeling techniques such as those used in three-dimension tomography or advanced MHD techniques, should allow the complete dynamical characterization of the heliospheric density, velocity, and magnetic field evolution from the Sun outward to 1 AU. This combination of data analysis and modeling should complete our understanding and ability to forecast solar disturbances, and to determine their potential role in the origin of geomagnetic storms.

Acknowledgments. I would like to thank my colleagues for their many valuable discussions and P.L. Hick in particular for contributing a significant portion of this manuscript. I thank Ch. Leinert, for his support of our analyses of the Helios photometer data and M. Kojima and G. Woan for a portion of the IPS data used in this review. The coronagraph image from Solwind was made available to me courtesy of R. Howard and the SMM image from the coronagraph group at the High Altitude Observatory. The kilometric dynamic spectrum from Ulysses is courtesy of J. L. Steinberg. The Helios data was processed and made available to us through the courtesy of the NSSDC. The work of B. Jackson, is supported at the University of California at San Diego by grant AFOSR-94-0070.

REFERENCES

Behannon, K. W., L. F. Burlaga, and A. Hewish, Structure and evolution of compound streams at ≤ 1 AU, *J. Geophys. Res., 96,* 21,213-21,225, 1991.

Borrini, G., J. T. Gosling, S. J. Bame, W. C. Feldman, and J. M. Wilcox, Solar wind helium and hydrogen structure near the heliospheric current sheet: A signal of coronal streamers at 1 AU, *J. Geophys. Res., 86,* 4565-4573, 1981.

Bougeret, J.-L., J. Fainberg, and R. G. Stone, Interplanetary radio storms II. Emission levels and solar wind speed in the range 0.05 - 0.8 AU, *Astron. Astrophys., 141,* 17-24, 1984.

Brueckner, G. E., R. A. Howard, M. J. Koomen, C. M. Korendyke, D. J. Michels, J. D. Moses, D. G. Socker, K. P. Dere, P. L. Lamy, A. Llebaria, M. V. Bout, R. Schwenn, G. M Simnett, D. K. Bedford, and C. J. Eyles, *Solar Phys., 162,* 357-402, 1995.

Burlaga, L. G., L. Klein, N. R. Sheeley, Jr., D. J. Michels, R. A. Howard, M. J. Koomen, R. Schwenn, and H. Rosenbauer, A magnetic cloud and a coronal mass ejection, *Geophys. Res. Lett., 9,* 1317-1320, 1982.

Chapman, R. W., NASA's search for the solar connection-II, *Sky and Tel., 58,* 223-227, 1979.

Coles, W. A., and J. J. Kaufman, Solar wind velocity estimation from multi-station IPS, *Radio Science, 13,* 591-597, 1978.

Crifo, F., J. P. Picat, and M. Cailloux, Coronal transients: loop or bubble?, *Solar Phys. 83,* 143-152, 1983.

Dessler, A. J., Solar wind and interplanetary magnetic field, *Rev. Geophys., 5,* 1-41, 1967.

Fisher, R., R. H. Lee, R. M. MacQueen, and A. I. Poland, New Mauna Loa coronagraph system, *Appl. Opt., 20,* 1094-1101, 1981.

Gapper, G. R., A. Hewish, A. Purvis, and P. J. Duffett-Smith, Observing interplanetary disturbances from the ground, *Nature, 296,* 633-636, 1982.

Gosling, J. T., Coronal mass ejections and magnetic flux ropes in interplanetary space, in *Physics of Magnetic Flux Ropes*, edited by C. T. Russell, E. R. Priest, and L. C. Lee, pp. 343-364, AGU Monograph 58, 1990.

Gosling, J. T., In situ observations of coronal mass ejections in interplanetary space, in *Eruptive Solar Flares, Lecture Notes in Physics, 399,* edited by Z. Svestka, B. V. Jackson and M. E. Machado, pp. 258-267, Springer-Verlag, Berlin, 1992.

Gosling J. T., The solar flare myth, *J. Geophys. Res., 98,* 18,937-18,949, 1993.

Gosling, J. T., and D. J. McComas, Field line draping about fast coronal mass ejecta: a source of strong out of the ecliptic magnetic fields, *Geophys. Res. Lett., 14,* 355-358, 1987.

Gosling, J. T., E. Hildner, R. M. MacQueen, R. H. Munro, A. I. Poland, and C. L. Ross, The speeds of coronal mass ejection events, *Solar Phys., 48,* 389-397, 1976.

Gosling, J. T., G. Borrini, J. R. Asbridge, S. J. Bame, W. C. Feldman, and R. T. Hansen, Coronal streamers in the solar wind at 1 AU, *J. Geophys. Res., 86,* 5438-5448, 1981.

Gosling, J. T., D. N. Baker, S. J. Bame, W. C. Feldman, R. D. Zwickl, and E. J. Smith, Bidirectional solar wind electron heat flux events, *J. Geophys. Res., 92*, 8519-8535, 1987.

Hewish, A., and S. Bravo, The sources of large-scale heliospheric disturbances, *Solar Phys., 106*, 185-200, 1986.

Hewish, A., P. F. Scott, and D. Wills, Interplanetary scintillation of small diameter radio sources, *Nature, 203*, 1214-1217, 1964.

Hick, P., and B. V. Jackson, Evidence of active region imprints on the solar wind structure, in *Solar Wind Eight*, edited by D. Winterhalter, J. T. Gosling, S. R. Habbal, W. S. Kurth, and M. Neugebauer, pp. 461-464, AIP Conference Proceedings 382, Woodbury, New York, 1996.

Hick, P. L, B. V. Jackson, S. Rappoport, G. Woan, G. Slater, K. Strong, and Y. Uchida, Synoptic IPS and Yohkoh soft X-ray observations, *Geophys. Res. Lett., 22*, 643-646, 1995.

Hildner, E., J. T. Gosling, R. M. MacQueen, R. H. Munro, A. I. Poland, and C. L. Ross, Frequency of coronal transients and solar activity, *Solar Phys. 48*, 127-135, 1976.

Hoeksema, J. T., J. M. Wilcox, and P. H. Scherrer, The structure of the heliospheric current sheet 1978-1982, *J. Geophys. Res., 88*, 9910-9918, 1983.

Houminer, Z., and A. Hewish, Correlation of interplanetary scintillation and spacecraft plasma density measurements, *Planetary and Space Sci., 22*, 1041-1042, 1974.

Howard, R. A., N. R. Sheeley, Jr., M. J. Koomen, and D. J. Michels, Coronal mass ejections: 1979-1981, *J. Geophys. Res., 90*, 8173-8191, 1985a.

Howard, R. A., N. R. Sheeley, Jr., D. J. Michels, and M. J. Koomen, The solar cycle dependence of coronal mass ejections, in *The Sun and the Heliosphere in Three Dimensions*, edited by R. G. Marsden, pp. 107-111, Rediel, Dordrecht, 1985b.

Hudson, H. S., Solar flares, microflares, nanoflares and coronal heating, *Solar Phys. 133*, 357-369, 1991.

Hudson, H. S., Solar soft X-ray antecedents of magnetic storms, in the *Chapman Conference on Magnetic Storms*, (this issue), 1997.

Hudson, H. S., B. M. Haisch, and K. T. Strong, Comment on "The solar flare myth," *J. Geophys. Res., 100*, 3473-3477, 1995.

Hundhausen, A. J., *Coronal Expansion and the Solar Wind*, 238 pp., Springer-Verlag, New York, 1972.

Hundhausen, A. J., An interplanetary view of coronal holes, in *Coronal Holes and High Speed Wind Streams*, edited by J. B. Ziker, pp. 225-329, Colorado Assoc. University Press, Colorado, 1977.

Hundhausen, A. J., J. T. Burkepile, and O. C. St. Cyr, Speeds of coronal mass ejections: SMM observations from 1980 and 1984-1989, *J. Geophys. Res., 99*, 6543-6552, 1994.

Jackson, B. V., IPS observations of the 14 August 1979 mass ejection transient, in *Proceedings of the Maynooth, Ireland Symposium on Solar/Interplanetary Intervals*, edited by M. A. Shea, D. F. Smart, and S. M. P. McKenna-Lawlor, pp. 169-173, Book Crafters, Chelsea, Michigan, 1984.

Jackson, B. V., Helios observations of the earthward-directed mass ejection of 27 November, 1979, *Sol. Phys., 95*, 363-370, 1985a.

Jackson, B. V., Imaging of coronal mass ejections by the Helios spacecraft, *Sol. Phys., 100*, 563-574, 1985b.

Jackson, B. V., Helios spacecraft photometer observations of elongated corotating structures in the interplanetary medium, *J. Geophys. Res., 96*, 11,307-11,318, 1991.

Jackson, B. V., Solar generated disturbances in the heliosphere, in *Solar Wind Seven*, edited by E. Marsch and R. Schwenn, pp. 623-634, Pergamon, Oxford, 1992.

Jackson, B. V., and H. R. Froehling, Three-dimensional reconstruction of coronal mass ejections, *Astron. Astrophys., 299*, 885-892, 1995.

Jackson, B. V., and P. L. Hick, Three dimensional reconstruction of coronal mass ejections, in the proceedings of the Third SOHO Workshop on Solar Dynamic Phenomena \& Solar Wind Consequences, pp. 199-202, *ESA SP-373*, 1994.

Jackson, B. V., and R. A. Howard, CME mass distribution derived from Solwind coronagraph observations, *Solar Phys., 148*, 359-370, 1993.

Jackson, B. V., and C. Leinert, Helios images of solar mass ejections, *J. Geophys. Res., 90*, 10,759-10,764, 1985.

Jackson, B. V., R. A. Howard, N. R. Sheeley, Jr., D. J. Michels, M. J. Koomen, and R. M. E. Illing, Helios spacecraft and earth perspective observations of three looplike solar mass ejections, *J. Geophys. Res., 90*, 5075-5081, 1985.

Jackson, B., R. Gold, and R. Altrock, The Solar Mass Ejection Imager, *Adv. Space Res., 11*, 377-381, 1991.

Jackson, B. V., D. F. Webb, P. L. Hick, and J. L. Nelson, Catalog of Helios $90°$ photometer events, 51 pp., *PL-TR-94-2040*, Hanscom AFB, Mass., 1994.

Jackson, B. V., A. Buffington, P. L. Hick, S. W. Kahler, R. C. Altrock, R. E. Gold and D. F. Webb, The Solar Mass Ejection Imager, in *Solar Wind Eight*, edited by D. Winterhalter, J. T. Gosling, S. R. Habbal, W. S. Kurth and M. Neugebauer, pp. 536-539, AIP Conference Proceedings 328, Woodbury, New York, 1996.

Jackson, B. V., P. L. Hick, M. Kojima,, and Y. Yokobe, Heliospheric tomography using interplanetary scintillation observations, in the proceedings of the EGS XXI meeting, the Hague, the Netherlands, 6-10 May, 1997 (in press).

Kahler, S. W., and D. V. Reames, Probing the magnetic topologies of magnetic clouds by means of solar energetic particles, *J. Geophys. Res., 96*, 9419-9424, 1991.

Klein, L. W., and L. F. Burlaga, Interplanetary magnetic clouds at 1 AU, *J. Geophys. Res., 87*, 613-624, 1982.

Koomen, M. J., C. R. Detwiler, G. E. Brueckner, H. W. Cooper, and R. Tousey, White light coronagraph in OSO-7, *Appl. Opt., 14*, 743-751, 1975.

Krieger, A. S., A. F. Timothy, and E. C. Roelof, A coronal hole and its identification as the source of a high velocity solar wind stream, *Solar Phys., 29*, 505-525, 1973.

Kundu, M., T. E. Gergely, P. J. Turner, and R. A. Howard, Direct evidence of type III electron streams propagating in coronal streamers, *Astrophys. J. Lett., 269*, 67-71, 1983.

Leinert, C., H. Link, E. Pitz, N. Salm, and D. Kluppelberg,

Helios zodiacal light experiment, *Raumfahrtforschung, 19*, 264-267, 1975.

MacQueen, R. M., The three-dimensional structure of 'loop-like' coronal mass ejections, *Solar Phys., 145*, 169-188, 1993.

MacQueen, R. M., J. A. Eddy, J. T. Gosling, E. Hildner, R. H. Munro, G. A. Newkirk, Jr., A. I. Poland, and C. L. Ross, The outer solar corona as observed from Skylab: preliminary results, *Astrophys. J. Lett., 187*, 85-88, 1974.

Montgomery, M. D, J. R. Asbridge, S. J. Bame, and W. C. Feldman, Solar wind electron temperature depressions following some interplanetary shock waves: evidence for magnetic merging?, *J. Geophys. Res., 79*, 3103-3110, 1974.

Munro R. H., and B. V. Jackson, Physical properties of a polar coronal hole from 2 to 5 R_o, *Astrophys. J., 213*, 874-886, 1977.

Ness, N. F., C. S. Scearce, and J. B. Seek, Initial results of the IMP 1 magnetic field experiment, *J. Geophys. Res., 69*, 3531-3569, 1964.

Neugebauer, M., and C. W. Snyder, Mariner 2 observations of the solar wind, *J. Geophys. Res., 71*, 4469-4484, 1966.

Parker, E. N., Dynamics of the interplanetary gas and magnetic fields, *Astrophys. J., 128*, 664-676, 1958.

Pilipp, W. G., H. Miggenrieder, M. D. Montgomery, K. H. Muhlhauser, H. Rosenbauer, R. Schwenn, and F. M. Neubauer, Unusual electron distribution functions in the solar wind derived from the Helios plasma experiment: double-strahl distributions and distributions with an extremely anisotropic core, *J. Geophys. Res., 92*, 1093-1101, 1987.

Pizzo, V. J., and D. G. Sime, Interpretation of interplanetary scintillation observations of large scale solar wind structures, presented at *Solar Wind Seven* held in Goslar, Germany 16-21 September, 1991.

Pneuman, G. W., and R. A. Kopp, Gas-magnetic field interactions in the solar corona, *Sol. Phys., 18*, 258-270, 1971.

Poland, A. I., Motions and mass changes of a persistent coronal streamer, *Sol. Phys., 57*, 141-153, 1978.

Richter, I., C. Leinert, and B. Planck, Search for short term variations of zodiacal light and optical detection of interplanetary plasma clouds, *Astron. and Astrophys., 110*, 115-120, 1982.

Rickett, B. J., Disturbances in the solar wind from IPS measurements, *Solar Phys., 43*, 237-247, 1975.

Rosenbauer, H., R. Schwenn, E. Marsch, B. Meyer, H. Miggenrieder, M. Montgomery, K. H. Muhlhauser, W. Pilipp, W. Voges, and S. K. Zink, A survey on initial results of the Helios plasma experiment, *J. Geophys., 42*, 561-580, 1977.

Schwenn, R., Large-scale structures of the interplanetary medium, in *Physics of the Inner Heliosphere, 1*, edited by R. Schwenn and E. Marsch, pp. 99-181, Springer-Verlag, Berlin, Germany, 1990.

Sheeley, N. R., Jr., R. A. Howard, M. J. Koomen, D. J. Michels, R. Schwenn, K. H. Muhlhauser, and H. Rosenbauer, Coronal mass ejections and interplanetary shocks, *J. Geophys. Res., 90*, 163-175, 1985.

Smart, D. F., M. A. Shea, and Y. Leblanc, Kilometric type II radiation observed by the Voyager spacecraft and shockwaves during April 1978, *EOS. TRANS. AGU, 66*, 1036, 1985.

Smith, E. J., and J. H. Wolfe, Observations of interaction regions and corotating shocks between one and five AU: Pioneers 10 and 11, *Geophys. Res. Lett., 3*, 137-140, 1976.

Steinberg, J.-L., Satellite observations of solar radio bursts, in *Radio Physics of the Sun*, edited by M. R. Kundu and T. E. Gergely, pp. 387-400, Reidel, Dordrecht, 1980.

Steinberg, J. L., S. Hoang, and G. A. Dulk, Evidence of scattering effects on the sizes of interplanetary type III radio bursts, *Astron. Astrophys., 150*, 205-216, 1985.

Stone, R. G. and 26 co-authors, The ISPM unified radio and plasma wave experiment, *ESA SP-1050*, 56 pp., 1983.

Tappin, S. J., Interplanetary scintillation and plasma density, *Planet. Space Sci., 35*, 271-283, 1987.

Tsurutani, B. T., W. D. Gonzales, F. Tang, S. I. Akasofu, and E. J. Smith, Origin of interplanetary southward magnetic fields responsible for major magnetic storms near solar maximum (1978-1979), *J. Geophys. Res., 93*, 8519-8531, 1988.

Tsurutani, B. T., W. D. Gonzales, F. Tang, and Y. T.Lee, Great magnetic storms, *Geophys. Res. Lett., 19*, 73-76, 1992.

Tsurutani, B. T., W. D. Gonzales, A. L. C. Gonzales, F. Tang, J. K. Arballo, M. Okada, Interplanetary origin of geomagnetic activity in the declining phase of the solar cycle, *J. Geophys. Res., 100*, 21,717-21,733, 1995.

Uchida, Y., A. McAllister, K. T. Strong, Y. Ogawara, T. Shimizu, R. Matsumoto, and H. S. Hudson, Continual expansion of the active-region corona observed by the Yohkoh Soft X-ray Telescope, *Publ. Astron. Soc. Japan, 44*, L155-160, 1992.

Watanabe, T., and T. Kakinuma, Radio-scintillation observations of interplanetary disturbances, *Adv. Space Res., 4*, 331-334, 1984.

Webb, D. F., and B. V. Jackson, The identification and characteristics of solar mass ejections observed in the heliosphere by the Helios-2 photometers, *J. Geophys. Res., 95*, 20,641-20,661, 1990.

Webb, D. F., and R. A. Howard, The solar cycle variation of coronal mass ejections and the solar wind mass flux, *J. Geophys. Res., 99*, 4201-4220, 1994.

Webb, D. F., R. A. Howard, and B. V. Jackson, Comparison of CME masses and kinetic energies near the sun and in the inner heliosphere, in *Solar Wind Eight*, edited by D. Winterhalter, J. T. Gosling, S. R. Habbal, W. S. Kurth and M. Neugebauer, pp. 540-543, AIP Conference Proceedings 382, Woodbury, New York, 1996.

Wilcox, J. M., and N. F. Ness, Quasi-stationary corotating structure in the interplanetary medium, *J. Geophys. Res., 70*, 5793-5805, 1965.

Wild, J. P., and L. L. McCready, Observations of the spectrum of high-intensity solar radiation at metre wavelengths - I. The apparatus and spectral types of solar bursts observed, *Australian J. Sci. Res., 3*, 387-398, 1950.

Wild, J. P., Some investigations of the solar corona: the first two years of observation with the Culgoora radioheliograph, *Proc. Astron. Soc. Australia, 1*, 365-370, 1970.

Woo, R., and R. Schwenn, Comparison of Doppler scintillation and in situ spacecraft plasma measurement of interplanetary disturbances, in *Solar Wind Seven*, edited by E. Marsch and R. Schwenn, pp. 685-688, Pergamon, Oxford, 1992.

Zirker, J. B., Coronal holes - an overview, in *Coronal Holes and High Speed Wind Streams*, edited by J. B Ziker, pp. 1-26, Colorado Assoc. University Press, Colorado, 1977.

Zwickl, R. D., The study of fluxuations in the solar wind density and their impact on IPS measurements, presentation at the SEIIM conference held in Colorado Springs, Colorado, 1988.

The Interplanetary Causes of Magnetic Storms: A Review

Bruce T. Tsurutani

*Space Physics Element, Jet Propulsion Laboratory,
California Institute of Technology, Pasadena, California*

Walter D. Gonzalez

Instituto Nacional Pesquisas Espaciais, Sao Jose dos Campos, San Paulo, Brazil

The physical mechanism for energy transfer from the solar wind to the magnetosphere is magnetic reconnection between the interplanetary field and the earth's field. During and a few years after solar maximum, the dominant interplanetary phenomena causing intense magnetic storms ($D_{ST} < -100$ nT) are the interplanetary manifestations of fast coronal mass ejections (CMEs). Two interplanetary regions are important for intense southward IMFs: the sheath region just behind the forward shock, and the ejecta material itself. Whereas the initial phase of a storm is caused by the increase in plasma ram pressure associated with the increase in density and speed at and behind the shock (accompanied by a sudden impulse [SI] at Earth), the storm main phase is due to southward IMFs. If the fields are southward in both of the sheath and solar ejecta, two-step main phase storms can result. The storm recovery phase begins when the IMF turns less southward, with delays of ~1-2 hours. The recovery phase has a decay time of ~10 hours and is physically due to a combination of several different energetic particle loss processes (Coulomb collisions, charge exchange and wave-particle interactions). During solar minimum, high speed streams from coronal holes dominate the interplanetary medium activity. The high-density, low-speed streams associated with the heliospheric current sheet (HCS) plasma sheet impinging upon the Earth's magnetosphere cause positive D_{ST} values (storm initial phases if followed by main phases). In the absence of shocks, SIs are infrequent during this phase of the solar cycle. High-field regions called Corotating Interaction Regions (CIRs) are created by the fast stream (emanating from the coronal hole) interaction with the HCS plasma sheet. However, because the B_z component is typically highly fluctuating within the CIRs, the main phases of the resultant magnetic storms typically have highly irregular profiles and are weaker. Storm recovery phases during this phase of the solar cycle are also quite different in that they can last from many days to weeks. The southward magnetic field (B_s) component of Alfvén waves in the high speed stream proper cause intermittent reconnection, intermittent substorm activity, and sporadic injections of plasma sheet energy into the outer portion of the ring current, prolonging its final decay to quiet day values. This continuous auroral activity is called High Intensity Long Duration Continuous AE Activity (HILDCAAs).

1. INTRODUCTION

The primary cause of magnetic storms are intense, long-duration southward interplanetary magnetic fields which interconnect with the earth's magnetic field and allow solar wind energy transport into the earth's magnetotail/magnetosphere (Gonzalez et al., 1994). It is the purpose of this paper to review the sources of such interplanetary magnetic fields distinguishing between solar maximum and the declining phases of the solar cycle.

The solar wind speed, V_{SW}, plays an equal role in the interplanetary cross tail electric field (-V_{SW} x B_S/c). However, it is found empirically that the solar wind speed is usually only a minor factor for the creation of storms. The reason for this is that the variability of the magnitude of the solar wind speed is much less than the variability of the magnitude of B_S.

1.1. *Solar Maximum*

During the most active phase of the solar cycle, solar maximum, the sun's activity is dominated by flares and disappearing filaments, and their concomitant Coronal Mass Ejections (CMEs). Coronal holes are present, but the holes are small and do not extend from the poles to the equator as often happens in the descending phase of the solar cycle. However, Gonzalez et al. (1996) and Bravo et al. (1996) have indicated possible roles for these small coronal holes. We refer the reader to these articles for further details.

The fast (>500 km s^{-1}) CMEs coming from the sun into interplanetary space are the solar/coronal features that contain high magnetic fields. Figure 1 is a schematic of the remnants of such a solar ejecta (driver gas) detected at 1 AU (each of the three main identifying features of CMEs observed close to the sun have not been identified at 1 AU; see Tsurutani and Gonzalez, 1995a for details). There are two principal regions of intense fields. If the speed differential between the remnants of the coronal ejecta and the slow, upstream solar wind is greater than the magnetosonic wave speed (50-70 km s^{-1}), a forward shock is formed. The larger the differential speed, the stronger the Mach number of the shock. The average interplanetary quiet field is 3-8 nT and shock compression (magnetic field jump) across the shock of this field is roughly proportional to the Mach number. Interplanetary shocks typically have Mach numbers of 2-3, so the interplanetary "sheath" fields downstream of the shock are typically up to 9-24 nT. In exceptional events, the speed differential is larger than Mach 4, and a maximum compression in the field of ~4 is attained.

The primary part of the driver gas might contain a so-called magnetic cloud (Burlaga et al. 1981; 1987; Klein and Burlaga, 1982; Lepping et al. 1990; Farrugia and Burlaga, 1993 a,b). The magnetic cloud is a region of slowly varying and strong magnetic fields (10-25 nT or higher) with exceptionally low proton beta, typically ~0.1 (Choe et al., 1992; Tsurutani and Gonzalez, 1995a; this is particular nicely shown in Farrugia et al., 1993a, Figure 4). The magnetic field often has a north-to-south (or vice versa) rotation to it (Figure 2) and is elongated along its axis, form-

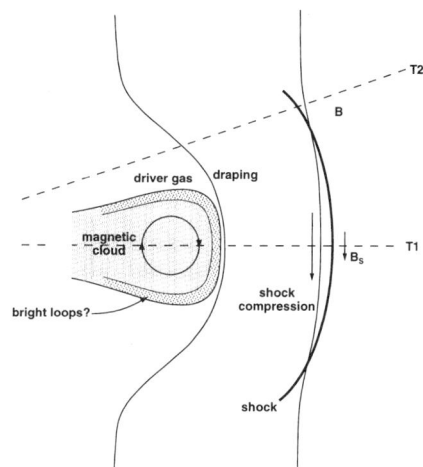

Solar Maximum: Types of Large B Fields

T1: Crossing at the center of the shock/magnetic cloud structure

T2: Crossing off-center of the shock/magnetic cloud structure (missing the driver gas)

Figure 1. Regions of intense interplanetary magnetic fields during solar maximum. T1 and T2 are two types of satellite crossings of the interplanetary structure.

ing a giant flux rope (Farrugia and Burlaga, this issue). Whether these fields remain connected to the sun or not is currently being debated.

Other three-dimensional shapes, such as spherical, toroidal or cylindrical forms, have been explored as well (Ivanov et al., 1989; Vandas et al., 1991, 1993; Farrugia et al., 1995). Simple configurations such as so-called "magnetic tongues" proposed by Gold (1962) have been sought in this study, but were not found in the ISEE-3 1978-79 data set.

At the present time we have not identified all of the component pieces of a CME at 1 AU as indicated in Figure 1. A "classic" CME is shown in Figure 3, courtesy of A. Hundhausen. This is a Solar Maximum Mission white-light coronagraph image. The time sequence goes from left to right. The three parts of a CME are illustrated in the left panel. Furthest from the sun are bright outer loops. Next is a dark region, and closest to the sun are bright twisted filaments. It has been speculated by Tsurutani and Gonzalez (1995a) that the magnetic cloud most probably corresponds to the central, dark region of the CME. This is because magnetic clouds are characterized by low ion temperatures (Farrugia and Burlaga, this issue). If the above argument is correct, then where are the loops and filaments? An intriguing possibility can be found in Figure 4, taken from Galvin et al. (1987). A magnetic cloud is present in the ISEE-3 data from 0830 to 1800 UT. It is characterized by high fields (peak of ~25 nT), a rotation from a southward direction to a northward direction (bottom panel), and a lack

Driver Gas Fields

a) First version of
   magnetic clouds
   Klein and Burlaga, 1982

b) Fluxropes
   Burlaga et al., 1990

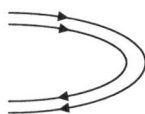

c) Magnetic tongues
   Gold, 1962

Figure 2. Types of solar ejecta magnetic fields.

of Alfvén waves and discontinuities (Tsurutani et al., 1988a; 1994). The plasma temperatures are quite low. The smooth fields allow bi-directional flow of electrons and ions which have been observed (Gosling et al., 1987; Marsden et al., 1987). Galvin et al. have emphasized the existence of an anomalous region from 0630-0830 UT just upstream of the magnetic cloud. This interval is characterized by higher density and temperature plasma, and enhanced He^{++}/H^+ values. There is also enhanced Fe (at temperatures from 1.8×10^6 K to ~3.5×10^6 K) in this region (not shown). The region is also bounded by magnetic field discontinuities at ~0630 and ~0830 UT. It is speculated that this plasma is the remnants of the bright loops of the CME. Such structures upstream of magnetic clouds are present 20-40% of the time at 1 AU.

1.2. *Magnetic Cloud Driven Storms*

A classic example of a magnetic storm driven by a magnetic cloud is shown in Figure 5. The forward shock is denoted by an "S" and a vertical dashed line in the Figure, and the start of the magnetic cloud by a second dashed vertical line. The preshocked solar wind speed is ~400 km s^{-1} and the post shock speed ~550 km s^{-1}. The magnetic field increased from ~6 nT to ~22 nT across the shock. Because B_Z ~ 0 in the sheath, there is no increased ring current activity associated with the sheath fields.

The plasma density increases from 5 cm^{-3} to > 40 cm^{-3} across the shock. Because of this density (and velocity) increase across the shock, the increased ram pressure exerted on the earth's magnetosphere, ρV_{sw}^2, causes a sudden compression of the magnetosphere and a positive jump in the horizontal component of the equatorial-region field. A positive jump in D_{ST} is noted at the time of the shock. This is a sudden impulse (SI) event. Since the SI is eventually followed by a storm main phase, it is called a storm sudden commencement or SSC (however, it has been argued [Joselyn and Tsurutani, 1990; Gonzalez et al., 1992] that this latter term is an artificial label because the physics of a SI [ram pressure increase] is independent of whether it is followed by a storm main phase or not).

The storm main phase (storm onset, or SO) occurs in near-coincidence with the sharp southward turning of the IMF at the magnetic cloud boundary. The delay is ~1 hour (Gonzalez et al., 1989). The storm main phase (decrease in D_{ST}) development is rapid and the decrease monotonic. In the example of Fig. 5, the peak D_{ST} value of -239 nT is reached ~two hours after the peak B_S value of ~-30 nT. It should be noted that the southward turning of the IMF was abrupt, and after the maximum B_S was reached, B_S was constant for several hours and the field then slowly and smoothly rotated to a northward direction.

The storm recovery phase is initiated by a gradual turning of the IMF to a northward direction from 1600 UT day 354 to 1400 UT day 355. The recovery starts as the field becomes less southward, is smooth and the 1/e time scale is a fraction of a day. Further discussions on the configuration and evolution of magnetic clouds and their geoeffectiveness can be found in a companion paper by Farrugia and Burlaga (this issue).

1.3. *Magnetic Storms Caused by Sheath Fields*

There are numerous mechanisms that lead to southward component fields in the sheath. A number of these are indicated schematically in Figure 6.

Two of the mechanisms lead to the intensification of magnetic fields, independent of whether they are oriented in a northward or a southward direction. They are shock compression b), discussed previously, and d) draping. In the former mechanism, the shock compresses both the magnetic field and plasma. In the latter mechanism (Midgley and Davis, 1963; Zwan and Wolf, 1976), draping of magnetic fields around a large object (in this case, the solar ejecta) leads to a squeezing of plasma out the ends of magnetic flux tubes. Although the dynamic pressure ($B^2/8\pi + \Sigma_i N_i kT_i$) is maintained across the whole sheath, draping leads to lower beta plasmas and thus higher field strengths. The so-called "plasma depletion layer" adjacent to the earth's magnetopause is a simple consequence of this effect, and should be present to some degree near the sheath stagnation points at all large objects where magnetic draping occurs.

Figure 7 illustrates the generation of magnetic storms by the shock compression mechanism. From day 245 until the shock on day 248, the B_Z value was fluctuating, but generally had a southward component. There is corresponding auroral electrojet (AE) activity as well as ring current

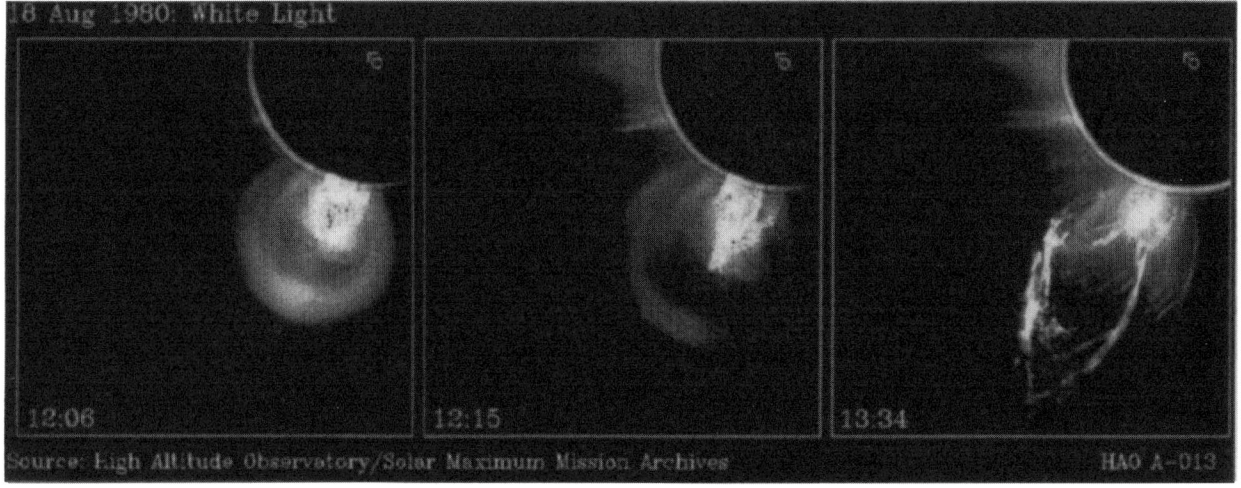

Figure 3. An example of a coronal mass ejection as seen in a white light coronograph image taken during the Solar Maximum Mission (courtesy of A. Hundhausen).

(D_{ST}) activity present. D_{ST} was ~ -30 nT from day 245 until the middle of day 247, and ~ -50 nT thereon until the shock. These D_{ST} values are relatively constant with little or no sign of the classic main phase/recovery phase signatures.

There is a short duration increase (small spike) in D_{ST} at and just after the shock due to solar wind ram pressure effects. This Sudden Impulse is the totality of the storm initial phase.

The B_z values in the sheath region behind the shock are fluctuating, but primarily directed southward from the shock until 1600 UT day 250. The peak B_S value of ~ -20 nT is reached at ~1200 UT day 249 and the peak D_{ST} of -280 nT several hours later. The mechanism for the southward component magnetic fields causing this storm are shock compression plus possible effects of draping.

Whether intense interplanetary fields are those of the sheath or the ejecta, the energy injection mechanism into the magnetosphere is the same. This is schematically shown in Figure 8. Interconnection of interplanetary fields and magnetospheric dayside fields leads to the enhanced reconnection of fields on the nightside with the concomitant deep injection of plasma sheet plasma in the nightside. The latter leads to the formation of the storm-time ring current. In general, the IMF structures leading to great ($D_{ST} < -250$ nT) and intense ($D_{ST} < -100$ nT) magnetic storms have features similar to the examples shown. The IMF B_S is intense and has a long duration. Gonzalez and Tsurutani (1987) have empirically found that interplanetary events with $E_{dawn-dusk} > 5$ mV/m (approximately $B_S > 10$ nT) with $T > 3$ hours lead to intense ($D_{ST} < -100$ nT) magnetic storms.

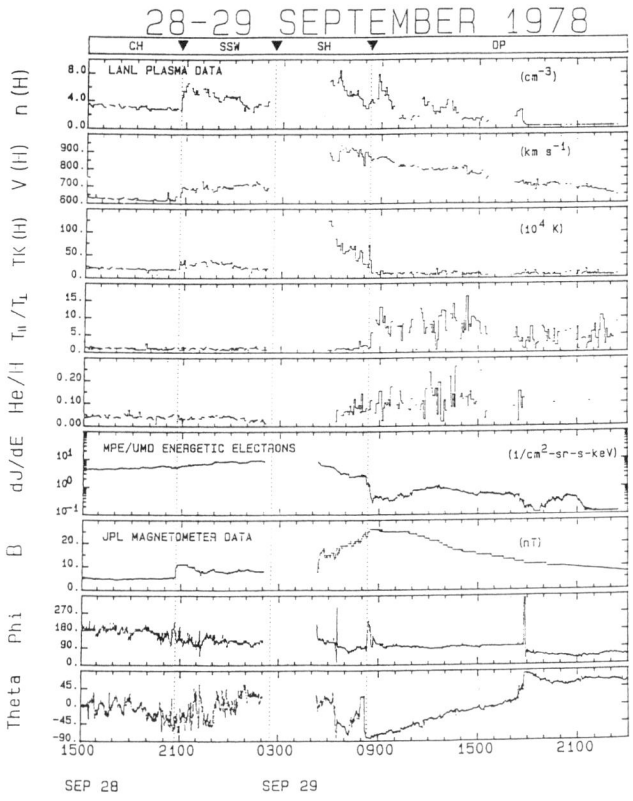

Figure 4. An example of possible remnants of the "bright loops" region (of a CME) followed by a magnetic cloud (taken from Galvin et al., 1987).

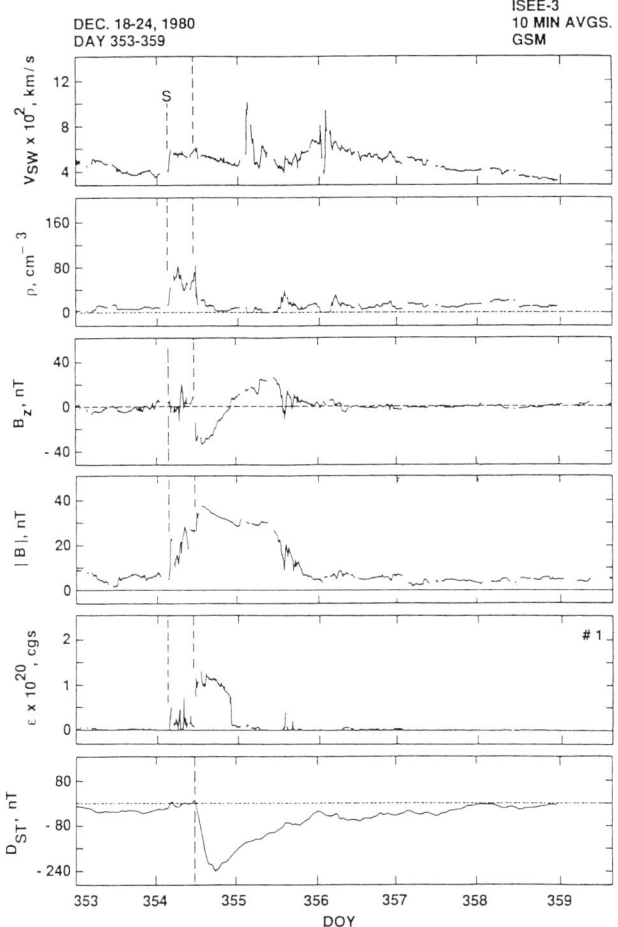

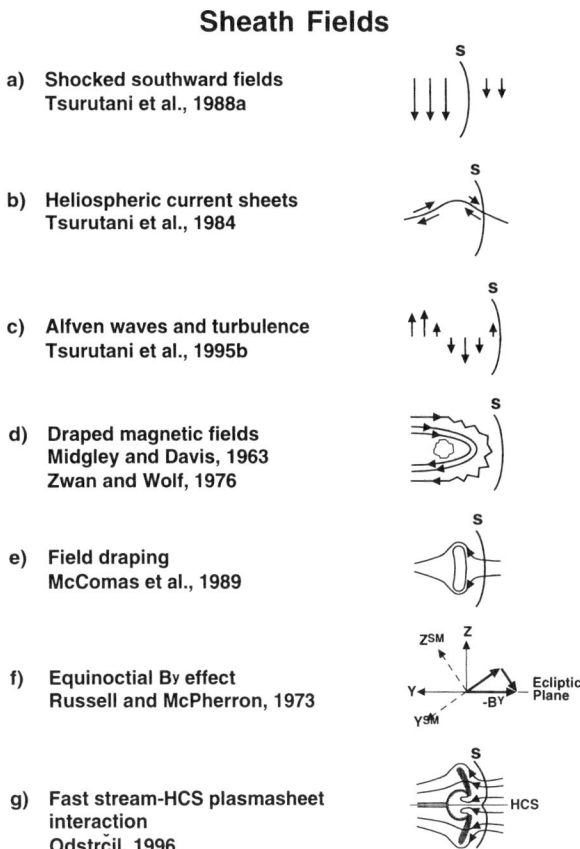

Figure 6. Types of "sheath" magnetic field structures.

Figure 5. A classical example of a magnetic storm driven by a magnetic cloud. The vertical dashed line labeled by a "S" indicates the presence of a fast forward shock. The vertical dashed line to the right indicates the start of the magnetic cloud.

In Tables 1 and 2 we give the causal connection between shocks/solar ejecta and storms of various levels of strength where we have defined the latter as follows: big: D_{ST} < -200 nT, intense: (-200 nT ≤ D_{ST} < -100 nT, moderate: -100 nT ≤ D_{ST} < -50 nT small: -50 nT ≤ D_{ST} < -30 nT magnetic storms. These come from prior work of the authors and from Gosling et al. (1991). Gosling et al. (1991) used Kp indices, and we have indicated the approximate D_{ST} values corresponding to these values. The Tables show that big storms have a 90% correspondence with fast solar eject events (with shocks), while small storms have only a 24% correspondence with fast solar ejecta.

Table 1 indicates that solar ejecta led by shocks do not always cause intense (D_{ST} ≤ -100 nT) magnetic storms. Studies using the ISEE-3 1978-1979 data indicate that only one out of every six solar ejecta (17%) are geoeffective in causing intense storms (Tsurutani et al., 1988b). From 57 fast solar ejecta events, it was found that some of the events did not have substantial B_S, others had large B_S values, but were highly fluctuating (about B_Z = 0 nT) in time. The important point is that they did not have B_S > 10 nT for T > 3 hours.

Table 3 gives the statistics for moderate magnetic storms. At these lower levels of storm intensity, one notes that the interplanetary causes are much more diverse. There are many mechanisms responsible for the causative B_S values. One such case (Alfvén fluctuations) were indicated in Figure 7 for the geomagnetic activity in the preshock interval. The general southward component (possibly intensified by the Russell-McPherron [1973] mechanism) and fluctuating B_Z led to D_{ST} ~ -50 nT.

1.4. Viscous Interaction

The earth's magnetopause can absorb solar wind energy through the fluid analogy of a viscous interaction (Axford and Hines, 1961). More specifically, mechanisms such as

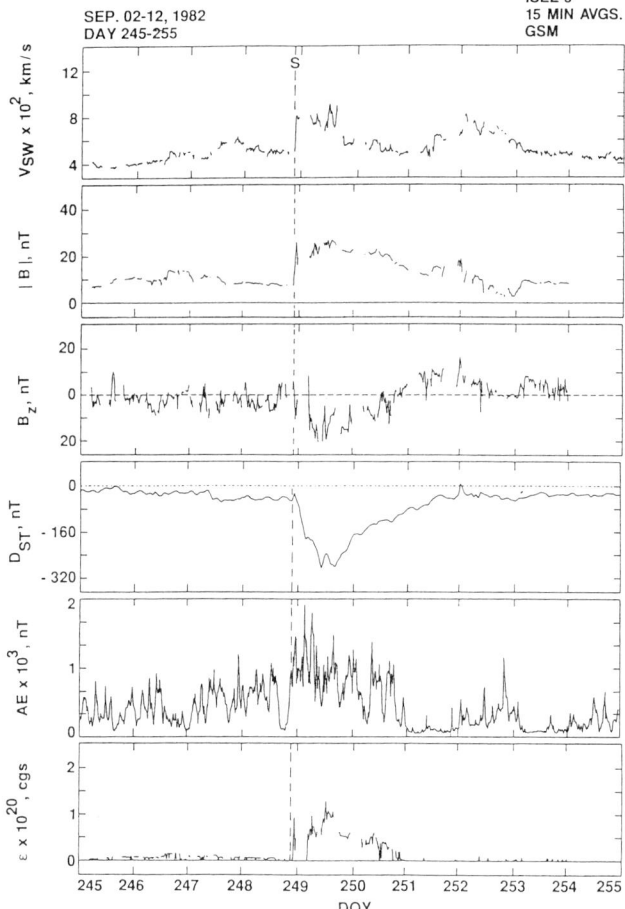

Figure 7. Example of a magnetic storm caused by shock compression of interplanetary B fields. The bottom panel contains the Perreault and Akasofu (1978) epsilon parameter.

the Kelvin Helmholtz instability (Parker, 1958, Tamao, 1965; Chen and Hasegawa, 1974; Southwood, 1968, 1974) or magnetosheath cross-field diffusion due to magnetopause boundary layer waves (Tsurutani and Thorne, 1982; Gendrin, 1983; Thorne and Tsurutani, 1991), are possible ways to inject solar wind energy into the magnetosphere.

An upper limit of the efficiency of solar wind energy access to the magnetosphere has been explored by examining intervals where the northern IMF component has characteristics: $B_n > 10$ nT and $T > 3$ hours. These interplanetary conditions allow reconnection between the IMF and terrestrial field tailward of the cusp (e.g., Dungey, 1961; Russell, 1972) hence justifying the statement that this is an upper limit calculation. The actual efficiency value might be lower. Without going through the (reasonably simple) details of the calculations, the conclusion is that ~ 1 to 4×10^{-3} of the solar wind ram energy is converted to magnetospheric energy in the form of auroral particles energy, Joule heating, or ring current particle energy (Tsurutani and Gonzalez, 1995b).

The efficiency of solar wind energy injection during magnetic reconnection events such as substorms and intense storms is 5-10% (Weiss et al., 1992, Gonzalez et al., 1989, respectively). The intercomparison of these numbers indicates that viscous interaction appears to be not more than 1/100th to 1/30 as efficient as magnetic reconnection. The highest solar wind speed event ever detected ($V_{SW} > 1500$ km s^{-1}, August, 1972) has also been studied for this effect. The efficiency of viscous interaction was found to have approximately the same value for this event as well (Tsurutani et al., 1992).

It should be noted that northward B_Z intervals satisfying the $B_Z > +10$ nT and $T > 3$ hours criteria are often found to be a portion of a magnetic cloud. Thus, since magnetic clouds often have south and then northward magnetic field orientations (or vice versa), clouds often cause magnetic storms followed by geomagnetic quiet (or vice versa).

1.5. *Descending Phase of the Solar Cycle*

In contrast to solar maximum, where coronal holes are not very important, during the descending phase of the solar cycle, coronal holes have major, even dominant effects on the interplanetary medium. Polar coronal holes extend from the polar regions down to the equator and sometimes even far past the equator (see Jackson, this issue). Coronal holes are low temperature regions above the sun, observed in soft x-rays (Timothy et al, 1975). They are areas of open magnetic field lines. Ulysses has shown that holes are regions of fast streams with velocities of 750-800 km s^{-1} (Phillips et al., 1994) and are dominated by large amplitude Alfvén waves (Tsurutani et al., 1994, 1996; Balogh et al., 1995; Smith et al., 1995a, b). The Alfvén waves are continuously present in the high velocity streams.

During the descending phase of the solar cycle, when the holes migrate down to lower latitude as "fingers", the streams emanating from the holes "corotate" at ~27 day intervals (as seen at the Earth), and thus plasma from these streams impinge on the Earth's magnetosphere at periodic intervals and cause recurrent geomagnetic storms (Burlaga and Lepping, 1977; Sheeley et al., 1976, 1977; Burlaga et al., 1978).

High speed streams emanating from coronal holes can create intense magnetic fields if the streams interact with streams of lower speeds (Belcher and Davis, 1971; Pizzo, 1985; Tsurutani et al., 1995c, d). A schematic of such an interaction is given in Figure 9. The magnetic fields of the slower speed stream are more curved due to the lower speeds, and the fields of the higher speed stream are more radial because of the higher speeds. The stream-stream interface (IF) is the boundary between the slow stream and fast stream plasmas and fields. Significant angular deflec-

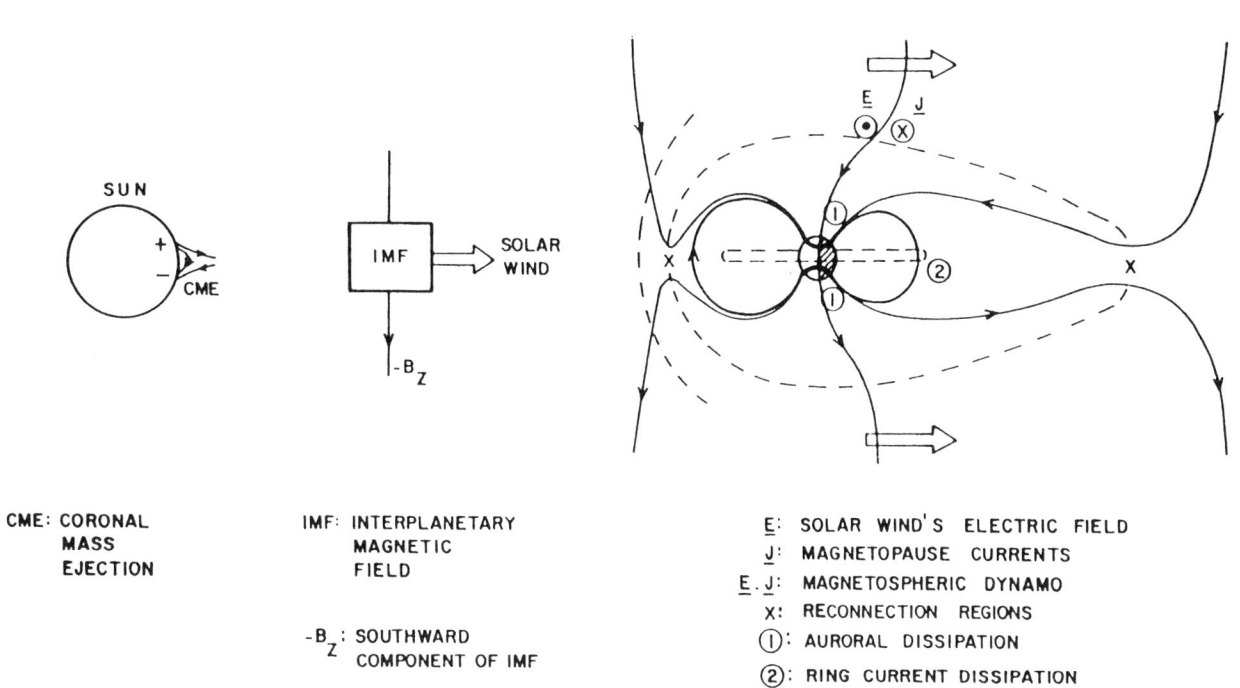

Figure 8. Schematic of interplanetary-magnetosphere coupling, showing the energy injection mechanism into the nightside magnetosphere.

tions in velocity can occur at or near this region (see Pizzo, 1985).

Antisunward of the IF are the compressed and accelerated slower speed plasma and fields. Behind the IF are the compressed and decelerated high speed stream plasma and fields. At large heliospheric distances (> 1.5 AU), where these corotating structures are well developed, they are bounded by fast forward (FS) and fast reverse (RS) shocks. This overall structure was first found in the Pioneer 10 and 11 data and were named Corotating Interaction Regions (CIRs) by Smith and Wolf (1976). See also Burlaga et al. (1985). As far as geomagnetic storms are concerned, the important feature of CIRs is that they are characterized by intense magnetic fields. The intensities can reach ~30 nT.

At 1 AU, CIRs are not fully developed. They almost never have forward shocks (this can and has been used as a reasonably reliable identifying feature) and usually do not have reverse shocks (~80% of the time). We therefore call these proto-CIRs (PCIR) in this paper.

An example of a PCIR and its consequential magnetic storm activity is shown in Figure 10. This event is typical of the events studied for the 1973-1975 epoch where two corotating streams (from two coronal holes) per solar cycle dominated interplanetary activity.

The unusually high plasma densities of > 50 cm^{-3} at the beginning of day 25 is intrinsic to the slow solar wind near the heliospheric current sheet (HCS), the region separating the north and south hemisphere heliospheric magnetic fields. This high density plasma has been called the HCS plasma sheet by Winterhalter et al. (1995). However, R. P. Lepping (personal communication 1996) notes that this plasma sheet may not always be present. The HCS is identified by a reversal in the Parker spiral direction by ~180° or a simultaneous reversal in the signs of both B_x and B_y. Such a reversal can be noted at ~2200 UT day 24.

The high density plasma of the HCS plasma sheet causes the "initial phase" of the magnetic storm found in the bottom panel. Note that this "phase" of the storm is caused by interplanetary conditions totally unlike those during solar maximum. Here the high densities are associated with a low velocity stream ($V_{SW} < 400$ km s^{-1}). Since the PCIRs typically do not have forward shocks at 1 AU, there will typically be a lack of a sudden impulse associated with this type of a storm.

The magnetic field of the PCIR increases gradually from about 0000 UT until 2000 UT day 25. A maximum value of ~25 nT is present from 1200 to 2000 UT. In this particular case, the PCIR is terminated by a reverse shock.

TABLE 1. ISEE-3 statistics (Aug 1978 - Dec 1979)

Storm Intensity	No. of Events	Definition	Association with Shocks (56) (supermagnetosonic speed ICMEs)
Intense	10	$D_{ST} < -100$ nT	80%
Moderate	40	-100 nT $\leq D_{ST} < -50$ nT	45%
Small	62	-50 nT $\leq D_{ST} < -30$ nT	24%

Shocks and Magnetic Storms
15% followed by intense storms
35% followed by moderate storms
30% followed by small storms
20% followed by no storms ($D_{ST} \geq -30$ nT)

TABLE 2. Statistics for Aug 1978 - Oct 1982

Storm Intensity	K_p Definition	D_{ST} Definition	Shock Association	ICME Association
Big	$8 \leq K_p \leq 9$	$D_{ST} \leq -200$ nT	100%	90%
Intense	$K_p = 7$	-200 nT $\leq D_{ST} < -100$ nT	80%	80%
Moderate	$5 \leq K_p \leq 6$	-100 nT $\leq D_{ST} < -50$ nT	40%	40%

TABLE 3. Interplanetary Association of Moderate Storms[a]

ISEE-3 (Aug 1978 - Dec 1979)
40% Shocks
23% High-speed streams without shocks
17% High-Low speed stream interactions
10% Noncompressive density enhancements (NCDEs)
10% Other (including Alfvénic fluctuations)

[a](-100 nT $\leq D_{ST} < -50$ nT)

The PCIR is responsible for the main phase of the magnetic storm. The reverse shock (~2000 UT day 25), across which the field decreases dramatically, leads to the start of the recovery phase of the magnetic storm with a delay of about 1 hour. We note, however, that the storm main phase is somewhat irregular in profile and the peak intensity is only $D_{ST} \sim -70$ nT (this is on the upper end of storm strength distribution during this phase of the solar cycle). The cause of this is in the character of B_Z within the PCIR. B_Z is highly fluctuating throughout the interval. There may be a net southward component within the PCIR, but this is accompanied by a much larger fluctuation amplitude.

Why are such fluctuations present? One possible answer is schematically shown in Figure 9. If B_Z fluctuations (Alfvén waves) are present in the high speed stream proper, then the deceleration and compression due to passage through the reverse shock could lead to amplification of such oscillations. Ulysses results (Tsurutani et al., 1995c) are consistent with such a scenario.

Figure 11 shows the geomagnetic activity during 1974 when there were two corotating streams (per 27 day solar rotation) present. The 3 storms where $D_{ST} < -100$ nT were caused by fields associated with fast solar ejecta events and not by the corotating streams. Thus, the corotating streams are far less geoeffective in creating intense or moderate magnetic storms.

A summary of the geoeffectiveness of PCIRs is given in Table 4. This was derived from a subset of the 1974 data set. Similar studies have been performed on the 1973 and 1975 data, with similar results.

MAXIMUM GEOMAGNETIC ACTIVITY DURING SOLAR MAXIMUM OR MINIMUM?

Although it is clear that there are far more large D_{ST} events during solar maximum than during solar minimum, the same cannot be said for auroral zone (AE) activity. For the period 1973-1975, the annual AE average (of the 2.5 min values) were: 247, 283 and 224 nT, respectively. For 1979-1981, the annual AE values were 221, 180 and 237 nT. The 283 nT value for 1974 was larger than any of the solar maximum years (Tsurutani et al., 1995c).

The interplanetary phenomenon causing this effect can be found in Figure 11. After each magnetic storm interval (sharp D_{ST} decrease), there are prolonged intervals of intense AE. These AE intensifications are directly correlated with the slow recovery of D_{ST}. In most of the events shown in the Figure, the D_{ST} index takes 10-20 days to recover to near-background values.

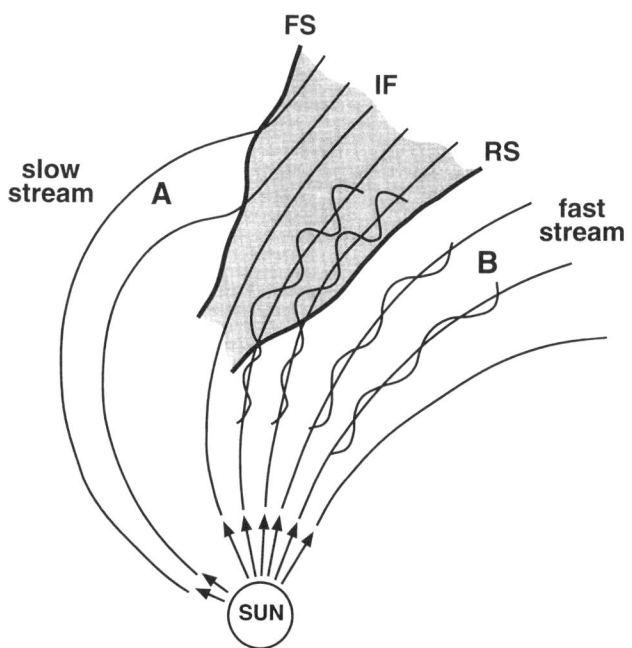

Figure 9. Schematic of the formation of corotating interaction regions (CIRs) during the descending phase of the solar cycle. The compression of plasma and magnetic field fluctuations are also shown. Taken from Tsurutani et al. (1995c).

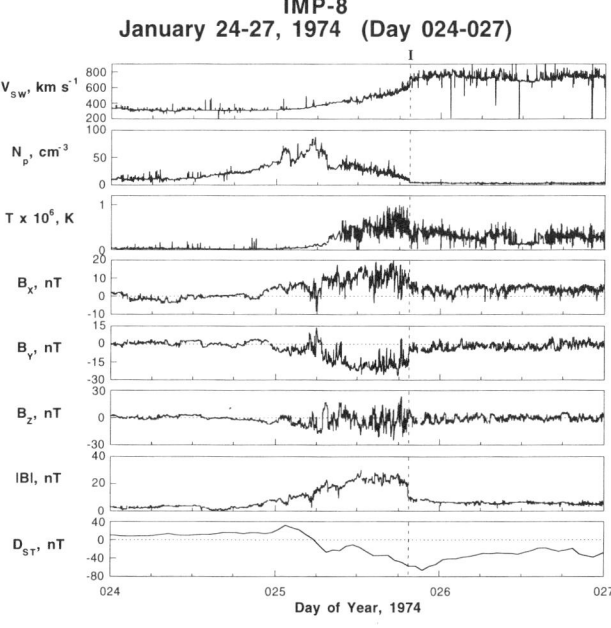

Figure 10. Example of a PCIR and associated geomagnetic activity, typical of 1973-1975. Taken from Tsurutani et al. (1995d).

Figure 12 illustrates a four day period of one of these storm recovery intervals. D_{ST} fluctuates at a value near -25 nT for the entire period with little or no sign of recovery. An intercomparison with the AE index indicates that there is a one-to-one relationship between AE increases and D_{ST} decreases. Thus one interpretation of this observation is that substorms (AE increases) are injecting fresh particles into the outer radiation belts, preventing the ring current from reaching quiet day values. However, it should be noted that plasma sheet current intensifications or earthward motions of the latter could cause such effects on the D_{ST} index as well. This problem will be investigated in the near future.

The cause of the continuous substorms in Figure 12 are the large amplitude B_Z fluctuations in the IMF. Although the average B_Z value is near zero, the large amplitude fluctuations provide very large B_S intervals and concomitant substorms through the reconnection process. The IMF fluctuations have been examined and have been shown to be Alfvén waves propagating outward from the sun in these coronal hole streams. The fluctuations are more or less continuous and the southward components of the larger period waves cause High Intensity Long Duration Continuous AE Activity (HILDCAAs) (Tsurutani and Gonzalez, 1987; Tsurutani et al, 1990).

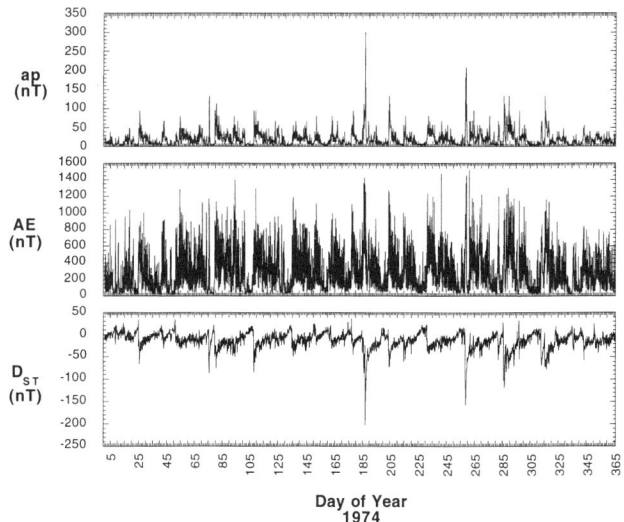

Figure 11. Geomagnetic activity indices for 1974.

TABLE 4. Geoeffectiveness of Proto-CIRs, IMP-8 Days 1-241, 1974, well-developed streams[a]

Storm Intensity	Definition	Geoeffectiveness
Intense	$D_{ST} < -100$ nT	0%
Moderate	-100 nT $\leq D_{ST} < -50$ nT	29%
Small	-50 nT $\leq D_{ST} < -30$ nT	29%
Negligible storm activity	-30 nT $\leq D_{ST}$	41%

[a] ($V_{SW} = 600\text{-}850$ km s^{-1})

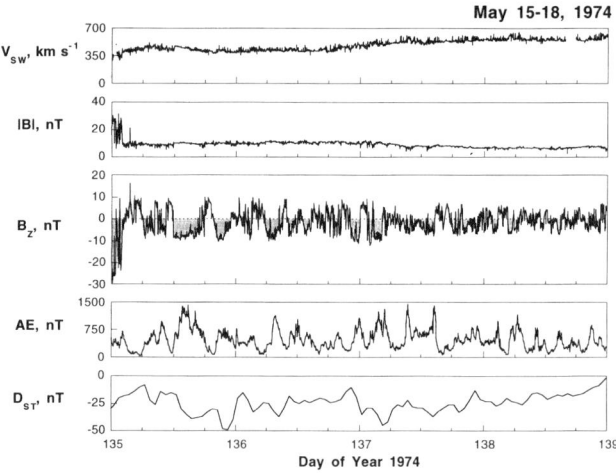

Figure 12. An example of a recovery phase of a magnetic storm during a HILDCAA interval.

CURRENT OUTSTANDING PROBLEMS

1. To predict the occurrence of a magnetic storm one needs to be able to predict three interplanetary parameters: V_{SW}, B_Z and the duration of B_S. The first parameter can be obtained with several days advanced warning by placing a coronagraph in space with a spacecraft/sun/earth angle of ~90°. A CME event occurrence plus its velocity can be obtained by the coronagraph measurements. We are currently not able to predict the latter two parameters. Unfortunately, these are the two most important parameters in determining the storm intensity.

2. There has been a great deal of focus on magnetic clouds because of their strong/weak interaction with the earth's magnetosphere (magnetic storms with the B_S portion/geomagnetic quiet during the B_N portion). A point that is often missed is that magnetic clouds are only present in one out of six fast ICMEs/driver gases (Tsurutani et al., 1988b). The reasons for the complex field configuration for "this more typical case" should be investigated and explained.

3. Because of the geoeffectiveness of fast ICMEs/driver gases and their sheaths, these solar/interplanetary phenomena have received the most attention. However, little is known about "slower" CMEs, those ejecta events that do not produced upstream shock waves. Questions we need to understand are: why aren't these geoeffective? Is the lack of sheath fields the cause of the lack of storms? Do the ICMEs/driver gases never/seldom have magnetic cloud field configurations? Or are the ICME/driver gas fields less intense, implying an intrinsic relationship between CME field strength and velocity?

Acknowledgments. We would like to acknowledge the excellent comments provided by two of the referees, C. J. Farrugia and R. P. Lepping. Portions of this work were performed at the Jet Propulsion Laboratory, California Institute of Technology, Pasadena, under contract with the National Aeronautics and Space Administration.

REFERENCES

Axford, W. I. and C. O. Hines, A unifying theory of high-latitude geophysical phenomena and geomagnetic storms, *Can. J. Phys., 39,* 1433, 1961.

Balogh, A., E. J. Smith, B. T. Tsurutani, D. J. Southwood, R. J. Forsyth, and T. S. Horbury, The heliospheric magnetic field over the south polar region of the sun, *Science, 268,* 1007, 1995.

Bravo, S. and J. A. L. Cruz-Abeyo, The spatial relation between active regions and coronal holes, and the occurrence of geomagnetic storms, *Chapman Conference on Magnetic Storms*, Pasadena, California, February 12-16, 1996.

Belcher, J. W. and L. Davis, Jr., Large amplitude Alfvén waves in the interplanetary medium, 2, *J. Geophys. Res., 76,* 3534, 1971.

Burlaga, L. F. and R. P. Lepping, The causes of recurrent geomagnetic storms, *Planet. Space Phy, 25,* 1151, 1977.

Burlaga, L. F., K. W. Behannon, S. F. Hansen, G. W. Pneumann and W. C. Feldman, Sources of magnetic fields in recurrent interplanetary streams, *J. Geophys. Res., 83,* 4177, 1978.

Burlaga, L. F., E. Sittler, F. Mariani, and R. Schwenn, Magnetic loop behind an interplanetary shock: Voyager, Helios and IMP-8 observations, *J. Geophys. Res., 86,* 6673, 1981.

Burlaga, L. F., V. Pizzo, A. Lazarus and P. Gazis, Stream dynamics between 1 AU and 2 AU: A comparison of observations and theory, *J. Geophys. Res., 90,* 7317, 1985.

Burlaga, L. F., K. W. Behannon, and L. W. Klein, Compound streams, magnetic clouds and major geomagnetic storms, *J. Geophys. Res., 92,* 5725, 1987.

Burlaga, L.F., R. P. Lepping and J. Jones, in *Physics of Flux Ropes,* ed. C. T. Russell, E. R. Priest and L. C.Lee, *AGU Monograph 58,* Wash. D.C., 373, 1990.

Chen, L. and A. Hasegawa, A theory of long-period magnetic pulsations, 1), Steady state excitation of field-line resonances, *J. Geophys. Res., 79,* 1024, 1974.

Choe, G. S., N. LaBelle-Hamer, B. T. Tsurutani and L. C. Lee, Identification of a driver gas boundary layer, *EOS, 73,* 485, 1992.

Dungey, J. W., Interplanetary magnetic field and the auroral zones, *Phys. Rec. Lett., 6,* 47, 1961.

Farrugia, C. J., L. F. Burlaga, V. A. Osherovich, I. G. Richardson, M. P. Freeman, R. P. Lepping and A J. Lazarus, A study of an expanding interplanetary magnetic cloud and its interaction with the Earth's magnetosphere: the interplanetary aspect, *J. Geophys. Res., 98,* 7621, 1993a.

Farrugia, C. J., I. G. Richardson, L. F. Burlaga, R. P. Lepping and V. A. Osherovich, Simultaneous observations of solar MeV particles in a magnetic cloud and in the earth's northern tail lobe: Implications for the global field line topologies of a magnetic cloud and for the entry of solar particles into the magnetosphere during cloud passage, *J. Geophys. Res., 98,* 15497, 1993b.

Farrugia, C. J., V. A. Osherovich and L. F. Burlaga, Magnetic flux rope versus the spheromak as models for interplanetary magnetic clouds, *J. Geophys. Res., 100,* 12293, 1995.

Farrugia, C. J. and L. F. Burlaga, Magnetic clouds and the quiet-storm effect at earth, in *Magnetic Storms,* edited by B. T. Tsurutani, W. D. Gonzalez and Y. Kamide, AGU Monograph, Wash. D.C., 1997.

Galvin, A. B., F. M. Ipavich, G. Gloeckler, D. Hovestadt, S. J. Bame, B. Kleckler, M. Scholer and B. T. Tsurutani, Solar wind ion charge status preceding a driver plasma, *J. Geophys. Res., 92,* 12069, 1987.

Gendrin, R., Magnetic turbulence and diffusion processes in the magnetopause boundary layer, *Geophys. Res. Lett.,* 769, 1983.

Gold, T., Magnetic storms, *Space Sci. Rev., 1,* 100, 1962.

Gonzalez, W. D. and B. T. Tsurutani, Criteria of interplanetary parameters causing intense magnetic storms (D < -100 nT), *Planet. Space Sci., 35,* 1101, 1987.

Gonzalez, W. D., B. T. Tsurutani, A.L.C. Gonzalez, E. J. Smith, F. Tang, and S.-I. Akasofu, Solar wind magnetosphere coupling during intense magnetic storms (1978-1979), *J. Geophys. Res., 94,* 8835, 1989.

Gonzalez, W. D., A. L. Clua de Gonzalez, O. Mendes, Jr., and B. T. Tsurutani, Difficulties in defining storm sudden commencements, *EOS, Trans. Amer. Geophys. Un., 73,* 180, 1992.

Gonzalez, W. D., J. A. Joselyn, Y. Kamide, H. W. Kroehl, G. Rostoker, B. T. Tsurutani, and V. M. Vasyliunas, What is a Geomagnetic Storm?, *J. Geophys. Res.* 99, 5771, 1994.

Gonzalez, W. D., B. T. Tsurutani, P. S. McIntosh, and A. L. Clua de Gonzalez, Coronal hole - active region - current sheet (CHARCS) association with intense interplanetary and geomagnetic activity, *Geophys. Res. Lett., 23,* 2577, 1996.

Gosling, J. T., D. N. Baker, S. J. Bame, W. C. Feldman and R. D. Zwickl, Bi-directional solar wind electron heat flux events, *J. Geophys. Res., 92,* 8519, 1987.

Gosling, J. T., S., J. Bame, D. J. McComas, and J. L. Phillips, Coronal mass ejections and large geomagnetic storms, *Geophys. Res. Lett., 127,* 901, 1990.

Gosling, J. T., D. J. McComas, J. L. Phillips and S. J. Bame, Geomagnetic activity associated with Earth passage of interplanetary shock disturbances and coronal mass ejections, *J. Geophys. Res., 96,* 7831, 1991.

Ivanov, K. G., A. F. Harschiladze, E. G. Eroshenko, and V. A. Styazhkin, Configuration, structure and dynamics of magnetic clouds from solar flares in light of measurements on board Vega 1 and Vega 2 in Jan.-Feb. 1986, *Solar Phys., 120,* 407, 1989.

Jackson, B. V., Heliospheric observations of solar disturbances and their potential role in the origin of storms, in *Magnetic Storms,* B. T. Tsurutani, W. D. Gonzalez and Y. Kamide, Amer. Geophys. Union Press, Washington D.C., 1997.

Joselyn, J. A. and B. T. Tsurutani, Geomagnetic sudden impulses and storm sudden commencements, *EOS, 71, 1808,* 1990.

Klein, L. W. and L. F. Burlaga, Interplanetary magnetic clouds at 1 A, *J. Geophys. Res., 87,* 613, 1982.

Lepping, R. P., J. A. Jones, and L. F. Burlaga, Magnetic field structure of interplanetary magnetic clouds at 1 AU, *J. Geophys. Res., 95,* 11957, 1990.

McComas, D. J., J. T. Gosling, S. J. Bame, E. J. Smith, and H. V. Cane, A test of magnetic field draping induced B_z perturbations ahead of fast coronal mass ejecta, *J. Geophys. Res., 94,* 1465, 1989.

Marsden, R. G., T. R. Sanderson, C. Tranquille and K.-P. Wenzel, ISEE-3 observations of low-energy proton bi-directional events and their relation to isolated interplanetary magnetic structures, *J. Geophys. Res., 92,* 11009, 1987.

Midgley, J. E. and L. Davis, Jr., Calculation by a moment technique of the perturbation of the geomagnetic field by the solar wind, *J. Geophys. Res., 68,* 5111, 1963.

Odstrcil, D., Numerical simulation of interplanetary plasma clouds propagating along the heliospheric plasma sheet, *Astrophys. Lett. Comm.* in press, 1997.

Parker, E. N. Interaction of solar wind with the geomagnetic field, *Phys. Fluids, 1,* 171, 1958.

Perreault, P. and S.-I. Akasofu, A study of geomagnetic storms, *Geophys. J. R., Astron. Soc. 54,* 547, 1978.

Phillips, J. L., A. Balogh, S. J. Bame, et al., Ulysses at 50° south: constant immersion in the high-speed solar wind, *Geophys. Res. Lett., 21,* 1105, 1994.

Phillips, J. L., S. J. Bame, W. C. Feldman, B. E. Goldstein, J. T. Gosling, C. M. Hammond, D. J. McComas, M. Neugebauer, E. E. Scime and S. T. Suess, Ulysses solar wind plasma observations at high southerly latitudes, *Science, 268,* 1030, 1995.

Pizzo, V. J., Interplanetary shocks on large scale: A retrospective on the last decade's theoretical efforts, in *Collisionless Shocks in the Heliosphere, Review of Current Research,* ed. by B. T. Tsurutani and R. G. Stone, *Geophys. Mon. Series, 35,* 51, Wash D.C., 1985.

Russell, C. T., The configuration of the magnetosphere, in *Critical Prob Magnet. Phys.,* edited by E. R. Dyer, 1, Nat. Acad. Sci., Wash. D.C., 1972.

Russell, C. T. And R. L. McPherron, Semiannual variation of geomagnetic activity, *J. Geophys. Res., 78,* 92, 1973.

Sheeley, N. R., Jr., J. W. Harvey, and W. C. Feldman, Coronal holes, solar wind streams and recurrent geomagnetic disturbances, 1973-1976, *Sol. Phys., 49,* 271, 1976.

Sheeley, N. R. Jr., J. R. Asbridge, S. J. Bame and J. W. Harvey, A pictoral comparison of interplanetary magnetic field polarity, solar wind speed, and geomagnetic disturbances index during the sunspot cycle, *Sol. Phys., 52,* 485, 1977.

Smith, E. J. and J. W. Wolfe, Observations of interaction regions and corotating shocks between one and five AU: Pioneers 10 and 11, *Geophys. Res. Lett., 3,* 137, 1976.

Smith, E. J., M. Neugebauer and B. T. Tsurutani, Ulysses observations of latitudinal gradients in the heliospheric magnetic field: Radial component and variances, *Space Sci. Rev., 72,* 165, 1995a.

Smith, E. J., A. Balogh, M. Neugebauer, and D. McComas, Ulysses observations of Alfvén waves in the southward northern solar hemisphere, *Geophys. Res. Lett., 22,* 3381, 1995b.

Southwood, D. J., The hydromagnetic stability of the magnetospheric boundary, *Planet. Space Sci., 16,* 587, 1968.

Southwood, D. J., Some features of field-line resonance in the magnetosphere, *Planet. Space Sci., 22,* 483, 1974.

Tamao, T., Transmission and coupling resonance of hydromagnetic disturbances in non-uniform Earth's magnetosphere, *Sci. Rep. Tohoku Univ. Ser., 5, 17,* 43, 1965.

Thorne, R. M. and B. T. Tsurutani, Wave-particle interactions in the magnetopause boundary layer, in *Physics of Space Plasmas (1990),* ed. by T. Chang, et al., Sci Publ. Inc., Cambridge, MA, 10, 119, 1991.

Timothy, A. F., A. S. Krieger and G. S. Vaiana, The structure and evolution of coronal holes, *Sol. Phys. 42,* 135, 1975.

Tsurutani, B. T. and R. M. Thorne, Diffusion processes in the magnetopause boundary layer, *Geophys. Res. Lett., 9,* 1247, 1982.

Tsurutani, B. T., C. T. Russell, J. H. King, R. D. Zwickl, and R. P. Lin, A kinky heliospheric current sheet: Causes of the CDAW6 substorms, *Geophys. Res. Lett., 11,* 339, 1984.

Tsurutani, B. T. and W. D. Gonzalez, The cause of high intensity long-duration continuous AE activity (HILDCAAs): Interplanetary Alfvén waves trains, *Planet. Space Sci., 35,* 405, 1987.

Tsurutani, B. T., W. D. Gonzalez, F. Tang, S.-I. Akasofu, and E. J. Smith, Origin of interplanetary southward magnetic fields responsible for major magnetic storms near solar maximum (1978-1979), *J. Geophys. Res., 93,* 8519, 1988a.

Tsurutani, B. T., B. E. Goldstein, W. D. Gonzalez, and F. Tang, Comment on "A new method of forecasting geomagnetic activity and proton showers", by A. Hewish and P. J. Duffet-Smith, *Planet. Space Sci., 36,* 205, 1988b.

Tsurutani, B. T., T. Gould, B. E. Goldstein, W. D. Gonzalez, and M. Sugiura, Interplanetary Alfvén waves and auroral (substorm) activity: IMP-8, *J. Geophys. Res., 95,* 2241, 1990.

Tsurutani, B. T., W. D. Gonzalez, F. Tang, Y. T. Lee, M. Okada, and D. Park, Reply to L. J. Lanzerotti: Solar wind ram pressure corrections and an estimation of the efficiency of viscous interaction, *Geophys. Res. Lett., 19,* 1993, 1992.

Tsurutani, B. T., C. M. Ho, E. J. Smith, M. Neugebauer, B. E. Goldstein, J. S. Mok, J. K. Arballo, A. Balogh, D. J. Southwood and W. C. Feldman, The relationship between interplanetary discontinuities and Alfvén waves: Ulysses observations, *Geophys. Res. Lett., 21,* 2267, 1994.

Tsurutani, B. T. and W. D. Gonzalez, The future of geomagnetic storm predictions: Implications from recent polar and interplanetary observations, *J. Atmos. Terr. Phys., 57,* 1369, 1995a.

Tsurutani, B. T. and W. D. Gonzalez, The efficiency of "viscous interaction" between the solar wind and the magnetosphere during intense northward IMF events, *Geophys. Res. Lett., 22,* 663, 1995b.

Tsurutani, B. T., C. M. Ho, J. K. Arballo, B. E. Goldstein, and A. Balogh, Large Amplitude IMF fluctuations in corotating interaction regions: Ulysses at midlatitudes, *Geophys. Res. Lett., 22,* 3397, 1995c.

Tsurutani, B. T., W. D. Gonzalez, A.L.C. Gonzalez, F. Tang, J. K. Arballo and M. Okada, Interplanetary original of geomagnetic activity in the declining phase of the solar cycle, *J. Geophys. Res. 100,* 21717, 1995d.

Tsurutani, B. T., B. E. Goldstein, C. M. Ho, M. Neugebauer, E. J. Smith, A Balogh and W. C. Feldman, Interplanetary discontinuities and Alfvén waves at high heliographic latitudes: Ulysses, *J. Geophys. Res., 101,* 11027, 1996.

Vandas, M., S. Fischer and A. Geranios, Spherical and cylindrical models of magnetic clouds and their comparison with spacecraft data, *Planet. Space Sci., 39,* 1147, 1991.

Vandas, M., S. Fischer, P. Pelant and A. Geranios, Spheroidal models of magnetic clouds and their comparison with spacecraft measurement. *J. Geophys. Res., 98*, 11467, 1993.

Weiss, L. A., P.H. Reiff, J. J. Moses, and B. D. Moore, Energy Dissipation in substorms, *Eur. Space Agency Spec. Publi., ESA-SP-335*, 309, 1992.

Winterhalter, D., E. J. Smith, M. E. Burton, N. Murphy and D. J. McComas, The heliospheric plasma sheet, *J. Geophys. Res., 99*, 6667, 1994.

Zwan, B. J. and R. A. Wolf, Depletion of the solar wind plasma near a planetary boundary, *J. Geophys. Res., 81*, 1636, 1976.

B. T. Tsurutani, Space Physics Element, Jet Propulsion Laboratory, California Institute of Technology, Pasadena, California 91109
email: btsurutani@jplsp.jpl.nasa.gov

Walter D. Gonzalez, Institutio Nacional Pesquisas Espaciais, Sao Jose dos Campos, San Paulo, Brazil
email: gonzalez@dge.inpe.br

Magnetic Clouds and the Quiet-Storm Effect at Earth

C. J. Farrugia

Institute for the Study of Earth, Oceans, and Space, University of New Hampshire, Durham

L. F. Burlaga and R. P. Lepping

NASA/Goddard Space Flight Center, Code 692, Greenbelt, Maryland

In this review, we discuss first magnetic clouds in the context of other interplanetary causes of geomagnetic storms. We then describe work on the global field line topology of magnetic clouds, focussing on information gained by the use of energetic particles. We then give a summary of theoretical and simulation work on the dynamics of magnetic clouds. In one approach, based on self-similar evolution of radially expanding magnetic flux ropes, the role of electrons is central. A section on the boundaries of magnetic clouds is followed by one on magnetic field line draping around these ejecta, including the formation of a magnetic barrier. In the aspect of the study dealing with the geomagnetic response to magnetic clouds, we discuss effects on the dayside magnetosheath; ionosphere; and nightside magnetosphere at geostationary orbit and beyond, utilizing primarily observations made during Earth's encounter with a magnetic cloud on January, 13 - 14, 1988. A case study is mentioned where solar energetic particles, injected into a magnetic cloud and then guided along its helical field lines, entered the magnetosphere through interconnection of the cloud's field lines with those of Earth. Simulation work on the geomagnetic response to magnetic clouds is briefly reviewed. We finally consider studies specifically correlating magnetic clouds, in isolation or as part of compound streams, with geomagnetic storm activity. Throughout, we indicate areas where further work is needed.

1. INTRODUCTION

Various studies have shown that a distinct class of transient solar ejecta is correlated with nonrecurrent storm activity at Earth: magnetic clouds. The point of view of the paper is that to properly appreciate this association it is important to review our present understanding of these configurations and to discuss storm activity within the broader context of magnetosheath and magnetspheric behavior during magnetic cloud passage.

2. MAGNETIC CLOUDS AND GEOMAGNETIC STORMS: PART I

A dominant mechanism coupling the momentum and energy of the solar wind to the magnetosphere and thus driving geomagnetic activity is reconnection between the interplanetary magnetic field (IMF) and the geomagnetic field at the dayside magnetopause [*Dungey*, 1961]. The resulting energy input depends on the strength and duration of a southward-directed IMF B_z component and on the solar wind bulk speed [*Baker et al.*, 1984, and references therein]. Indeed, studying major storms (-220 nT $\leq$ Dst $\leq$ -100 nT) for a 16-month period near solar maximum, *Gonzalez and Tsurutani* [1987] found that a dawn-dusk electric field $\geq 5mVm^{-1}$, i.e., approximately IMF $B_z \leq -10$ nT, maintained for long periods ($\geq$ 3 h) was a necessary and sufficient condition to generate these storms. The interplanetary sources of negative IMF B_z are many (see *Tsurutani and Gonzalez* [1996], and references therein). However, a common finding is that a substantial fraction of all storms is caused by a class of transient solar ejecta called magnetic clouds. Thus 5 out of 10 major magnetic storms in the years 1978 - 1979 [*Tsurutani et al.*, 1988a] and 2 of the 5 largest magnetic storms during 1971 - 1986 with -325 nT $\leq$ Dst $\leq$ -250 nT [*Tsurutani et al.*, 1992] were due to magnetic clouds. The very first study of magnetic clouds as we understand them today [*Burlaga et al.*, 1981] associated the cloud studied there (January, 1978) with a major storm disturbance.

Magnetic clouds are defined by: (a) enhanced magnetic field strengths, (b) a large and smooth rotation of the magnetic field vector over a period of order 1 day (at 1 AU), and (c) low proton temperatures, T_p [*Burlaga et al.*, 1981; *Burlaga*, 1991, 1995]. Figure 1 shows an example of a much-studied magnetic cloud. The above definition identifies magnetic clouds as a distinct class of ejecta with a clear physical interpretation and signature. They have also been considered as a proper subset of interplanetary transients known collectively as coronal mass ejections (CME's) and thus it is appropriate to review at this point work on CME's and geomagnetic storms. (On a matter of nomenclature: we shall henceforth use the term "solar ejecta" for the interplanetary manifestation of CME's and reserve the term "CME" for the phenomenon observed near the Sun in coronagraphs.)

This work was done by *Gosling* and co-workers [1990, 1991], who studied storm activity in relation to solar ejecta for the 4-year period 1978 - 1982. The solar ejecta were identified by bidirectional flows of suprathermal electrons (Energy $\geq$ 80 eV; *Gosling et al.*, [1987]), and the storm strength was gauged by the Kp index. An overwhelming majority of large magnetic storms were found to be associated with the following 3 mechanisms, either acting alone or in combination: the solar ejecta itself, the compression of a negative IMF B_z component

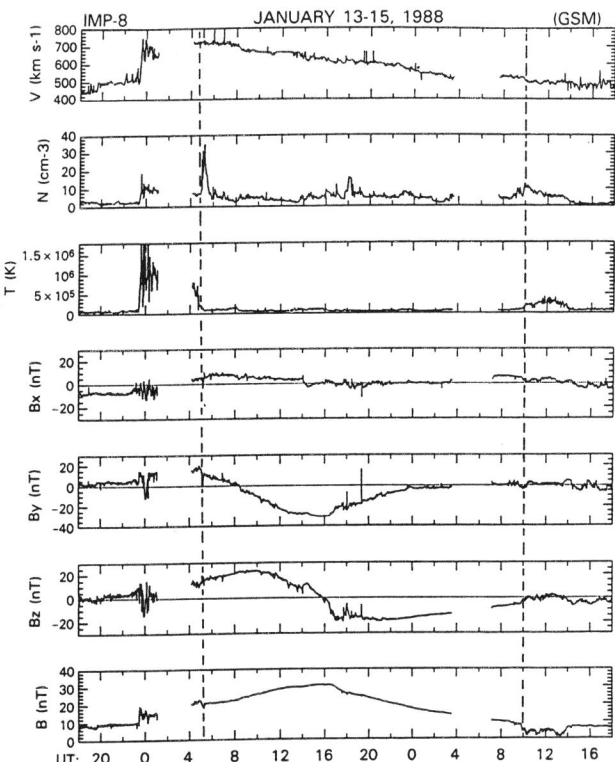

Figure 1. Observations of a magnetic cloud by IMP 8 on January 14 - 15, 1988. The time interval shown is from 18 UT, January 13 to 18 UT, January 15, 1988. The panels display the bulk flow speed, V, density, n, temperature, T, the B_x, B_y, B_z components of the field (GSM coordinates) and the total field, B. The magnetic cloud interval is shown between vertical guidelines. The passage of this cloud caused a major magnetic storm, see Figure 7 below. (After *Farrugia et al.*, 1992).

by the shock which fast solar ejecta drive, or IMF draping around the ejecta. The correspondence is, however, not one-to-one, and the authors found that only about one in six shock/solar ejecta events gave rise to large magnetic storms. (On this point see also the paper by *Hewish and Duffet-Smith* [1987], and Comment by *Tsurutani et al.* [1988b]) The least geoeffective solar ejecta were those which did not drive shocks or which lacked a prolonged southward B_z.

3. THE STATIC FLUX ROPE MODEL OF MAGNETIC CLOUDS

According to one model which has gained wide acceptance, magnetic clouds are magnetic flux ropes (see, e.g., the reviews by *Burlaga* [1991, 1995], and *Gosling* [1990], and references therein). In 1983 Goldstein considered magnetic clouds as equilibrium force-free configurations of cylindrical symmetry, i.e., as solutions of curl $\mathbf{B} = \alpha(r)\mathbf{B}$, where $\mathbf{B}$ is a two-component magnetic field, $\mathbf{B} = (0, B_\phi(r), B_z(r))$, and $\alpha(r)$ is an arbitrary function of the radial distance from the symmetry axis of the tube, r. *Burlaga* [1988] showed that magnetic cloud signatures at 1 AU can be explained well by assuming these ejecta to be approximately cylindrically-symmetric, force-free fields of constant α for which the *Lundquist* [1950] solution is appropriate. *Lepping et al.* [1990] carried out a systematic least-squares fit of this model to many magnetic cloud field signatures and found that, whereas many magnetic cloud flux ropes have only a small inclination to the ecliptic, some are highly inclined to it. The orientation of the magnetic cloud axes is relevant to discussions of the cloud-storm association (see section 14).

4. ON THE GLOBAL TOPOLOGY OF MAGNETIC CLOUDS

The global magnetic field line topology of magnetic clouds is a topic of lively debate (and, in the context of the earlier, so-called "plasma clouds", one of long standing, see *Burlaga*, [1991, 1995], and references therein.). We shall just discuss briefly energetic particle evidence in favour of a bent magnetic flux rope rooted at both ends to the Sun. That magnetic cloud flux ropes are bent in the large (to be carefully distinguished from their connection or otherwise to the Sun) was inferred by *Burlaga et al.* [1990] from a study of data from 5 well-separated spacecraft.

Observations of the intensity and flow properties of solar energetic particles made in association with magnetic clouds provide useful information on the global topology of these ejecta. The basic idea was put forward by *Kahler and Reames* [1991]. They argued that if energetic particles from a solar event arrive promptly at a spacecraft engulfed in a magnetic cloud which existed before the solar event, then the magnetic field lines of the magnetic cloud cannot be detached from the Sun. If the magnetic field lines formed closed loops detached from the Sun, such energetic particles would have to diffuse rapidly across field lines, a theoretical time scale which is long in comparison with their transit time past the cloud. Several such prompt onsets have been studied [*Kahler and Reames*, 1991; *Farrugia et al.*, 1993a,b; *Richardson and Cane*, 1995], and these observations have all been interpreted successfully in terms of a bent flux rope model rooted at its ends to the Sun's surface. Typically, a prompt onset is signalled by a sharp rise in the intensities of particles of energies in the several MeV range, accompanied by a strong unidirectional flow which can be either from the east or from the west. This unidirectional flow changes to bidirectional streaming once the particles mirror inside the cloud closer to the Sun [*Richardson and Reames*, 1993].

A case study has been made where a prompt onset was observed by two spacecraft just before a magnetic cloud engulfed one of them [*Farrugia et al.*, 1993b]. Once inside the ejecta, intensity levels of < 20 MeV protons dropped with respect to values on the second spacecraft, and took hours to recover. Intensity levels of ~ 1 MeV electrons were, however, not affected. It was found that the observed delay of arrival of the protons was consistent with that expected if the protons were injected near the Sun into one leg of a Lundquist flux rope and travelled along its helical field lines to the spacecraft, a trajectory longer than that along open interplanetary lines. (For example, the extra delay for 5 MeV protons travelling along field lines at $0.9\,R$ from the rope's axis was estimated to be of order 3 hours.) On the other hand, ~ 1 MeV electrons are highly relativistic and would populate the flux rope in a matter of minutes. We shall return to this multi-spacecraft case study below (section 10).

5. TEMPORAL EVOLUTION OF MAGNETIC CLOUDS

Magnetic clouds expand as they propagate antisunward [*Klein and Burlaga*, 1982]. One must thus go beyond static models. One defining characteristic of magnetic clouds is based on the plasma (low T_p). One must thus go beyond force-free fields, where the magnetic and thermodynamic structures are decoupled.

In one approach to the modelling of magnetic cloud evolution, *Chen and Garren* [1993] build on earlier work by *Chen* [1989, 1990] and describe the dynamics of solar flux loops which erupt when toroidal current is injected. A polytropic equation of state is assumed with a quasi-isothermal polytropic index, $\gamma = 1.1$. The model invokes high thermal conductivity parallel to the field to account for observed temperatures at 1 AU, and thus the loop is required to remain connected to the Sun.

Another approach to the study of the evolution of magnetic clouds treats these configurations as self-similarly expanding flux ropes of cylindrical symmetry. The force-free concept still plays an important role in this theory. In this work, the electrons are essential for the dynamics. Model predictions on the temporal evolution of magnetic clouds have been successfully compared with various data. Space limitations do not allow us to describe this model and the data comparisons in detail, for which the reader is referred to *Osherovich et al.*, [1993a, b, c, 1995], *Farrugia et al.*, [1992, 1993a, c, 1995a] and *Farrugia and Burlaga*, [1994]. We shall just look at the role of the electrons.

While studying the effect of gas pressure on the evolution of magnetic clouds, *Osherovich et al.* [1995] found that the major contributor to the gas pressure gradient is the electrons. *Osherovich et al.* [1993c] and *Farrugia and Burlaga* [1994] found further that in magnetic clouds, the electron temperature T_e anticorrelates with the density n and thus T_e has as much structure as n. In general, T_e is enhanced in magnetic clouds, and *Osherovich et al.* [1993c] predicted that continued expansion should further increase the temperature ratio T_e/T_p. *Osherovich et al.* [1993c] and *Farrugia and Burlaga* [1994] argued further that the energetics of the electrons may be represented by a polytrope of index $\gamma_e < 1$.

The $T_e - n$ anticorrelation has also been found at large heliospheric distances. *Fainberg et al.* [1995] show data for a solar ejecta seen by Ulysses at 4.64 AU and S 32.5° on June 9 - 13, 1993 and which has been identified as a magnetic cloud [*Gosling et al.*, 1994]. A detailed $T_e - n$ anticorrelation is apparent. The T_e/T_p temperature ratio reaches values of 10 - 20, i.e., higher than typical values at 1 AU ($\sim$ 6-10). In view of these observational results, it is clearly essential that simulations of magnetic clouds reproduce the $T_e - n$ anticorrelation.

An important finding is that the electron distribution inside the cloud is non-Maxwellian. While the core electrons dominate outside the cloud, the halo electrons contribute equally to the pressure inside [*Fainberg et al.*, 1996]. Deviations from Maxwellian distributions in low density plasma have been investigated by *Scudder* [1992] in connection with the so-called "velocity filtration" mechanism for heating the solar corona, and it was shown there that such deviations can lead to $\gamma < 1$.

One result of this approach to magnetic cloud evolution is that asymmetries often seen in the B-signature (i.e., peak field strengths shifted towards the leading edge and a $B(t)$ profile generally having stronger fields towards the leading edge, see bottom panel of Figure 1) may be largely the result of radial expansion, i.e., a kinematic effect [*Farrugia et al.*, 1992, 1995; *Osherovich et al.*, 1993a]. There is little doubt that asymmetries do also arise out of interaction with the interplanetary medium (e.g., a cloud running into a slower stream or being compressed from the rear by a faster stream; e.g., *Gosling* [1990], *Burlaga* [1995]). And, indeed, more effort should be directed to studying the interaction of magnetic clouds with other interplanetary flows. However, the following seems to us to be a fair point to make: if the asymmetries in the field strength profile are due mainly to interaction with the medium, then with the passage of time these asymmetries should increase. If, on the other hand, the effect is mainly geometric, then they should diminish. The observations at large radial distances made by Ulysses might be useful in this context.

Magnetic cloud propagation was simulated also by *Vandas et al.*, [1995, 1996a,b] using the 2.5-dimensional MHD code of *Detman et al.* [1991]. The approach is different from those outlined above. For example, the polytropic index is 5/3. *Vandas et al.* [1996b] obtain good agreement with the analytical results of *Osherovich et al.* [1993a] for the decrease of B_{max} with heliospheric distance without requiring a value of $\gamma < 1$.

There are now a number of models on the thermodynamics and temporal evolution of magnetic clouds. We have already mentioned three, and to these must be added the recent work of *Kumar and Rust* [1996] which appeared as we were going to press. It is important to resolve the different points of view espoused in these various approaches. A central issue is whether the energetics of the magnetic cloud plasma is adequately represented by a polytropic relation for the electrons with index ~ 0.5, or whether it is better represented by a polytrope for the protons. If the latter is the case, should the polytropic index be $\sim$1 or 5/3 ? The polytropic index can be obtained from observations once the symmetries in the models are also considered. In an attempt to resolve the matter, one should tread the old empirical path of confronting model predictions with observations. For example, the $T_e - n$ anticorrelation noted earlier is first and foremost an observation which all detailed modeling should account for. Signatures of evolution should be present in any given cloud observation since the duration of such an observation constitutes a substantial fraction of the cloud's "lifetime". The availability of in situ measurements on magnetic clouds at various radial distances from the Sun (i.e., various "ages") offers an even wider scope for comparison of models with data.

6. THE BOUNDARIES OF MAGNETIC CLOUDS

Determination of magnetic cloud boundaries is as yet uncertain. Solar ejecta may be characterized by a number of signatures [*Zwickl et al.*, 1983; see also review by *Gosling*, 1990], although not all of these may be present in a given observation. Bidirectional solar wind electron heat fluxes are often regarded as a good signature of the extent of solar ejecta [*Gosling et al.*, 1987]. However, the frequent disagreement of the boundaries thus determined with those obtained from bidirectional flows of low-energy (< 1 MeV) protons over the same periods [*Marsden et al.*, 1987] may suggest that at times this technique does not identify the entire ejecta material. (See also discussion in *Tsurutani and Gonzalez* [1996].) It has long been suggested that abnormally low T_p is a good indication of solar ejecta [*Gosling et al.*, 1973], and recently *Richardson et al.* [1995] proposed that T_p depressions are a more comprehensive and reliable indication of the presence of ejecta material than bidirectional streaming of electron heat fluxes. Specializing to magnetic clouds, it would appear from section 5 that the quantity T_e/T_p may be used to good purpose here. Examples have been given where a value $T_e/T_p = 1$ coincided with the boundary inferred from other signatures [*Osherovich et al.*, 1993c; *Farrugia et al.*, 1994], suggesting that the boundaries of magnetic clouds might also be thermal boundaries. See Figures 2a, b. The aforementioned departure of the electron distribution from Maxwellian, too, may be a useful boundary signature, and one that appears to be consistent with an identification based on the T_e/T_p temperature ratio [*Fainberg et al.*, 1995].

Besides the location, the nature of the front boundary is also important in as much as this serves as one of the boundary conditions for the solution of the set of MHD equations describing flow around magnetic clouds [*Erkaev et al.*, 1995]. Examples show that this boundary might have an elaborate field and plasma structure. (See Figure 3.) Techniques developed in the study of the structures of low- and high-shear magnetopauses could be used to study magnetic cloud boundaries further. Summing up this section, it is apparent that the determination of cloud boundaries and their nature is an urgent research problem.

7. MAGNETIC FIELD LINE DRAPING AROUND, AND MAGNETIC BARRIERS AHEAD OF, MAGNETIC CLOUDS

Gosling and McComas [1987] were the first to study IMF draping about solar ejecta and to link it with geomagnetic storm activity. And there are examples of storms, major and even "great", attributable to this mechanism, either acting alone or together with field compression at the shock and/or intervals of negative B_z in the ejecta proper [*Tsurutani et al.*, 1988a, 1992; *Gosling et al.*, 1990, 1991]. Draping arises when magnetic field lines are stretched as two infinitely conducting magnetoplasmas travel past each other. It is expected that the draping effect should increase with increasing relative speed [*Gosling and McComas*, 1987]. Draping of the IMF about a solar ejecta occurs irrespective of whether the ejecta drives a shock or not.

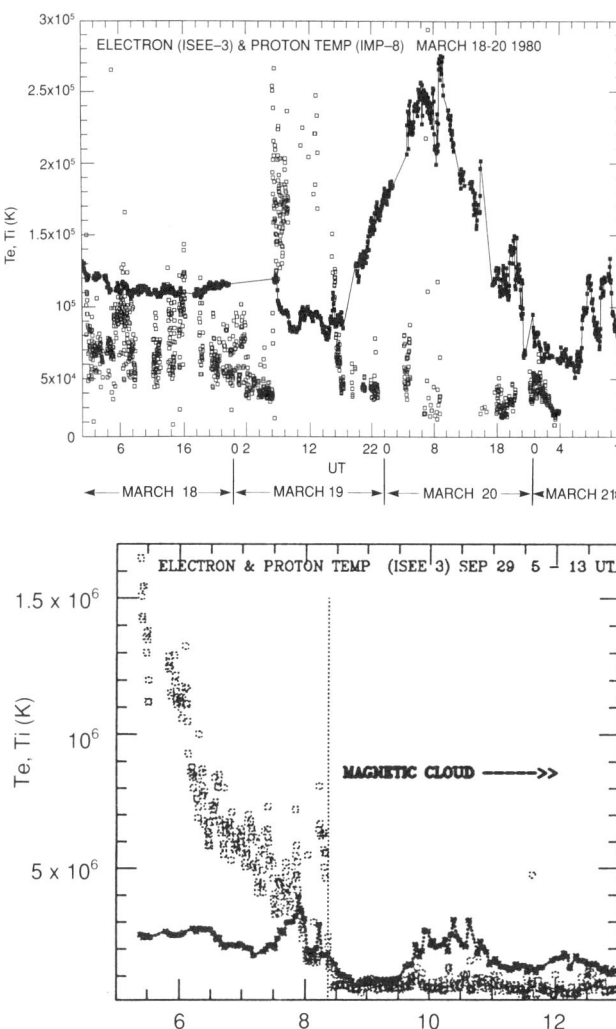

Figure 2a,b. At the front boundary of magnetic clouds, there is often a cross over from $T_e < T_p$ in the sheath to $T_e > T_p$ in the cloud. Two examples from the work of *Osherovich et al.*, 1993c (Figure 2a) and that *Farrugia et al.*, 1994 (Figure 2b) are shown. In both, $T_p > T_e$ in the sheath and $T_e > T_p$ in the cloud. In these examples, the front boundary determined from the condition $T_p = T_e$ agrees well with that determined independently.

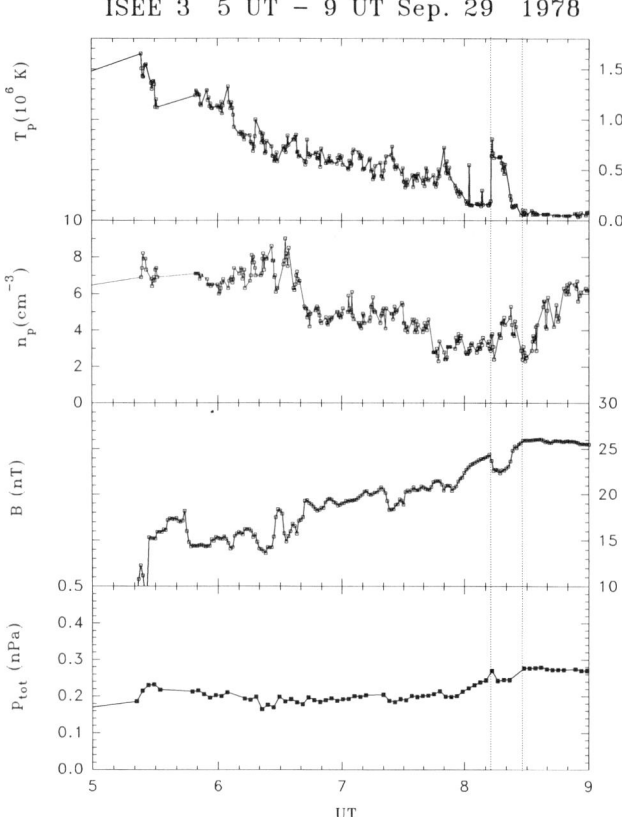

Figure 3. A possible magnetic barrier ahead of the September 29 - 30, 1978 magnetic cloud, the same as in Figure 2b. Shown are variations of parameters T_p, n, B, and P_p+P_b in the sheath region ahead of the cloud. The structure of the front boundary of the magnetic cloud is shown between vertical dotted lines. (After *Farrugia et al.*, 1994).

Gosling and McComas [1987] describe qualitatively how the draping process can give rise on occasion to large southward B_z fields in the sheath region behind the shock.

In practice, comparisons of predictions of field line draping around ejecta with observations is frustrated by the generally turbulent and fluctuating sheath field, the lack of precise knowledge of the spacecraft path relative to the ejecta, the shape of the ejecta, etc. [*McComas et al.*, 1989]. To these may be added the difficulty of comparing quantitative, steady-state MHD flow models around ejecta [e.g., *Erkaev et al.*, 1995] with data since the ambient solar wind can hardly be expected to remain steady during the long spacecraft traversal of the sheath ($\sim$ 12 h on average).

In discussing draping around interplanetary ejecta, the analogy with IMF draping around the magneto-sphere is often drawn. (One must handle this analogy with care, however, since draping depends on the dimensionality of the problem: e.g., draping about a straight cylinder is a two-dimensional problem while draping around the magnetosphere is inherently three-dimensional.) Theoretical studies of MHD flow past the magnetosphere predict the occasional presence of a "magnetic barrier" (aliter: "depletion layer") adjacent to the sunward side of the magnetopause [*Midgley and Davis*, 1963; *Lees*, 1964; *Zwan and Wolf*, 1976; *Erkaev*, 1988]. Spacecraft passes through the low-latitude, dayside magnetosheath next to the magnetopause routinely reveal the presence of a depletion layer when the local magnetic shear across the magnetopause, angle ψ, is small ($\leq 30°$) [*Phan et al.*, 1994, and references therein], though a depletion layer should also be present irrespective of magnetic shear when the solar wind Alfven Mach number, M_A, is low [*Farrugia et al.*, 1995b, 1996]. In this layer, the IMF exerts a strong influence on the flow, resulting, e.g., in an enhancement of the magnetic field strength, B, and a simultaneous decrease in density, n, as the low-latitude magnetopause is approached. Under normal solar wind conditions (Upstream Alfven Mach number, M_A, $\sim$ 8 - 12), the magnetic barrier is observationally $\sim$ 0.3 Re thick in the Earth-Sun direction [*Phan et al.*, 1994], but it widens as M_A decreases [*Farrugia et al.*, 1995b, 1996]. Reconnection weakens depletion.

Flow in the sheath ahead of magnetic clouds is generally under lower upstream M_A conditions (4 - 5) than in the terrestrial magnetosheath. The quantitative MHD model of flow about magnetic clouds driving shocks [*Erkaev et al.*, 1995] predicts that a magnetic barrier, when present, should amount to a substantial fraction of the cloud radius. The calculation in various limiting cases shows quantitatively how the IMF is enhanced in the magnetic barrier region.

Are there magnetic barriers ahead of magnetic clouds? Are they thick ? Figure 3 shows data in the sheath region in front of the magnetic cloud on September 29, 1978: T_p, n, $P_b + P_g$ are shown plotted for the interval 5 - 9 UT (29) [*Farrugia et al.*, 1994]. In the sheath, the electron pressure is unimportant in this case [*Farrugia et al.*, 1994]. The front boundary is crossed at $\sim$ 0828 UT. Sheath passage takes $\sim$ 5 h and there are data available for the $\sim$ 3 hours next to the cloud. As the spacecraft approaches the cloud, n and T_p decrease, B increases and the plasma beta falls quickly to values below unity. These properties define a magnetic barrier. (See *Tsurutani and Gonzalez* [1996] for an alternative interpretation.) Assuming steady conditions, the spatial extent of this layer is of the order of one-half the total sheath thickness. Similar features are seen in the sheath region ahead of the magnetic cloud seen by Helios on June 20, 1980 (Figure 1 in *Burlaga* [1991]). Quantity M_A is ~ 4.5 here. The magnetic barrier is very evident and occupies most of the sheath extent, where the plasma beta is ≤ 1.

Despite these clear cases, it is puzzling that magnetic cloud magnetic barriers are not discovered in the data as a matter of course, as they are at Earth. This issue requires further study. Should a systematic search - an exercise which presupposes good knowledge of the location of the ejecta boundaries - reveal that magnetic barriers in front of magnetic clouds are thin or, worse, generally absent, then this would be an indication that draping around these ejecta is fundamentally different from that around planetary magnetospheres. In addition, an MHD theory of flow around magnetic clouds not driving shocks and modelled as cylindrical flux ropes needs to be elaborated since the one in existence is only in a kinematic approximation [*Farrugia et al.*, 1987; *Walthour et al.*, 1993]

8. EARTH ENCOUNTERS MAGNETIC CLOUDS

The study of Earth-magnetic cloud encounters leads to further insight into the larger problem of solar wind-magnetospheric interactions for the following reasons. During their long (~ 1 day) passage at Earth, magnetic cloud parameters are steady over typical response times of the ionosphere - magnetosphere system [*Freeman et al.*, 1990, and references therein], but change greatly over the long duration of cloud passage. Extreme values of interplanetary parameters are reached, which are hardly otherwise ever sampled. Depending on the orientation of the flux rope, the forcing of the magnetosphere by the cloud can range from very strong to very weak during a single encounter, allowing the active as well as the quiescent magnetosphere to be studied. The empirical criteria for major storms (section 2) are routinely met in magnetic clouds. Magnetospheric dynamics can be related to geometrical properties of the cloud: its inclination to the ecliptic, the absolute and relative strengths of the axial and azimuthal components (B_{ax}, B_{az}), the pitch of the helical field structure ($= r\, B_{ax}(r)/B_{az}(r)$), etc. Global simulations of magnetospheric dynamics with magnetic cloud $B(t)$ and $V(t)$ variations as input are particularly instructive [*Chen et al.*, 1995, and references therein] since they offer the possibility of modelling the causal chain between specific solar disturbances and their attendant magnetospheric manifestations.

9. EFFECTS ON THE MAGNETOSHEATH

During cloud passage, the normally turbulent magnetosheath is smooth and steady with little fluctuation [*Lepping et al.*, 1991; *Farrugia and Burlaga*, 1994]. *Lepping et al.* [1991] studied a very large magnetic cloud observed simultaneously on opposite sides of the bow shock and concluded that the cloud retained its structure almost unchanged by its interaction with the bow shock.

Unusual properties of the magnetosheath during cloud passage have been related to the low Alfven Mach number, M_A, of magnetic clouds, on the one hand, and to the electron behavior in magnetic clouds, on the other (see above). To solve the MHD equations describing the flow of the shocked solar wind past the magnetosphere, boundary conditions at the magnetopause and at the bow shock have to be stipulated. For the former, the key parameter is the local magnetic shear, ψ. Many studies have shown that transfer processes at the magnetopause depend strongly on ψ. At the bow shock, the key parameter is the magnetosonic Mach number. The sonic Mach number in magnetic clouds is typical of usual conditions, while the Alfven Mach number, M_A, is lower (~ 3 vs ~ 8-12). The MHD equations then imply that the flow in the terrestrial magnetosheath is controlled by the magnetic cloud's magnetic field, leading, e.g., to wide magnetic barriers. Since during cloud passage the parameter ψ varies over a wide range, one can also examine the effect of this parameter, modelling the magnetopause as a tangential (rotational) discontinuity if ψ is low (high).

Such studies have been made. *Farrugia et al.*, [1995b, 1996] subdivided the ~ 30-hour-long passage of the January, 1988 cloud (Figure 1) into 2-h intervals, where conditions are steady, and worked out the structure of the subsolar magnetosheath in each. Parameter $M_A \sim 3$ except for the last interval (~ 7). For low ψ and $M_A \sim 3$, they found a wide magnetosheath and a magnetic barrier region generally extending up to the bow shock. The profiles of plasma and field parameters show considerable pile-up near the magnetopause, and there is hardly any jump of B across the low-shear magnetopause. For low ψ and $M_A \sim 7$, the magnetic barrier is much thinner. For high ψ and $M_A \sim 3$, the thickness of the magnetosheath decreased and the B-enhancement and n-depletion were much more pronounced. However, the magnetic barrier was still wide. Finally, for high ψ and $M_A \sim 7$, the magnetic barrier disappeared. Since magnetosheath flow is of the stagnation-line type under low shear [*Sonnerup*, 1974; *Phan et al.*, 1994; *Farrugia*

98 MAGNETIC CLOUDS AND STORMS

et al., 1995b], one can study this type of flow pattern systematically since during cloud passage it should persist uninterruptedly for many hours. This represents a considerable advance in magnetosheath flow studies when one recalls how elusive observational identification, let alone detailed study, of stagnation line flow proved to be in the past.

The behavior of electrons in magnetic clouds (sections 5 and 6) leads to other unusual magnetosheath properties. The condition $T_e \gg T_p$ is favourable for the generation of ion acoustic waves in clouds [*Osherovich et al.*, 1993c; *Farrugia and Burlaga*, 1994], which have also been observed [*Burlaga et al.*, 1980; *D. Gurnett*, private communications, 1992, 1993; *Fainberg et al.*, 1995]. In one case example [September 29, 1978], correspondingly large T_e/T_p ratios in the Earth's magnetosheath (~ 6) were noted [*Farrugia and Burlaga*, 1994], and these authors suggested that every cloud passage at Earth should be accompanied by ion acoustic emissions in the magnetosheath. This prediction should be checked.

10. MAGNETIC CLOUDS AND THE ENTRY OF SOLAR ENERGETIC PARTICLES INTO THE EARTH'S MAGNETOSPHERE

According to the connected, bent flux rope model, solar energetic particles (SEP's) can gain access to the terrestrial magnetosphere guided by magnetic cloud field lines. Since the "legs" of the cloud are rooted to the Sun, SEP's can be injected into the cloud. Near earth, they can enter into the magnetosphere through interconnection of cloud and magnetospheric field lines if the cloud field points south. Using dual-spacecraft observations, made simultaneously inside the magnetic cloud (by ISEE 3) and inside the northern tail lobe (by IMP 8), it was possible to obtain direct confirmation of this scenario [*Farrugia et al.*, 1993b].

Figure 4, top panels, shows the magnetic cloud field, with the continuous and dashed vertical lines representing the times when the front boundary swept over ISEE 3 and IMP 8, respectively. ISEE 3 was near the L1 libration point. Due to a variable cloud dynamic pressure, IMP 8 was alternately in the magnetosheath and the northern tail lobe. During the interval of interest here, it was crossing the lobe in a dusk-dawn direction. The bottom panels show flow directions for few-MeV ions ($\theta = 0°$, 180°, 90° represent flow from and towards the Sun, and flow from the west, respectively). Several particle onsets were seen during cloud passage. We focus on the interval from 12 UT (16) to 18 UT (16) when the

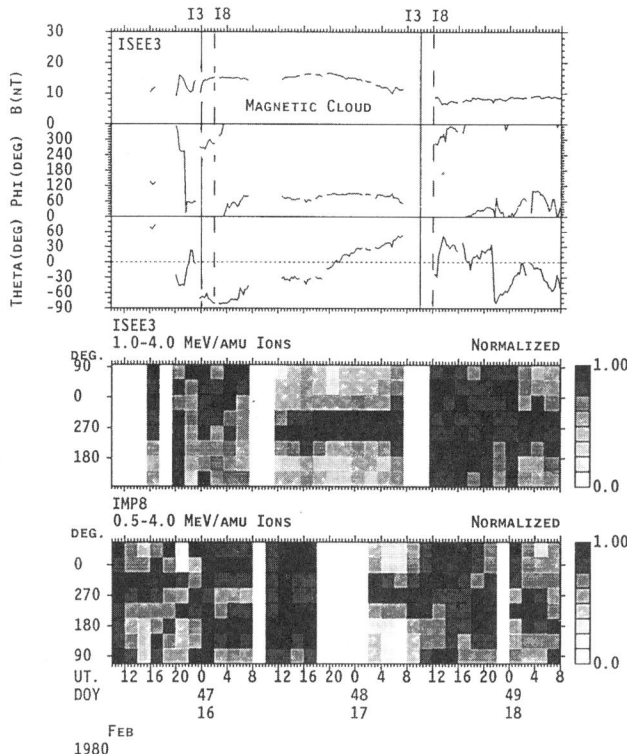

Figure 4. Flow information for few MeV ions on ISEE 3 (I3; bottom panel) and IMP 8 (I8) during the passage of a magnetic cloud (top three panels) at the two spacecraft on February 16 - 17, 1980. For further details, see text.

cloud field pointed south (average $B_z \sim -8$ nT). ISEE 3, inside the cloud, observed strong streaming from the west. Simultaneously, IMP 8, inside the lobe, observed tail-aligned bidirectional flows.

11. EFFECTS ON THE IONOSPHERE

We now discuss ionospheric flow patterns in both hemispheres prior to and during the major geomagnetic storm caused by the January 1988 magnetic cloud (Figure 1). The independent studies of *Freeman et al.* [1993] and *Knipp et al.* [1993] employed a large number of data sets and different techniques which, coupled with the stability of cloud parameters, allowed inferences to be drawn of unprecedented range and accuracy. The magnetic cloud was oriented with its axis in the ecliptic and pointing roughly east-west [*Farrugia et al.*, 1992]. Cloud passage was thus divided into an 11-h period of GSM $B_z > 0$ followed by an 19-h period of $B_z < 0$, where B_z is the azimuthal component of the flux rope field (Figure 1). The axial field ($\approx$ GSM B_y in this orientation) provided a further large east-west component (minimum ~ -31 nT). *Freeman et al.* [1993] discussed convection for the whole duration of cloud passage in terms of cloud parameters. *Knipp et al.* [1993] concentrated on the earlier, $B_z > 0$ interval and interpreted results in terms of a general merging model. The scope and richness of the observations defy easy condensation here, so we shall just summarize some of the main conclusions.

Freeman et al. [1993] contrasted peak flows in the two hemispheres under strong positive and negative cloud B_z. For cloud $B_z \gg 0$, the southern hemisphere convection was characterized by a two-cell convection pattern of the reverse type [*Maezawa*, 1976], confined to latitudes $\geq 75°$. The strength of the flows, of order $\sim 1 km s^{-1}$, was comparable to that seen later when cloud $B_z < 0$. A maximum cross-polar cap potential of $\sim$80 kV was reached. For cloud $B_z \gg 0$, and under near winter solstice conditions, the northern hemisphere flows were weak and irregular and varied on short spatial scales. In both hemispheres and for cloud $B_z \gg 0$, a weak two-cell standard component was also apparent, which Freeman et al. attributed to viscous interaction. When cloud $B_z \ll 0$, interhemisphere flow differences disappeared and in both a two-cell standard pattern was present which extended to as low as 50° in latitude. Peak cross-cap potentials $\sim$180 kV were measured. (The results for positive cloud B_z were also arrived at by *Knipp et al.* [1993; see also references therein]. These workers were able to study all ionospheric patterns ever observed under northward IMF conditions.)

Using radar data, *Freeman et al.* [1993] monitored the transition from reverse to standard convection. They found that this was brought about by the strong axial cloud field component (B_y), which provided the magnetic shear at the magnetopause, and it occurred some hours before cloud B_z turned negative. A major transition in flow configurations occurring a few hours before the cloud B_z changed polarity was also monitored by *Knipp et al.* [1993]. These latter workers also made the important suggestion that for interplanetary $B_z > 0$, $|By/Bz| \approx 1$ gives a reliable indication when standard replaces reverse ionospheric circulation, what we would relate to the relative strengths of the axial and azimuthal components of the magnetic cloud field.

Freeman et al. [1993] also studied the variation of the polar cap area, suitably defined, with time. They distinguished between long- (hours) and short- ($\leq$ 1h) term changes. They attributed the former to temporal gradients of the cloud B_z component and noted that when B_z changed rapidly (i.e., when the cloud's axis swept over Earth and B_z went negative), the polar cap area increased dramatically by a factor of $\sim$ 2.4, but when $B_z < 0$ and increasing slowly for $\sim$ 15 h at the rear of the cloud (Figure 4), there was a long-term variation of only $\sim 2°$ in angular radius. (A complementary observation is that of *Lepping et al.* [1991] who, for another cloud, found that auroral activity increased by a factor of 10 within 1 h of the cloud field turning negative.) On short time scales, the polar cap area changed by $\sim 3°$ in latitude, which Freeman et al. interpreted in terms of unbalanced day and night reconnection rates, specifically, an increase in area during the observed growth phase of substorms, when flux is opened on the dayside and accumulated before release on the nightside.

The large axial cloud component (B_y) induced various asymmetries in the flow. Freeman et al. found that a switch in B_y polarity from positive to negative under $B_z \gg 0$ conditions resulted in the dawnside antisunward cell of the reverse two-cell pattern going from being suppressed by the duskside cell to dominating over it. When cloud $B_z < 0$, and $B_y \ll 0$, a one-cell convection component with an anticlockwise sense of rotation (when viewed from above the magnetic pole) was present in the southern hemisphere with, correspondingly, a concentration of antisunward flow to the dawnside. This east-west asymmetry results from a cloud B_y-related asymmetry in the Maxwell stresses on newly-reconnected field lines. During cloud $B_z < 0$ (and $B_y < 0$), the convection boundary in the northern hemisphere was consistently up to 10° lower in latitude at dusk than at dawn, an asymmetry which persisted even when $B_y \approx 0$. *Freeman et al* [1993] propose this to be a novel asymmetry perhaps related to the abnormally large magnetic fields in the cloud. This observation deserves further study.

A final point from Freeman et al.'s work is these authors' study of the maximum potential difference across convection patterns (V_{max}), which characterizes the strength of convection, as a function of cloud B_z. The latter quantity varied by $\sim$ 20 nT on both sides of zero. For cloud $B_z > 0$, values of V_{max} across the reverse pattern in the southern hemisphere are of order 80 kV, high values also obtained in the *Knipp et al.* [1993] study. For cloud $B_z < 0$, V_{max} changes quasi-linearly with cloud B_z with no evidence of saturation. These points are illustrated in Figure 5.

12. EFFECTS ON THE NIGHTSIDE MAGNETOSPHERE AT AND BEYOND GEOSTATIONARY ORBIT

Farrugia et al. [1993d] used spacecraft observations on field and particles, together with ground magne-

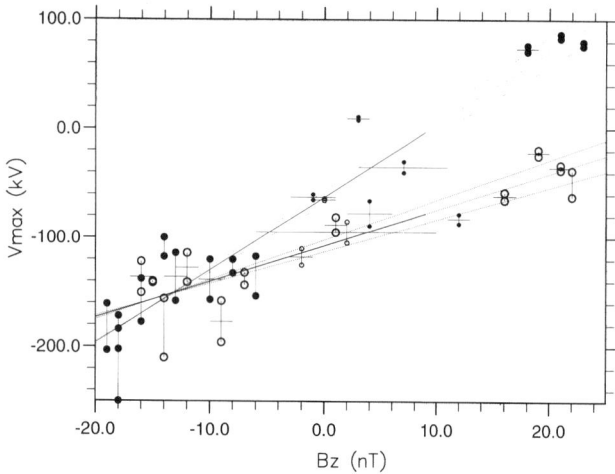

Figure 5. The maximum electric potential drop (V_{max}) across convection patterns in the southern (closed symbols) and northern (open symbols) hemispheres as a function of the B_z component of the January 1988 magnetic cloud shown in Figure 1. The error bars on V_{max} refer to two empirical ways of obtaining this parameter. Small circles refer to data outside the cloud. Separate regression lines are drawn for the two hemispheres. (After *Freeman et al.*, 1993).

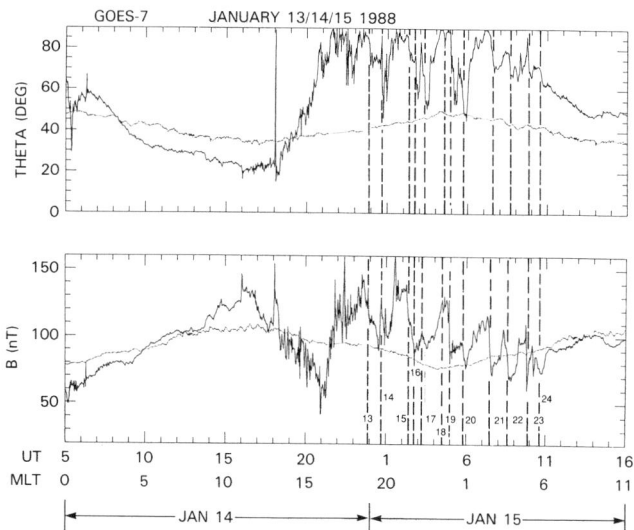

Figure 6. Goes 7 observations for the period 5 UT, January 14, to 16 UT January, 15, 1988, showing the inclination of the field to the dipole (angle theta, top panel, where $\theta = 0°$ indicates an approximately dipolar field), and the total field at geostationary orbit. The lighter trace shows the geomagnetic field measured on January 13, a geomagnetically quiet day. (After *Farrugia et al.*, 1993d.)

tograms, to examine the geomagnetic activity on the nightside generated by the January 1988 magnetic cloud (Figure 1), relating the energy input into the magnetosphere to the magnetospheric output. They addressed 3 issues: (a) the state of the magnetosphere during the early, $B_z \gg 0$ interval; (b) substorm activity accompanying the storm during the second part of cloud passage; and (c) the transition from (a) to (b). They also made a juncture with theories on deterministic nonlinear dynamics of the magnetosphere concerning the loading-unloading capability of the magnetosphere [*Klimas et al.*, 1994, and references therein].

We may illustrate some of these points by Figure 6, which shows the geomagnetic field at geostationary heights observed by the Goes 7 spacecraft. The thin line in the top panels refers to January 13, a geomagnetic quiet day included for comparison.

(a) Prior to ~ 13 UT, when reverse convection was replaced by standard convection (above), the diurnal variation of the geomagnetic field was enhanced: with respect to January 13, the field is more compressed on the dayside and more distended on the nightside (Figure 7). This is because the cloud dynamic pressure was higher. In addition, the field is punctuated by impulsive changes due to arrival of dynamic pressure fronts.

These cause sudden impulses (SI's) on the ground, and *Farrugia et al.* [1993d] were the first to examine the geomagnetic effect of discontinuities within magnetic clouds. In view of the wide range of effects dynamic

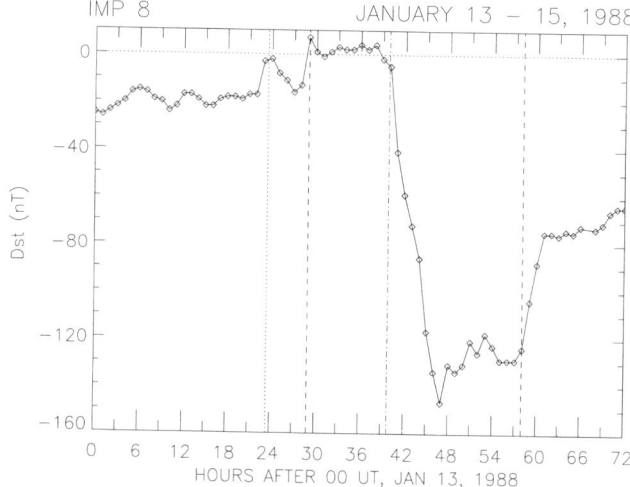

Figure 7. Hourly Dst values for the period January 13 - 15, 1988. The vertical guidelines mark, from left to right: arrival of an interplanetary shock driven by the magnetic cloud, the front boundary of the cloud, the B_z negative transition inside the cloud, and the rear boundary of the cloud, respectively.

pressure variations have on magnetospheric dynamics, it appears that not even during the earlier, $B_z \gg 0$ part of cloud passage did the magnetosphere attain a 'ground state'.

(b) On January 15, both Goes 7 was on the nightside, and therefore ideally placed to observe substorm activity. Substorm expansion phase onsets are characterized by a sudden dipolarization of the field (decreases in angle θ in Figure 6), and by a rise or fall of B, whichever change of the latter makes the field more dipolar. There are many substorm onsets in this period (shown by vertical guidelines). Indeed, the magnetic field at Goes 7 alternates between a tail-like field typical of a substorm growth phase and a more dipolar configuration typical, in turn, of a substorm expansion phase, with little evidence of a recovery phase. This behavior is unlike what typically happens in isolated substorms and is probably due to the strong forcing of the magnetosphere by the magnetic cloud. The authors identified in all 23 substorms during cloud passage with an average inter-substorm time interval of $\sim$ 50 min. This is much shorter than the corresponding average obtained from statistical studies of isolated substorms [*Borowsky et al.*, 1993]. Whether the high substorm recurrence rate during cloud passage is due to the continuity of the forcing and/or its (high) level is a question which should be investigated further.

The quasi-periodic response of the magnetosphere stands in sharp contrast to the cloud input. The cloud input may be measured by the so-called "epsilon" and "VBs" parameters. The epsilon parameter gives the integrated Poynting flux into the magnetosphere [*Perreault and Akasofu*, 1978], and the VBs parameter, defined as equal to $-VB_z$ if $B_z < 0$, and zero otherwise, is a measure of the solar wind rectified electric field (see *Baker et al.* [1985], and references therein). The input rises sharply to a peak as B_z reaches maximum negative values, and decreases slowly to zero as B_z increases steadily to zero for many hours thereafter. (See Figure 6 in *Farrugia et al.* [1993d]). The contrast between input and output confirms that the magnetosphere-ionosphere system has an intrinsic storage-release capacity [e.g. *Bargatze et al.*, 1984; *Baker et al.*, 1984, and references therein] and is not just directly-driven. This conclusion, coupled with the observed high substorm recurrence rate under strong forcing, agrees with recent theoretical ideas [*Klimas et al*, 1994] and simulation work [*Freeman and Farrugia*, 1995].

A large geomagnetic storm accompanied the long substorm sequence. Figure 7 shows hourly values of the Dst index for the period January 13 - January 15, 1988.

The vertical guidelines mark, from left to right, the arrival at IMP 8 of the interplanetary shock driven by the cloud, the front boundary of the cloud, the B_z transition inside the cloud, and the rear cloud boundary, respectively. Slightly positive Dst's are registered for most of the early, cloud $B_z > 0$ interval. This is due to the compression of the magnetosphere by the enhanced dynamic pressure inside the cloud (see above). A systematic decrease starts around 14 UT, January 14, 1988, when cloud B_z is still > 0. As cloud B_z becomes negative, a deep depression in Dst develops, with minimum values of -147 nT being reached a few hours after minimum B_z is attained in the cloud. A strong recovery starts when the rear boundary of the cloud passes Earth. The start of the main phase of this geomagnetic storm coincided approximately with the start of substorm activity (see Figure 2a, *Farrugia et al.*, 1993d).

(c) As noted in section 12, at 13 UT (14) various data sets indicated that standard convection replaced reverse convection. The growth phase of the first substorm in the subsequent long substorm sequence was due to dayside reconnection under the strong axial component of the cloud field (B_y). Thus the start of the main phase of the storm, too, was due to this component.

13. MAGNETIC CLOUDS AND GEOMAGNETIC STORMS: PART II

We finally turn to studies specifically addressing the relation between magnetic clouds and geomagnetic storms. [We note in parenthesis that the largest magnetic clouds do not necessarily have the most intense field strengths, since the field strength depends on expansion, as noted earlier. Thus any association large cloud - large storm would be misleading.] Using the Kp index as storm activity indicator, *Burlaga et al.* [1981] reported one such association. A correlation of magnetic clouds with storms of various intensities (valid at the 99 % confidence level) was established by *Wilson* [1987], who employed a superposed epoch analysis technique to study the period 1972 - 1978, using the set of magnetic clouds listed by *Klein and Burlaga* [1982]. The intensity and onset of storm activity were related to the polarity of the cloud B_z component. The author subdivides clouds into those where northward field is followed by southward field (NS) or vice versa (SN clouds). The strength of the storm, measured by the Dst index, was not found to be significantly different for the two sets of clouds, but the timing of the storm main phase onset depended on the phase of the cloud field, usually beginning when the cloud B_z turned south. Storm recovery started as soon as cloud B_z turned north. *Wilson* [1987] also found that a minority of clouds, despite having negative B_z (which in all cases was, however, > -10 nT) of long duration, were not effective in generating storms.

Zhang and Burlaga [1988] extended this analysis to the 1978 - 1982 period near solar maximum. They found that clouds can cause disturbances with Dst ~ -100 nT. In the 19 clouds sampled, the authors distinguished between SN and NS clouds, and according to whether the clouds drove a shock or not. There were 6 NS clouds and 13 SN clouds, with 3 in the former and 1 in the latter category not being associated with shocks. Again, storm onset was found to correlate with the phase of the cloud: For SN clouds, Dst values decreased near the time of arrival of the cloud, whereas for NS clouds the Dst index did not decrease significantly until later in cloud passage, when the magnetic field turned south. There were some examples of NS clouds where storm main phase onset did not occur until several hours after the southward turning of the cloud B_z.

SN and NS clouds had a minimum Dst of -125 nT and -91 nT, respectively. Zhang and Burlaga argue that this difference is not due to the strength of the B_z component, which was the same for both categories. Rather, SN clouds cause larger storms because of their higher bulk flow speed on average. It is worth investigating why clouds driving shocks are generally more geoeffective than those not driving shocks. This is related to Zhang and Burlaga's observation, but it is not the same. Zhang and Burlaga's results were corroborated in a further study by *Wilson* [1990]. An example of an SN cloud causing a large geomagnetic storm is shown in Figure 8.

The above mentioned studies concentrated on the geomagnetic effectiveness of clouds with a long northward followed by a long southward field excursion, or vice versa. The associated flux tubes generally have their axes in the ecliptic. However, even clouds with a substantial tilt to the ecliptic plane can be very effective in causing geomagnetic activity. An example of a cloud highly inclined to the ecliptic ($\sim 80°$) occurred on August 27-28, 1978 [*Burlaga*, 1988]. Data for the period August 26 - August 29, 1978 are shown in Figure 9, in the same format as Figure 8. The peak Dst activity this cloud elicited is comparable to that of the December 1980 cloud (-220 vs -240 nT). Even so, the two clouds are different in many respects: the August 1978 cloud has a lesser maximum B, a less negative B_z but of longer duration, and a slower average bulk speed. The largest negative gradient in Dst(t), occurs ~ 7 h after the last B_z negative turning, though there is some Dst activity before then.

Thus far, we have considered the geomagnetic effect of isolated clouds, without taking account of the streams they are embedded in. However, the geomagnetic effect is often enhanced if magnetic clouds are in fast flows because the fields can be amplified by interaction with a shock and or fast flow. This emerges from the study of *Burlaga et al.* [1987].

The two largest storms in 1979 (gauged by Ap index), which occurred on April 3 and April 25, were associated with compound streams. In the former, the compound stream consisted of a magnetic cloud that was overtaking other ejecta. In the latter, the compound stream consisted of a magnetic cloud that was interacting with a corotating stream. Even without magnetic clouds, compound streams may cause large storm disturbances: the largest storm in the period 1968 - 1986 (July 13, 1982) was associated with a compound stream.

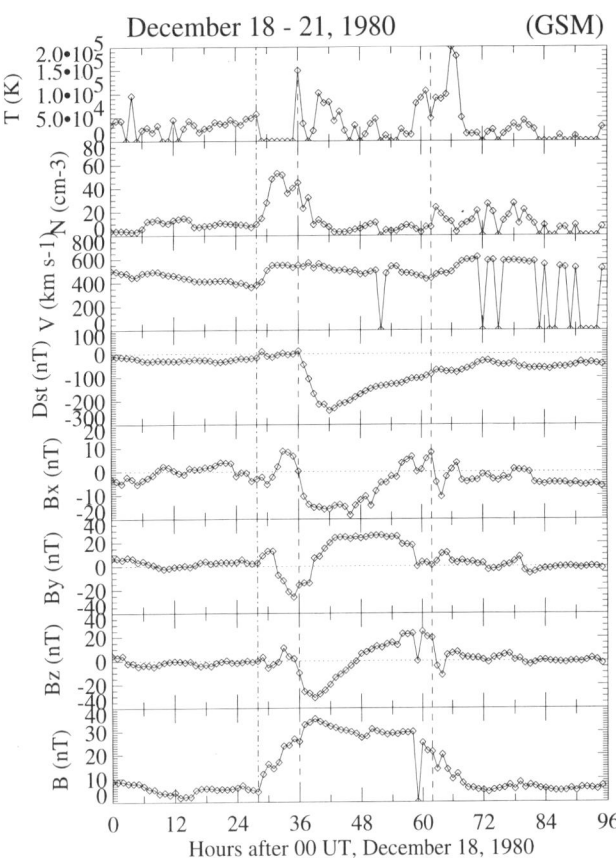

Figure 8. Plasma and field data (1 hour averages) for a magnetic cloud causing a major geomagnetic storm. The panels show, from top to bottom, the proton temperature, density and bulk speed, the Dst index, the GSM B_x, B_y, B_z components, and the total field strength.

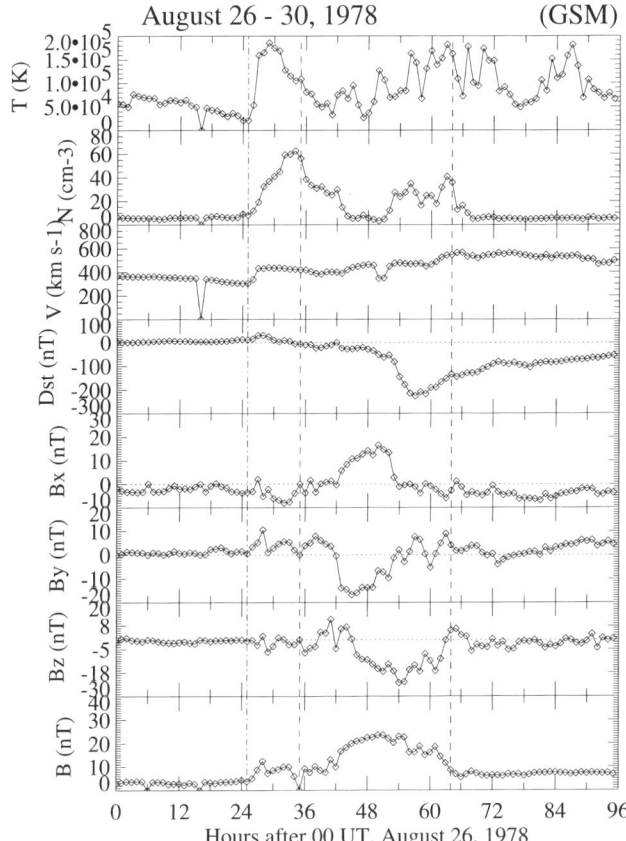

Figure 9. Similar to Figure 8, but for the magnetic cloud in August, 1978. This magnetic cloud, which is highly inclined to the ecliptic, caused a major magnetic storm comparable in magnitude to that shown in Figure 8.

Burlaga et al. [1987] further find that most large storms between 1972 and 1983 were associated with compound streams and/or magnetic clouds.

A potentially very important recent development is *Chen et al's* [1995] three-dimensional simulation of the geomagnetic response produced by a magnetic cloud. The model magnetic cloud is that obtained from the eruptive solar loop model of Chen and Garren (section 5). The cloud input is coupled to an ideal MHD model of the magnetosphere-ionosphere system [*Fedder et al.*, 1995] and references therein]. Chen et al. calculate the variation of the cross cap potential, the Joule heating rate and the electron energy precipitation integrated over the polar cap, spatial distribution and relative intensity of precipitation energy. They also compute synthetic equatorial Dst* and auroral AE* indices. The model shows clearly that geomagnetic activity picked up while cloud $B_z >0$ but $B_y \approx B_z$, as observed during the January 1988 cloud.

Acknowledgments We thank Mervyn Freeman and V. A. Osherovich for helpful comments on the manuscript. We are grateful to two of the referees, W. D. Gonzalez and B. T. Tsurutani for their many useful suggestions. CJF thanks the American Geophysical Union for its financial support. Figures 4 and 5 are courtesy of Ian Richardson and Mervyn Freeman, respectively. This work is supported in part by NASA Grant NAG5-2834.

REFERENCES

Baker, D. N., S.-I. Akasofu, W. Baumjohann, et al., Substorms in the magnetosphere, in *Solar Terrestrial Physics: Present and Future*, NASA Ref. Publ. 1120, edited by D. M. Butler and K. Papadopoulos, p.8-1, NASA, Washington, D. C., 1984.

Bargatze, L. F., D. N. Baker, R. L. McPherron, and E. W. Hones, Magnetospheric impulse response for many levels of geomagnetic activity, *J. Geophys. Res.*, *90*, 6387, 1985.

Borovsky, J. E., R. J. Nemzek, and R. D. Belian, The occurrence rate of magnetospheric substorm onsets: Random and periodic substorms, *J. Geophys. Res.*, *98*, 3807, 1993.

Burlaga, L. F., *Interplanetary Magnetohydrodynamics*, Oxford University Press, New York, 1995.

Burlaga, L. F., Magnetic Clouds, Chapter 5 in *Physics of the Inner Heliosphere, Vol 2*, edited by R. Schwenn and E. Marsch, p.1, Springer-Verlag, Berlin-Heidelberg, 1991.

Burlaga, L. F., Magnetic clouds: Constant alpha force-free configurations, *J. Geophys. Res.*, *93*, 7217, 1988.

Burlaga, L. F., R. Lepping, and J. Jones, Global configuration of a magnetic cloud, in *Physics of Flux Ropes*, edited by C. T. Russell, E. R. Priest, and L. C. Lee, p. 373, AGU Geophysical Monograph 58, American Geophysical Union, Washington, D. C., 1990.

Burlaga, L. F., K. W. Behannon, and L. W. Klein, Compound streams, magnetic clouds and major geomagnetic storms, *J. Geophys. Res.*, *92*, 5725, 1987.

Burlaga, L. F., E. Sittler, F. Mariani, and R. Schwenn, Magnetic loop behind an interplanetary shock: Voyager, Helios and IMP 8 observations, *J. Geophys. Res.*, *86*, 6673, 1981.

Burlaga, L. F., R. Lepping, R. Weber, et al., Interplanetary particles and fields, November 22 to December 6, 1977: Helios, Voyager and IMP observations between 0.6 AU and 1.6 AU, *J. Geophys. Res.*, *85*, 2227, 1980.

Chen, J., Dynamics, catastrophe and magnetic energy release of toroidal solar current loops, in *Physics of Magnetic Flux Ropes, Geophys. Monogr. Ser.*, vol. 58, edited by C. T. Russell, E. R. Priest, and L. C. Lee, p. 269, AGU Washington, D.C., 1990.

Chen, J., Effects of toroidal forces in current loops embedded in a background plasma, *Astrophys. J.*, *338*, 453, 1989.

Chen, J., and D. A. Garren, Interplanetary magnetic clouds: Topology and driving mechanisms, *Geophys. Res. Lett.*, *20*, 2319, 1993.

Chen, J., S. Slinker, J. A. Fedder, and J. G. Lyon, Simulation of geomagnetic storms during the passage of magnetic clouds, *Geophys. Res. Lett.*, *22*, 1749, 1995.

Detman, T. R., M. Dryer, T. Yeh, S. M. Han, S. T. Wu, and D. J. McComas, A time-dependent, three-dimensional MHD numerical study of interplanetary magnetic field draping around plasmoids in the solar wind, *J. Geophys. Res.*, *96*, 9531, 1991.

Dungey, J. W., Interplanetary magnetic field and the auroral zones, *Phys. Rev. Lett.*, *6*, 47, 1961.

Erkaev, N. V., Results of the investigation of MHD flow around the magnetosphere, *Geomagn. Aeron. 28*, 455, 1988.

Erkaev, N. V., C. J. Farrugia, L. F. Burlaga, et al., Ideal MHD flow behind interplanetary shocks driven by magnetic clouds, *J. Geophys. Res.*, *100*, 19,919, 1995

Fainberg, J., V. A. Osherovich, R. G. Stone, et al., Ulysses observations of electron and proton components in a magnetic cloud and related wave activity, in: *Solar Wind Eight*, AIP Conference Proceedings 382, edited by D. Winterhalter, J. Gosling, S. R. Habbal, W. S. Kurth, M. Neugebauer, p. 554, 1995.

Farrugia, C. J., N. V. Erkaev, H. K. Biernat, and L. F. Burlaga, Dependence of magnetosheath properties on solar wind Alfven Mach number and magnetic shear across the magnetopause, in *Proceedings of the International Workshop The Solar Wind-Magnetosphere System 2*, ed. by S. Bauer, H. K. Biernat, F. Ladreiter, and C. J. Farrugia, Austrian Academy of Sciences Press, in press, 1996

Farrugia, C.J., L.F. Burlaga, V.A. Osherovich, and R.P. Lepping, The magnetic flux rope versus the spheromak as models for interplanetary magnetic clouds, *J. Geophys. Res.*, *100*, 12,293, 1995a

Farrugia, C. J., N. V. Erkaev, H. K. Biernat, and L. F. Burlaga, Anomalous magnetosheath properties during Earth passage of an interplanetary magnetic cloud, *J. Geophys. Res.*, *100*, 19,245, 1995b

Farrugia, C. J., and L. F. Burlaga, A fast-moving magnetic cloud and features of its interaction with the dayside magnetosheath, in: *The Solar Wind-Magnetosphere System*, edited by H. K. Biernat, G. A. Bachmaier, S. Bauer, and R. P. Rijnbeek, Austrian Academy of Sciences Press, Vienna, p. 33, 1994.

Farrugia, C.J., R.J. Fitzenreiter, L.F. Burlaga, et al., Observations in the sheath region ahead of magnetic clouds and in the dayside magnetosheath during cloud passage, *Adv. Space Res.*, Vol 14, (7)105, 1994

Farrugia, C. J., L. F. Burlaga, V. Osherovich, I. G. Richardson, et al., A study of an expanding interplanetary magnetic cloud and its interaction with the the Earth's magnetosphere: The interplanetary aspect, *J. Geophys. Res.*, *98*, 7621, 1993a

Farrugia, C.J., I.G. Richardson, L.F. Burlaga, et al., Simultaneous observations of Solar MeV particles in a magnetic cloud and in the Earth's northern tail lobe: Implications for the global field line topology of magnetic clouds and entry of solar particles into tail lobe during cloud passage, *J. Geophys. Res.*, *98*, 15,497, 1993b

Farrugia, C. J., V. A. Osherovich, and L. F. Burlaga, Elements of the magnetohydrodynamics of interplanetary magnetic clouds, in *Current Topics in Astrophysical and Fusion Plasma Research*, edited by M. F. Heyn, W. Kernbichler, and H. K. Biernat, dbv-Verlag, p.254, Technical University of Graz press, 1993c.

Farrugia, C. J., M. P. Freeman, L. F. Burlaga, et al., The Earth's magnetosphere under continued forcing: Substorm activity during the passage of an interplanetary magnetic cloud, *J. Geophys. Res.*, *98*, 7657, 1993d.

Farrugia, C.J., L.F. Burlaga, V.A. Osherovich, and R.P. Lepping, A comparative study of dynamically expanding force-free, constant-alpha magnetic configurations with applications to magnetic clouds, in *Solar Wind Seven*, eds. E. Marsch and R. Schwenn, Pergamon Press, p.611, 1992

Farrugia, C. J., R. C. Elphic, D. J. Southwood, and S. W. H. Cowley, Field and flow perturbations outside the reconnected field line region in flux transfer events: theory, *Planet. Space. Sci.*, , *35*, 227, 1987.

Fedder, J. A., S. P. Slinker, J. G. Lyon, and R. D. Elphinstone, Global numerical simulations of the growth phase and expansion onset of a substorm observed by Viking, *J. Geophys. Res.*, *100*, 19,083, 1995.

Freeman, M. P., and C. J. Farrugia, A statistical study of the possible effects of solar wind variability on the recurrence rate of substorms, *J. Geophys. Res.*, *100*, 23,607, 1995

Freeman, M. P., C. J. Farrugia, L. F. Burlaga, et al., The interaction of a magnetic cloud with Earth: Ionospheric convection in the northern and southern hemispheres for a wide range of quasi-steady interplanetary magnetic field conditions, *J. Geophys. Res.*, *98*, 7633, 1993

Freeman, M.P., C.J. Farrugia, S.W.H. Cowley, The response of dayside ionospheric convection to the Y-component of the magnetosheath magnetic field:a case study, *Planet. Space. Sci.*, *38*, 13, 1990

Goldstein, H., On the field configuration in magnetic clouds, in *Solar Wind Five*, edited by M. Neugebauer, p. 731, NASA Conf. Publ. 2280, Washington, D. C., 1983.

Gonzalez, W. D., and B. T. Tsurutani, Criteria of interplanetary parameters causing intense magnetic storms (Dst < -100 nT), Planet. *Planet. Space. Sci.*, *35*, 1101, 1987.

Gosling, J. T., Coronal Mass ejections and magnetic flux ropes in interplanetary space, in *Physics of Magnetic Flux Ropes*, edited by C. T. Russell, E. R. Priest and L. C. Lee, p. 344, AGU Geophysical Monograph 58, American Geophysical Union, Washington, DC, 1990.

Gosling, J. T., S. J. Bame, D. J. McComas, et al., A forward-reverse shock pair in the solar wind driven by over-expansion of a coronal mass ejection: Ulysses observations, *Geophys. Res. Lett.*, *21*, 237, 1994.

Gosling, J. T., D. J. McComas, J. L. Phillips, and S. J. Bame, Geomagnetic activity associated with Earth passage of interplanetary shock disturbances and coronal mass ejections, *J. Geophys. Res.*, *96*, 7831, 1991.

Gosling, J. T., S. J. Bame, D. J. McComas, and J. L. Phillips, Coronal mass ejections and large geomagnetic storms, *Geophys. Res. Lett.*, *17*, 901, 1990.

Gosling, J. T., and D. J. McComas, Field line draping about fast coronal mass ejecta: A source of strong out-of-ecliptic interplanetary magnetic fields, *Geophys. Res. Lett.*, *14*, 355, 1987.

Gosling, J. T., D. N. Baker, S. J. Bame et al., Bidirectional solar wind electron heat flux events, *J. Geophys. Res.*, *92*, 8519, 1987.

Gosling, J. T., V. Pizzo, and S. J. Bame, Anomously low proton temperatures in the solar wind following interplanetary shock waves: Evidence for magnetic bottles ?, *J. Geophys. Res.*, *78*, 2001, 1973.

Hewish, A., and P. J. Duffet-Smith, A new method of forecasting geomagnetic activity and proton showers, *Planet. Space. Sci.*, *35*, 487, 1987.

Kahler, S. W., and D. V. Reaves, Probing magnetic topologies of magnetic clouds by means of solar energetic particles, *J. Geophys. Res.*, *96*, 9419, 1991.

Klein, L. W., and L. F. Burlaga, Interplanetary magnetic clouds at 1 AU, *J. Geophys. Res.*, *87*, 613, 1982.

Klimas, A. J., D. N. Baker, D. Vasiliadis, and D. A. Roberts, Substorm recurrence during steady and variable solar wind driving: Evidence for a normal mode in the unloading dynamics of the magnetosphere, *J. Geophys. Res.*, *99*, 14,855, 1994.

Knipp, D. J., B. A. Emery, A. D. Richmond, et al., Ionospheric Convection response to slow, strong variations in a northward interplanetary magnetic field: A case study for January 14, 1988, *J. Geophys. Res.*, *98*, 19,273, 1993.

Kumar, A., and R. M. Rust, Interplanetary magnetic clouds, helicity conservation, and current- core flux ropes, *J. Geophys. Res.*, *101*, 15,667, 1996.

Lees, L., Interaction between the solar wind plasma and the geomagnetic cavity, *AIAA. J.*, *2*, 1576, 1964.

Lepping, R. P., L. F. Burlaga, B. T. Tsurutani, et al., The interaction of a very large interplanetary magnetic cloud with the magnetosphere and with cosmic rays, *J. Geophys. Res.*, *96*, 9425, 1991.

Lepping, R. P. J. A. Jones, and L. F. Burlaga, Magnetic field structure of interplanetary magnetic clouds at 1 AU, *J. Geophys. Res.*, *95*, 11,957, 1990.

Lundquist, S., Magnetohydrostatic fields, *Ark. Fys.*, *2*, 361, 1950.

Maezawa, K., Magnetospheric convection induced by the positive and negative Z components of the interplanetary magnetic field: Quantitative analysis using polar cap magnetic records, *J. Geophys. Res.*, *81*, 2289, 1976.

Marsden, R. G., T. R. Sanderson, C. Tranquille, and K.-P. Wenzel, ISEE-3 observations of low energy proton bidirectional events and their relation to isolated interplanetary magnetic structures, *J. Geophys. Res.*, *92*, 11,009, 1987.

McComas, D. J., J. T. Gosling, S. J. Bame, et al., A test of magnetic field draping induced Bz perturbations ahead of fast coronal mass ejections, *J. Geophys. Res.*, *94*, 1465, 1989.

Midgley, J. E., and L. Davis, Calculation by a moment technique of the perturbation of the geomagnetic field by the solar wind, *J. Geophys. Res.*, *68*, 5111, 1963.

Osherovich, V.A., C.J. Farrugia, and L.F. Burlaga, The nonlinear evolution of magnetic flux ropes: 2. Finite beta plasma, *J. Geophys. Res.*, *100*, 12307, 1995.

Osherovich, V.A., C.J. Farrugia, and L.F. Burlaga, Dynamics of aging magnetic clouds, *Adv. Space Res.*, *13*, 6(6), 57, 1993a.

Osherovich, V.A., C.J. Farrugia, and L.F. Burlaga, The nonlinear evolution of magnetic flux ropes: 1. The low beta limit, *J. Geophys. Res.*, *98*, 13,225, 1993b.

Osherovich, V.A., C.J. Farrugia, L.F. Burlaga, et al., Polytropic relationship for magnetic clouds, *J. Geophys. Res.*, *98*, 15,331, 1993c.

Perreault, P., and S.-I. Akasofu, A study of geomagnetic storms, *Geophys. J. R. Astron. Soc.*, *54*, 547, 1978.

Phan, T. -D., G. Paschmann, W. Baumjohann, and N. Sckopke, The magnetosheath region adjacent to the dayside magnetopause: AMPTE/IRM observations, *J. Geophys. Res.*, *99*, 121, 1994.

Richardson, I. G., and H. V. Cane, Regions of abnormally low proton temperature in the solar wind (1965-1991), and their association with ejecta, *J. Geophys. Res.*, *100*, 397, 1995.

Richardson, I. G., and D. V. Reames, Bidirectional $\sim$ 1MeV/ amu ion intervals in 1973-1991 observed by the Goddard Space Flight Center instruments on IMP 8 and ISEE 3/ICE, *Astrophys. J. (Supp.)*, *85*, 411, 1993.

Scudder, J. D., On the causes of temperature change in inhomogeneous low-density astrophysical plasmas, *Astrophys. J.*, *398*, 299, 1992.

Sonnerup, B. U. O., The reconnecting magnetopause, in: *Magnetospheric Physics,* edited by B. M. McCormac, p.23, D. Reidel, Norwell, Mass., 1974.

Tsurutani, B. T., and W. D. Gonzalez, The interplanetary causes of magnetic storms: *A review, submitted to Magnetic Storms,* edited by B. T. Tsurutani, W. D. Gonzalez, and Y. Kamide, AGU Monograph, Washington, D. C., 1996.

Tsurutani, B. T., W. D. Gonzalez, F. Tang, and Y. T. Lee, Great magnetic storms, *Geophys. Res. Lett.*, *19*, 73, 1992.

Tsurutani, B. T., W. D. Gonzalez, F. Tang et al., Origin of interplanetary southward magnetic fields responsible for major magnetic storms near solar maximum (1978-1979), *J. Geophys. Res.*, *93*, 8519, 1988a.

Tsurutani, B. T., W. D. Gonzalez, and F. Tang, Comment on "A new method of forecasting geomagnetic activity and proton showers" by A. Hewish and P. J. Duffet-Smith, *Planet. Space. Sci.*, *2*, 205, 1988b.

Vandas, M., S. Fischer, M. Dryer, et al., Simulation of magnetic cloud propagation in the inner heliosphere in two dimensions, 2, A loop parallel to the ecliptic plane and the role of helicity, *J. Geophys. Res.*, *101,* 2505, 1996a.

Vandas, M., S. Fischer, M. Dryer, Z. Smith, and T. Detman, Parameteric study of loop-like magnetic cloud propagation, *J. Geophys. Res.*, *101,* 15,645, 1996b.

Vandas, M., S. Fischer, M. Dryer, et al., Simulation of magnetic cloud propagation in the inner heliosphere in two dimensions, 1, A loop perpendicular to the ecliptic plane, *J. Geophys. Res.*, *100,* 12,285, 1995.

Walthour, D. W., B. U. O. Sonnerup, G. Paschmann, et al., Remote sensing of two-dimensional magnetopause structures, *J. Geophys. Res.*, *98,* 1489, 1993.

Wilson, R. M., Geomagnetic response to magnetic clouds, *Planet. Space. Sci.*, *35,* 329, 1987.

Wilson, R. M., On the behavior of the Dst geomagnetic index in the vicinity of magnetic cloud passage at Earth, *J. Geophys. Res.*, *95,* 215, 1990.

Zhang, G., and L. F. Burlaga, Magnetic clouds, geomagnetic disturbances, and cosmic ray decreases, *J. Geophys. Res.*, *95,* 2511, 1988.

Zwan, B. J., and R. A. Wolf, Depletion of solar wind near a planetary boundary, *J. Geophys. Res.*, *81,* 1636, 1976.

Zwickl, R. D., J. R. Asbridge, S. J. Bame, W. C. Feldman, J. T. Gosling, and E. J. Smith, Plasma properties of driver gas following interplanetary shocks observed by ISEE 3, *Solar Wind Five, NASA Conf. Publ.*, CP-2280, 711, 1983.

C. J. Farrugia, Institute for the Study of Earth, Oceans, and Space, University of New Hampshire, Durham, N. H. 03824. (e-mail: ferrugia@unhedi1.sr.unh.edu)

L. F. B urlaga and R. P. Lepping, NASA/Goddard Space Flight Center, Code 692, Greenbelt, MD, 20771

The Role of Magnetosphere–Ionosphere Coupling in Magnetic Storm Dynamics

Ioannis A. Daglis

Institute of Ionospheric and Space Research, National Observatory of Athens, Penteli, Greece

The magnetic storm is the prime dynamic process in Geospace. It interconnects, in a uniquely global manner, the solar wind, the magnetosphere, the ionosphere, the upper atmosphere, and, occasionally through large induced currents, the Earth's surface. Here we concentrate on the role and the importance of the magnetosphere-ionosphere coupling in the evolution of the ring current, as manifested through storm-time observations of the inner magnetosphere by CRRES. The observations show that during the main phase of great storms, i.e., during times of very intense ring currents, the abundance and energy density of ionospheric-origin ions (O^+ in particular) in the inner magnetosphere are extraordinarily high. O^+ alone provides more than 40% of the particle energy density, compared with a level of ~20% during small to moderate storms, and a level of less than 10% during quiet times. Hence, the ring current growth is associated with dramatic compositional changes in the near-Earth magnetosphere. Considering the domination of O^+ and taking into account that a fraction of H^+ is also of ionospheric origin, it is conceivable that the cause of the intense ring current during large storms is terrestrial, although the energy source is unambiguously of solar origin. The overwhelming dominance of O^+ during the observed storms probably is also related to the fact that CRRES operated during solar maximum, since previous studies have shown a solar-cycle dependence of ionospheric outflow.

INTRODUCTION

Along with man's increasing capability to perturb his environment, we have witnessed severe effects of solar-terrestrial variability on human environment [*Wilcox*, 1976; *Lanzerotti*, 1994]. Of particular interest in this framework are great magnetic storms, which are associated with hazards on technological systems [e.g., *Lanzerotti*, 1992]. The frequency of occurrence of storms is much smaller than that of magnetospheric substorms, which constitute the elementary dynamic process in geospace. However, in terms of global disturbances, the magnetic storm is the prime dynamic process in the geospace.

A major feature of the magnetic storm is the injection, transport and loss of charged particles that constitute the ring current. The characteristics of the ring current actually determine the gross characteristics of the storm. Obviously, the comprehension of the processes influencing the buildup and decay of the ring current is most important. This study addresses the supply of ionospheric-origin ions (particularly O^+ ions) to the inner magnetosphere during storms, and discusses their role in the evolution of the storm-time ring current.

Before the AMPTE (Active Magnetospheric Particle Tracer Explorers) mission, the composition of the bulk ring current (i.e. within the energy range ~20-300 keV) was unknown [e.g., *Williams*, 1983]. The charge–energy–mass (CHEM) spectrometer [*Gloeckler et al.*, 1985a] on board AMPTE/CCE was the first experiment to investigate the near-Earth magnetosphere with multi-species ion measurements extending in the higher-energy (≥20 keV) range. However, AMPTE operated during solar minimum, and had the chance to observe only one great storm.

The great storm in February 1986 was studied in detail by *Hamilton et al.* [1988]. *Hamilton et al.* [1988] showed that O^+ dominated near the storm's maximum phase, with 47% of the energy density compared with 36% contributed by H^+. Furthermore, they estimated that 67-80% of the ring current density near the maximum of the storm was of ionospheric origin (since also a fraction of H^+ and He^+ is of ionospheric origin). Accordingly, the authors suggested a major ionospheric ring current component near the maximum phase of great storms.

In the following section storm observations by the Combined Release and Radiation Effects Satellite (CRRES) during solar maximum will be presented. One of the storms included here is the great storm of March 24, 1991. This storm attracted high interest among a plethora of scientists, not only because its size was very large, but also because some extraordinary features appeared. For example, an intense new radiation belt of relativistic electrons was created almost instantly (within a few minutes after the SSC) and lasted for several months [*Blake et al.*, 1992]. Observations regarding the ring current composition of this storm were first presented by *Wilken et al.* [1992a]. *Wilken et al.* [1992a] used measurements of the Magnetospheric Ion Composition Spectrometer (MICS) on board CRRES [*Wilken et al.*, 1992b] to compare the effects of two magnetic storms and one strong compression of the magnetosphere on the state of the inner magnetospheric ion population. The two storms (on March 24, 1991, and on June 5, 1991) were large, with *Dst* values lower than −200 nT in both cases. The compression followed an increase (by a factor of five) in solar wind density on March 21, 1991. To compare the effects of the three events, averages of energy spectra and pitch angle distributions of the five species H^+, O^+, He^{++}, He^+ and O^{++} were taken over the L-range 5 to 6 in the premidnight magnetosphere. It was shown that the basic difference between the compression and the two storms was that the former simply energized the preexisting population (presumably via betatron acceleration), while the latter brought new particles (especially of ionospheric origin) into the inner magnetosphere. All three events had in common that the contribution of the ionospheric-origin O^+ ions to the total energy density increased relative to its quiet state contribution. The increase however was less for the compression event than for the two storms: starting from a quiet-state contribution of <10%, the peak O^+ contribution was ~25% in the compression event, ~50% in the June storm and ~75% in the March storm.

STORM OBSERVATIONS

Following we shall present CRRES observations regarding the relative contribution of the two major ion species H^+ and O^+ to the total ion energy density during five storms in 1991. The set includes a weak to moderate storm, with a minimum *Dst* of −80 nT, and four large storms that occurred in March, June and July 1991, with minima *Dst* from −180 nT to −300 nT.

Starting with the smallest storm of the group (February 1, 1991), we can see the time profiles of compositional changes and of the *Dst* index in Figure 1. The top two panels show the contribution of the two major ion species H^+ and O^+ to the total energy density of the energetic ion population in the outer ring current, that is in the L-range 5 to 7 R_E. The reason for considering only the outer ring current in this study, is the operation of the particle identifier (PID) of MICS at lower altitudes (i.e. below L~5). PID is a fast analog processor evaluating the sensor information, with the purpose of preventing high proton fluxes from overloading the relatively slow analog-to-digital conversion of the sensor signals [*Wilken et al.*, 1992b]. The by-product is that the mass / mass per charge matrix countrates used in this study are affected by the PID, in the sense that the H^+ rates are close to zero. Hence the actual relative contribution of the main ion species cannot be calculated from the matrix countrates below L~5.

The contribution to the total energy density is calculated as follows. The energy density of all ion species is calculated over the range 50-430 keV. The lower energy threshold for complete identification by MICS differs slightly among the various ion species (20 keV for H^+, 50 keV for O^+, 60 keV for He^{++} and 80 keV for O^{++}), but since the energy range 50-430 is common for the major ion species H^+ and O^+, it is the most appropriate for the calculations. After calculating the energy density for each ion species and the total measured ion energy density, the ratio of the energy density of H^+ and O^+ over (i.e. their contribution to) the total energy density is calculated. It is expected that the O^+ matrix count rates contain also some ionospheric N^+, which can be 7-20% of O^+ [*Gloeckler and Hamilton*, 1987].

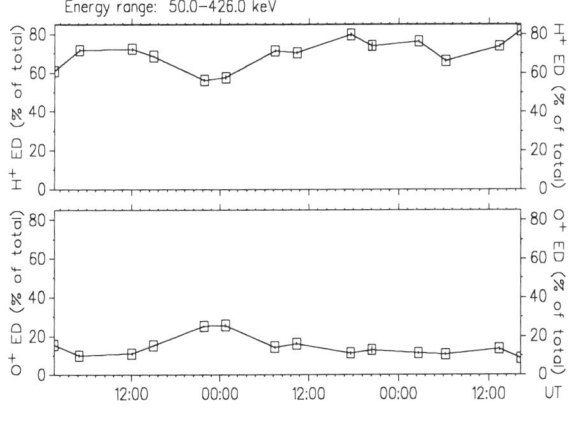

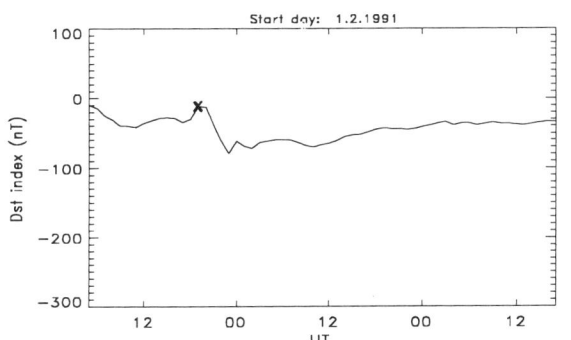

Figure 1. The weak to moderate storm of February 1, 1991: The diagram shows the time profile of the contribution of the two major ion species H^+ and O^+ to the total energy density of the energetic ion population in the outer ring current region, and the time profile of the Dst index. Squares with error bars mark the actual data points on the particle panels, while a cross marks the SSC occurrence on the Dst panel. The main feature to be observed is the concurrent increase of the $|Dst|$ level and of the O^+ contribution to the total ion energy density.

Since the variations in one specific region are considered (outer ring current, $L=5-7$), and the CRRES orbital period is 10 h, the values shown in the diagrams are separated by 1 to 6 hours. As expected [*Williams*, 1983] and as demonstrated by the AMPTE mission [*Krimigis et al.*, 1985], the bulk of the storm-time ring current energy is contained within the energy range 20-300 keV. Furthermore, the dominant contribution to the total energy density comes from the H^+ ions, with O^+ being the only other species to become occasionally important or dominant [*Gloeckler et al.*, 1985b; *Gloeckler and Hamilton*, 1987; *Daglis et al.*, 1993, 1994]. Accordingly, the energy range covered by CRRES-MICS is perfectly suitable for studies of the energy budget of the storm-time ring current. The only "deficiency" of MICS is the underestimation of the O^+ contribution to the total energy density.

In the case of the moderate storm of February 1, 1991, we see in Figure 1 that before the storm, H^+ is the dominant ion species. Before the SSC (at 1842 UT), we have four passes of CRRES through the outer ring current region (two passes outbound and two passes inbound), with O^+ contributing less than 10% of the total energy density. Right after the SSC, we have a jump of the O^+ contribution to 20% concurrently with a drop of Dst to its lowest value (–80 nT) during this storm. During the following passes, the O^+ contribution falls off to its initial value, while Dst recovers. It is remarkable that a slight drop of Dst at 1100 UT on February 2 occurs concurrently with a slight increase of the O^+ contribution. Both variations are short-lived.

Figure 2 shows the time profiles of the H^+ and O^+ contribution to the total energy density and of the Dst

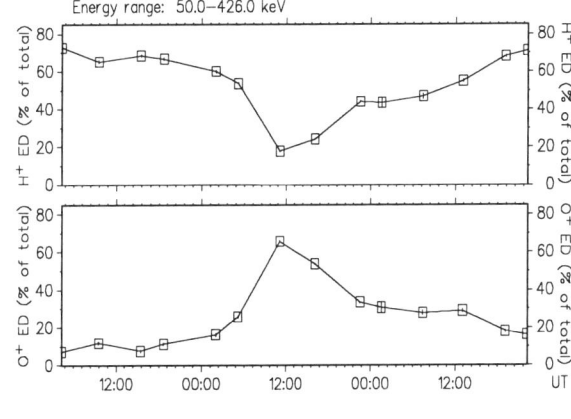

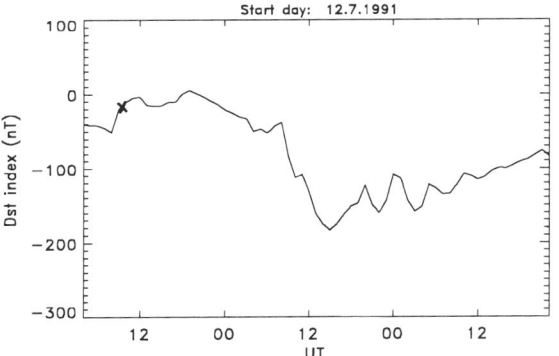

Figure 2. Same format as Figure 1 for the July 13, 1991, storm. The main features are the concurrent increase of the $|Dst|$ level and of the O^+ contribution to the total ion energy density, and the dominance of O^+ near the storm maximum.

index for the July 13, 1991, storm. This storm is considerably larger than the February 1 storm, since Dst reaches a minimum of −180 nT, to be compared with a minimum of −80 nT in the February storm. The picture is qualitatively similar to the small storm: the O^+ contribution is initially below 10%, and after the SSC at 0924 UT increases rapidly, to drop again during storm recovery. However, O^+ becomes dominant in this storm, reaching a contribution level of 65%. It is remarkable that the peak of O^+ contribution to the total energy density precedes the peak in $|Dst|$. This indicates that the ionospheric feeding of the inner plasma sheet attenuates before the maximum of the storm main phase. The newly injected ionospheric-origin ions convect earthward and enhance the ring current, which is mainly responsible for the $|Dst|$ peak.

The time profiles of the H^+ and O^+ contribution to the total energy density and of the Dst index for the July 9, 1991, storm are shown in Figure 3. This storm is slightly

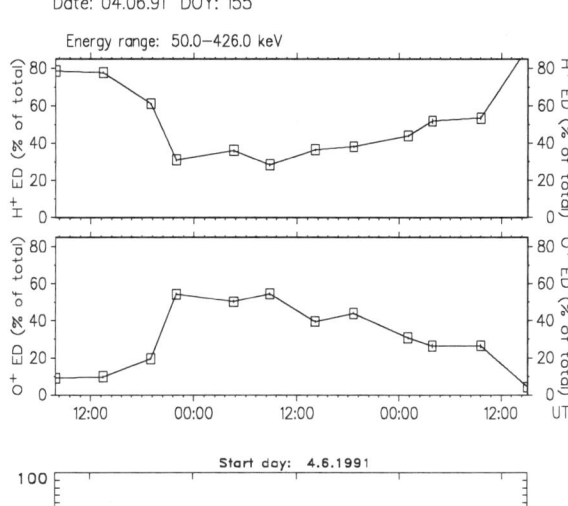

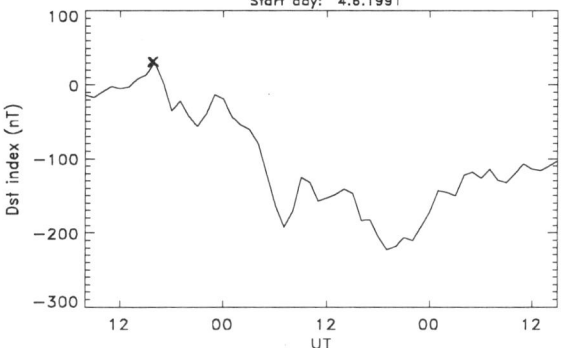

Figure 4. Same format as Figure 1 for the June 4, 1991, storm. The profiles of Dst and O^+ contribution are somehow different than in the July storms: there are two deep minima in Dst and a prolonged high level of O^+ contribution to the total energy density.

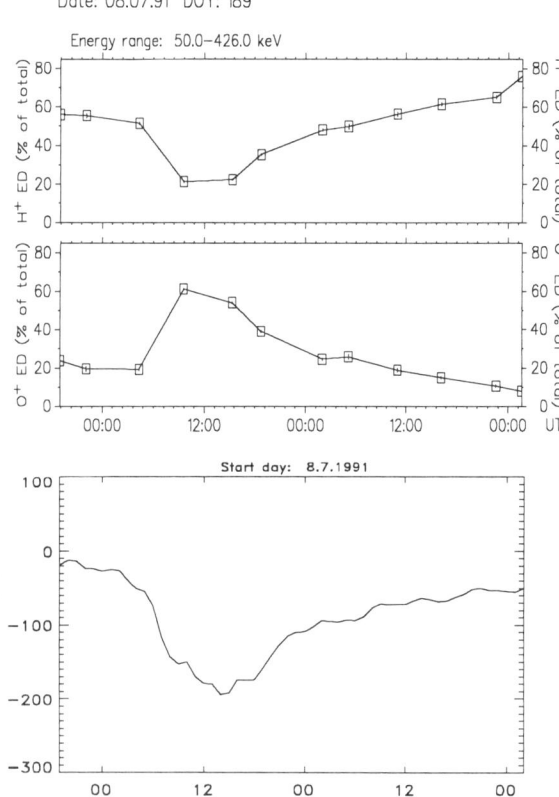

Figure 3. Same format as Figure 1 for the July 9, 1991, storm. As in the July 13 storm, $|Dst|$ and O^+ contribution rise concurrently, and O^+ dominates near storm maximum.

larger than the July 13 one, with Dst reaching −190 nT. The qualitative picture of the event is similar to the former two events: the initially low O^+ contribution to the total energy density (~10%) increases along with $|Dst|$, reaching 61%. The levels reached by both Dst and the O^+ contribution are similar to the ones in the July 13 storm. Again the peak of O^+ contribution to the total energy density precedes the peak in $|Dst|$.

The June 4 storm, presented in Figure 4, exhibits a somewhat different picture than the two large storms in July. First of all, there are three consecutive minima in Dst. Although the peak $|Dst|$ (225 nT) is higher than in the two July storms, the peak level of O^+ contribution to the total energy density in the outer ring current region is lower (55%). However, O^+ is the dominant ion species in the outer ring current during most of the storm main phase. The O^+ contribution remains enhanced and above

30%, an exceptionally high level [*Daglis et al.*, 1993], for more than 24 hours. This means that the ionospheric ion source feeds the inner plasma sheet for a long time interval, leading to the enhancement of the ring current with the newly injected ions that move earthward and follow closed drift paths. Regarding the timing of the peak of O^+ contribution relatively to the peak in $|Dst|$, it is not as clear as in the former storms, because of the complicated profile of Dst. The overall trend is that high levels of O^+ contribution precede the two deep minima in Dst.

Finally, the great storm of March 24, 1991, is presented in Figure 5. Dst reaches a deep minimum of -300 nT in this storm, and the O^+ contribution to the total energy density in the outer ring current climbs to more than 66%. Figure 5 shows the by now familiar picture of concurrent O^+ and $|Dst|$ increase. The bottom panel of Figure 5 shows the preliminary AE index, which was provided by Alexander G. Yahnin (Polar Geophysical Institute, Apatity, Russia). The March storm exhibits a peculiar two-phase profile, both in Dst and in the level of O^+ contribution. Following the SSC (at 0341 UT) we can see an increase of the O^+ contribution from the $\sim$10% level to the $\sim$40% level, along with a drop of Dst to ~-100 nT. A period of transient Dst recovery and O^+ decrease follows, and then we enter the main phase of the storm, with both Dst and O^+ reaching their peaks. A similar pattern was observed in the great storm of February 1986 [*Hamilton et al.*, 1988]. Furthermore, it can be seen in Figure 5 that the O^+ contribution remains at an extraordinarily high level (above 40%) for a very extended time period (more than 30 h). It should be noted that O^+ has a stronger earthward gradient than H^+, resulting in an earthward increase of the O^+ contribution. Over the L-range 5 to 6 for example (the values plotted in Figure 5 are averages over L=5-7), the O^+ contribution exceeds 70% [*Wilken et al.*, 1992a].

It can be seen that, just as in the two July storms, the peaks of O^+ contribution preceded the peaks in $|Dst|$. The bottom panel of Figure 5 shows the time profile of the preliminary AE index. It demonstrates the well known fact that the O^+ abundance and energy density in the inner plasma sheet increase during periods of enhanced auroral activity [*Lennartsson and Shelley*, 1986; *Daglis et al.*, 1994]. Furthermore, it indicates that a series of substorm expansions initiates and sustains the enhanced ionospheric feeding of the inner plasma sheet [*Daglis and Axford*, 1996], which contributes to the enhancement of the ring current.

It has been predicted that a higher abundance of O^+ in the plasma sheet will permit substorm breakup at relatively low L-values [*Rothwell et al.*, 1988]. Lower-energy (<30 keV) O^+ ions that are injected directly to the inner ($L<6$) midnight-duskside magnetosphere [*Kaye et al.*, 1981; *Strangeway and Johnson*, 1983], will be energized very efficiently by substorm induced electric fields [e.g., *Delcourt et al.*, 1990]. Hence, the combination of, or moreover the feedback between O^+ injections and substorm-breakups successively proceeding to the duskside and to lower L-shells [*Baker et al.*, 1985], will substantially contribute to a rapid enhancement of the ring current.

All storms presented here show an remarkably persistent feature: O^+ and Dst change in parallel. Although the number of storms presented here precludes statistical significance, the trend is unmistakable: the enhancement of the ring current is concurrent with an increased ionospheric feeding of the inner plasma sheet. It is very

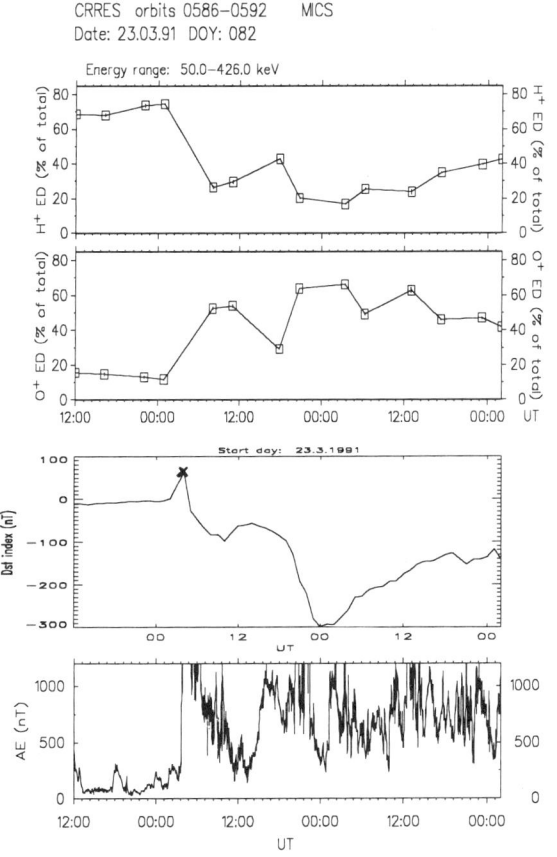

Figure 5. Same format as Figure 1 for the great storm of March 24, 1991, plus an extra panel in the bottom showing the AE time profile. The Dst first reached a minimum of -100 nT and later a deep minimum of -300 nT during this storm, while O^+ became the dominant ion contributing more than 66% of the total ion energy density in the outer ring current. The peaks of O^+ contribution preceded the peaks in $|Dst|$, just as in the July storms.

remarkable that even the two-phase profile of Dst during the March 1991 storm is mirrored in the O^+ contribution profile Figure 5. Such a two-phase profile in Dst has been previously observed in several other large storms [*Tsurutani and Gonzalez, 1996*], however there are no measurements of the bulk ring current composition available for those storms. In the case of the February 1986 storm, the two-phase profile in Dst and O^+ contribution was also observed [*Hamilton et al., 1988*].

SIGNIFICANCE OF THE STORM TIME O^+ DOMINANCE

The outstanding storm feature that is manifested by the CRRES observations of four large and one moderate storm is the concurrent increase of |Dst| and of O^+ (and consequently ionospheric) contribution to the total particle energy density. Furthermore, it is manifested that O^+ is the dominant ion species during the main phase of large storms. The fact that CRRES operated during solar maximum may indicate that such a dominance is related to the solar-cycle dependence of ionospheric outflow [*Yau et al., 1988*] and occurs only during solar maximum large storms, while at solar minimum it occurs only during great storms [*Hamilton et al., 1988*].

Taking into account that a fraction of H^+ (about 30% in the storm-time outer ring current [*Gloeckler and Hamilton, 1987*]) is also of ionospheric origin, it is clear that, although the energy source of storms is unambiguously of solar origin, the majority of the storm-time ring current particles are of terrestrial origin. This indicates that large storms are closely connected with the ionospheric response to enhanced solar wind-magnetosphere coupling, in other words: the intensification of the ring current leading to very low Dst levels is attributed to the contribution of outflowing ionospheric ions. The Sun contributes the energy and the Earth ionosphere contributes the majority of the particles for the formation of the ring current during large storms. It is noteworthy that preliminary quasi-static modeling of the differences between H^+ and O^+ ions on the ring current intensification by *Wodnicka* [1989], showed that energetic (>10 keV) O^+ ions result in a stronger ring current (larger |Dst|) and in a faster initial ring current decay, as compared to H^+. A follow-up study [*Wodnicka, 1991*] showed that smaller initial pitch angles, lower radial injection distances and higher initial energies of O^+ would further increase |Dst|.

The overwhelming dominance of O^+ during the observed storms probably is also related to the fact that CRRES operated during solar maximum, since previous studies have shown a solar-cycle dependence of ionospheric outflow [*Yau et al., 1988*] and of the relative abundance of lower energy (<16 keV) O^+ in the magnetosphere [*Young et al., 1982; Lennartsson, 1989; Stokholm et al., 1989*]. A larger database of storm observations with composition measurements of the bulk ring current would permit to address the question if the ionospheric response is a prerequisite for very strong ring currents (that is great storms). It is conceivable, that very intense ring currents responsible for very low Dst levels are only created when the ionospheric response to the solar wind-magnetosphere coupling is of sufficient strength and temporal extent. The CRRES (and previous AMPTE) observations have shown that the enhanced magnetosphere-ionosphere coupling, in form of ion feeding of the inner plasma sheet, provides the additional new population associated with the storm main phase. However, in order to fully assess the role of magnetosphere-ionosphere coupling, one does not only need compositional measurements in the inner magnetosphere, but also complete information on the solar and interplanetary conditions preceding and accompanying large storms.

It has been empirically shown that intense storms (Dst ≤ –100 nT) are primarily caused by $B_z \leq -10$ nT fields with duration greater than 3 hours [*Gonzalez and Tsurutani, 1987; Tsurutani et al., 1992; Tsurutani and Gonzalez, 1996*]. Unfortunately there was no complete solar wind information for the present set of large storms in 1991. Nevertheless, it is clear that one has to investigate if similar solar wind conditions (i.e., the occurrence of sufficiently strong southward interplanetary magnetic fields, for a sufficiently long time) lead to similar-size storms independently of the extent of ionospheric outflow, or, on the contrary, if a larger abundance of ionospheric ions accompanies larger |Dst| levels despite similar solar wind conditions. The latter would speak for the explosive ionospheric response as a prerequisite of large storms.

An increased relative abundance of O^+ ions in the inner magnetosphere does not only have the effect of storm-time ring current enhancement. The resulting mass loading is important for wave growth, propagation and absorption [*Kozyra et al., 1984; Moore, 1991; Thorne and Horne, 1994*]. Increased O^+ abundance also influences the decay rate of the ring current, since the charge-exchange lifetime of O^+ is considerably shorter than the H^+ lifetime for ring current energies (≥ 40 keV, *Smith and Bewtra* [1978]). This implies that O^+-dominated ring current will decay faster, at least initially. Such a fast initial ring current decay, associated with a large O^+ component during the storm main phase, has been indeed observed in the February 1986 storm [*Hamilton et al., 1988*]. The same trend can be seen in the storms of July 9 and of March 24,

1991 (Figures 3 and 5), where the initially fast recovery of Dst is concurrent with an initially fast drop of the O^+ contribution to the total energy density.

There has been the question of whether or not substorm occurrence is essential for the storm-time ring current growth [*Kamide*, 1979, 1992; *Kamide et al.*, 1996]. This issue may be related to the connection of substorms with ionospheric outflow, since the ring current growth is concurrent with increased abundance of ionospheric-origin ions in the inner equatorial magnetosphere. The association of strong substorms (as observed during storms) and enhanced ionospheric ion abundance and energy density in the inner plasma sheet has been shown by *Daglis et al.* [1992] and *Daglis et al.* [1994] on the basis of a large set of substorm observations by AMPTE/CCE. A recent study of substorms observed by CRRES confirmed the AMPTE/CCE results [*Daglis et al.*, 1996]. Clues to this issue can be provided by studies addressing the processes of ionospheric ion acceleration and extraction into the magnetosphere [e.g., *Yau et al.*, 1988]. Viking observations of ionospheric outflow and associated electric fields [*Lundin et al.*, 1987; *Hultqvist et al.*, 1988; *Lundin et al.*, 1990] prompted relevant modeling and simulation studies, which showed that outflowing ionospheric ions are accelerated very efficiently by low-frequency large-amplitude electric field fluctuations [*Lundin and Hultqvist*, 1989; *Hultqvist*, 1996]. Since such electric fields occur during intense auroral activity (G. Marklund, personal communication), it is expected that this type of acceleration of ionospheric ions at low altitudes operates during substorm expansion. Consequently, a higher abundance of ionospheric ions (O^+ in particular) in the inner magnetosphere is expected during substorm expansion, in accordance with the results of *Daglis et al.* [1994].

A further important aspect of storm/substorm relationship through the increased O^+ abundance may be the feedback between O^+ injections and substorm breakups (and associated induced electric fields) moving progressively duskward and earthward [*Baker et al.*, 1985; *Rothwell et al.*, 1988].

It has been predicted that a higher abundance of O^+ in the plasma sheet will permit substorm breakup at relatively low L-values [*Rothwell et al.*, 1988]. Lower-energy (<30 keV) O^+ ions that are injected directly to the inner ($L < 6$) midnight-duskside magnetosphere [*Kaye et al.*, 1981; *Strangeway and Johnson*, 1983], will be energized very efficiently by substorm induced electric fields [e.g., *Delcourt et al.*, 1990]. Hence, the combination of, or moreover, the feedback between substorm-breakups and O^+ injections successively proceeding to the duskside and to lower L shells [*Baker et al.*, 1985], will substantially contribute to a rapid enhancement of the ring current.

However, although studies on both high and low altitude observations have shown that the outflow of ionospheric O^+ and the energy density of O^+ in the inner magnetosphere are closely correlated with the auroral activity, the ring current grows more efficiently during the main phase of storms than during non-main phase periods with the auroral electrojets having the same strength [*Gonzalez et al.*, 1994]. The answer to this paradox should be the persistence and long duration of enhanced auroral activity resulting in a prolonged ionospheric outflow during the storm main phase. A prolonged ionospheric outflow has been suggested by *Daglis and Axford* [1996] to account for the continuing rise in O^+ energy density in the inner plasma sheet, in contrast with the one-step-rise of H^+ and He^{++} energy density during substorm expansion. Of course there are also some other factors influencing the extent of ionospheric outflow and of the O^+ abundance in the inner magnetosphere [*Daglis et al.*, 1995]. The ionospheric outflow is for example influenced by the solar cycle, that is the solar UV radiation [*Young et al.*, 1982; *Yau et al.*, 1985], by the neutral atmosphere variability [*Moore*, 1984], by the solar wind velocity [*Lundin et al.*, 1995]. It seems however that the most important influence is the occurrence of substorms and the associated high-latitude activity.

Another answer to this paradox (i.e., ring current growing more efficiently during the main phase of storms, with AE levels seemingly similar to non-main phase periods) may simply be the inherent deficiencies of AE indices. Since the auroral electrojets move equatorwards during large storms, the AE ground magnetogram stations will not provide the actual value of the electrojet intensity, since they will miss a part of the electrojets, which will be larger the greater the storm is. That means that the greater the storm, the worse the underestimate of the auroral electrojet intensity will be.

Recently, *Sun and Akasofu* [1996] proposed an alternative explanation of the non-equivalency of main-phase and non-main-phase auroral electrojet activity levels with regard to the ring current growth. *Sun and Akasofu* [1996] suggested that the AE / AL indices do not represent well the efficiency of ionospheric ion injection; they presented a new index AF, which monitors time variations of the total field-aligned currents and correlates with the Dst index better than AE or AL do. *Sun and Akasofu* [1996] suggest that the AF index is a good measure for the efficiency of the injection of ionospheric ions to the storm-time ring current (it should be reminded here that regions of outflowing ionospheric ions correlate with regions of field-aligned currents into the ionosphere [*Yau et al.*, 1984; *Hultqvist et al.*, 1988; *Lundin and Hultqvist*, 1989]. The good correlation between *AE* and *Dst* is a further indication of the importance of the ionospheric ion feeding of the inner plasma sheet for the ring current growth during storms.

SUMMARY

Storm observations by CRRES have shown that at solar maximum O^+ is the dominant ion species during the main phase of large storms. Combined with the fact that a fraction of H^+ is also of ionospheric origin, the observations imply that the cause of intense ring currents during large storms is terrestrial, although the energy source is the solar wind. Furthermore, CRRES observations show that the increases of $|Dst|$ and of the O^+ (ionospheric) contribution to the total particle energy density in the inner magnetosphere are concurrent. Both features demonstrate the importance of the magnetosphere-ionosphere coupling for the development of intense ring currents during large storms.

It is conceivable that intense ring currents responsible for the global geomagnetic effects that characterise large storms, are only created when the magnetosphere-ionosphere coupling is of suffcient duration and size (in terms of ion outflux). If this were the case, the high abundance of ionospheric ions (in particular heavy O^+ ions) in the inner magnetosphere would be a prerequisite for great storms. However, for a complete assessment of the role of storm-time magnetosphere-ionosphere coupling, combined information on both the magnetosphere and the solar and interplanetary conditions during storms is needed. The numerous spacecraft monitoring the Earth's magnetosphere, the interplanetary space and the Sun itself during the International Solar-Terrestrial Physics (ISTP) Program will provide an unprecedented potential for investigations of the cause/effect chain connecting solar events through the solar wind with geospace processes. In this framework, the issue of the importance of magnetosphere-ionosphere coupling for the evolution of magnetic storms (in terms of both ring current growth and decay rate) has the opportunity to be thoroughly investigated.

Acknowledgments. Parts of this work were conducted at the Max-Planck-Institut für Aeronomie (MPAe) with support from the Deutsche Agentur für Raumfahrtangelegenheiten (German Space Agency) under project 50 OC 95022. The author wishes to thank Bengt Hultqvist, Rickard Lundin, and Göran Marklund for valuable discussions. Congratulations are due to the conveners Walter Gonzalez, Yohsuke Kamide and Bruce Tsurutani, for organizing the highly successful Chapman Conference on Magnetic Storms. The CRRES-MICS experiment was a joint effort of the Aerospace Corporation (Los Angeles, USA), the MPAe (Katlenburg-Lindau, Germany) as the lead institution, the Rutherford Appleton Laboratory (Didcot, England) and the University of Bergen (Bergen, Norway). Principal Investigator of CRRES-MICS is Berend Wilken (MPAe).

REFERENCES

Baker D. N., T. A. Fritz, W. Lennartsson, B. Wilken, H. W. Kroehl, and J. Birn, The role of heavy ions in the localization of substorm disturbances on March 22, 1979: CDAW 6, *J. Geophys. Res.*, *90*, 1273-1281, 1985.

Blake, J. B., et al., Injection of electrons and protons with energies of tens of MeV into L < 3 on 24 March 1991, *Geophys. Res. Lett.*, *19*, 821-824, 1992.

Daglis, I. A., and W. I. Axford, Fast ionospheric response to enhanced activity in geospace: Ion feeding of the inner magnetotail, *J. Geophys. Res.*, *101*, 5047-5065, 1996.

Daglis, I. A., E. T. Sarris, G. Kremser, and B. Wilken, On the solar wind-magnetosphere-ionosphere coupling: AMPTE/CCE particle data and the *AE* indices, in *Study of the Solar-Terrestrial System, ESA SP-346*, ed. J. J. Hunt and R. Reinhard, pp. 193-198, ESA, Paris, 1992.

Daglis, I. A., E. T. Sarris, and B. Wilken, AMPTE/CCE observations of the ion population at geosynchronous altitudes, *Ann. Geophys.*, *11*, 685-696, 1993.

Daglis, I. A., S. Livi, E. T. Sarris, and B. Wilken, Energy density of ionospheric and solar wind origin ions in the near-Earth magnetotail during substorms, *J. Geophys. Res.*, *99*, 5691-5703, 1994.

Daglis, I. A., W. I. Axford, S. Livi, and B. Wilken, Factors regulating the supply of ionospheric ions to the magnetosphere during geomagnetically active times, *Eos Trans. AGU*, *76* (46), 526, 1995.

Daglis, I. A., W. I. Axford, S. Livi, B. Wilken, M. Grande, and F. Søraas, Auroral ionospheric ion feeding of the inner plasma sheet during substorms, *J. Geomagn. Geoelectr.*, *48*, 729-739, 1996.

Delcourt, D. C., J.-A. Sauvaud, and T. E. Moore, Cleft contribution to ring current formation, *J. Geophys. Res.*, *95*, 20,937-20,943, 1990.

Gloeckler, G., and D. C. Hamilton, AMPTE ion composition results, *Phys. Scr.*, *T18*, 73-84, 1987.

Gloeckler, G., et al., The charge-energy-mass (CHEM) spectrometer for 0.3 to 300 keV/*e* ions on the AMPTE/CCE, *IEEE Trans. Geosci. Remote Sens.*, *GE-23*, 234-240, 1985a.

Gloeckler, G., B. Wilken, W. Stüdemann, F. M. Ipavich, D. Hovestadt, D. C. Hamilton, and G. Kremser, First composition measurements of the bulk of the storm time ring current (1 to 300 keV/*e*) with AMPTE/CCE, *Geophys. Res. Lett.*, *12*, 325-328, 1985b.

Gonzalez, W. D., and B. T. Tsurutani, Criteria of interplanetary parameters causing intense magnetic storms ($Dst < -100nT$), *Planet. Space Sci.*, *35*, 1101-1109, 1987.

Gonzalez, W. D., J. A. Joselyn, Y. Kamide, H. W. Kroehl, G. Rostoker, B. T. Tsurutani, and V. M. Vasyliunas, What is a geomagnetic storm?, *J. Geophys. Res.*, *99*, 5771-5792, 1994.

Hamilton, D. C., G. Gloeckler, F. M. Ipavich, W. Stüdemann, B. Wilken, and G. Kremser, Ring current development during the great geomagnetic storm of February 1986, *J. Geophys. Res.*, *93*, 14,343-14,355, 1988.

Hultqvist, B., On the acceleration of positive ions by high-latitude, large amplitude electric field fluctuations, *J. Geophys. Res.*, in press, 1996.

Hultqvist, B., R. Lundin, K. Stasiewicz, L. Block, P.-A. Lindqvist, G. Gustafsson, H. Koskinen, A. Bahnsen, T. A. Potemra, and L. J. Zanetti, Simultaneous observations of upward moving field-aligned electrons and ions on auroral zone field lines, *J. Geophys. Res.*, *93*, 9765-9776, 1988.

Kamide, Y., Relationship between substorms and storms, in *Dynamics of the magnetosphere*, edited by S.-I. Akasofu, pp. 425-443, D. Reidel, Boston, Mass., 1979.

Kamide, Y., Is substorm occurrence a necessary condition for a magnetic storm?, *J. Geomagn. Geoelectr.*, *44*, 109, 1992.

Kamide, Y., et al., Current understanding of magnetic storms: Storm/substorm relationships, *J. Geophys. Res.*, submitted, 1996.

Kaye, S. M., R. G. Johnson, R. D. Sharp, and E. G. Shelley, Observations of transient H^+ and O^+ bursts in the equatorial magnetosphere, *J. Geophys. Res.*, *86*, 1335-1344, 1981.

Kozyra, J. U., T. E. Cravens, A. F. Nagy, E. G. Fontheim, and R. S. B. Ong, Effects of energetic heavy ions on electromagnetic ion cyclotron wave generation in the plasmapause region, *J. Geophys. Res.*, *99*, 2217-2233, 1984.

Krimigis, S. M., G. Gloeckler, R. W. McEntire, T. A. Potemra, F. L. Scarf, and E. G. Shelley, Magnetic storm of 4 September 1985: A synthesis of ring current spectra and energy densities measured with AMPTE-CCE, *Geophys. Res. Lett.*, *12*, 329-332, 1985.

Lanzerotti, L. J., Comment on "Great magnetic storms" by Tsurutani et al., *Geophys. Res. Lett.*, *19*, 1991-1992, 1992.

Lanzerotti, L. J., Impacts of solar-terrestrial processes on technological systems, in *Solar-Terrestrial Energy Program*, COSPAR Colloquia Series Vol. 5, edited by D. N. Baker, V. O. Papitashvili and M. J. Teague, pp. 547-555, Pergamon Press, London, 1994.

Lennartsson, W., Energetic (0.1- to 16-keV/e) magnetospheric ion composition at different levels of solar $F10.7$, *J. Geophys. Res.*, *94*, 3600-3610, 1989.

Lennartsson, W., and E. G. Shelley, Survey of 0.1- to 16-keV/e plasma sheet ion composition, *J. Geophys. Res.*, *91*, 3061-3076, 1986.

Lundin, R., and B. Hultqvist, Ionospheric plasma escape by high altitude electric fields: magnetic moment pumping, *J. Geophys. Res.*, *94*, 6665-6680, 1989.

Lundin, R., M. Yamauchi, J. Woch, and G. Marklund, Boundary layer polarization and voltage in the 14 MLT region, *J. Geophys. Res.*, *100*, 7587-7597, 1995.

Moore, T. E., Superthermal ionospheric outflows, *Rev. Geophys.*, *22*, 264-274, 1984.

Moore, T. E., Origin of magnetospheric plasma, *U.S. Natl. Rep. Int. Union Geod. Geophys. 1987-1991*, *Rev. Geophys.*, *29*, 1039-1048, 1991.

Rothwell, P. L., L. P. Block, M. B. Silevitch, and C.-G. Fälthammar, A new model for substorm onsets: The pre-breakup and triggering regimes, *Geophys. Res. Lett.*, *15*, 1279-1282, 1988.

Smith, P. H., and N. K. Bewtra, Charge exchange lifetimes for ring current ions, *Space Sci. Rev.*, *22*, 301-318, 1978.

Stokholm, M., H. Balsiger, J. Geiss, H. Rosenbauer, and D. T. Young, Variations of the magnetospheric ion number densities near geostationary orbit with solar activity, *Ann. Geophys.*, *7*, 69, 1989.

Strangeway, R. J., and R. G. Johnson, Mass composition of substorm-related energetic ion dispersion events, *J. Geophys. Res.*, *88*, 2057-2064, 1983.

Sun, W., and S.-I. Akasofu, A new devised index AF as monitoring the efficiency of the injection of energetic particles associated with substorms, Paper presented at the Chapman Conference on Magnetic Storms, Pasadena, 12-16 February 1996. Book of abstracts, p. 21, 1996.

Thorne, R. M., and R. B. Horne, Energy transfer between energetic ring current H^+ and O^+ by electromagnetic ion cyclotron waves, *J. Geophys. Res.*, *99*, 17,275-17,282, 1994.

Tsurutani, B. T., and W. D. Gonzalez, The interplanetary causes of magnetic storms: A review, this volume, 1996.

Tsurutani, B. T., W. D. Gonzalez, F. Tang, and Y. T. Lee, Great magnetic storms, *Geophys. Res. Lett.*, *19*, 73-76, 1992.

Wilcox, J. M., Solar structure and terrestrial weather, *Science*, *192*, 745, 1976.

Wilken, B., I. A. Daglis, and S. Livi, Observations of geomagnetic storms by the CRRES satellite, *Eos Trans. AGU*, *73* (43), 457, 1992a.

Wilken B., W. Weiß, D. Hall, M. Grande, F. Søraas, and J. F. Fennell, Magnetospheric Ion Composition Spectrometer onboard the CRRES spacecraft, *J. of Spacecraft and Rockets*, *29*, 585-591, 1992b.

Williams, D. J., The Earth's ring current: Causes, generation, and decay, *Space Sci. Rev.*, *34*, 223-234, 1983.

Wodnicka, E. B., The magnetic storm main phase modelling, *Planet. Space Sci.*, *37*, 525-534, 1989.

Wodnicka, E. B., What does the magnetic storm development depend on?, *Planet. Space Sci.*, *39*, 1163-1170, 1991.

Yau, A. W., B. A. Whalen, W. K. Peterson, and E. G. Shelley, Distribution of upflowing ionospheric ions in the high-altitude polar cap and auroral ionosphere, *J. Geophys. Res.*, *89*, 5507-5522, 1984.

Yau, A. W., P. H. Beckwith, W. K. Peterson, and E. G. Shelley, Long-term (solar cycle) and seasonal variations of upflowing ionospheric ion events at DE-1 altitudes, *J. Geophys. Res.*, *90*, 6395-6407, 1985.

Yau, A. W., W. K. Peterson, and E. G. Shelley, Quantitative parameterization of energetic ion outflow, in *Modeling Magnetospheric Plasma*, Geophys. Monogr. Ser., vol. 44, edited by T. E. Moore and J. H. Waite Jr., pp. 211-217, AGU, Washington, D. C., 1988.

Young, D. T., H. Balsiger, and J. Geiss, Correlations of magnetospheric ion composition with geomagnetic and solar activity, *J. Geophys Res.*, *87*, 9077-9096, 1982.

I. A. Daglis, Institute of Ionospheric and Space Research, National Observatory of Athens, Metaxa & Vas. Pavlou Str., GR–152 36 Palea Penteli, Greece.

Dynamics of the Magnetotail During Magnetic Storms: Review of ISEE 3 and GEOTAIL Observations

Susumu Kokubun

Solar-Terrestrial Environment Laboratory, Nagoya University, Toyokawa, Japan

This paper reviews ISEE 3 and GEOTAIL observations of the distant magnetotail during geomagnetically disturbed intervals. Although the probability of encountering the magnetotail is reduced because non-radial solar wind flow associated with interacting solar wind streams moves a compressed tail away from the nominal position during disturbed intervals, a series of temporal entries of spacecraft into the tail at positions of larger than 100 R_E downstream from the earth are often observed during the storm main phase. In such cases the magnitude of lobe magnetic field becomes at times more than 30 nT. The magnetic field of as large as 53 nT was observed in the distant tail lobe by GEOTAIL at -182 R_E on March 9, 1993, when a moderate magnetic storm was in progress. The characteristic time scale of magnetospheric period is approximately 20 minutes. These entries into the magnetotail are usually preceded by a strong southward component of sheath magnetic field. It is confirmed that pressure balance is approximately holds at times of magnetopause crossings during enhanced magnetosheath conditions. This indicates that the strong lobe magnetic fields are due to the compression associated with the increase of solar wind thermal and magnetic pressures. Observations of large lobe field events suggest that the addition of magnetic flux to the magnetotail caused by the merging process near the earth during magnetic storms constrain the magnetotail radius from decreasing the nominal value significantly even at distances beyond ~ 150 R_E, although the magnetotail is in a compressed state due to the increase in exterior pressure.

INTRODUCTION

In their extensive review, Gonzalez et al. [1994] define a geomagnetic storm as an interval of time when a sufficiently intense and long-lasting interplanetary convection electric field leads, through a substantial energization in the magnetosphere-ionosphere system, to an intense ring current strong enough to exceed some key threshold of the quantifying storm time Dst index. The principal defining property is the creation of an enhanced ring current, formed by ions and electrons in the 10-300 keV energy range, located usually between 2 to 7 R_E. The ring current produces a magnetic field disturbance, which is opposite to the Earth's dipole field inside the ring current. Outside the ring current, it produces an enhancement of magnetic flux equivalent to an increase of the Earth's dipole moment. Oguti [1995] has estimated this effect to the magnetotail and the magnetopause position by a simplified model of a equivalent dipole moment of the ring current. It is suggested that a 100-200 nT Dst cause a 10 - 30 % increase of the tail radius.

It has been known that the auroral oval expands equatorward from its normal position associated with the development of the ring current [e.g., Akasofu and Chapman, 1963; Meng, 1984]. The expansion of auroral oval, and thus the polar cap size, during magnetic storms is closely related to a long-lasting interplanetary convection electric field. Thus, an increase of the open magnetic flux would cause a magnetic flux increase in the magnetotail. Observations in the distant magnetotail will give useful information on: to what extent the additional magnetic flux reaches.

ISEE 3 observations of the distant magnetotail have revealed that the lobe field magnitude is approximately 9 nT at ISEE 3 apogee distances near 230 R_E [Slavin et al., 1985]. It has also been reported that the tail has "ceased" expanding in radius at 230 R_E [Baker et al., 1987], indicating the tail boundary be aligned parallel to the solar wind flow, at least on average. Furthermore, Fairfield [1993] has shown that the magnetotail configuration is controlled not only by solar wind flow and temperature, but also by the interplanetary magnetic field (IMF) and has suggested that very northward IMF can eliminate the extended tail. In fact, GEOTAIL observations support this suggestion that a very different configuration occurs when the IMF remains northward for periods of at least several hours [Fairfield et al., 1996].

The effect of IMF By has also been studied extensively [e.g. Tsurutani et al., 1984: Sibeck et al., 1985a and b; Sibeck et al., 1986; Owen et al., 1995]. When the IMF By is positive, the sense of torque would twist the north lobe toward the dawn flank and toward the dusk flank for the negative By [Cowley, 1981]. ISEE 3 observations have been shown to be consistent with a sense predicted from the tail twisting models [e.g. Sibeck et al., 1986].

Previous studies on the plasma sheet dynamics by using ISEE 3 and GEOTAIL data have focused mainly on substorm-related phenomena, such as plasmoids and traveling compression region (TCR) [e.g. Baker et al., 1987; Moldwin and Hughes, 1992; Slavin et al., 1993; Nagai et al., 1994]. Few studies have, however, been made on dynamics of the distant tail during magnetic storms. This may be partly because the probability of encountering the magnetotail is reduced during enhanced solar wind intervals, even when the spacecraft is in the nominal center of the magnetotail [Fairfield, 1993]. Since the presence of substorms is certainly not a sufficient condition for the development of a magnetic storm [Gonzalez et al., 1994], studies of dynamical behavior of the magnetotail during magnetic storms are needed to understand the physical process occurring in the distant tail..

This paper reviews recent GEOTAIL observations during magnetic storms, as well as ISEE 3 observations.

ISEE 3 OBSERVATIONS DURING DISTURBED INTERVALS

In the analysis of the ISEE 3 observations Fairfield [1993] has presented simultaneous IMF 8 and ISEE 3 data for five geomagnetically disturbed intervals. During these intervals ISEE 3 remained mostly within the magnetosheath, even when it was very near the center of an average tail as shown Figure 1. Figure 1 shows the relative amount of time that ISEE 3 spent in various regions on 40 days between July 4 and August 13, 1983. In the bottom of the figure by Fairfield [1993] the Dst index variation is added. ISEE 3 was near apogee from X = -236 to X = -215 and was within ±10 R_E from an aberrated magnetotail axis (3.5°). The spacecraft moved gradually from south to north. During the interval shown, five periods stand out because of their high percentage of time in the magnetosheath: July 12-13, July 16, July 23-24, August 2 and August 7-8. Figure 2 shows hourly average solar wind parameters from IMP 8 for the interval of Figure 1. Vertical dashed lines roughly delineate the four intervals within which ISEE 3 was mostly in the magnetosheath. These data indicate that the missing intervals of the magnetotail are associated with high Kp and with high pressures and anomalous solar wind flow angles.

Fairfield [1993] suggested that non-radial solar wind flow associated with interacting solar wind streams moves a compressed tail due to enhanced external pressure away from the nominal position at these times. These effects along with the well-known association of high-velocity streams with geomagnetic activity account for the general association of enhanced Kp and ISEE 3 in the magnetosheath seen in Figure 2. It is further pointed out that during several few-hour intervals of strongly northward interplanetary magnetic field (IMF) within these periods the solar wind is more radial and cannot explain the entry of the spacecraft into the magnetosheath. It is suggested that the magnetotail became narrower under northward IMF [Fairfield, 1993; Fairfield et al., 1996]

A comparison with the Dst index also indicates that the minimum probability of magnetospheric residence is observed in the positive phase of Dst index, as is seen in Figure 1. This is explained well in terms of the combined effect of an external pressure increase and the northward IMF on the decrease in tail radius. We also note in Figure 1 that the probability of the residence of the spacecraft in the magnetosphere tends to increase rapidly toward the nominal value, associated with the development of Dst. It is interesting in this connection to examine whether non-radial solar wind flow and the associated enhanced thermal pressure can account for this tendency.

Since the magnetosphere is often in a compressed state due to enhanced solar wind during the main phase of magnetic storms, a rapid increase in the probability of magnetospheric residence appears to be controlled not only by solar wind parameters, but also by the direction of IMF. As will be shown in the next section, GEOTAIL observations suggests that additional magnetic flux caused by the merging process near the earth associated with a sustained IMF Bz significantly

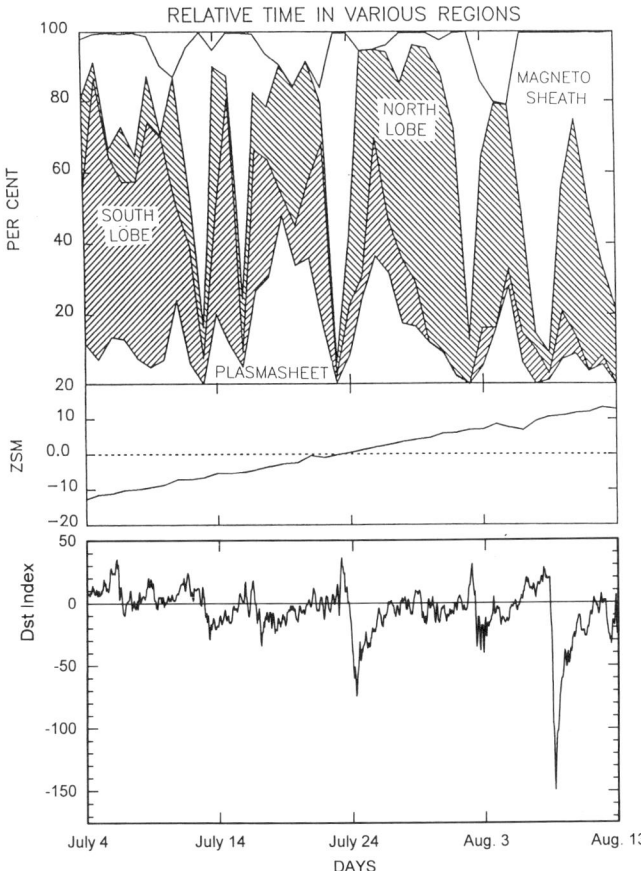

Figure 1. Percent time of ISEE 3 spent in various magnetospheric regions on days between July 4 and August 3, 1983. The area above the top line indicates the percent of time the region could not be determined. The other regions are as indicated. The lower middle panel indicates the daily average solar magnetospheric Z coordinate of the spacecraft. The average position is a good predictor of the relative time in various regions except for five intervals that are predominantly magnetosheath [Fairfield, 1993]. In the bottom panel is plotted the Dst variation.

constrain the magnetotail radius from decreasing the nominal value during magnetic storms.

Although the dynamical behavior of the plasma sheet in the distant tail has been studied by a number of workers by using ISEE 3 data [e.g. Baker, et al., 1987; Moldwin and Hughes, 1992; Owen, et al., 1995; Slavin et al., 1993], few studies have been made on the plasma sheet dynamics in relation to magnetic storms. Ho and Tsurutani [1996] have recently studied the distant tail behavior during three magnetic storms with a peak Dst ranging from -150 to -220 nT. They have reported important features that very strong earthward flows (up to 1200 km/s) were observed with ISSE 3 during the storm recovery phase.

In studies of CDAW 8 events by Fairfield et al. [1989] and Richardson et al. [1989] we can find observations of the magnetotail at GSM coordinates (-108, -1, 4) R_E during a moderate magnetic storm. Figure 3 shows ISEE 3 and AE data for CDAW 8 interval B (0500-1400 UT, March 25, 1983) [Richardson et al., 1989]. Figure 4 represents the Dst variation on March 24-25, 1983. Prior to a storm sudden commencement (ssc) at 0544, the AE index had remained at levels of more than ~500 nT associated with a slow decrease in Dst index. The Dst decreased rapidly, reaching ~ -100 nT 3 hours after the ssc. During the CDAW 8 interval the Dst index was at levels of -90 ~ -100 nT. A compression of the tail lobe field

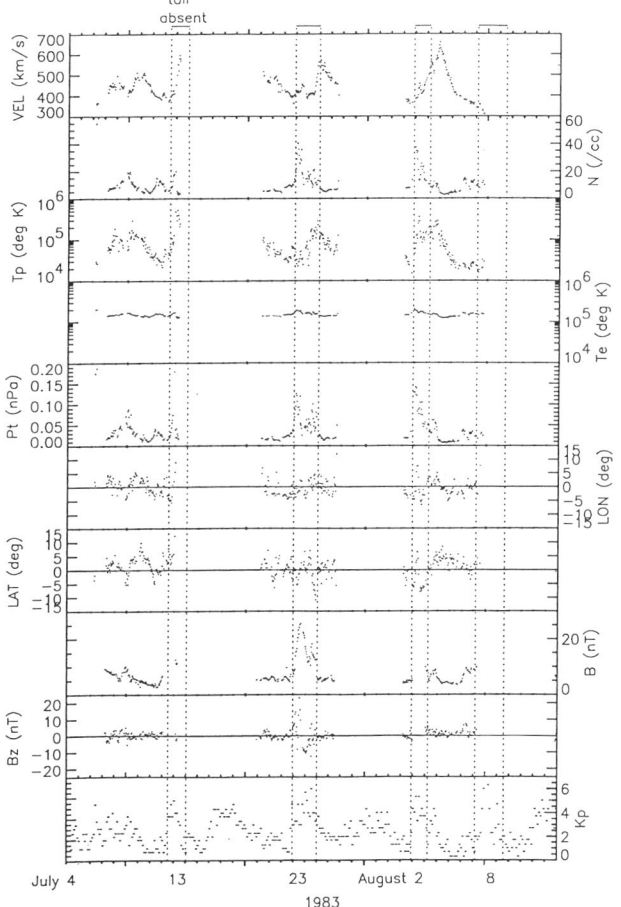

Figure 2. Hourly average solar wind parameters from IMP 8 for the interval of Figure 1. Magnetospheric intervals are roughly designated by vertical lines. The measured values of velocity, density, and temperature are shown in the top three panels. The fourth panel indicates a value of electron temperature inferred from measured density. Subsequent panels display the thermal pressure, the longitudinal and latitudinal angles of the solar wind flow, the magnetic field strength, and its Bz component, and Kp index [From Fairfield, 1993].

was observed to propagate down to the tail at approximately the solar wind speed, following the ssc [Fairfield, et al., 1989]. At ISEE 3, the lobe magnetic field increased from 12 nT to 21 nT during this interval and remained elevated around ~20 nT in the following day.

Fairfield et al. [1989] have identified seven separated electrojet intensifications in ground magnetogram, as indicated in the top part of Figure 3. Richardson et al. [1989] have discussed signatures of plasmoid passage and other features related to reconfiguration and energy losses in the near-Earth magnetotail produced by these intensifications in detail. We have not mentioned their results in detail here, but their conclusion from the analysis of the CDAW 8 interval is important:

Plasmoids are observed in the distant tail following disturbance enhancements, the time of their appearance being generally consistent with disconnection from the near-Earth region at the time of the enhancement. Their structure is consistent with the "neutral line model" [e.g., Hones, 1979; Baker, et al., 1987; Richardson and Cowley; 1985; Richardson et al., 1987]. However, not all enhancements in geomagnetic activity result in the observations of plasmoids. In particular, the CDAW 8 data suggest that during extended intervals of strong activity, the neutral line may reside in the near-Earth tail on an essentially continuous basis and that some disturbance enhancement may then relate to an increase in the reconnection rate at a pre-existing neutral line, rather than to new neutral line, and plasmoid formation.

In Figure 3 we note that ISEE 3 spent two hours from 1100, moving back and forth between the north tail lobe and the magnetosheath several times. Figure 5 shows more clearly transitions between the magnetotail and the magnetosheath [Fairfield et. al., 1989]. The time scale of these transition is 10 ~ 20 minutes. The Bz component of sheath magnetic field was negative in this period, as expected from a continuous ground magnetic activity. These occasional spacecraft residences in the magnetotail or in the magnetosheath have been examined by Ho and Tsurutani [1996] and are also seen in the data during disturbed intervals presented by Fairfield [1993]. We will see the same feature in GEOTAIL observations, as will be presented in the next section.

GEOTAIL OBSERVATIONS

GEOTAIL was controlled by the lunar double swingby maneuvers until October, 1994 since its launch in July, 1992 [Nishida, 1994]. Although the solar activity was in a declining phase for this period, GEOTAIL observed the magnetotail for more than 20 storm periods during the two years [Kokubun et al., 1996]. We will review GEOTAIL observations during magnetic storms in this section.

Yamamoto et al. [1994] have shown that the average profile of the magnetic field strength in the distant tail lobe is confirmed to fit approximately to the power law, $B(X) = 125 \times |X|^{-0.53}$ ($X \leq -130$ R_E), derived from the ISEE 3 observation by Slavin et al., [1985]. Tsurutani et al. [1986] have reported that the lobe field magnitude increases with increasing geomagnetic activity. Yamamoto et al. [1994] have also pointed out that GEOTAIL occasionally observes large lobe magnetic fields of more than 20 nT associated with magnetic storms. Figure 6 illustrates hourly average magnetic fields obtained with GEOTAIL along with Dst variations in March, 1993. We can see a general association of peaks in magnetic field data with developments of Dst variations. Almost all peaks in magnetic field data are confirmed to observed in the tail lobe. As will be shown later, the largest magnitude of 53 nT in the tail lobe was recorded on March 9, 1993 when a moderately large magnetic storm was in progress [Kokubun et al,. 1994].

Kokubun et al., [1996] have examined large lobe field events of more than 20 nT as observed with GEOTAIL. They found that large lobe field were mostly observed during storm main phases under enhanced solar wind conditions. Figure 7 shows nine examples of Dst index variations corresponding to intervals of large field events. In this figure intervals when the magnitude of lobe magnetic field exceeded 20 nT are indicated by thick lines. It is, however, noted that GEOTAIL did not always stay in high field tail lobe during the whole interval indicated by the thick line. Most events were observed in association with a temporal residence of the spacecraft in the magnetotail. Durations of large field events in the magnetotail is found to range from several minutes to several hours. This observation is also consistent with the probability of magnetospheric residence of ISEE 3 as seen in Figure 1.

Figure 8 shows magnetic field obtained by GEOTAIL and IMP 8 on March 8-9, 1993 [Kokubun, et al., 1996]. In Figure 8 time for GEOTAIL data is shifted ahead by 33 minutes to account for an apparent convection time from IMP 8 to GEOTAIL. A storm sudden commencement (ssc) occurred at 2137 on March 8. The corresponding interplanetary shock was identified at 2134 in the IMP-8 magnetic field data. IMP-8 was located at (7.0, 16.1, 30.1) R_E in GSM coordinates at that time. A sudden change of sheath magnetic field was observed with GEOTAIL at a distance of $X_{GSM} = -182.5$ R_E at 2208:10. In comparing the magnetic field data from GEOTAIL and IMP-8 we note that the variation in sheath field observed by GEOTAIL were strikingly similar to IMF magnetic variation, although the field magnitude in the distant sheath was slightly smaller than that of IMF. A series of GEOTAIL entries into the magnetotail were observed in intervals of 0112-29, 0147-0210, 0228-0238, 0406-15, 0440-53 and 0529-0714, as indicated by dashed lines in Figure 8. These tail entries of the spacecraft are evidently identified by large deviations of GEOTAIL magnetic field data from IMF

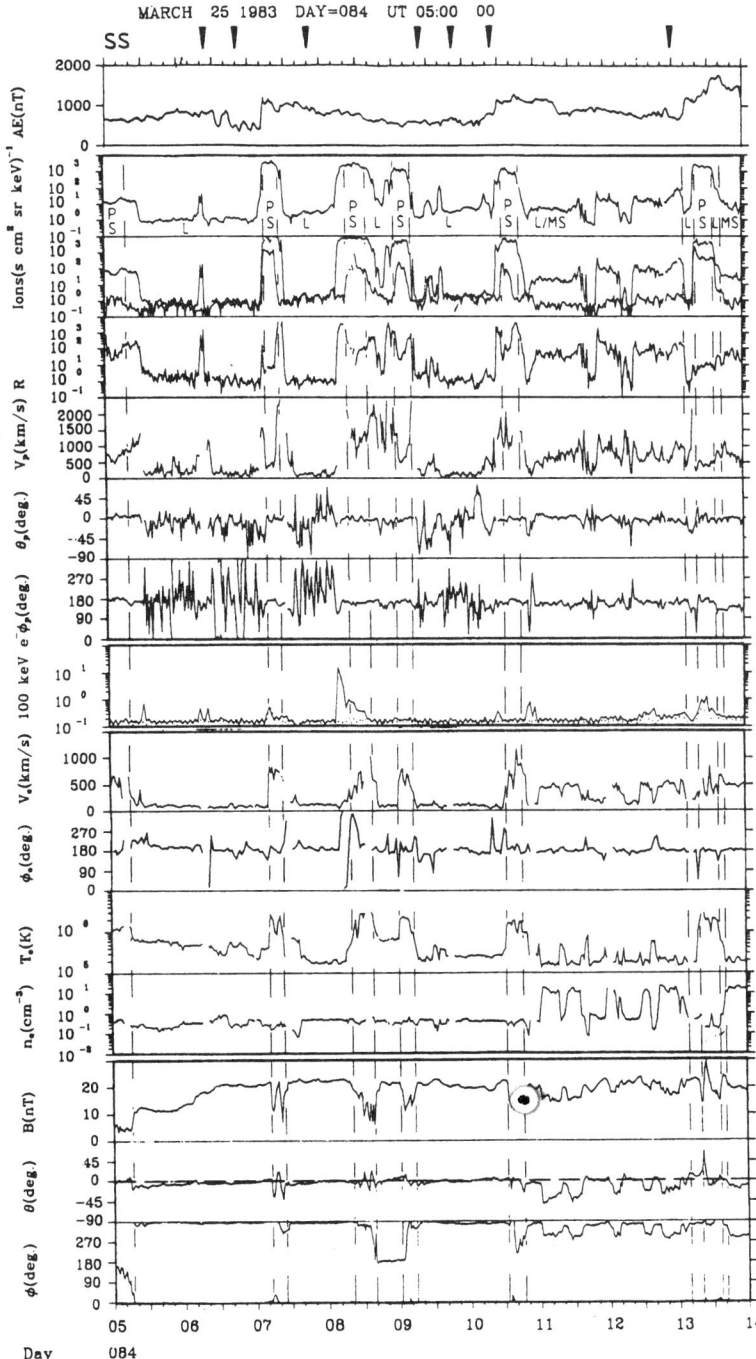

Figure 3. ISEE 3 and AE data for CDAW 8 interval B (0500-1400 UT, March 25, 1983) (spacecraft location (-108, -1, 4) R_E). The top panel shows the AE index (nT), together with indications of the times of enhancement of geomagnetic activity, given by the arrows. The second panel shows energetic particle anisotropy spectrometer data, from the top, the 35 to 56-keV direction averaged ion intensity, the tailward and earthward directed 35 to 56-keV ion intensities and their ratio (R), and the ion bulk flow velocity and polar and azimuthal directions. The electron plasma ecliptic plane flow speed and azimuth, temperature and density are show in the third panel. The bottom panel indicates the magnetic field strength and GSM direction. The regions encountered by ISEE 3, plasma sheet (PS), lobe (L), and magnetosheath (MS), are indicated in the second panel[From Richardson et al., 1989].

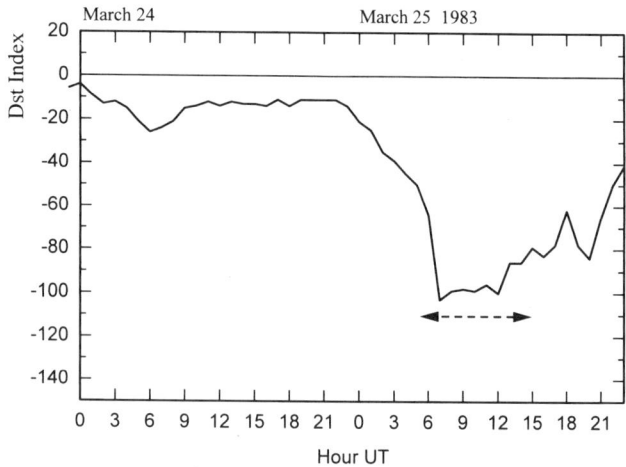

Figure 4. The Dst index on March 24 - 25, 1983.

values. These entries, except the last one, occurred during an interval of negative By of ~ -20 nT. The largest value of the lobe magnetic field, 53.0 nT, in the distant tail epoch of the GEOTAIL observations was recorded at 0121:03 during the first lobe entry on March 9. It is important to note that the B_z components of both the IMF and sheath magnetic fields were southward during certain intervals before the lobe entries of GEOTAIL.

Ground magnetic activities are illustrated in Figure 9. In the bottom of the figure are overlapping plots of horizontal components from auroral and subauroral latitude stations. Magnitudes of horizontal disturbances in the polar cap, denoted by PC in Figure 9, roughly represent the sunward current in the central polar cap. The Dst index was positive for approximately two hour after the ssc and then decreased from ~0h to ~6h on March 9 associated with a large enhancement of AE activity. The series of tail entries occurred during a

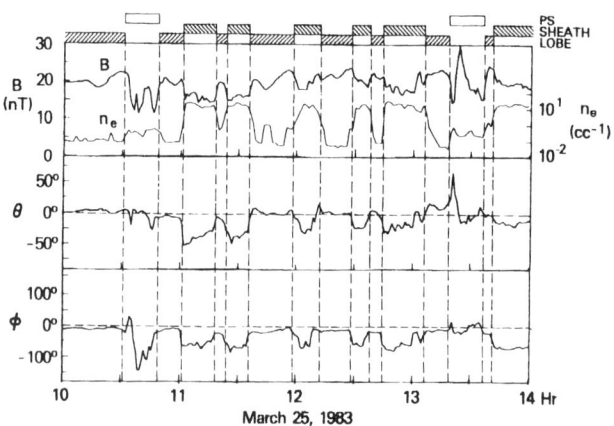

Figure 5. ISEE 3 magnetic field data from 1000 to 1400 UT on March 25, 1983. Vertical lines indicate transitions between the plasma sheet, the tail lobe, and the magnetosheath [From Fairfield at al., 1989].

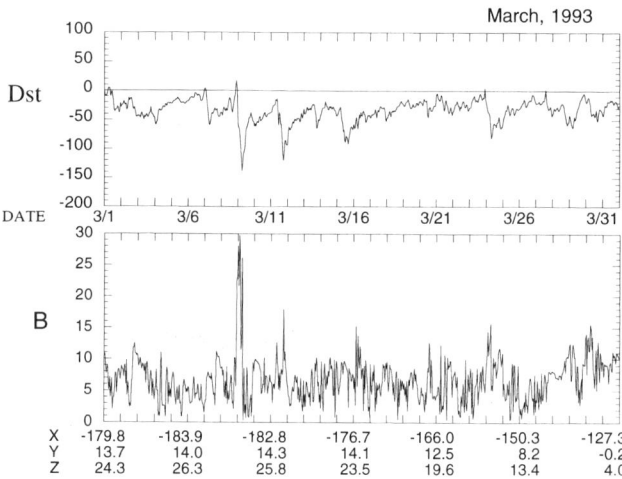

Figure 6. Hourly average values of the magnetic field strength from GEOTAIL and Dst variations in March, 1993. Most peaks of more than 10 nT in magnetic field strength correspond to minima in Dst.

growing stage of ring current activity associated with the enhancement of AE activity. The first three magnetospheric intervals of durations of 10 - 23 minutes occurred during a period of continuous electrojet activity. The fourth and fifth

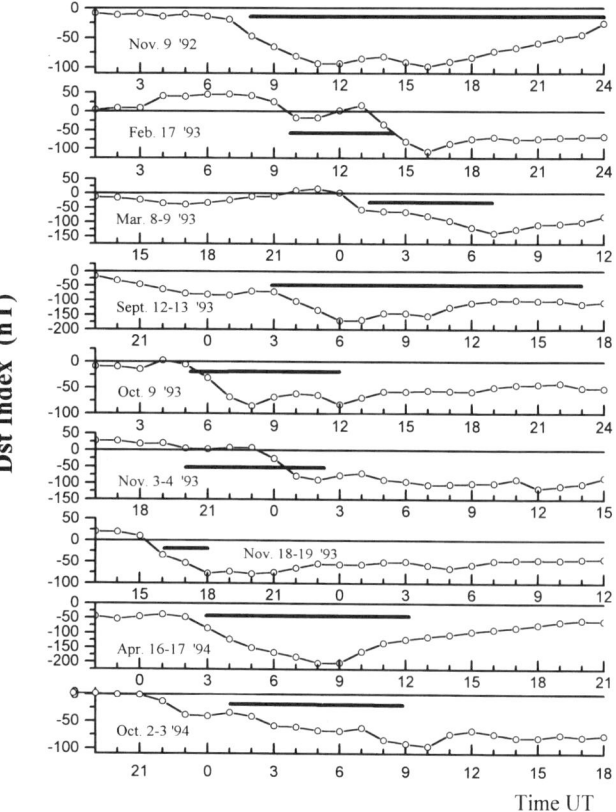

Figure 7. Dst indices and intervals of large lobe fields indicated by thick lines as observed with GEOTAIL [Kokubun et al., 1996].

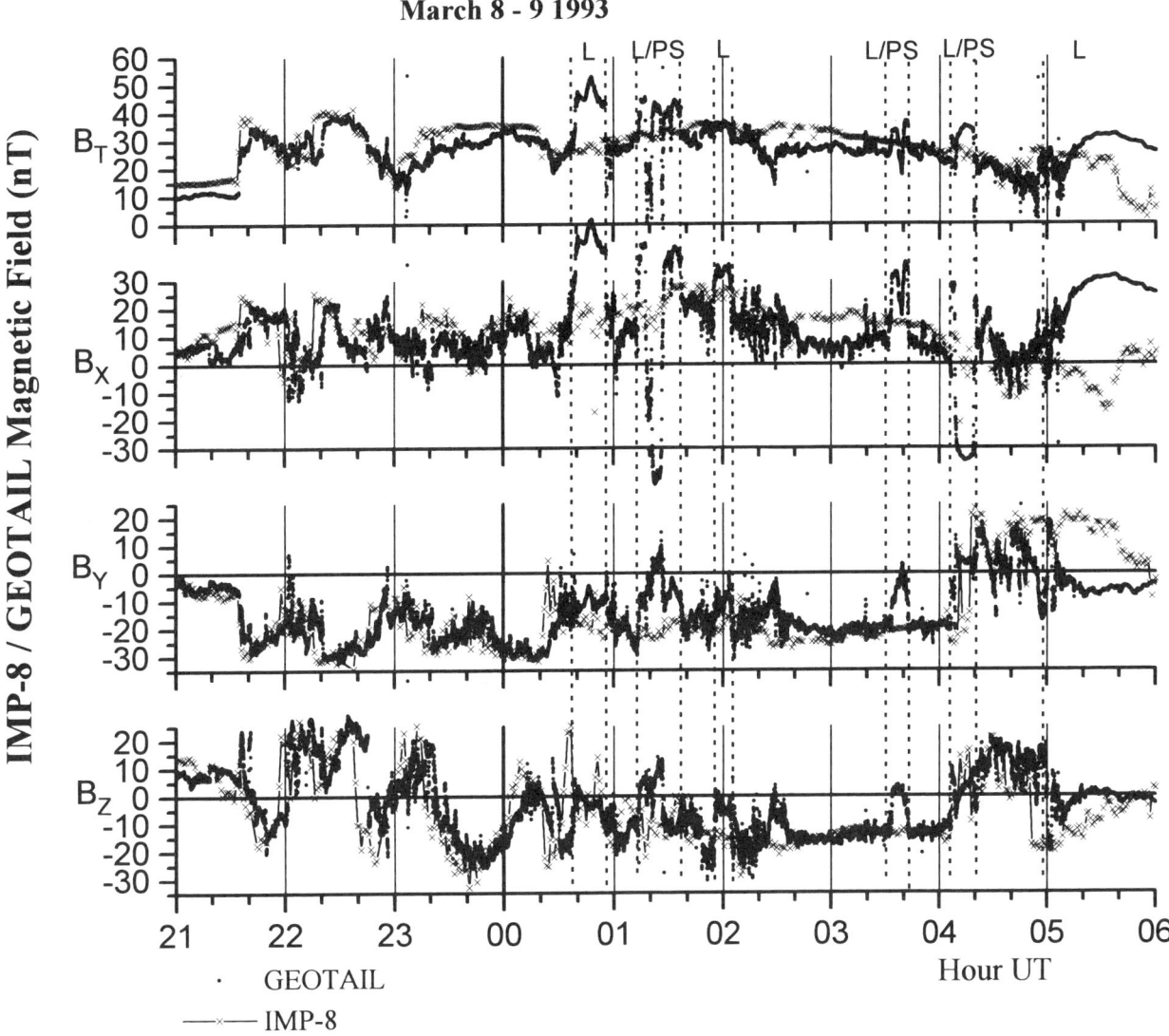

Figure 8. Geotail and IMP-8 magnetic field data in GSM coordinates on March 8 - 9, 1993. Time for GEOTAIL data is shifted by 33 minutes taking into account an convection speed of IMF variations to the location of GEOTAIL. Dashed lines indicate magnetopause crossings of the GEOTAIL spacecraft. L and PS denote that GEOTAIL observed the lobe and the plasma sheet, respectively [Kokubun et al., 1996].

intervals were observed under a similar magnetic condition, but of smaller AL activity. Smallness in AL may be due to the lack of station in the north Atlantic. The last entry appears to follow the substorm onset around 05h.

Figure 10 displays plasma data, the flow speed, the flow angles to the GSM tail axis, the ion temperature, the density, the dynamic pressure ($N_p m V^2$), and the total pressure ($2N_p k T_p + B_T^2/2\mu_0$) [Frank et al., 1994], together with magnetic field data for the five hour interval, when large lobe fields of more than 30 nT were intermittently observed. Here, it is assumed in the total pressure estimation that ion and electron temperatures are equal, because electron moments were not available at the time of analysis. High density ($20 \sim 35/cm^3$) and fast tailward flows ($V_x = \sim 600$ km/s) were observed in the magnetosheath throughout this interval. The signatures of the tail lobe and the plasma sheet are clearly identified by the plasma data. Plasma sheet crossings from the north lobe to the south lobe were observed even though the spacecraft was located far from the nominal neutral sheet in an aberrated GSM coordinates.

Figure 10 also reveals that the total pressure is almost continuous across the magnetopause and is unusually high, being

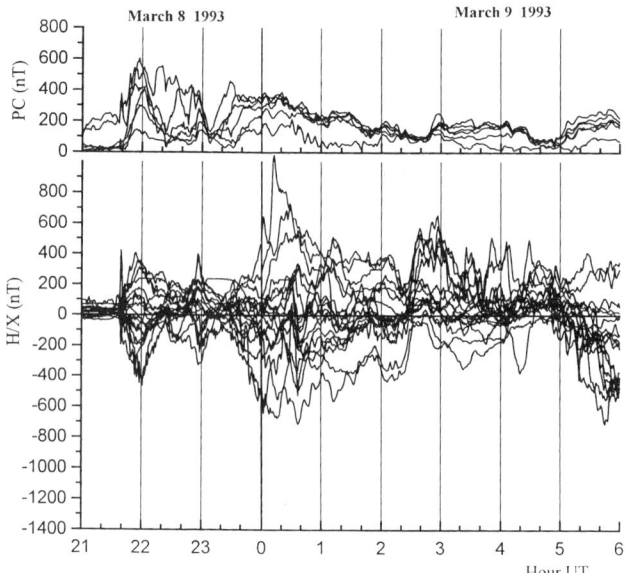

Figure 9. Overlapping plots of magnetic field data from polar regions for the interval of Figure 8. PC in the top panel indicates the horizontal magnitude of deviation from the quiet day mean at Resolute Bay (83.1, 297.4 in geomagnetic coordinates), Cambridge Bay (76.7, 300), Baker Lake (73.6, 320.0), Mold Bay (79.7,260.9), Thule (88.3, 14.1), and Godhavn (79.1, 34.6). In the bottom panel are plotted magnetic field data from 19 stations from 60° to 70° in geomagnetic latitude.

approximately 30 times typical values of 0.02-0.03 nPa in the solar wind [Fairfield, 1993]. Thus, the magnetotail should be in a compressed state compared with the average condition. Assuming that a circular tail is confined by the magnetosheath pressure and that the average tail flux is conserved, we find that tail radius (R) has a 1/4 power dependence on the external pressure (P) [Fairfield, 1993]:

$$R = (P_0/P)^{1/4} R_0 \quad (1)$$

Observed external pressures are as large as 0.52 - 0.99 nPa for eleven crossings of the magnetotail shown in Figure 10. From these values we obtain tail radius of 12.5 - 10.6 R_E for the average tail radius ($R_0 = 25 R_E$) and external pressure ($P_0 = 0.032$ nPa : corresponding to the average lobe magnetic field of 9.1 nT). Kokubun et al., [1996] have estimated the tail radius, assuming the tail axis is parallel to the solar wind/sheath flow, as done by Shodhan [1996] and Nakamura et al., [1996]. Tail radii at times of magnetopause crossings estimated from the same procedure range from 14 to 42 R_E for GEOTAIL plasma data and from 10 to 23 R_E for IMP 8 plasma data. These values are considerably larger than those from the flux conservation as mentioned before. However, tail radii estimated from GEOTAIL and IMP 8 data are not necessarily consistent with each other. Since simulations of

Frank et al, [1996] suggest that the Y and Z flow velocities near the boundary depend on position, the sheath flow may not a good substitute for the solar wind when the By component is large as in case of March 9 event.

Kokubun et al., [1996] have reported the following feature observed during magnetospheric intervals: A common feature of flow direction in the magnetospheric intervals, except C and F, is that plasma flows were slightly northward after the entry and were southward before the sheath entry. Also, flows were duskward during magnetospheric intervals. It is very likely that the tail boundary moved outward-then-inward during these intervals, suggesting a wavy magnetopause motion.

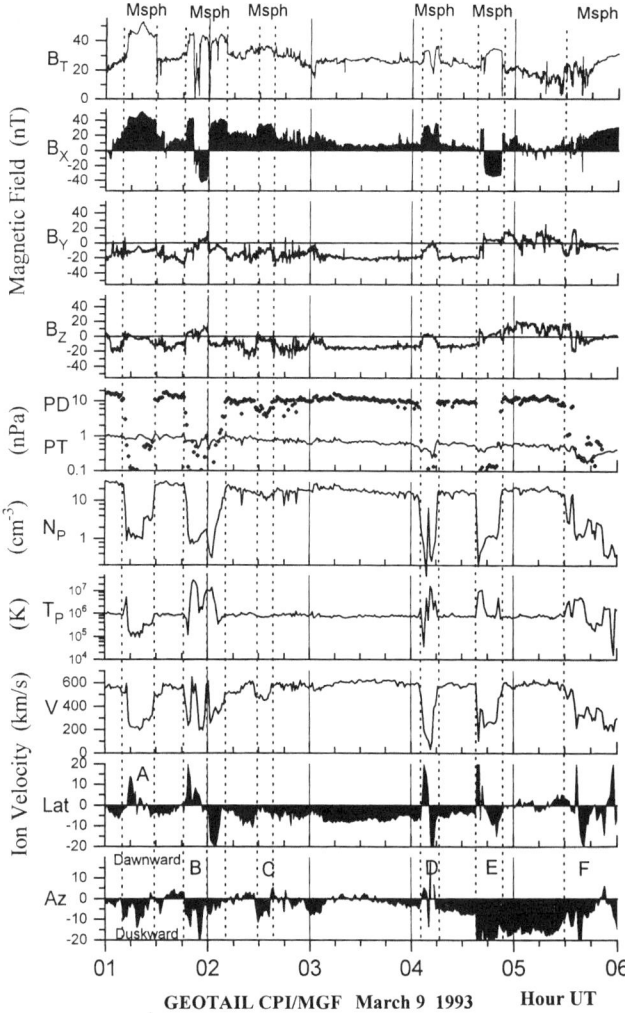

Figure 10. Magnetic field data (~3 sec.) from 0100 to 0500 UT on March 9, 1993 are plotted along with plasma parameters, density (N_p), temperature (T_p), bulk speed (V), latitude and azimuthal angles from $-X_{GSM}$ axis, from the CPI hot plasma detector [Frank et al., 1994]. Dynamic pressure, $N_p mV^2$ (PD) and total pressure, $2N_p kT_p + B_T^2/2\mu$ (PT) are also shown [Kokubun et al., 1996].

The next example is the observation of the largest magnetic storm for the deep tail epoch of GEOTAIL observations on April 17, 1994 [Kokubun, et al., 1996]. Figure 11 shows ground magnetic activities in the polar region (IMP 8 data are not available for this interval). The Dst index began to decrease around 01h and reached the minimum, ~280 nT, around 0740 [Iyemori et al., 1995]. Large electrojet activities of more than 1000 nT were observed at stations located around 60° in geomagnetic latitude associated with the development of ring current. We also note significantly large activity of 200 - 400 nT in the polar cap.

In Figure 12 are plotted low energy particle and magnetic field data for a 15 hour interval during this magnetic storm sequence. Although GEOTAIL was near the central part of the nominal magnetotail, it was in the magnetosheath for more than half of this interval. Note that the Bz component is dominant in the sheath and that the By component is small in the first half of this interval. The magnetospheric signatures were observed in the following periods: 0010-0121, 0202-0212, 0225-0248, 0302-0307, 0314-0416, 0433-0439, 0612-0638, 1012-1240, and 1307-1322. After the last entry the spacecraft was in the sheath for about seven hours. The Bz component in the sheath was strongly northward for this interval (see Figure 14).

During magnetospheric intervals GEOTAIL observed the south lobe and/or the north lobe. An interesting feature to be noted here is a systematic tendency of plasma flow directions in the magnetotail. When the spacecraft was in the southern lobe, plasma flows had northward components. In the period of 0612-0638, GEOTAIL first entered the plasma sheet of high temperature and tailward flows and then moved to the northern lobe. Southward and dawnward components of plasma flows predominate in the northern lobe. Similar to the March 9 cases, magnetosheric periods were preceded by the southward magnetosheath magnetic fields. It is seen that the magnitude of the Z-component of the sheath magnetic field gradually increased, corresponding to the development of Dst and substorms.

In order to examine how entries of the spacecraft relate to directional changes in sheath flow, calculation was made for distances from the expected center of the magnetotail, assuming the tail axis is parallel to the sheath flow. Figure 13 represents the radial distance of GEOTAIL from the estimated tail axis and the GEOTAIL location in Y-Z plane based on sheath flows when the spacecraft was in the magnetosheath. In the bottom panel is also shown the radius of the tail calculated from the magnetosheath pressure, assuming the tail flux conservation. The interesting feature to be noted is that radial distances show a change with a time scale of two hours or more, indicating the windsock effect [Shodhan et al., 1996]. Spacecraft entries were observed when radial distances range in 20 - 30 R_E. Radial distances at times of magnetopause crossings are 5 R_E or more compared with tail radii estimated from the flux conservation. Other features to be noted are relation between the spacecraft location in the magnetotail and the By component of magnetic field. GEOTAIL was mostly in the north lobe in period of 0612-0638, although the spacecraft was in the south of estimated ecliptic plane (Zc = -3 ~ -7 R_E). In periods before and after this interval By components in the sheath were negative. The observed north lobe encounter is expected from the magnetotail twisting model. The model predicts that the sense of the torque from the negative By is to raise the south lobe above the dawnside ecliptic plane and lower the north lobe below the duskside ecliptic plane [e.g. Sibeck et al., 1986]. The south lobe encounter in the interval of 1012-1240 is also explained in terms of the By effect. In this interval the spacecraft appears to be close to the ecliptic plane. By components before and after this interval were strongly positive. Thus, the south lobe encounter is consistent with the twisting model.

Since electrojet activity is too complicated to identify individual substorm onsets, it is difficult to relate individual substorm onsets to magnetospheric entries of GEOTAIL. The following correspondences between ground magnetic

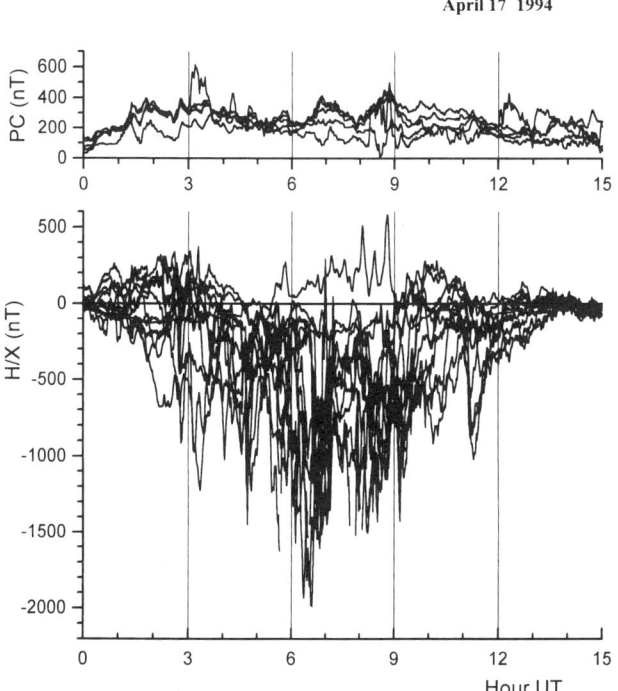

Figure 11. Ground magnetic data from 0000 to 1500 UT on April 17, 1994. PC in the top panel indicates variations of the horizontal magnitude from Eureka (88.9, 318 in geomagnetic coordinates), Resolute Bay, Cambridge Bay, Baker Lake and Ny Ålesund (76.0, 112.3). In the bottom panel are plotted magnetic field data from 12 stations in auroral and subauroral regions [Kokubun et al., 1996].

126 DYNAMICS OF MAGNETOTAIL

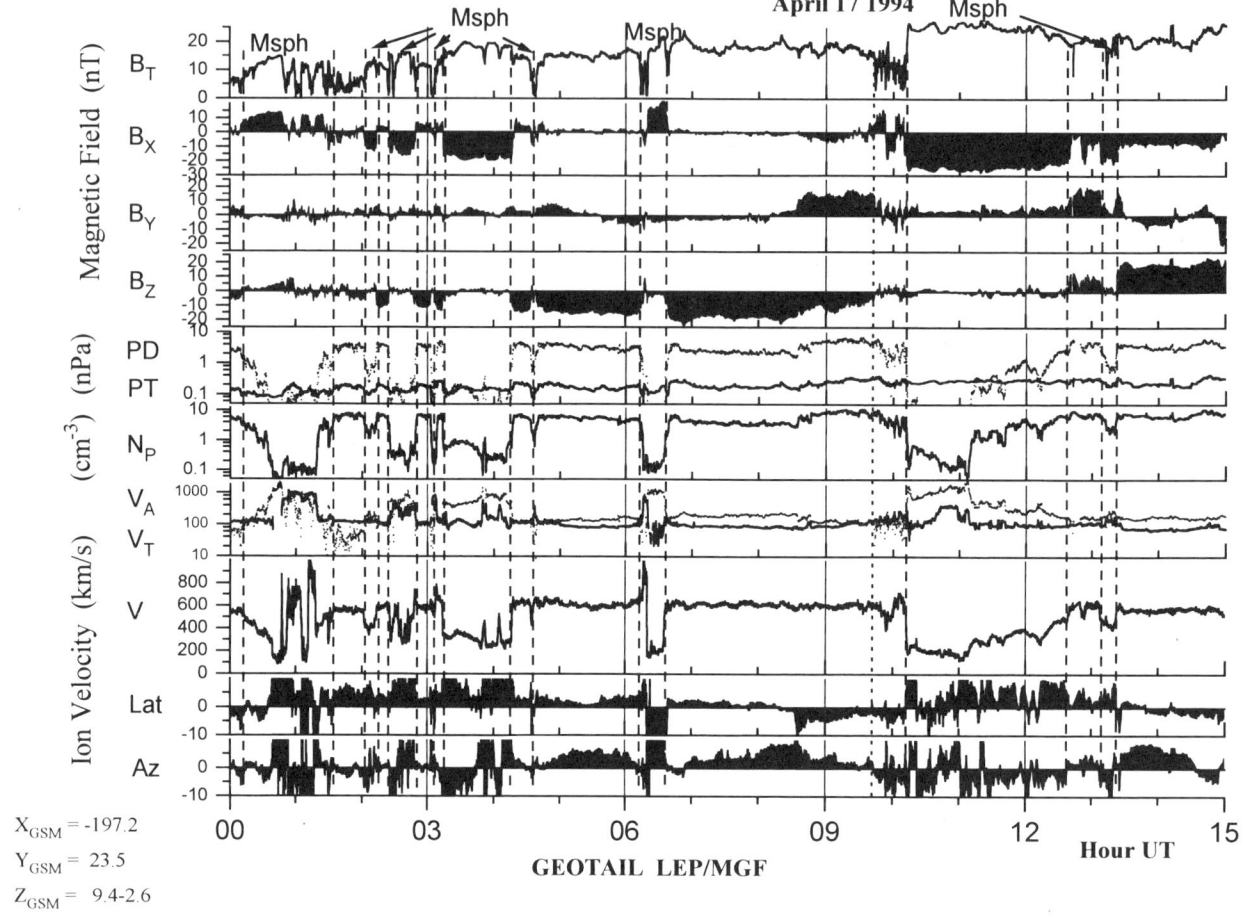

Figure 12. Magnetic field and plasma parameters from the low energy particle experiment [Mukai et al., 1994] on April 17, 1994 are plotted for 15 hour interval in GSM coordinates. The GEOTAIL spacecraft was at positions around X_{GSM} = -197 R_E. Time resolutions are ~3 seconds for magnetic field data and ~12 seconds (four spin periods) for plasma data, respectively. V_A (dot) and V_T (line) in the fourth panel from the bottom represent Alfven velocity and thermal velocity (km/s) [Kokubun et al., 1996].

activities and excursions into the magnetotail are, however, noted here: several excursions from 0202 to 0433 would occur associated with the development of electrojets after about 0130, since the tail orientation change is considered to be a slowly varying phenomenon with a time scale of more than two hours [Shodhan, et al., 1996]. The temporal entry after 0612 may correspond to maximum electrojet activity around 06 h.

A sudden rotation of the sheath magnetic field occurred associated with an increase in ion density at 0834. If this discontinuity propagates with a speed of approximately 600 km/s as measured at GEOTAIL in the magnetosheath, it would pass by the earth around 08 h. A sudden change is seen in the bottom panel of Figure 11 in association with an increase in the polar cap activity. This suggests that the lobe entry from 1012 would relate to magnetic activity after 08 h. Dst and substorm activities recovered rapidly associated with the northward excursion of Bz component as observed by GEOTAIL after the last entry.

Figure 14 shows changes of radial distance in the 8 hour period from 1300 on April 17 in a format different from Figure 13. The sheath magnetic field was strongly northward for about seven hours and the sheath pressure was larger than that observed before this interval. The radial distance inferred from the sheath flow changes with a time scale of 1 ~ 2 hours, indicating the windsock effect. Although the radial distance occasionally became equivalent to the tail radius estimated from the tail flux conservation, only three short entries of the spacecraft for 20 minutes after 1935 were observed under the condition of strongly northward sheath field. This feature

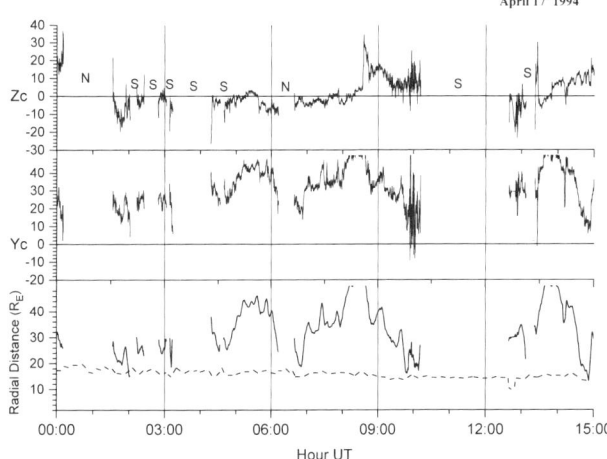

Figure 13. Radial distances from the center of the magnetotail, and spacecraft locations in Y-Z plane, assuming the tail axis is parallel to the sheath flow measured with LEP. In the bottom panel is shown the radius of the tail (dashed line) calculated from the flux conservation. N and S indicate spacecraft residences in the north and south lobes, respectively [Kokubun et al., 1996].

supports the suggestion by Fairfield [1993] and Fairfield et al. [1996] that the magnetotail has less open flux when the IMF is very northward for periods of at least several hours.

SUMMARY AND DISCUSSION

The solar wind flow direction should be an important factor in determining whether or not the spacecraft observes the tail lobe at a distance of larger than the average cross-sectional radius of the tail [Fairfield, 1993]. The size of the tail radius is another crucial parameter. Fairfield [1993] and Nakamura et al. [1996] have concluded that solar wind variations, which cause magnetic storm activity, significantly change the orientation as well as the size of the distant tail so that the magnetosheath can be observed even near the average center of the magnetotail. Although the probability of encountering the magnetotail is controlled by directional changes of the solar wind, an increase in the exterior pressure and the IMF during enhanced solar wind intervals, temporal and successive entries of the spacecraft into the magnetotail at positions of larger than 100 R_E downstream from the earth are often observed associated with the development of the ring current during magnetic storms. The following features are noted for these temporal entries:

1) Spacecraft entries into the magnetotail are often preceded by the negative sheath magnetic field, as expected from the fact that magnetic storms occur in association with sustained intervals of the IMF negative Bz component.

2) The magnitude of the lobe magnetic field during these magnetospheric residences is usually two or more times the average lobe magnetic field of ~9 nT. The largest magnitude observed by GEOTAIL was 53 nT. It appears that the pressure balance holds at times of the magnetopause crossings during enhanced sheath conditions.

3) The time scale of magnetospheric intervals is typically ~20 minutes when the spacecraft is near the magnetopause at distances beyond 100 R_E from the earth. This value was obtained as the median for 73 magnetospheric intervals of the GEOTAIL spacecraft at distances beyond 100 R_E during magnetic storms. The tail radius at times of the magnetopause crossings in these cases is 5 R_E or more, compared with the tail radius inferred from the tail flux conservation. This suggests that the dayside merging of southward IMF contributes to an increase of magnetic flux in the distant tail down to about -200 R_E during disturbed periods.

Tsurutani et al. [1986] and Schindler et al. [1989] have extended the Coroniti and Kennel [1972] model to predict the effect of variation in solar wind ram pressure and magnetic merging at the magnetopause on configuration changes of the distant magnetotail. Their results show that high solar wind streams would place the flaring termination distance beyond the ISEE 3 orbit of 240 R_E. According to calculations by Tsurutani et al. [1986], assuming standard solar wind parameters, the Coroniti and Kennel model predicts substantial field increase inside 100 R_E, but essentially negligible changes at large distances > 140 R_E. Since the flaring termination distance X^*, however, depends on solar wind parameters, the magnetic flux of one tail lobe and the lobe magnetic field strength at $X = X^*$, the calculation by Tsurutani et al.[1986] could not apply to disturbed conditions. As has been discussed by Coroniti and Kennel [1972], the increased solar wind

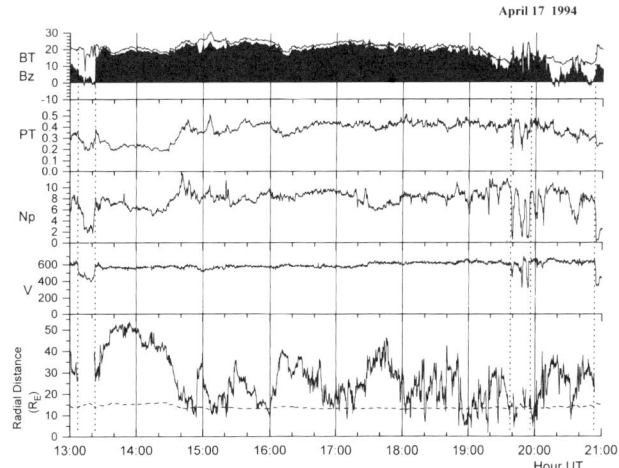

Figure 14. Variations of radial distance are plotted along with plasma, total pressure, density, and bulk flow speed, and the Z component and the magnetic field magnitude for an 8 hour interval on April 17, 1994. The sheath magnetic field was predominantly northward in this interval.

flaring stress and enhanced field-cutting rates at the dayside magnetopause associated with IMF Bz would cause an earthward movement of the whole tail structure and a large increase in magnetic field of the distant tail during magnetic storms. In the following we will illustrate this feature by using observed parameters with GEOTAIL.

An approximate expression for X* is given by Schindler et al. [1989], including the solar wind magnetic pressure and the lobe plasma pressure

$$X^* = X_0 - (4\mu_0/\pi)^{1/2} N/B^{*3/2} [0.6 - (B^*/B_0)^{3/2}/3] \quad (2)$$

where

$$N = VF^{1/2} \quad F = \pi R_0^2 B_0/2$$

with $V = (\rho_{sw} V_{sw}^2)^{1/2}$. Here X_0, R_0, and B_0 are X coordinate, the tail radius and the lobe magnetic field associated with an appropriate near-Earth tail cross section. For cases of March 9, 1993, and April 17, 1994, V^2 and B* are approximately estimated to be 12 nPa and 3 nPa, and 40 nT and 18 nT, respectively. We choose $X_0 = -15$ R_E, $R_0 = 17$ R_E for the former case, and $R_0 = 23$ R_E for the latter case, referring to the empirical models of the near-Earth magnetotail by Roelof and Sibeck [1993] and Petrinec and Russell [1996]. As for the magnetic flux of one tail lobe, we assume $F = 10^9$ Wb/m, which corresponds to the polar cap magnetic flux of co-latitude θp = 21°. Thus, magnetic fields at $X_0 = -15$ R_E are 54 nT and 30 nT. Previous observations [Ohtani and Kokubun, 1990; Nakai et al., 1991] indicate that the lobe field strength around $X = -15$ R_E is in range of 30 ~ 40 nT. Nakai et al. [1991] have also shown that the field strength increases with increases of the solar wind dynamic pressure and |AL|. Therefore, the assumption for the magnetic flux is not inconsistent with near-earth observations. From the equation (2), we obtain $X^* = -47$ R_E and -79 R_E for these two sets of parameters, respectively. We also obtain $R^* = 20$ and 30 R_E. These estimated radii are consistent with observations. The Coroniti and Kennel model, thus, predicts that the spacecraft was in the non-flaring region of the distant tail in these cases.

Shodhan et al., [1996] have examined three possible mechanisms that might cause multiple crossings of the GEOTAIL spacecraft near 170 R_E downstream from the earth: windsock, wrenching, and breathing. They have concluded: (1) Two mechanisms, windsock and breathing, caused the multiple mantle crossing observed during a relatively quiet period on March 18 -19, 1993. (2) Breathing causes variations at the time scale of tens of minutes; windsock causes variations at the time scale of hours. (3) Multiple crossings are little influenced by changes in the IMF orientation (the wrenching mechanism). GEOTAIL observations during magnetic storms, as reviewed in the previous section, indicate the existence of similar time scales for the windsock effect and probably for the breathing mechanism due to internally driven expansions and contractions of the tail cross section [Siscoe et al., 1994]. Although the breathing mechanism might be associated with substorms, observations of plasmoids and TCRs do not account for the time scale of ~20 minutes. Comprehensive surveys of plasmoids and TCRs in the distant tail from ISEE 3 data show that typical plasmoid and TCR last for 228s and 158s, respectively [Moldwin and Hughes, 1992; Slavin et al., 1993].

In this connection it is worthwhile to note that a very high correlation with near-Earth substorm signatures and plasmoids was not found in the midtail (X ~ -80 to -110 RE) during periods of intense geomagnetic activity [Nishida et al., 1988; Fairfield et al., 1989; Richardson et al., 1989]. Moldwin and Hughes [1993] also reported that during periods of intense geomagnetic activity, plasmoids are observed in the deep tail but at a lower frequency than the number of individual substorm onsets or intensifications. Schindler et al. [1989] have suggested from theoretical consideration that the distant neutral line is located at the flaring termination distance. The flaring termination may move earthward as far as ~50 R_E during enhanced solar wind conditions, as discussed before based on the Coroniti and Kennel theory [1972]. A lower frequency of plasmoid detection in the distant tail during periods of intense geomagnetic activity would be due to a large shift of the distant neutral line toward the earth.

In recent years a number of global magnetohydrodynamic (MHD) simulation models have been developed for stating interactions of solar wind with the magnetosphere [Usadi et al., 1993; Walker et al., 1993; Ogino et al , 1994; Frank et al., 1995]. Among them, Frank et al. [1995] have reported the first direct comparison of in situ observations of plasma and magnetic fields in the distant magnetotail with a time-dependent MHD simulation. In this study simultaneous observations of the solar wind ions and the IMF with the IMP 8 spacecraft upstream from the earth during a moderately active period, when IMF By and Bz components were fluctuating with a time scale of 30 minutes, were used as the driving input for a global MHD model. These observations were obtained with GEOTAIL at a downstream distance of 81 R_E near the dawnside magnetopause. The magnetopause positions encountered at this time were also examined by using energetic particle data [Williams et al., 1994]. They found that the unusual magnetopause encounters of time scales between 30 - 70 minutes could be explained in part by solar wind aberration effects, suggesting that additional processes, such as magnetospheric breathing modes, are required to fully explain these observations. Observations for storm intervals also suggest that internal processes related to substorms and/or continuous geomagnetic activity are needed to account for intermittent and temporal entries of the spacecraft into the magnetotail of which time scale is typically 20 minutes. In order to understand the large-scale structure and complicated dynamics of the magnetotail based on a single point measurement during disturbed conditions, comparison with global MHD models should further be proceeded.

Acknowledgement. The author is grateful to R. Nakamura for her help of ground magnetic data processing and to Y. Kamide for useful comments on manuscript. The author would like to thank all members of GEOTAIL MGF and LEP teams. This work was supported by a grand-in-aid for scientific research project 05452082, Ministry of Education Science, Sports and Culture, Japan.

REFERENCES

Akasofu, S.-I., and S. Chapman, The lower limit of latitude (US sector) of northern quiet auroral arcs, and its relation to Dst (H), *J. Atmos. Terr. Phys., 25,* 9-12, 1963.

Baker, D. N., R. C. Anderson, R. D. Zwickl, and J. A. Slavin, Average plasma and magnetic field variations in the distant magnetotail associated with near-Earth substorm effects, *J. Geophys. Res., 92,* 71-81, 1987.

Coroniti, F. V., and C. F. Kennel, Changes in magnetospheric configuration during the substorm growth phase, *J. Geophys. Res., 77,* 3361-3370, 1972.

Cowley, S. W. H., Magnetospheric asymmetries associated with the Y-component of the IMF, *Planet. Space Sci., 29,* 79-96, 1981.

Fairfield, D. H., Solar wind control of the distant magnetotail: ISEE 3, *J. Geophys. Res., 98,* 21,265-21,276, 1993.

Fairfield, D. H., D. N. Baker, J. D. Craven, R. C. Elphic, J. G. Fennell, L. A. Frank, I. G. Richardson, H. G. Singer, J. A. Slavin, B. T. Tsurutani, and R. D. Zwickl, Substorms, plasmoid, flux rope, and magnetotail flux loss on March 25 1983: CDAW 8, *J. Geophys. Res., 94,* 15,135-15,152, 1989.

Fairfield, D. H., R. P. Lepping, L. A. Frank, K. L. Ackerson, W. R. Paterson, S. Kokubun, T. Yamamoto, K. Tsuruda, and M. Nakamura, Geotail observations of an unusual magnetotail under very northward IMF conditions, *J. Geomag. Geoelectr.,48,* 473-487, 1996.

Frank, L. A., K. L. Ackerson, W. R. Paterson, J. A. Lee, M. R. English, and G. L. Picket, The comprehensive plasma instrumentation (CPI) for the GEOTAIL spacecraft, *J. Geomag. Geoelectr., 46,* 23-37, 1994.

Frank, L. A., M. Ashour-Abdalla, J. Berchem, J. Raeder, W. R. Paterson, S. Kokubun, T. Yamamoto, R. P. Lepping, F. V. Coroniti, D. H. Fairfield, and K. L Ackerson, Observations of plasma and magnetic fields in Earth's distant magnetotail: Comparison with a global MHD model, *J. Geophys. Res., 100,* 19,177-19,190, 1995.

Gonzalez, W. D., J. A Joselyn, Y. Kamide, H. W. Kroehl, G. Rostoker, B. T. Tsurutani, and V. M. Vasyliunas, What is a geomagnetic storm, *J. Geophys. Res., 99,* 5771-5792, 1994.

Ho, C. M., and B. T. Tsurutani, The distant tail behavior during high speed solar wind streams and magnetic storms, submitted to *J. Geophys. Res.*, 1996.

Hones, E. W., Jr., Transient phenomena in the magnetotail their relation to substorms, *Space Sci. Rev., 23,* 393, 1979.

Iyemori, T., T. Araki, T. Kamei, and M. Takeda, Mid-latitude geomagnetic indices ASY and SYM, No. 5, Kyoto University, April 1995

Kokubun, S., Y. Kamide, R. Nakamura, T. Yamamoto, K. Tsuruda, T. Mukai, A. Nishida, L. A. Frank, and W. R. Paterson, Structure of the distant magnetotail during the main phase of magnetic storm, *EOS, 75,* 538, 1994.

Kokubun, S., L. A. Frank, K. Hayashi, Y. Kamide, R. P. Lepping, T. Mukai, R. Nakamura, W. R. Paterson, T. Yamamoto, and K. Yumoto, Large field events in the distant magnetotail during magnetic storms, *J. Geomag. Geoelectr., 48,* 561-575, 1996.

Meng, C.-I., Dynamic variation of the auroral oval during intense magnetic storms, *J. Geophys. Res., 89,* 227-235, 1984.

Moldwin, M. B., and W. J. Hughes, On the formation and evolution of plasmoids: A survey of ISEE 3 geotail data, *J. Geophys. Res., 97,* 19,259-19282, 1992.

Moldwin, M. B., and W. J. Hughes, Geomagnetic substorm association of plasmoids, *J. Geophys. Res., 98,* 81-88, 1993.

Mukai, T., S. Machida, Y. Saito, M. Hirahara, T. Terasawa, N. Kaya, T, Obara, M. Ejiri, and A. Nishida, The low energy particle (LEP) experiment onboard the GEOTAIL Satellite, *J. Geomag. Geoelectr., 46,* 669-692, 1994.

Nakai, H., Y. Kamide, and C. T. Russell, Influences of solar wind parameters and geomagnetic activity on the tail lobe magnetic field: A statistical study, *J. Geophys. Res., 96,* 5511-5523, 1991.

Nagai, T., K. Takahashi, H. Kawano, T. Yamamoto, S. Kokubun, and A. Nishida, Initial GEOTAIL survey of magnetic substorm signatures in the magnetotail, *Gephys. Res. Lett., 21,* 2991-2994, 1994.

Nakamura, R., S. Kokubun, Y. Kamide, T. Yamamoto, L. A. Frank, and W. R. Paterson, Magnetosheath observations near the center of the magnetotail during a weak geomagnetic storm on January 25, 1993, *J. Geomag. Geoelectr, 48,* 577-588, 1996.

Nishida, A., The Geotail Mission, *Geophys. Res. Lett., 21,* 2871-2873, 1994.

Nishida, A., J., Bame, D. N. Baker, G. Gloeckler, M. Scholer, E. J. Smith, T. Terasawa, and B. T. Tsurutani, Assessment of the boundary layer model of the magnetospheric substorm, J. Geophys. Res., 93, 5579-5588, 1988.

Ogino, T., R. J. Walker, and M. Ashour-Abdalla, A global magnetohydrodynamic simulation of the response of the magnetosphere to a northward turning of the interplanetary magnetic field, *J. Geophys. Res., 99,* 11,027-11,042, 1994.

Oguti, T., Magnetosphere inflation due to equatorial ring current, *J. Geomag. Geoelectr., 47,* 347-352, 1995.

Ohtani, S., and S. Kokubun, IMP 8 magnetic observations of the high-latitude tail boundary: Location and force balance, *J. Geophys. Res., 95,* 20,759-20,769, 1990.

Owen, C. J., J. A. Slavin, I. G. Richardson, N. Murphy, and R. J. Hunds, Average motion, structure and orientation of the distant magnetotail determined from remote sensing of the edge of the plasma sheet boundary layer with E > 35 keV ions, *J. Geophys. Res., 100,* 185-204, 1995.

Petrinec, S. M., and C. T. Russell, Near-Earth magnetotail shape and size as determined from the magnetopause flaring angle, *J. Geophys. Res., 101,* 137-152, 1996.

Roelof, E. C., and D. G Sibeck, Magnetopause shape as a bivariate function of interplanetary magnetic field Bz and solar wind dynamic pressure, *J. Geophys. Res., 98,* 21,421-21,450, 1993.

Richardson, I. G., and S. W. H. Cowley, Plasmoid-associated energetic ion bursts in the deep geomagnetic tail: Properties of the boundary layer, *J. Geophys. Res., 90,* 12,133-12,158, 1985.

Richardson, I. G., S. W. H. Cowley, E. W. Hones, Jr., and S. J. Bame, Plasmoid-associated energetic ion bursts in the deep geomagnetic tail: Properties of plasmoids and the post plasmoid plasma sheet, *J. Geophys. Res., 92.* 9997-10,013, 1987.

Richardson, I. G., C. J. Owen, S. W. H. Cowley, A. B. Galvin, T. R. Sanderson, M. Scholer, L. A. Slavin, and R. D. Zwickl, ISEE 3 observations during the CDAW 8 intervals: Case studies of the distant geomagnetic tail covering a wide range of geomagnetic activity, *J. Geophys. Res., 94,* 15,189-15,220, 1989.

Schindler, K., D. N. Baker, J. Birn, E. W. Hones, Jr., J. A. Slavin, and A. B. Galvin, Analysis of an extended period of earthward plasma sheet flow at ~220 R_E: CDAW 8, *J. Geophys. Res., 94,* 15,177-15,188, 1989.

Shodhan, S., G. L. Siscoe, L. A. Frank, K. L. Ackerson, and W. R. Paterson, Boundary oscillation at Geotail: Windsock, breathing, and wrenching, *J. Geophys. Res., 101,* 2577-2586, 1996.

Sibeck, D. A., G. L. Siscoe, J. A. Slavin, E. J. Smith, B. T. Tsurutani, and R. P. Lepping, The distant magnetotail's response to a strong interplanetary magnetic field B_y: Twisting, flattening, and field line bending, *J. Geophys. Res., 90,* 4011-4019, 1985a.

Sibeck, D. A., G. L. Siscoe, J. A. Slavin, and R. P. Lepping, Major flattening of the distant geomagnetic tail, *J. Geophys. Res., 90,* 4223-4237, 1985b.

Sibeck, D. A., J. A. Slavin, E. J. Smith, and B. T. Tsurutani, Twisting of the geomagnetic tail, in *Solar Wind-Magnetosphere Coupling,* edited by Y. Kamide and J. A. Slavin, P731-738, Terra Scientific Pub., Tokyo, 1986

Siscoe, G. L., L. A. Frank, K .L. Ackerson, and W. R. Paterson, Properties of the mantle-like boundary layer: Geotail data compared with a mantle model, *Geophys. Res. Lett., 21,* 2975-2978, 1994.

Slavin, J. A., E. J. Smith, D. A. Sibeck, D. N. Baker, R. D. Zwickl, and S.-I. Akasofu, An ISEE 3 study of average and substorm conditions in the distant magnetotail, *J. Geophys. Res., 90,* 10,875-10,895, 1985.

Slavin, J. A., M. F. Smith, E. L. Mazur, D. N. Baker. E. W. Hones, Jr., T. Iyemori, and E. W. Greenstadt, ISEE 3 observations of traveling compression region in the earth's magnetotail, *J. Geophys. Res., 98,* 15,425-15,446, 1993.

Tsurutani, B. T., D. E. Jones, R. P. Lepping, E. J. Smith, and D. G. Sibeck, The relationship between the IMF By and the distant tail (150 - 238 Re), *Geophys. Res. Lett., 11,* 1082-1085, 1984.

Tsurutani, B. T., B. E. Goldstein, and M. E. Burton, and D. E. Jones, A review of the ISEE-3 geotail magnetic field results, *Planet. Space Sci., 34,* 931-960, 1986.

Usadi, A., A. Kageyama, K. Watanabe, and T. Sato, A global simulation of the magnetosphere with a long tail: Southward and northward interplanetary magnetic field, *J. Geophys. Res., 98,* 7503-7517, 1993.

Walker, R. J., T. Ogino, J. Raeder, and M. Ashour-Abdalla, A global magnetohydrodynamic simulation of the magnetosphere when the interplanetary magnetic field is southward: The onset of magnetotail reconnection, *J. Geophys. Res., 98,* 17,235-17,249, 1993.

Williams, D. J., A. T. Y. Lui, R. W. McEntire, V. Angelopoulos, C. Jacquey, S. P. Christon, L. A. Frank, K. L. Ackerson, W. R. Paterson, S. Kokubun, T. Yamamoto, and D. H. Fairfield, Magnetopause encounters in the magnetotail at distances of ~80 R_e, *Geophys. Res. Lett. 21,* 3007-3010, 1994.

Yamamoto, T., K. Shiokawa, and S. Kokubun, Magnetic field structures of the magnetotail as observed by Geotail, *Geophys. Res. Lett., 21,* 2875-2878, 1994.

Susumu Kokubun, Solar-Terrestrial Environment Laboratory, Nagoya University, 3-13 Honohara, Toyokawa 442, Japan.

The Role of Substorms in the Generation of Magnetic Storms

R. L. McPherron

*Institute of Geophysics and Planetary Physics and Department of Earth and Space Sciences
University of California Los Angele*

Abstract. A typical magnetic storm is characterized by an initial phase, a main phase and a recovery phase. The initial phase is caused by increased solar wind dynamic pressure, the main phase by injection into and energization of particles in the radiation belts, and the recovery phase by charge exchange loss of these particles. The effect of the radiation belt particles is roughly equivalent to that of a geocentric ring of current of radius 3-4 Re which decreases the horizontal component of the earth's surface field by as much as 500 nT. The magnitude of this effect is generally indexed by the Dst index. The Dessler-Parker-Sckopke relation shows that in the absence of other effects Dst is directly proportional to the total energy of the radiation belt particles. Thus if Dst is corrected for other effects its time rate of change is a direct measure of the rate at which energy is flowing into or out of the ring current. Various models have been developed to explain the rate of energy input via solar wind coupling and the rate of loss by charge exchange. Statistical studies have parameterized these models and the resulting differential equation solved to predict the time history of Dst as a function of solar wind coupling. It is noteworthy that these models are remarkably accurate, and that they do not involve any measure of substorm activity such as the AL index. In fact, when the rectified solar wind electric field is used to predict both Dst and AL, the prediction residuals for these two indices are completely uncorrelated. This result suggests that the effect of particles injected and energized by the expansion phase of substorms is undetectable in the pressure corrected Dst index. It thus seems more likely that it is the global convection electric field that injects and energizes ring current particles. Thus, despite the fact that substorms occur throughout the main phase development of the ring current, the inductive electric field of the expansion phase is not likely to be the primary source of ring current particles or energy. However, it almost certainly plays a role in trapping the particles transported by the global convection field.

INTRODUCTION

Magnetic storms are produced by short term increases in the number of particles in the earth's radiation belts. Intense magnetospheric substorms occur while magnetic storms are developing. Particles are observed to be injected into the outer edge of the ring current during the expansion phase of substorms. These correlations have led to the development of the substorm injection hypothesis, i. e. that magnetic storms are caused by successive substorm expansions. Collapse of a tail-like field to a more dipolar configuration during the expansion phase is the basic mechanism assumed in many models of this process.

Some recent studies cast doubt on this hypothesis [*McPherron et al.*, 1986,1988;.*Chen et al.*, 1993, 1994; *Iyemori and Rao*, 1996; *Siscoe and Petschek*, 1996]. Prediction filters that relate the solar wind to the Dst index of storm strength show that injection occurs too quickly for the substorm expansion phase to be the cause of the injection. The solar wind is able to account for most of the variation in the strength of the ring current without recourse to indices of substorm processes. A superposed epoch analysis of the Dst index shows that the rate of decrease of Dst decreases, not increases, at expansion onset. These results can be understood in terms of the generalized virial theorem provided most of the energy in a tail field collapse goes into other processes than the thermal energy of radiation belt particles.

In this paper we review these results and suggest an alternative process for ring current injection. Our primary conclusion is that global convection driven by dayside reconnection is the agent that transports and energizes the particles injected into the radiation belts during storms. However, it is essential that there be fluctuations in this field that will move particles on open drift paths to closed drift paths. Natural variations in the solar wind, waves in the magnetosphere, and substorm expansions probably all contribute to this trapping process. However, the signature of substorm expansion does not appear to be evident in the Dst index used to monitor storm strength.

What Is A Magnetic Storm?

Magnetic storms were originally defined as intervals of extraordinary fluctuations in the surface magnetic field [*Chapman*, 1962], however, they are most evident in the aurora. During storms the aurora becomes brilliant and active and expands to lower latitudes. On the ground beneath the aurora there are intense, rapid magnetic field variations caused by concentrated currents flowing through the aurora. At midlatitudes, well away from the aurora and currents, the horizontal component of the earth's magnetic field is significantly reduced by anywhere from 50 to 500 nT. The reduction occurs everywhere on the earth and can be approximated by an axial field parallel to the earth's dipole axis. Maps of this magnetic perturbation suggest that it is caused by a ring of current flowing around the earth. This current is produced by the charge dependent drift of radiation belt particles. The reduction of the surface field is a consequence of both this drift and the dipole moment created by the gyration of a particle around field lines.

A magnetic storm is therefore caused by the trapping, energization, injection, and loss of particles in the radiation belts. The major outstanding question is what physical process is responsible for their energization and injection. It is widely accepted that it is the inductive electric field of the substorm expansion phase. We believe the process is more complicated than this. We suggest that it is the global convection field that is the main transport and energization mechanism. However, fluctuations in global convection induced by changes in the rate of dayside magnetic reconnection, low frequency waves, and substorm intensifications (dipolarizations) are likely to be responsible for trapping the convected particles into the ring current.

Measuring The Strength of a Storm with the Dst Index

The strength of the ring current is generally measured by the Dst (disturbance storm time) index. Since our argument depends on the properties of this index it is necessary to first review how it is defined and calculated from observations. We show that Dst is likely to have systematic and random errors as well as contributions from a number of current systems besides the ring current.

A good index of the strength of a ring current is the axial magnetic field it produces at its center. On the earth this must be approximated by measurements from a station on the surface. If the station is at the magnetic equator then the H component measures the axial field. Elsewhere the H component must be normalized by the cosine of the station's latitude to obtain the equivalent axial field. Since measurements from a single station may contain random measurement errors or systematic errors such as effects of the partial ring current it is better to average many stations. Random errors will be reduced in the average as will be systematic errors if they have an antisymmetric distribution around the earth. Thus the Dst index is defined by

$$D_{st} = \frac{1}{N}\sum_{I=1}^{N} \frac{\Delta H_I}{\cos\theta_i} \quad (1)$$

where ΔH_I is the effect of the ring current at station I with latitude θ_I. Unfortunately, the H component measured at any station contains effects of many other current systems. At a given instant H is given by

$$H = H_0 + H_{SQ} + H_{MP} + H_{SR} + H_{PR} + H_T + H_{SS}$$

where the subscripts indicate respectively the main field secular variation (*0*), solar quiet daily variation (*SQ*), magnetopause current (*MP*), symmetric ring current (*SR*), partial ring current (*PR*), tail current (*T*), and substorm current wedge (*SS*). The main field is the dominant term, but it changes slowly with time. The SQ variation is next in importance. On a quiet day

$$H^Q = H_0 + H_{SQ} + H_{MP}^Q + H_{SR}^Q$$

The secular variation can be estimated by assuming the SQ variation vanishes at midnight and then polynomial fitting midnight readings of H on quiet days.

$$\hat{H}_0 = H_0^Q(00) + H_{MP}^Q(00) + H_{SR}^Q(00)$$

The daily SQ variation is estimated by two dimensional Fourier fits for day of year and time of day throughout a solar cycle subject to the constraint of being zero at midnight. The daily deviation used in calculating Dst will then be

$$\Delta H_I = H^D - \hat{H}_0 - \hat{H}_{SQ}$$

where H^D is the H component measured on a disturbed day. Substituting we obtain

$$D_{st} = \langle H_{SR}^D \rangle + \langle H_{MP}^D \rangle - \langle [H_{MP}^Q + H_{SR}^Q] \rangle$$
$$+ \langle [H_{SQ}^D - H_{SQ}^Q] \rangle + \langle [H_T^D + H_{PR}^D + H_{SS}^D] \rangle$$

where $\langle \rangle$ imply average over the longitudinal distribution of stations. The last two terms are assumed to average to zero. The estimate of the secular variation H_0 is biased by the presence of the quiet magnetopause and ring currents. Also, the diurnal variation H_{SQ} at each station exhibits variations from day to day due to changes in wind patterns and ionospheric composition possibly caused by solar emissions or prior magnetic activity. In the standard Dst index the five quietest days in each month are used to simultaneously estimate both the absolute value of H_0 and H_{SQ}. If the month is unusually disturbed the quiet day subtracted from H will be more negative than it should be so that Dst index during this month will have a systematic positive bias.

In most storm studies the Dst index is corrected to remove effects of the magnetopause and quiet time ring current. The magnetopause effect on a disturbed day is easily estimated by assuming it is an infinite planar sheet current that just cancels the earth's field outside the subsolar point. Since this is determined by a balance of magnetic and dynamic pressure (P_{dyn}) we obtain

$$\frac{B^2}{2\mu_0} = k\rho v^2 \quad \text{or} \quad B = \sqrt{2\mu_0 k\rho v^2} \tag{2}$$

Thus the perturbation from the magnetopause current scales as the square root of the dynamic pressure. The effects of the quiet time magnetopause and ring current may be approximated by an empirical constant

$$c = \langle H_{MP}^Q + H_{SR}^Q \rangle \tag{3}$$

This constant must be estimated from the data and certain assumptions about the behavior of the ring current. Finally, the corrected Dst index is defined as the effect of the disturbed ring current alone

$$D_{ST}^* = \langle H_{SR}^D \rangle$$

Then the corrected Dst index is given by

$$D_{st}^* = D_{st} - b\sqrt{P_{dyn}} + c \tag{4}$$

Note that the scale factor b must also be estimated from data since the tail current location also scales with dynamic pressure and its perturbation partially balances the effect of the magnetopause.

The corrected Dst index still does not necessarily represent the effect of the ring current alone. During the substorm growth phase the tail current grows in strength and moves earthward decreasing the vertical magnetic field in the equatorial plane. Field-aligned currents driven by global convection may also contribute to this effect. Early in a substorm the partial ring current begins to develop with its center near dusk. This current has the same geometry as the substorm current wedge, but with opposite sense and greater width [Horning et al., 1974]. During the expansion phase the substorm current wedge forms near midnight. For both current systems the magnetic perturbation well outside the wedge has the opposite sign to that inside (see Figure 7 of Clauer et al., 1974). However, stations outside are further from the currents so that the perturbations outside are smaller than inside. Thus the interior regions dominate the ground signatures and so both current systems contribute to the Dst index. However, the residual positive effect of the substorm current wedge may be compensated by the negative effect of the partial ring current. These considerations explain why virtually all previous studies of the ring current have ignored these two effects and used the pressure corrected Dst index as a measure of the strength of the ring current.

The effect of the two current systems is made more complicated in practice by the fact that hourly averages from only four stations are used to calculate the standard Dst index. With only four stations the current wedge will make a positive contribution to Dst only if one of the Dst stations happens to be inside the wedge. Stations outside the wedge will make small negative contributions. The same logic applies to the partial ring current except that its angular extent is greater and it is more probable that there will be a residual negative contribution. The process of constructing hourly averages of H also decreases the effect of the narrow substorm current wedge. The typical midlatitude positive bay has a duration of about one hour so that the hour average typically reduces the effect of the wedge on Dst by at least 50%. The partial ring current has a longer life time than the current wedge and so when only it is present Dst will be more negative than it should be.

The effect of the tail current on the Dst index is usually ignored. Simple calculations with a sheet current model of the tail current (and magnetopause return currents) show that the tail effect at quiet times is negative and of order -10 nT. With very disturbed conditions this decreases to about -30 nT. A more complex model of the tail current includes the Region 1 current system as well. As this system grows in strength it decreases the magnetic field just inside the subsolar magnetopause so that pressure balance is maintained as the magnetopause is eroded by reconnection. The magnitude of this effect clearly depends on the size of the magnetosphere determined by solar wind dynamic pressure and how much flux has been eroded. Since the magnetopause is seldom closer than synchronous orbit where the ambient field is 100 nT we guess that the total effect at the earth must be a fraction of this amount.

The preceding arguments suggest that the Dst index is more likely to be biased negative than positive. However, at times the fortuitous location of a station will make a positive contribution to Dst. Positive errors may also be present because the quiet day variation is not properly removed on a particular day. The accumulated error from all of these effects is probably of order 10-30 nT and will appear as random noise in the data, provided the substorm events used in any statistical analysis are uniformly distributed in universal time. Thus time series analysis techniques and superposed epoch analyses of the Dst index should still be able to resolve features of the index that are highly correlated with the solar wind or with the time of onset.

THE DESSLER-PARKER-SCKOPKE RELATION

The physical basis for using the Dst index to study magnetic storms is the Dessler-Parker-Sckopke (DPS) relation [*Dessler and Parker*, 1959; *Sckopke*, 1966]. In its simplest form this relation states that the magnetic perturbation caused by ring current particles is directly proportional to their total energy. This relation can be derived in the following manner. For an equatorial ring current of strength I the magnetic perturbation at the center of the ring is given by $\delta B_r = \mu_0 I / 2\pi r$ where r is the radius of the ring. For a single particle drifting near the equatorial plane the current produced by drift is $I = qv_d / 2\pi r$ where v_d is the average drift velocity. In a dipole field the equatorial drift velocity is given by $v_d = -3E_\perp r^2 / qM$ with q the charge, M the earth's dipole moment, and $E_\perp$ the gyrational energy of the particle. Combining these relations one finds $\delta B_1 = 3\mu_0 E_\perp / 4\pi M$. The effect of gyration as well as drift must be considered. A single gyrating particle creates an infinitesimal current loop or dipole with its moment antiparallel to the earth's field. This dipole creates a positive axial field at the center of the earth given by $\delta B_2 = (\mu_0 / 4\pi)(\mu / r^3)$ where μ is the dipole moment. The dipole moment of a gyrating particle is given by $\mu = E_\perp / B = (r^3 / M) E_\perp$. Thus $\delta B_2 = \mu_0 E_\perp / 4\pi M$. Summing the two contributions we obtain $\delta B = \mu_0 E_\perp / 4\pi M$. If we write this as a fraction of the earth's equatorial surface field we finally obtain

$$\delta B / B_0 = -2E_\perp / 3U_M \qquad (5)$$

where U_M is the total energy in the dipole field outside the earth.

WHAT IS A MAGNETOSPHERIC SUBSTORM?

The concept of a substorm grew out of studies of magnetic storms. *Chapman* [1962] stated:

"A magnetic storm consists of sporadic and intermittent polar disturbances, the lifetimes being usually one or more hours. These I call polar substorms."

Two years later Akasofu generalized this concept by noting that polar substorms were caused by the growth and decay of the auroral electrojets in association with a systematic expansion of the aurora away from midnight. He called his phenomenological model of the auroral development the *auroral substorm* [*Akasofu*, 1964], and similarly the model of electrojet development the *polar magnetic substorm* [*Akasofu et al.*, 1965]. *Coroniti et al.* [1968] and *Akasofu* [1968] generalized the substorm concept still further recognizing that the substorm was magnetospheric wide in scope and hence the term *magnetospheric substorm* was more appropriate. Subsequently *Rostoker et al.* [1980] and *Rostoker et al.* [1987] attempted to provide a more precise definition of a magnetospheric substorm. *Rostoker et al.* [1980] wrote:

"A magnetospheric substorm is a transient process initiated on the night side of the Earth in which a significant amount of energy derived from the solar wind-magnetosphere interaction is deposited in the auroral ionosphere and magnetosphere."

A fundamental part of the substorm definition is the division of isolated substorms into three distinct phases: growth, expansion, and recovery. The growth phase is one in which energy extracted from the solar wind is primarily stored in the magnetotail while smaller amounts are dissipated in the ring current and ionosphere. The expansion phase is a phase of explosive energy release in which particles are injected into synchronous orbit, the tail-like field becomes more dipolar, and the aurora and the electrojets build up in the ionosphere. The recovery phase is the interval during which a quiet time configuration of field, particles, aurora and ionospheric currents is reestablished.

THE SUBSTORM INJECTION HYPOTHESIS

Chapman's original definition of substorm visualized the substorm as a building block for magnetic storms. Substorms were simply little storms. Thus a magnetic storm was nothing more than the sum of many substorms,

$$\text{Storm} = \sum \text{Substorms}$$

The observational basis for this belief was the fact that no magnetic storm has ever been seen in which substorms are not present during the development of the main phase, i.e. during the injection and energization of the ring current. However, a visual comparison of plots of the AL and Dst indices reveals that substorms occur at times when there is

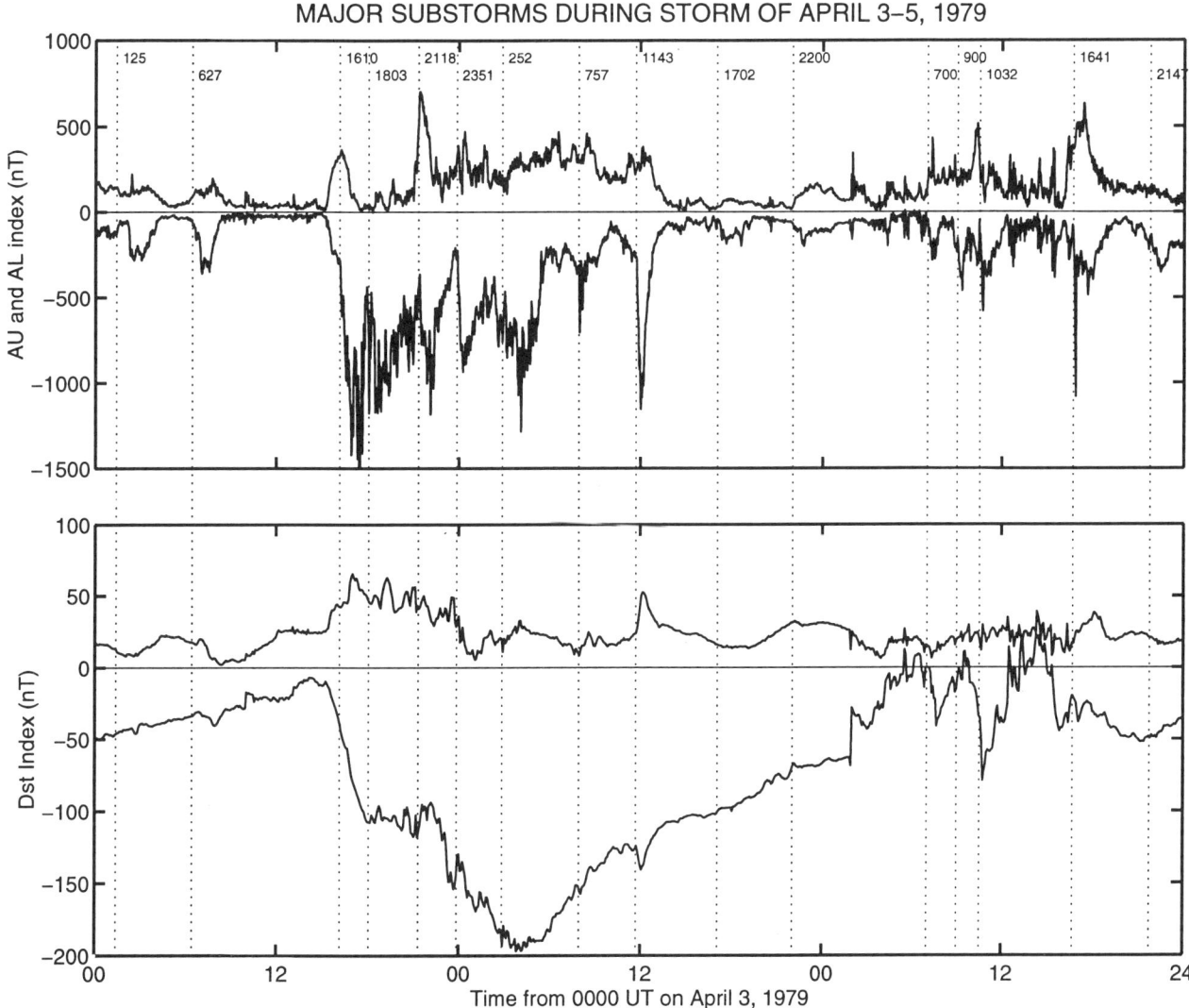

Figure 1. The auroral zone and ring current indices (AU,AL, Dst, Asym) for a magnetic storm in April 1979. Vertical dotted lines indicate substorm onsets determined by ground magnetograms, synchronous particles and fields, and tail data.

no ring current development or the ring current is recovering. We show one such example in Figure 1 discussed below. This implies that substorms are something more than a component of magnetic storms.

The association of substorms with storm main phase, combined with the observation that particles are injected into synchronous orbit during the substorm expansion phase have led many people to believe that it is substorm expansions that create the ring current. In fact, many investigators routinely use the word substorm to mean substorm expansion phase, neglecting the other two phases of the substorm. As we will demonstrate there is little evidence to support this view. Instead it is more likely that the substorm growth phase is the primary agent for ring current injection.

Additional support for the belief that substorms are the cause of storms can be found in the original papers defining the auroral electrojet (AE) indices of auroral zone magnetic activity [*Davis and Sugiura*, 1966; *Davis and Parthasarathy*, 1967]. As is well known, these indices are calculated from a longitudinal chain of about 12 auroral zone stations. For each station the monthly mean *H* is subtracted from the *H* trace. The deviations in *H* from all stations are plotted with a common base line. The upper envelope of the collection of traces is called the AU index, the lower envelope the AL index, and the separation of envelopes is called AE. Their mean value, AO, is analogous to the Dst index. Physically AU is roughly proportional to the current density in the eastward electrojet, and AL to that in the westward electrojet. The eastward electrojet is located premidnight under the aurora and the westward electrojet post midnight. *Davis and Parthasarathy* [1967] fund that the instantaneous Dst index could be well approximated by the linear superposition of the preceding 10 hours of the AE index. This obviously implies that there is a strong correlation between the AE index and Dst.

Kamide and Fukushima [1971] subsequently investigated this relation in greater detail. They concluded that Dst could be predicted by integrating the injection rate. For the injection rate they used an exponential decay to weight the instantaneous AE index. Their predicted Dst index virtually duplicated the observed index (see Figure 8 in *Gonzalez et al.*, 1994). This result would appear to provide strong support for the hypothesis of substorm injection into the ring current. Ten years later *Akasofu* [1981] demonstrated that in weak storms the Dst index is nearly a linear function of the AE index. *Wrenn* [1989] showed that Dst can be predicted as a linear multiple of a recursively filtered ap index. His filter constant was close to 10 hours, the same as found in the original work with the AE index. Most recently, *Cade et al.* [1995] demonstrated that Dst is linearly proportional to a recursively filtered ($\tau \sim 9.4$ h) AL index.

SOLAR WIND CONTROL OF THE DST INDEX

While the foregoing results may seem persuasive it must be remembered that there is an even larger body of work that demonstrates that it is the solar wind that injects the particles that create the ring current. *Russell et al.* [1974] clearly showed that the ring current develops in response to a southward turning of the interplanetary magnetic field. In particular, the strength of a storm appears to depend on the magnitude of the southward component and the velocity of the solar wind. *Burton et al.* [1975] developed a model which assumed that magnetic reconnection is the process responsible for injection into the ring current. In their model the ring current injection rate is linearly proportional to the low pass filtered, rectified, dawn-dusk solar wind electric field delayed 20 minutes. By correcting the Dst index as discussed above they were able to use observed solar wind dynamic pressure and electric field to predict the 2.5 minute Dst index with rather high accuracy. As their paper is the starting point for following work, we briefly outline the physical assumptions of their model.

The development of the model begins with the DPS relation discussed above

$$\frac{D_{st}^*(t)}{B_0} = \frac{2E(t)}{3U_M}$$

where the ring current perturbation at the center of the earth has been replaced by the corrected Dst index. If we take the time rate of change of this relation we obtain

$$\frac{dD_{st}^*}{dt} = \frac{2B_0}{3U_M}\frac{dE}{dt}$$

The time rate of change of energy in the ring current is a balance between energy input and energy output, thus

$$\frac{dE}{dt} = U(t) - \frac{E(t)}{\tau}$$

where $U(t)$ is the rate at which energy is injected, and $E(t)/\tau$ is the rate at which it is lost. This model embodies the assumption that the primary ring current loss process is charge exchange for which the loss rate depends on the number of particles present, or roughly on the total energy of the ring current. Substituting this expression for the power into the ring current we obtain

$$\frac{dD_{st}^*}{dt} = \frac{2B_0}{3U_M}\left[U(t) - \frac{E(t)}{\tau}\right] \text{ or}$$

$$\frac{dD_{st}^*}{dt} = Q(t) - \frac{D_{st}^*(t)}{\tau} \qquad (6)$$

Here $Q(T)$ is the Dst injection function defined by

$$Q(t) = \frac{2B_0}{3U_M}U(t) = \left[\frac{dD_{st}^*}{dt} + \frac{D_{st}^*}{\tau}\right]$$

In the *Burton et al.* model, $Q(t)$ was taken to be

$$Q(t) = aF_L * \left[v(t-t_d)\left(B_s(t-t_d) - B_z^0\right)U(\theta - \pi/2)\right] \qquad (7)$$

where F_L is a low pass filter, t_d is the time delay, B_z^0 is a threshold, and U is the unit step function. Note when the injection function is zero the corrected Dst decays to zero with a time constant τ.

A somewhat different formulation of this problem was made by *Perreault and Akasofu* [1978] using the concept of energy coupling. They noted that Joule heating by ionospheric currents and particle precipitation are two additional sinks for solar wind energy input to the magnetosphere. They estimated these as functions of the AE index and derived an expression for the total power, U_t, dissipated in the magnetosphere.

$$U_t = \frac{3}{2}\frac{U_M}{B_0}\left(\frac{dD_{st}^*}{dt} + \frac{D_{st}^*}{\tau}\right) + kAE$$

They then assumed that the rate at which energy is input to the magnetosphere is some fraction of the interplanetary Poynting vector depending on the clock angle θ of the IMF around the earth-Sun line. They called this energy coupling function epsilon (ε)

$$\varepsilon = l_0^2 v B^2 \sin^4(\theta/2)$$

where $(l_0)^2$ is an area on the magnetopause through which energy enters. Note that the *Perreault and Akasofu* formulation can not be used to predict either Dst or AE because the manner in which energy is partitioned between the two sinks is not specified. *Akasofu* [1981] subsequently argued that the wave form of ε is much closer to that of U_t when the decay time, τ, is made a function of the input ε such that it becomes small when the input is large.

The *Burton et al.* [1975] model has been reexamined in several subsequent papers. *Feldstein et al.* [1984] argued that there is an additional constant input to the ring current beside that depending on the solar wind electric field. This input maintains a constant quiet ring current which can be accounted for by the empirical constant c defined above. In addition they argue that the decay constant τ depends on the level of Dst such that it becomes shorter as Dst becomes more negative. *Pudovkin et al.* [1985] came to a quite different conclusion. They claim that the dynamic pressure scale factor b depends on the strength of the ring current (although they used a constant). In addition they argued that *Burton et al.* were incorrect in their assumption that the disturbed ring current decays completely in the absence of an input. Finally, they claim that the decay time depends on both the phase of the storm and the strength of the ring current. During the main phase they find a short and constant value independent of Dst while in the recovery phase they find recovery times increase as the minimum Dst becomes more negative. A comprehensive review of past work on this problem is given in *Feldstein* [1992].

CHANGES IN DST AT SUBSTORM ONSET

If the substorm injection hypothesis is correct one might expect to see an increase in the energy of ring current particles immediately after expansion phase onset. According to the DPS theorem this implies a decrease in the Dst index. Does this happen? Figure 1 shows indices for a magnetic storm in April 1979 with dotted lines indicating the times of major expansion onsets. It should be noted that Dst began to decrease at the same time as a substorm growth phase began in the auroral zone, almost an hour before the first expansion onset at 1610 UT on April 3. At the time of this onset the slope of Dst remained constant suggesting the expansion phase had no effect on the ring current. Furthermore, there is no apparent correlation of Dst with any of the subsequent onsets except one during the storm recovery at 1143 UT. The data in this figure suggest two conclusions. First, the Dst index responds on a time scale short compared to the duration of a substorm growth phase. Second, for some substorms there is no obvious change in the Dst index at expansion onset.

The first of these facts has been known for some time. Figure 2 shows the prediction filter obtained by the author [*McPherron et al.*, 1986] relating VBs in the solar wind to the ring current injection function, $Q(t)$ defined above. The filter is a Gaussian pulse centered at 30 minutes delay. Ten minutes of this delay is due to solar wind propagation from the upstream monitor to the magnetopause. Convolution of this filter with the rectified solar wind electric field both low passes the time series and delays it by 10 minutes to the magnetopause, and 20 minutes within the magnetosphere. This is exactly the prescription used by *Burton et al*. The filter implies that a pulse in the solar wind electric field at the magnetopause will decrease the Dst index 20 minutes later. This is much too short a time to correspond to the onset of a substorm expansion which typically occurs after a 55 minute growth phase.

The failure to observe the expected change in Dst at expansion onset has been demonstrated more quantitatively in a recent paper by *Iyemori and Rao* [1996]. In this paper the authors used one minute time resolution Dst indices derived from six stations to study changes in Dst after an expansion onset. Onsets were defined by the beginning of midlatitude Pi 2 bursts. Substorms were divided into two groups depending on whether they occurred during the main or recovery phase of a storm. The traces of the Dst index during each substorm were then superposed with expansion onset as epoch zero. Their results show that in the main phase of storms the magnitude of the slope of superposed Dst decreases after substorm onset rather than increases as might be expected from the substorm injection hypothesis. In the recovery phase the effect is even more dramatic. Dst appears to increase (become more positive). In both cases the results seem to imply that rather than increase the rate of ring current injection, substorm expansions decrease it!

One possible criticism of these results is that the Dst index used in the Iyemori and Rao study contains effects of the tail current, the substorm current wedge and the partial ring current as well as the effect of the symmetric ring current. As we showed above no attempt is made to remove these effects in either the derivation or correction of the Dst index because it is usually assumed that they cancel in the longitudinal average. Whether this is true depends on the detailed time history of the various current systems.

At substorm expansion onset the substorm current wedge begins to grow and causes a positive H perturbation inside the wedge and a negative H perturbation outside the wedge [*Horning et al.*, 1974]. Averaging the axial component of the ground perturbation around the earth gives a positive contribution to Dst because the negative side lobes of the current wedge do not cancel the positive lobe in the midnight sector. Of course during the

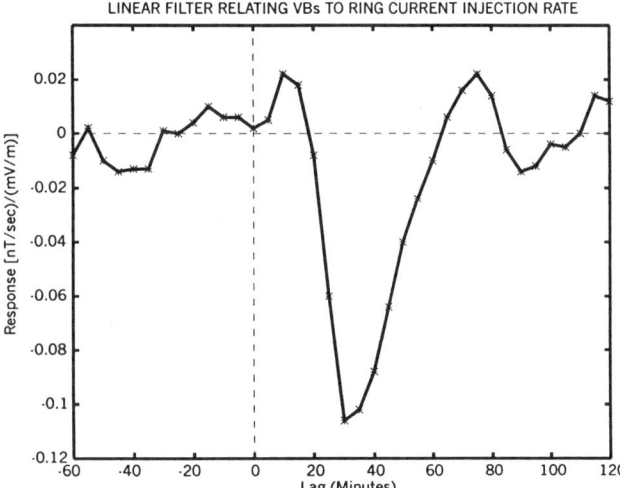

Figure 2. High time resolution prediction filter for the ring current injection function, Q(t), with rectified solar wind electric field as input [*McPherron et al.*, 1986].

expansion phase the partial ring current is also a current wedge growing with opposite sense, and it makes a negative contribution to Dst. The magnitude of this contribution depends on the angular extent of the partial ring current. For an 180° wedge the ground perturbation is antisymmetric and averages to zero. More likely it is less than this and so typically reduces the effect of the substorm current wedge. The questions therefore are how long does the positive Dst bias from the substorm current wedge persist, and is its effect large enough to overcome the effects of the changes in the symmetric ring current?

The duration of the expansion phase is typically less than an hour. However, this is only the time for the substorm current wedge to fully develop. The current wedge decays during the recovery phase which may last another 90 minutes. During the same interval the partial ring current either decays or is converted to a symmetric ring current. As these two current systems decrease, the Dst index should approach a value given by the symmetric ring current alone. Since this will happen in 2-3 hours, a time much shorter than the decay time of the ring current, it might be expected that the Dst index at the end of the substorm would be more negative. However, the effects found by Iyemori and Rao persist for many hours. In fact, their superposition of the asymmetry indices for *H* and *D* show that the effect of the expansion phase probably lasts for about an hour. Thus, even though the Dst index used in the Iyemori and Rao study is clearly affected by other current systems we feel that their conclusion is likely to be correct. As shown next, a recent theoretical argument explains why this result is expected. Finally, in the following section we show that linear prediction filter results also support their conclusion.

In a recent paper, *Siscoe and Petschek* [1996] consider the substorm injection hypothesis and the results of *Iyemori and Rao* using the generalized virial theorem [*Siscoe*, 1970]. This theorem is a generalization of the DPS relation discussed above. The generalization states that

$$\frac{B_m(0)}{B_0} = \frac{2K + M - \oiint \vec{R} \cdot \hat{n} dA}{3U_M} \quad (8)$$

where U_M is the total energy of the earth's dipole field outside the earth, K is the total thermal energy of particles within a volume of space, M is the magnetic energy due to ring, magnetopause and tail currents, and the vector R gives the flux of energy across the surface of the volume. Since substorm expansions frequently occur after the IMF turns northward and energy input to the magnetosphere ceases ($R = 0$), the substorm injection hypothesis requires that magnetic energy M be converted to particle energy K in such a way as to make $B_m(0)$ larger. This is obviously the case if M is completely converted to K, since the factor of two multiplying K guarantees that the perturbation at the center of the earth will increase. However, if more than half the magnetic energy is partitioned into other sinks such as Joule heating and particle precipitation (not to mention plasmoids), $B_m(0)$ will actually get smaller, i.e. Dst will appear to recover. *Siscoe and Petschek* review earlier empirical results, numerical simulations, and analytical estimates and conclude that the preponderance of evidence supports the view that more than half the magnetic energy in the tail is converted to ionospheric heat during a substorm expansion, and that therefore the *Iyemori and Rao* result is expected. In addition they discuss a lumped circuit model for the substorm injection hypothesis and show for this model it is impossible for more than half the energy to go into the ring current.

DO LINEAR PREDICTION MODELS SUPPORT THE SUBSTORM INJECTION HYPOTHESIS?

The *Burton et al.* [1975] work described earlier established that the ring current injection rate $Q(t)$ is well approximated by a function of the rectified solar wind electric field. *Fay et al.* [1986] obtained a virtually identical result using the technique of linear prediction filtering. In this section we extend the linear prediction analysis using hourly averages and find additional evidence suggesting that the substorm expansion is not the cause of ring current injection.

We begin with the well known fact that both the Dst and AL indices can be predicted by the solar wind electric field [*Fay, et al.*, 1986; *McPherron et al.*, 1988]. Because both currents are caused by the same process (magnetic reconnection with the IMF), the two indices are closely correlated and it will therefore appear that substorms are responsible for creation of the ring current. We will demonstrate first that almost all the variance in the Dst index (85%) can be accounted for by solar wind dynamic pressure and electric field. Removal of the predictable part of the Dst time series leaves a residual that can probably be explained by noise in the index. If this is true, there is no need to postulate additional injection during the substorm expansion phase.

The AL index is much less predictable than Dst, with only 60% of the hourly variance accounted for by a linear filter. For 2.5 minute AL data only 45% of the variance is predictable. The unpredictable residual in high resolution data is largest near expansion phase onset suggesting that the onset is not strongly correlated with the solar wind. This would be expected if the expansion phase is an internal magnetospheric process in which stored energy is suddenly unloaded. In fact, the substorm injection hypothesis states that it is this unloading that causes injection into the ring current. If this were the case we would expect the Dst index produced by these particles to also be weakly correlated with the solar wind. Thus it is possible that the portion of Dst that is unpredictable by the solar wind is directly related to the substorm expansion. We investigate this by examining the correlation between the AL prediction residual and the Dst prediction residual, where the residual is with respect to the solar wind prediction. We will show there is no significant correlation implying that the mutual relationship between Dst and AL is explained by the solar wind electric field.

The linear model states that output of a linear system can be written as the convolution of two filters, one with the input to the system and one with the output from the system.

$$O(t) = A * O(t) + B * I(t)$$

The filter B is a finite impulse response (FIR) filter while the filter A is an infinite impulse response filter (IIR). Alternative names are convolutional filter and recursive filter respectively. Note that the recursive filter A "feeds back" the output. The equation may be rewritten in terms of summations as

$$O(n) = \sum_{i=2}^{na} A(i)O(n+1-i) + \sum_{i=1}^{nb} B(i)I(n+1-i)$$

where na and nb are the number of coefficients in each filter.

There are $na + nb$ unknowns in these two filters. If we consider successive time points in the output series, $n, n+1, \ldots n+N$, we can write down a set of algebraic equations relating the output at one time to input and output at earlier times. Use of a sufficient number of output times will give an over determined set of equations that can be solved by least square techniques for the filter coefficients $\{A_i$ and $B_i\}$. We utilize this technique in what follows to examine the substorm injection hypothesis.

Let the solar wind coupling function be $C(t)$. This function may have different forms depending on what is thought to be the cause of the enhanced magnetic activity during magnetic storms. We investigate this aspect of the problem by using six previously suggested functions for solar wind coupling. The function with the highest prediction efficiency turns out to be

$$C(t) = \left(\frac{v^2}{v_0}\right)\sqrt{B_y^2 + B_z^2} \sin^4(\theta/2) \quad (9)$$

where the velocity dependence is normalized by the most probable solar wind velocity to give the expression units of electric field. The radical gives the magnitude of the IMF transverse to the earth-Sun line, and the trigonometric function of the clock angle of the IMF gives the angular gating caused by magnetic reconnection.

We use this coupling function to determine the pure convolution filters (B) that give corrected Dst, AL, and the ring current injection $Q(t)$ as a function of this input. We also calculate a filter relating AL to Dst. Next we calculate the prediction residuals by subtracting the observed indices from the predictions, and then determine filters relating the residuals. We do this for both Res(AL) to Res(Dst), and for Res(AL) to Res($Q(t)$).

It should be emphasized that it is the corrected Dst index

$$D_{st}^* = D_{st} - b\sqrt{P_{dyn}} + c$$

that is used in these calculations, and that this quantity involves undetermined constants. The form of the coupling function, the parameters b, c, and τ, as well as the filter B must be determined from the data. This inverse problem is clearly non linear. We have solved it in the following manner. Substitute for corrected Dst in the differential equation describing the development of the ring current, obtaining

$$\frac{d}{dt}\left[D_{st} - b\sqrt{P_{dyn}} + c\right] = Q(t) - \frac{1}{\tau}\left[D_{st} - b\sqrt{P_{dyn}} + c\right]$$

Rearrange terms and assume the injection function is linearly proportional to the coupling function, $Q(t) = a * C(t)$

$$\left[\frac{dD_{st}}{dt} + \frac{D_{st}}{\tau}\right] = aC(t) + b\left[\frac{d\sqrt{P_{dyn}}}{dt} + \frac{\sqrt{P_{dyn}}}{\tau}\right] - c\left(\frac{1}{\tau}\right) \quad (10)$$

The results shown in Figure 1 indicate that the injection function is the convolution of a short filter (~20 minutes) with the solar wind electric field at the magnetopause. For hourly averages the convolution thus becomes a simple multiplication. If the form of $C(t)$ and the value of τ are assumed this equation becomes a simple multiple regression between the dependent variable on the left and three dependent variables on the right. Using all data for the year 1979 we have found the optimum parameters a, b, c for five possible coupling functions and eighteen different values of τ. The prediction efficiencies obtained are summarized in Figure 3. The coupling function with the highest prediction efficiency is proportional to V^2B as discussed above. The function VB_s differs very little from this function. Other suggested functions are significantly less efficient predictors. The value of τ which maximizes the prediction efficiency for the best function is 6 hours. For the remaining calculations we have used this coupling function and value of τ.

The set of filters plotted in Figure 4 were obtained using the coupling function and value of decay time found above. The top panel shows the dynamic pressure filter which accounts for ~9% of the variance in Dst. Note that the filter is not a perfect impulse as it should be. The response prior to zero lag is probably a consequence of not correcting hourly average solar wind parameters for propagation from the monitor (IMP-8) which is sometimes downstream of the subsolar magnetopause. Negative coefficients in the filter are artifacts of trends in the solar wind parameters that were not entirely removed prior to calculating the correlation functions used to determine the filters. The second panel shows the AL filter which accounts for 62% of the variance in AL. The filter demonstrates that AL responds in the same hour as a change in the coupling function and that the effect of this change persists for at least three hours. Note that the AL prediction efficiency for hourly average AL (62%) is much higher than is obtained when 2.5 minute data are used (~45%). The third panel presents the coupling to Dst filter which accounts for 76% of the Dst variance. The rapid rise of this filter represents ring current injection while the long exponential tail predicts its decay. The shape of the filter is a consequence of choosing a pure FIR filter to represent the solar wind interaction with the magnetosphere. If we use both a FIR and an IIR filter we can account for injection in a single convolutional coefficient and decay in a single recursive coefficient. Together the solar wind pressure and coupling filters predict 85% of the Dst variance. The bottom panel presents the AL to Dst filter. Over 70% of the Dst variance is predicted by the AL index. This fact is the primary basis for the hypothesis that substorms inject particles into the ring current. Note, however, that the filter is not causal. The Dst index responds before the AL impulse at zero lag. Such behavior suggests that some other agent, in this case the solar wind, is responsible for the mutual relation. It should be mentioned that the filters for AL and Dst are virtually identical to those obtained by *Iyemori* [1979] who also used hourly averages for input and output.

A comparison of the observed and predicted Dst index during a typical magnetic storm is presented in Figure 5. The original index has been corrected by removal of the dynamic pressure and quiet ring corrections prior to convolving the filter with the coupling function. Note also

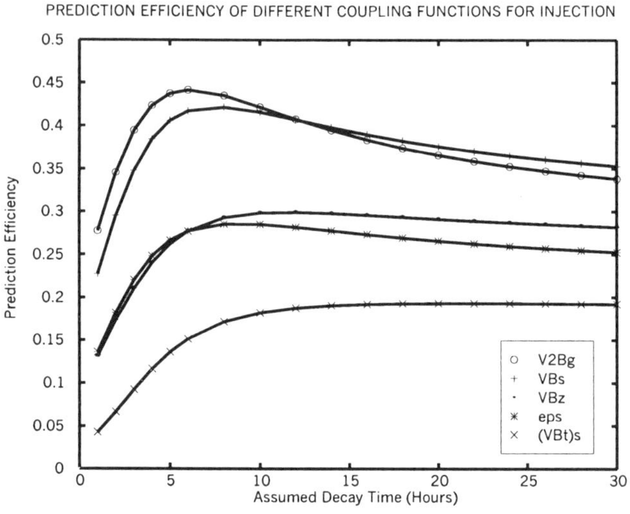

Figure 3. Efficiency of various coupling function in predicting the ring current injection function as a function of assumed ring current decay time.

that Dst is well above the baseline before and after the storm. This emphasizes the point that the filters predict the variance in the index, but not the DC value or trends. The quality of the prediction is very high, since 9% of the variance was removed by the dynamic pressure correction (not shown) and another 76% removed by the coupling function.

We have also used the various inputs to predict the Dst injection function, Q(t)

$$Q(t) = \left[\frac{dD'_{st}}{dt} + \frac{D'_{st}}{\tau} \right] \quad (11)$$

where the prime indicates correction for solar wind dynamic pressure. This function depends on the time rate of change of Dst which must be estimated numerically. We have used a five point operator to smooth the derivative thus averaging the slope over five hours. Despite the smoothing this operation emphasizes high frequency noise in the time series and the prediction efficiencies are substantially lower (49% using coupling and 41% using AL index as input). The filters obtained are displayed in Figure 6. Both the coupling and AL filters are pulses near zero lag. The AL filter is significantly acausal indicating that changes in Dst occur prior to changes in AL. The primary result is shown in the third panel. The residual filter between coupling residuals of AL and Dst accounts for only 2.4% of the variance in the Dst residual. This indicates that almost all of the mutual relation between AL and injection into the ring current is explained by the solar wind coupling function.

The prediction efficiencies obtained for various assumed relations are summarized in Table 1. Column 1 of the

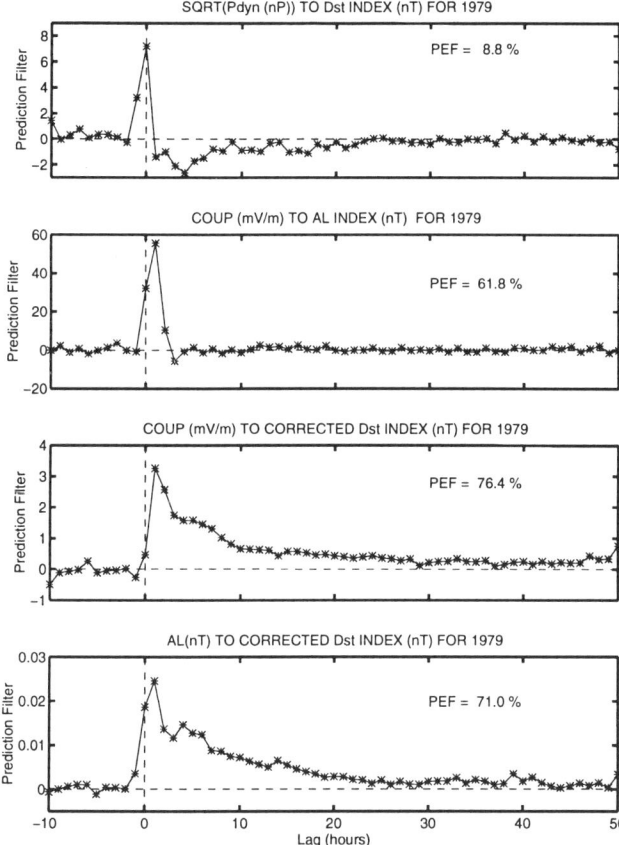

Figure 4. Prediction filters relating the solar wind dynamic pressure and coupling function to various output functions. From the top down the filters include the dynamic pressure to Dst filter, coupling to AL index, coupling to Dst index, and AL index to Dst index.

table shows the input and output times series used in the analysis, and column 2 shows the prediction efficiency for the assumed relation. For example, the first row shows that 9% of the Dst variance is caused by solar wind dynamic pressure. The second row demonstrates that the coupling function defined in equation (9) accounts for another 76% of the variance. Thus a total of 85% of the hourly average Dst variance is accounted for by the pressure and coupling function. In contrast, only 61% of hourly average AL variance can be predicted with the coupling function (pressure makes no contribution). Since the solar wind does well in predicting AL and Dst one would expect them to be closely related. This is verified by the fact that AL predicts 71% of the variance in the pressure corrected Dst index. However, when we remove the predictable parts of Dst and AL and examine the correlation of the residuals we find that only a very small fraction of the variance in the two residuals is correlated (10%).

A better test of the injection hypothesis is to examine the relation between various inputs and the Dst injection function

$$\left[\frac{dD_{st}}{dt} + \frac{D_{st}}{\tau}\right]$$

defined in equation 11. Table 1 shows that the solar wind coupling function predicts only 49% of this quantity. This is significantly worse than for its coupling to Dst, but understandable since the derivative operator emphasizes high frequencies where noise in the index is significant. The AL index does almost as well, predicting 41% of the injection variance. However, when we remove the parts of the AL index and the injection function predictable by the solar wind electric field, the residual AL predicts only 2% of the residual injection variance. This implies that essentially none of the residual variance in the injection function is related to parts of AL that are uncorrelated with the solar wind electric field.

Discussion

Substorms and magnetic storms are both caused by southward interplanetary magnetic field. A short interval of southward field causes an isolated substorm with three phases: growth, expansion and recovery. This southward IMF also controls injection into the ring current and hence a decrease in Dst. However, if the IMF is weak and the southward interval short, the change is Dst will be too small to be observable in the hourly average Dst index. Pulsating intervals of southward IMF will cause repeated substorms, and repeated weak ring current injection, but the ring current decays during intervals of northward IMF, and no storm develops. Long intervals of strong southward IMF cause intense substorms and continuous ring current injection leading to the main phase of a magnetic storm. The Dst index decreases until the rate of ring current decay increases to match the injection rate. At this time the main phase reaches its minimum value unless the injection rate becomes stronger. When the IMF turns northward injection ceases and the ring current decays. If the IMF subsequently turns southward for a brief interval there will be additional ring current injection which usually has the effect of slowing the apparent recovery rate. If the injection is strong enough it may overwhelm the recovery and drive Dst downward once again.

The foregoing behavior of the ring current is almost completely predictable using the simple differential equation discussed previously

$$\frac{dD_{st}^*}{dt} = a*C(t) - \frac{D_{st}^*(t)}{\tau}$$

The best coupling function, $C(t)$, is very nearly the rectified solar wind electric field, VB_s, but, the AL index can be used as its proxy with some degradation in the

142 RELATION OF SUBSTORMS TO STORMS

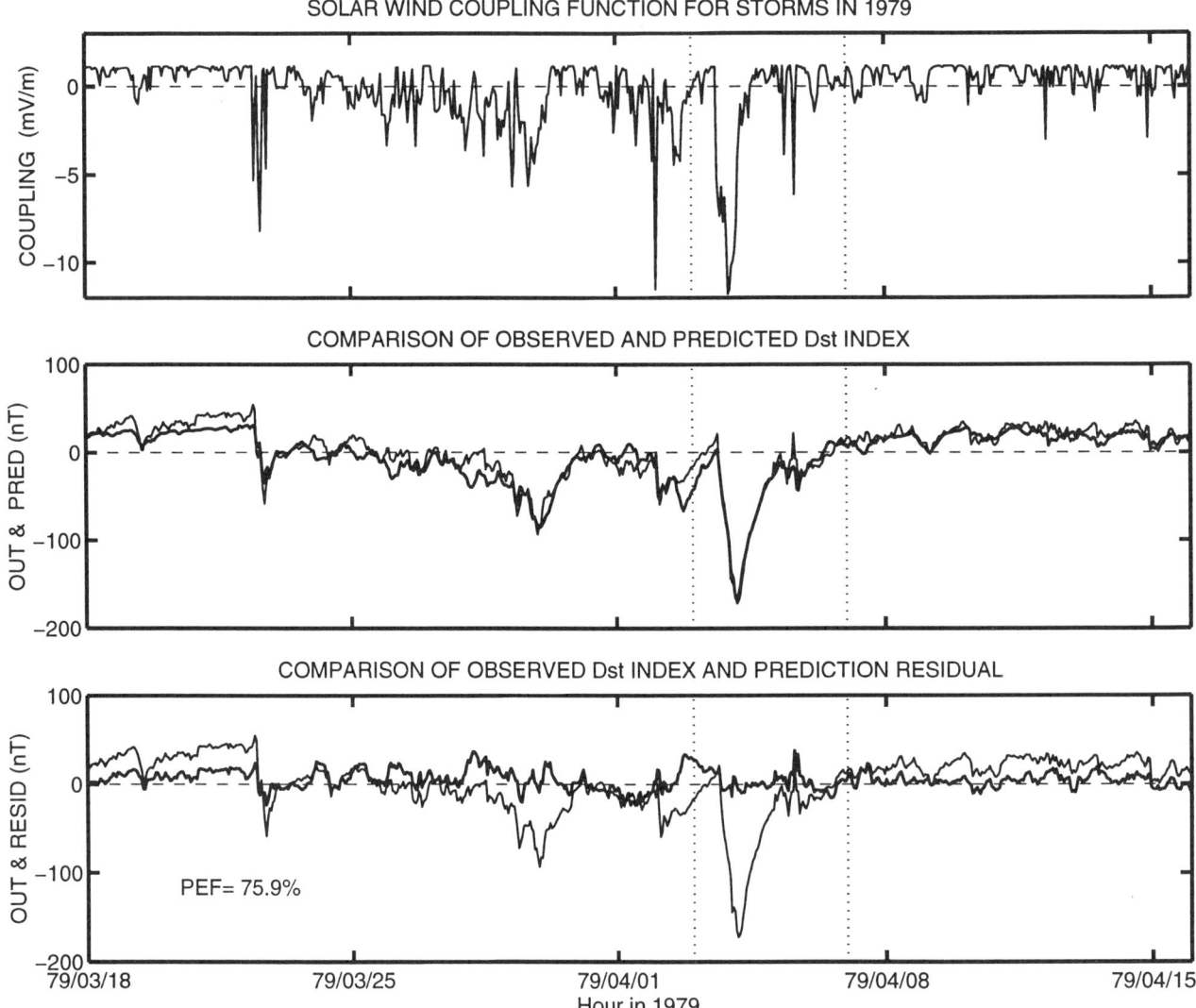

Figure 5. Prediction of the Dst index by solar wind coupling function for a moderate magnetic storm in April 1979. The three panels show respectively the coupling function, the observed and predicted Dst index, and a comparison of the observed index with its prediction residual. Vertical line denote the interval shown in higher time resolution in Figure 1.

quality of the prediction. If substorms are the direct cause of ring current injection it would be expected that AL would be a better predictor of injection than VB_s. The fact that it is not suggests that substorms and storms are both direct consequences of the solar wind electric field. In fact, a, the high time resolution prediction filter in this equation was shown in Figure 2 to be a Gaussian pulse centered at 20 minutes delay. This delay is the same as the delay in the onset of convection and magnetic disturbances that result from dayside reconnection [*Todd et al.*, 1988]. It is much shorter than the typical 55 minutes duration of a substorm growth phase, and hence unlikely to be associated with the onset of nightside reconnection.

It should be remembered that the high resolution prediction filter between solar wind electric field and the AL index has two peaks, one at 20 minutes, and one at 60 minutes [*Bargatze et al.*, 1985; *Blanchard and McPherron*, 1995]. The 20-minute peak has been attributed to the westward electrojet driven by dayside reconnection (the DP-2 current system), and the 60-minute peak to the premidnight westward electrojet driven by nightside reconnection (the DP-1 system). Convolution of the solar wind electric field with the first peak produces the disturbance in AL known as the substorm growth phase, while convolution with the second peak produces the expansion phase effects. The fact that there are two peaks

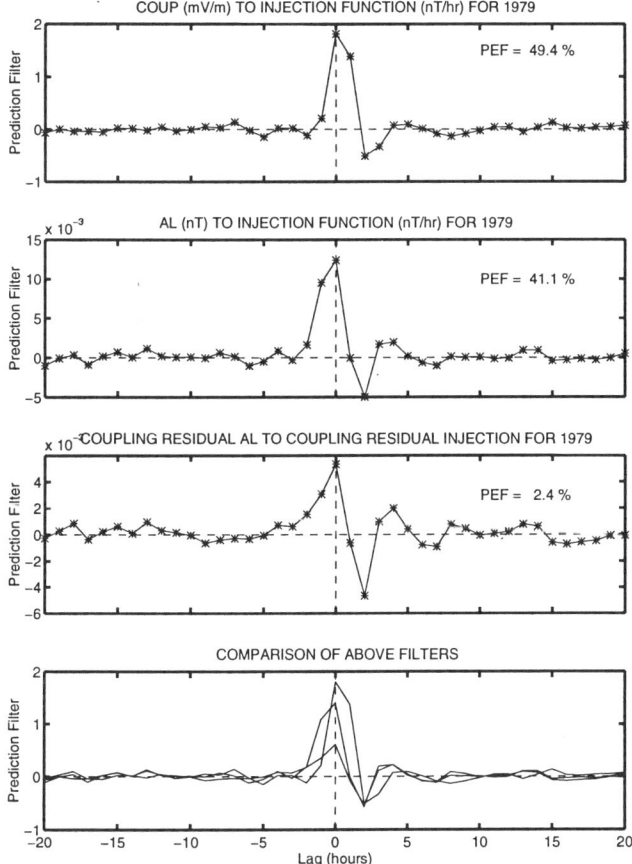

Figure 6. Prediction filters relating various inputs to the ring current injection function. From the top down the four panels show the coupling filter, the AL filter, the filter relating the coupling-AL residual to the coupling-Dst residual, and a comparison of the three filters.

in the AL filter implies that both DP-1 and DP-2 are driven by the solar wind electric field, although with quite different delays. The single peak at 20 minutes in the Dst injection filter (corrected for propagation) implies that ring current injection is driven by the solar wind electric field, but it occurs during the substorm growth phase.

Our result supports the conclusion derived from the generalized virial theorem, Dst is not decreased (made more negative) by converting magnetic energy of the tail lobes into particle energy. When reconnection coverts tail lobe energy to particle energy, more than half the energy stored in the lobes goes into other processes (plasmoids, Joule heating, particle precipitation, waves) and Dst increases (becomes less negative). If this is true, what causes Dst to decrease? In terms of the virial theorem it must be the term, R, that was assumed negligible, and which gives the flow of energy into the magnetosphere.

Presumably this is the Poynting vector associated with field lines opened on the day side by magnetic reconnection.

What physical process is responsible for ring current injection? The linear prediction analysis suggests it is global convection driven by magnetic reconnection. Our results imply that a dawn-dusk magnetospheric electric field transports particles up the tail and within 20 minutes injects them into the ring current at a rate proportional to the strength of the convection electric field. This is too quickly for injection models that depend on collapse of the tail field (e.g. *Mauk* [1986] and *Liu and Rostoker* [1995]). This is a surprising result. If the electric field remains constant the particles will not be trapped. Instead they drift to the dayside magnetopause and are lost into the magnetosheath. Only when the convection electric field relaxes will some of the convecting particles be trapped into circular drift paths creating a ring current.

Convection is the basis for one mechanism of ring current injection proposed by *Lyons and Williams* [1980]. In their model trapped particles are assumed to be present at distances through and somewhat beyond the typical ring current (L~4). They apply a dawn-dusk electric field to the inner magnetosphere and allow these particles to drift for an interval comparable to a typical substorm (~3 h). On the night side they drift inward gaining energy, while on the day side they drift outward losing energy. Particles with drift period equal to the duration of convection or much shorter experience no net change in energy or location because on the average they experience as much outward as inward drift. Particles whose half drift period exactly matches the duration of convection will gain the most energy provided they start across the night side just as convection begins. At L = 3 in a dipole field this will be 40 keV or lower energy particles. Since there are always more particles at low energy than high, the number of particles moved to high energy on the night side more than compensates for the high energy particles moved to low energy on the day side. Thus there is a net increase in the total energy of particles along a drift path. The authors note that the increase in ring current strength in this model will depend on both the strength and duration of the electric field, as well as the preexisting population of particles in and outside the ring current region. They propose that either convection from the tail or injection from the ionosphere are suitable candidates for creating the preexisting population.

Lyons and Schulz [1989] proposed a closely related mechanism to account for the higher energy particles injected into the ring current. They begin with the assumption that some process has created a preexisting population of medium energy particles in the ring current region (e.g. convection as described above). They then consider the effects of fluctuating electric fields on these trapped particles. For fluctuations of a given period there will be some particles whose half drift period is comparable. These particles will drift inward on the night

Table 1. Prediction Efficiencies Obtained With Various Input-Output Functions ($\Delta t = 1$ Hr).

INPUT - OUTPUT	PREDICTION EFFICIENCY
Pdyn - Dst	9%
Coup - Dst'	76%
Coup - AL	61%
Coup - Injection	49%
AL - Dst'	71%
AL - Injection	41%
Res(AL) - Res(Dst')	10%
Res(AL) - Res(Injection)	2%

side and outward on the day side. Since there are more particles at larger distances this produces a net inward radial diffusion, and net gain in ring current energy. They conclude that the transport rate should be more dependent on the rms power in the fluctuations than on the DC electric field. This process should also depend on the preexisting population, and the distribution function for particles near the outer edge of the ring current.

The two mechanisms discussed above have been evaluated numerically by *Chen et al.* [1993, 1994] using an ensemble of 100 simulated magnetic storms. The storms were modeled by superposing transient impulses on a steady, Volland-Stern shielded convection electric field (50 keV). Each impulse consisted of a sudden rise and 20-minute exponential decay. The amplitudes of the impulses was taken from a random Gaussian distribution with 200 keV mean and 50 keV standard deviation. The times between impulses were chosen from a random distribution with the constraint of a 10-min dead time following the onset of each impulse so as to avoid extreme values of the convection potential. A set of 12 particles was then followed backward in time from different locations in a symmetric ring located at 3 R_e. The magnetic moment of the particles was varied to determine the origin of different energy particles.

The authors found that particles with final energy below 110 keV are primarily convected to L = 3 along open drift paths originating in the tail, and are then trapped. Particles above 280 keV are radially diffused from closed drift shells both inside and outside L = 3. At intermediate energies, the highest energies typically found in the ring current, some particle have convective access while others are radially diffused. Trapping of the particles from open drift paths is a result of fortuitous changes in position of the sepratrix between open and closed drift paths, primarily on the dusk side as the convection electric field varies in response to the impulses. Particle trapping into the ring current depends on the fluctuations in the electric field. If the field is steady, particles will be energized and drift closer to the earth, but will not be trapped until the convection electric field relaxes.

We emphasize that the linear prediction filter results show that a steady convection electric field continuously injects ring current particles until they are lost as fast as they enter. How can a steady electric field trap these particles? One possibility is that this electric field is never steady. Natural variations in the solar wind, and reconnection on the dayside magnetopause may introduce fluctuations into the magnetospheric electric field that continuously scatter particles from open to closed drift paths. Large amplitude waves in the dusk sector might also accomplish this. A more likely source of variation is the substorm expansion phase and its associated inductive electric fields. However, as we have noted above, the expansion phase is typically delayed by 55 minutes or more, and successive major onsets are separated by about three hours. This would suggest that ring current injection should consist of clearly identifiable bursts much like that observed by synchronous satellites during substorms rather that the continuous injection indicted by the prediction filters.

It should be admitted, however, that the repetition rate of substorms is probably more frequent than one every three hours during major storms. Furthermore, most substorms consist of a superposition of intensifications separated by about 20 minutes [*Rostoker et al.*, 1987]. Often one or more of these is observed in the substorm growth phase and several are seen in the expansion phase, or occasionally can be found in the recovery phase. Whether these can effect the global electric field enough to trap particles into the ring current seems to be a moot point. The *Chen et al.* model makes rather extreme assumptions in this regard assuming global impulses of up to 400 keV following each other within 10 minutes.

Again we must return to the central question, why is the ring current injection rate directly proportional to the convection electric field? The only answer we can suggest is that global convection carries particles from the tail toward the ring current. An increase in convection carries more particles closer to the earth and gives them more energy. Secondary processes that are not observable in the Dst index convert the available particles from open to closed drift paths. The fluctuating electric fields associated with the frequent intensifications of substorms is a likely candidate for this conversion process. The Dst index must primarily measure the flux of particles regardless of whether they are on open or closed drift paths in order to respond as quickly as it does. It must be unable to distinguish when particles are converted to closed drift paths since there is no hint of bursty particle injection at substorm expansion onset.

CONCLUSIONS

Magnetospheric substorms can occur at any time regardless of whether a magnetic storm is in progress. However, substorms are more frequent and stronger during the main phase of magnetic storms. Indices of substorm activity like AL are good predictors of the index of storm strength, Dst. It is well known that the substorm expansion phase causes particle injection into synchronous orbit. These facts have led many investigators to conclude that the ring current is created by successive substorm expansion phase injections. Several models have been developed to accomplish this. Most of these depend on dipolarization of the field after expansion onset. Tail like field lines collapse earthward, carrying particles to regions of higher field strength where the field lines are shorter. Conservation of the first two adiabatic invariants then requires that the total energy of the particles increase. The changes in magnetic field allow particles initially on trapped drift paths to be placed on closed drift paths.

A visual examination of plots of 5-min resolution AL and Dst indices does not support the expansion phase injection model. There is little apparent correlation between decreases in the Dst index and the times of substorm onsets. This observation is supported by a superposed epoch analysis of Dst relative to substorm onset. During the main phase of storms the rate of decrease of Dst decreases after expansion onset. In the recovery phase Dst actually recovers. A consideration of the generalized virial theorem shows why this behavior might be expected. In the generalized theorem the change in Dst is proportional to twice the thermal energy of particles plus the magnetic energy of currents in a volume of integration. If all stored magnetic energy is converted to particle energy Dst should decrease. However, the partitioning of magnetic energy into plasmoids, Joule heating in the ionosphere, particle precipitation leaves less than half the energy for the ring current, and Dst must actually increase or at least decrease less rapidly.

What then can cause ring current injection? The generalized virial theorem suggests that it must be related to the usually neglected surface integral that gives the rate at which energy is flowing into the volume of integration. This energy is transported by global convection. The dependence of ring current injection on convection can often be seen at the beginning of storms in high resolution indices. Dst begins to decrease well before the first substorm expansion. A linear prediction filter analysis of the ring current injection rate corrected for solar wind propagation and ring current decay is a Gaussian pulse centered at 20 minutes delay. This delay is generally attributed to the formation of the DP-2 convection system driven by dayside reconnection. It is far too quick to be attributed to the substorm expansion phase.

The high correlation observed between substorm and storm indices is a consequence of the fact that both processes are controlled by the solar wind electric field. Two output signals generated through quasi-linear processes by a single input will be highly correlated. In fact one output can be used as a proxy for the real input in a prediction scheme. When one removes the solar wind controlled part of AL and Dst, the Dst residual is very small, suggesting that the solar wind alone can predict ring current injection. Furthermore, the correlation of the residuals is negligible implying that the effect of internal magnetospheric processes (not correlated with solar wind), such as substorm expansions, is indistinguishable in the injection function.

A recent model of ring current injection combines both direct convection of particles from the tail to the ring current and radial diffusion of preexisting particles. This model utilizes fluctuations in the convection electric field to move particles from open to closed drift paths. However, the linear prediction filter results imply that a steady convection electric field injects ring current. This empirical result suggests that changes in the Dst index are dominated by the flux and energy of particles drifting Sunward on open drift paths. Of course there must be some process that scatters these particles from open to closed drift paths but it appears to be undetected in the linear filter analysis of Dst. The process must be relatively efficient and continuous otherwise the injection rate would not depend so strongly on the solar wind electric field. We suggest this process is a combination of inherent fluctuations in the solar wind electric field, waves in the magnetosphere, and inductive electric fields caused by substorm expansion phases.

What then is the answer to the question posed in the title to this paper, "What is the relation of substorms to storms?" We suggest the primary role of substorms is to fluctuate the magnetospheric electric and magnetic fields so that particles on open drift paths are scattered to closed drift paths. Otherwise, the global convection electric field is the primary agent that transports particles to the ring current region and increases their energy. Substorms also play a role in injecting ionospheric ions into the magnetosphere where they are subsequently energized and transported by global convection.

What questions are suggested by our analysis and conclusions? Our discussion of the Dst index and the current systems that contribute to it suggests that Dst is probably an inadequate measure of the strength of the symmetric ring current, particularly when the IMF is strongly southward. The magnetopause current, the quiet ring current, the tail current, the Region 1 current, the substorm current wedge, and the partial ring current all contribute to the H component of midlatitude observatories that are used in its generation. The standard correction to Dst only eliminates the magnetopause and quiet time effects. It is generally assumed that the remaining effects cancel each other and are removed in the average over a longitudinal distribution of stations. However, the use of hourly averages from only four stations makes this rather unlikely in specific events. In fact, the apparent very rapid recovery of Dst immediately after a northward turning of the IMF suggests it is these effects disappearing from the index that causes the increase. If this is the case, the average effect of all of these currents is to bias Dst negative. However if positive and negative errors in Dst are equally likely at all phases of a substorm, time series analysis techniques such as superposed epoch analysis and linear prediction filters will still be able to detect the portion of the Dst index variation that is correlated with the solar wind or substorm onset. It is this fact that gives us confidence in our conclusion that ring current injection as a consequence of the substorm expansion phase can not be detected in the currently available Dst index.

None-the less we strongly recommend that future studies of the ring current utilize a high time resolution ring current index generated from many stations. The best approach would be to remove the dynamic pressure and quiet day in both H and D from each station separately. Then the residual variations in the horizontal field should be modeled with simple line current models of the symmetric ring, the partial ring, and the substorm current wedge. Together these models would require at least eight parameters. Two component data from 10-15 stations would be sufficient to make a least square fit to the model parameters. The temporal behavior of the fit to the symmetric ring current magnitude could then be used with greater confidence to investigate the question addressed in this paper. Whether these models can be made to converge and provide smoothly varying parameters is an interesting question.

Another question suggested by our analysis is whether the ring current injection rate depends on the magnitude of fluctuations in the solar wind electric field. Also, can it be shown that injection occurs even when the solar wind is steady for long periods of time? If such examples can be found, do ground and synchronous data suggest that the internal magnetospheric fields are fluctuating? For example, does ring current injection occur during convection bays? If so, what is responsible for trapping particles in the ring current? It has been suggested [*Kamide*, personal communication] that the superposed epoch results discussed above are flawed by the fact that they do not take into account the contributions to Dst of other currents than the ring current. It would be interesting to do a linear prediction analysis on a high time resolution ring current index obtained from models of the tail current, substorm current wedge, and the partial ring current. These are the three effects generally assumed to cancel in the Dst analysis.

Acknowledgments: This work was initiated while the author was a visiting professor at the Solar-Terrestrial Environment Laboratory in Toyokawa, Japan. We thank Professors S. Kokubun and Y. Kamide, and the laboratory for their support. The work was completed while the author was a visiting professor at the Department of Physics of the University of Newcastle. We thank Prof. B. Fraser and the University for their support. Partial financial support for this work was provided by grants from the National Science Foundation, ATM-95-02124 and ATM-13667, and the National Aeronautics and Space Administration, NAG5-1167.

REFERENCES

Akasofu, S.-I., The development of the auroral substorm, *Planet. Space Sci.*, *12*(4), 273-282, 1964.

Akasofu, S.-I, *Polar and Magnetospheric Substorms*, D. Reidel Pub. Co., Dordrecth, Holland, 1968.

Akasofu, S.I., Relationships between the AE and Dst indices during geomagnetic storms, *J. Geophys. Res.*, *86*, 4820, 1981.

Akasofu, S.-I., S. Chapman, and C.-I. Meng, The polar electrojet, *J. Atmos. Terr. Phys.*, *27*(11/12), 1275-1305, 1965.

Bargatze, L.F., D.N. Baker, R.L., McPherron, and E.W. Hones, Magnetospheric impulse response for many levels of geomagnetic activity, *J. Geophys. Res.*, *90*(A7), 6387-6394, 1985.

Blanchard, G.T. and R.L. McPherron, Analysis of the linear response function relating AL to VBs for individual substorms, *J. Geophys. Res.*, *100*(A10), 19,155-19,165, 1995.

Burton, R.K., R.L. McPherron, and C.T. Russell, An empirical relationship between interplanetary conditions and Dst, *J. Geophys. Res.*, *80*(31), 4204-4214, 1975.

Cade III, W.B., J.J. Sojka, and L. Zhu, A correlative comparison of the ring current and auroral electrojets using geomagnetic indices, *J. Geophys. Res.*, *100*(A1), 97-105, 1995.

Chapman, S., Earth storms: retrospect and prospect, 17, *J. Phys. Soc. Jpn.*, 6, 1962.

Chen, M.W., M. Schulz, L.R. Lyons, and D.J. Gorney, Stormtime transport of ring current and radiation belt ions, *J. Geophys. Res.*, *98*, 3835-3849, 1993.

Chen, M.W., M. Schulz, and L.R. Lyons, Simulations of phase space distributions of stormtime proton ring current, *J. Geophys. Res.*, *99*, 5745-5759, 1994.

Clauer, C.R., and R.L. McPherron, Mapping the local time development of magnetospheric substorms using mid-latitude magnetic observations, *J. Geophys. Res.*, *79*(19), 2811-2820, 1974.

Coroniti, F.V., R.L. McPherron, and G.K. Parks, Studies of the magnetospheric substorm: 3. Concept of the magnetospheric substorm and its relation to electron precipitation and micropulsations, *J. Geophys. Res.*, *73*(5), 1715-1722, 1968.

Davis, T.N., and M. Sugiura, Auroral electrojet activity index AE and its universal time variations, *J. Geophys. Res.*, *71*(3), 785-801, 1966.

Davis, T.N., and R. Parthasarathy, The relationship between polar magnetic activity DP and growth of the geomagnetic ring current, *J. Geophys. Res.*, *72*(23), 5825-5836, 1967.

Dessler, A.J., and E.N. Parker, Hydromagnetic theory of magnetic storms, *J. Geophys. Res.*, *64*(12), 2239-2259, 1959.

Fay, R.A., C.R. Garrity, R.L. McPherron, and L.F. Bargatze, Prediction filters for the Dst index and the polar cap potential, 111-117, Terra Scientific Publishing Co. (Tokyo), 1986.

Feldstein, Y.I., Modeling of the magnetic field of magnetospheric ring current as a function of interplanetary medium parameters, *Space Sci. Revs.*, *59*, 83, 1992.

Feldstein, Y.I., V. Yu. Pisarsky, N.M. Rudneva, and A. Grafe, Ring current simulation in connection with interplanetary space conditions, *Planet. Space Sci.*, *32*(8), 975-984, 1984.

Gonzalez, W.D., J.A. Joselyn, Y. Kamide, H.W. Kroehl, G. Rostoker, B.T. Tsurutani, and V.M. Vasyliunas, What is a geomagnetic storm?, *J. Geophys. Res.*, *99*(A4), 5771-5792, 1994.

Horning, B.L., R.L. McPherron, and D.D. Jackson, Application of linear inverse theory to a line current model of substorm current systems, *J. Geophys. Res.*, *79*(34), 5202-5210, 1974.

Iyemori, T. and D.R.K. Rao, Decay of the Dst field of geomagnetic disturbances after substorm onset and its implication to storm-substorm relation, *Ann. Geophys.*, *14*, 608-618, 1996.

Iyemori, T., H. Maeda, and T. Kamei, Impulse response of geomagnetic indices to interplanetary magnetic field, *J. Geomag. Geoelectr.*, *31*, 1-9, 1979.

Kamide, Y., and N. Fukushima, Analysis of magnetic storms with DR-indices for equatorial ring current field, *Rep. Ionos. Space Res. Jpn.*, *26*, 79, 1972.

Liu, W.W. and G. Rostoker, Energetic ring current particles generated by recurring substorm cycles, *J. Geophys. Res.*, *100*(A11), 21,897-21,910, 1995.

Lyons, L., and D.J. Williams, A source for the geomagnetic storm main phase ring current, *J. Geophys. Res.*, *85*, 523, 1980.

Lyons, L.R. and M. Schulz, Access of energetic particles to storm time ring current through enhanced radial diffusion, *J. Geophys. Res.*, *94*(A5), 5491-5496, 1989.

Mauk. B.H., Quantitative modeling of the "convection surge" mechanism of ion acceleration, *J. Geophys. Res.*, *91*(A12), 13,423-13,431, 1986.

McPherron, R.L., Baker, D.N., and Bargatze L.F., Linear filters as a method of real time prediction of geomagnetic activity, pp 85-92., Terra Scientific Publishing Co. (Tokyo), 1986.

McPherron, R.L., D.N. Baker, L.F. Bargatze, C.R. Clauer, and R.E. Holzer, IMF control of geomagnetic activity, *Adv. Space Res.*, *8*(9), 71-86, 1988.

Perreault, P. and S.-I. Akasofu, A study of geomagnetic storms, *Geophys. J. R. Astr. Soc.*, *54*, 547-573, 1978.

Pudovkin, M.I., S.A. Zaitseva, L.Z. Sizova, Growth rate and decay of magnetospheric ring current, *Planet. Space Sci.*, *33*(10), 1097-1102, 1985.

Rostoker, G., S.-I. Akasofu, J. Foster, R.A. Greenwald, Y. Kamide, K. Kawasaki, A.T.Y. Lui, R.L. McPherron, and C.T. Russell, Magnetospheric substorms - definition and signatures, *J. Geophys. Res.*, *85*(A4), 1663-1668, 1980.

Rostoker, G., S.-I. Akasofu, W. Baumjohann, Y. Kamide, and R.L. McPherron, The roles of direct input of energy from the solar wind and unloading of stored magnetotail energy in driving magnetospheric substorms, *Space Science Reviews*, *46*, 93-111, 1987.

Russell, C.T., R.L. McPherron, and R.K. Burton, On the cause of geomagnetic storms, *J. Geophys. Res.*, *79*(7), 1105-1109, 1974.

Siscoe, G. L., and H. E. Petschek, On storm weakening during substorm expansion phase, Ann. Geophys, in press, 1996.

Siscoe, G.L., The virial theorem applied to magnetospheric dynamics, *J. Geophys. Res*, *75*(28), 5340-5350, 1970.Siscoe, G.L., and H.E. Petschek, Why substorms might not be substorms, *Geophys. Res. Lett.*, submitted, February, 1996.

Todd, H., S.W. Cowley, M. Lockwood, M. Willis, and H. Luhr, Response time of the high-latitude dayside ionosphere to sudden changes in the north-south component of the IMF, *Planet. Space Sci.*, *36*, 1415-1428, 1988, .

Wrenn, G.L., Persistence of the ring current, *Geophys. Res. Lett.*, *16*, 891, 1989

Robert L. McPherron, Institute of Geophysics and Planetary Physics and Department of Earth and Space Sciences, University of California, Los Angeles, Los Angeles CA 90095-1567. (e-mail: rmcpherron@igpp.ucla.edu)

Physics of Magnetic Storms

Gordon Rostoker, Erena Friedrich and Matthew Dobbs

Department of Physics, University of Alberta, Edmonton, Alberta Canada

Abstract: Magnetic storms are defined by the presence of a ring current which comes into existence through the acceleration of particles during episodes of strongly enhanced input of energy from the solar wind into the magnetosphere. The main phase of a storm, during which the primary ring current growth takes place is typically accompanied by sustained substorm expansive phase activity, leading to the suggestion that substorm perturbations play a role in ring current growth. In this paper we shall show that substorm expansive phases do indeed play an important role in energization of ring current particles; however they contribute only a small portion of the energy which is typical for ring current particles. It appears that the dynamical changes in the near-Earth tail magnetic field which occur during substorm expansive phases are not effective in ring current generation regardless of the proximity of the locale of the initiation of the expansive phase with respect to the Earth. Rather, the substorm expansive phase involves a breakdown of the shielding electric field and, in this way, substorms cause the locale of future expansive phases to migrate further earthward. This, in turn, permits plasma sheet ions to penetrate closer to the Earth and become energized adiabatically to the rather high energies typical of the particles that contribute significantly to the ring current. Cyclical stretching and dipolarization of the near-Earth tail magnetic field can energize plasma sheet ions to the extent that the incremental energy provided by the convection electric field may lead to energies in excess of 100 keV for ring current particles and enhance the lifetime of the ring current itself.

1. INTRODUCTION

Of all magnetic perturbations of the Earth's main field, the one that has been known the longest and studied most intensively in the pre-satellite era has been the magnetic storm. In its classic form (Figure 1), it is manifested by a sudden increase in the north-south (H) component of the low latitude magnetic field (the ssc), followed some time later by a depression in that component developing over a time span of one to a few hours and concluding with a decay which may extend over several days. The size of the ssc ranges from a few nT to over 100 nT while the main phase of storms may reach strengths of a few hundred nT. The ssc is understood to be a consequence of enhanced solar wind dynamic pressure, and in modern times, is not considered to be a necessary component of the storm (cf. Akasofu, 1965). Thus the storm is defined purely by the growth and subsequent decay of the depression in the H-component of the low latitude magnetic field.

The strength of magnetic storm is usually measured by the magnitude of the Dst index. This index, introduced by *Sugiura* [1964] for the study of IGY data, was developed as a measure of the symmetric component of the ring current. The H-component of the surface magnetic field at low latitudes is measured at several stations distributed in longitude. After subtraction of the quiet time baseline (removing the effects of

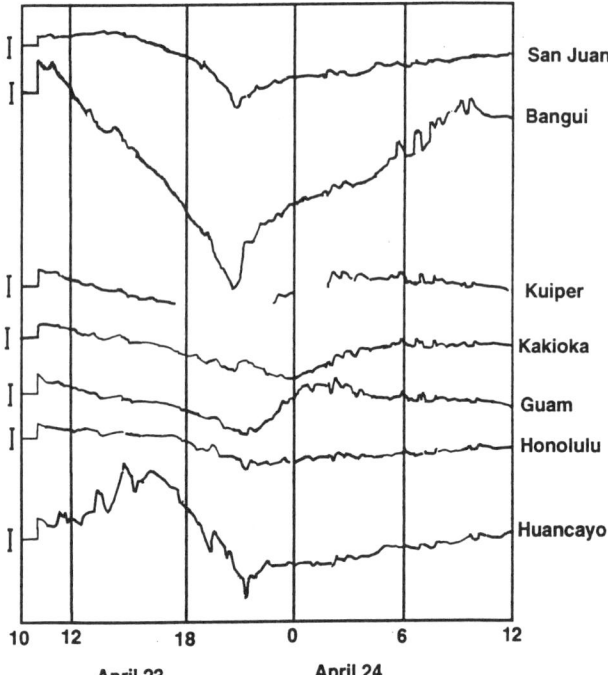

Fig. 1. Low latitude H-component magnetograms taken during the course of a magnetic storm on April 23-24, 1959, at a number of stations located at different longitudes. The vertical bars on the left side of the figure indicate 50 nT. There is evidently a large long lived negative H-component perturbation at all stations, however there is clearly a local time variation in in the character of the disturbance field. The Dst index is an attempt to identify the component of the disturbance field which is not dependent on local time.

Sq), the average value of the H-components of the contributing stations is calculated and multiplied by the secant of the average latitude of the contributing stations to define Dst (cf. Rostoker, 1972). Ideally, the index does not contain contributions from current systems which are not azimuthally symmetrical. Unfortunately, the fact that data from very few (viz. as little as four) stations are used in the computation of the index permits contributions from other current systems to leak into the index (viz. the asymmetric ring current, the magnetotail current as well as contributions to the H-component from field-aligned currents). Thus every value of Dst may be thought of as the value due to the symmetric ring current together with some "error" associated with the contributions of other current systems. We shall provide a measure of the size of the expected "error" later in this paper.

When researchers study magnetic storms, they typically choose events for which the main phase features a Dst of tens to hundreds of nT. Curiously, there is no well defined limit for the lower limit of Dst for which one can say a storm does or does not exist. There are operational definitions for the purposes of predictions, with ~50 nT being the lower limit [*Joselyn and Tsurutani*, 1990]. However, there is no physical reason why any particular magnitude of Dst should be chosen as a lower limit. Thus, when one says that a particular group of substorms is or is not associated with an episode of ring current growth, no study has yet been performed to establish that fact based on a quantitative measure which establishes the presence or absence of a storm time ring current. Later in this presentation, we shall try to look at this question of what kind of lower limit can be set above which the presence of a storm time ring current can be established.

We should now like to address the question of what constitutes a magnetospheric substorm, so that we can better understand how the development of a storm time ring current might be related to substorms in general. It is now reasonably well accepted that a substorm involves two distinctive processes.

The first of these is the directly driven process in which energy from the solar wind that enters the magnetosphere is immediately deposited in the high latitude ionosphere with the only delay being the Alfvén propagation time from the magnetospheric boundary layers to the ionosphere (cf. Akasofu, 1979; Rostoker et al., 1987). The electric current manifestations of the directly driven process are the eastward and westward electrojets flowing from near noon across the dusk and dawn meridians, respectively (Figure 2a). These primarily Hall currents flow in the auroral oval and are colocated in the ionosphere with the Birkeland currents (cf. Zmuda and Armstrong, 1974) which flow into and out of the auroral ionosphere in anti-parallel sheets linked by primarily Pedersen meridonal ionospheric currents. The Region 1 and Region 2 Birkeland currents (cf. Iijima and Potemra, 1976) are offset such that there is net downward field aligned current across the noon sector and net upward field-aligned current across the the midnight sector. These large scale currents vary rather slowly, but are responsible for the dissipation of a considerable amount of the energy which enters the magnetosphere from the solar wind.

The second of the the substorm processes is the storage/release process in which some of the energy from the solar wind is stored in the magnetotail as the magnetic energy of the tail lobe and as the drift energy of the earthward convecting plasma in the tail plasma sheet. From time to time, this energy is released in what is termed expansive phase activity. (It is extremely important to note here that most researchers, when they use the term substorm, are actually referring to the expansive phase of the substorm. In this paper we shall try to be explicit in referring to the phenomenon accompanying the auroral breakup (cf. Akasofu, 1964) as the expansive phase.) It is, in fact, a major problem for the substorm researcher to distinguish between directly driven activity and substorm expansive phase activity using magnetometer data since both involve changes in westward electrojet strengths.) Expansive phases are often triggered by decreases in energy input from the solar wind to the magnetosphere, commonly induced by a weakening of a southward IMF or a turning towards the north of the IMF (cf. Caan et al., 1977; Rostoker, 1983). The storage process normally occurs concurrent with growth of the directly driven current systems after the start of increased

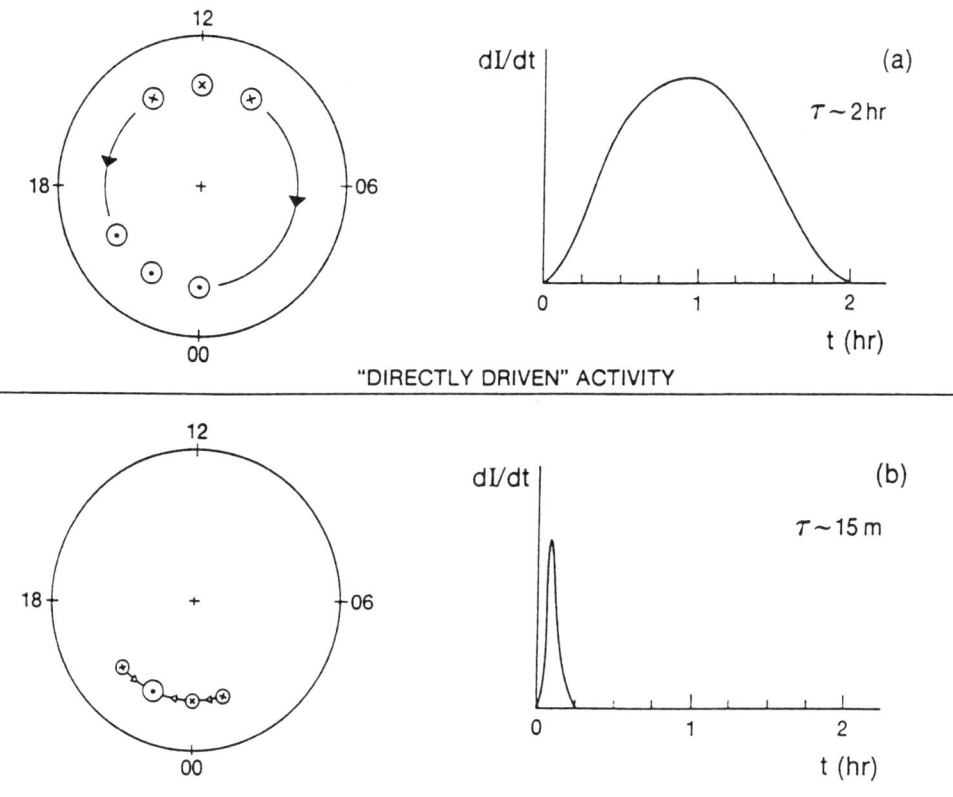

Fig. 2. Cartoon showing the two components of substorm activity. Fig. 2a shows the large scale eastward and westward electrojets associated with directly driven activity while 2b shows the localized field-aligned and ionospheric currents associated with an expansive phase intensification (after Rostoker, 1991). A substorm expansive phase involves the development of many small localized current elements such as that shown in 2b, and the ensemble of these structures contributes significantly to the overall substorm magnetic perturbation pattern.

energy input into the magnetosphere (typically as the result of a southward turning of the IMF). The expansive phase involves the sudden growth of an azimuthally confined segment of westward electrojet in the late evening sector, beginning near the equatorward edge of the auroral oval [*Samson et al.*, 1992]. As expansive phase activity develops after onset, the activity spreads poleward in discrete steps [*Kisabeth and Rostoker*, 1974]. After the maximum poleward movement has taken place, activity may continue for some time afterwards both at the poleward edge of the expanded oval and near its equatorward border. When the substorm moves into its recovery phase, the activity is normally restricted to the poleward border of the oval. However, the magnetic disturbances at the poleward border can be quite strong although the substorm might be said to be in its recovery phase (cf. Lyons et al., 1990).

2. CONCEPTS OF HOW SUBSTORMS MIGHT INFLUENCE THE GROWTH OF THE STORM TIME RING CURRENT

The question of how substorms relate to the development of a magnetic storm has long been a subject of intense discussion. *Kamide* [1979] has explored the suggestion by *Akasofu* [1968] that a magnetic storm simply was the sum of polar magnetic substorms together with the ring current (which is the unique feature of storm). In fact, *Kamide* [1992] specifically explored the question of whether or not a substorm is a necessary component of a magnetic storm. In light of the fact that the symmetric ring current signature (quantified by the index Dst) appears to be the unique identifier of a magnetic storm, we shall devote this paper to an effort to understand the nature of the physical processes through which the ring current is created.

A storm involves the energization of a radially localized population of protons to energies in excess of ~100keV in an L-shell range which is earthward of the typical position of the inner edge of the plasma sheet. During the development of the main phase, significant amounts of oxygen are transported out of the ionosphere into the near-Earth plasma sheet (cf. Lennartsson and Shelley, 1986). The distinguishing feature of the stormtime ring current is the length of time required for its decay compared to normal decay times of substorm related currents. As mentioned earlier (cf. Figure 2a), the large scale electrojets vary on a time scale of many tens of minutes. In fact, *Rostoker et al.* [1991] have shown that for sudden northward turnings of the IMF followed by extended periods of northward IMF, the decay times of the auroral electrojets is of the order of an hour. The wedgelets which characterize the intensifications taking place during the expansive phase tend to have decay time of the order of ~15 min (cf. Figure 2b). In contrast, the ring current decay times range from a few hours (associated with its energetic oxygen component) to several tens of hours (associated with its energetic proton component).

In trying to understand how ring current particles might be energized to the levels of ~100-200 keV which have been identified as the most important in terms of the ion population identified as the current carriers, one is initially tempted to look to the rapid time variations of the magnetic field associated with substorm expansive phase activity. This first approach seems natural since *Akasofu and Chapman* [1961] have identified auroral oval substorm expansive phase activity as a unique feature of the period of time in which ring current growth takes place. However, one quickly realizes that there are many substorm expansive phases which seem to have no particular response in terms of ring current growth. This could mean one of two things. On one hand, it may suggest that the expansive phases *per se* have no major impact in terms of energizing ring current particles. This would be consistent with the suggestion by *Burton et al.* [1975] that the energization is simply a consequence of the convection electric field moving plasma sheet particles closer to the earth. For such a scenario, the energization would be adiabatic reflecting the combined effects of betatron and Fermi acceleration (cf. Hines, 1963). Another possibility is that substorm expansive phases which are triggered too far from the Earth cause acceleration of particles which are unable to make complete drift paths around the Earth and hence cannot form a long lived symmetric ring current. Only if the expansive phases are triggered relatively close to the Earth (viz. the inner edge of the tail current has moved sufficiently earthward) will the energized particles be able to make complete circuits around the Earth and hence form a symmetric ring current. This has been suggested by *Rostoker* [1994a, 1996a] and we shall explore the latter possibility in this paper.

3. RESPONSE OF THE RING CURRENT INDEX DST TO EQUATORWARD MOVEMENT OF THE OVAL

Substorm magnetic field variations with magnitudes of tens of nT are often observed at latitudes many hundreds of kilometers equatorward of the equatorward edge of the auroral oval. Typically, the perturbations are in the horizontal (H and D) components and are due to the distant effect of field-aligned currents associated with the substorm current wedge, the directly driven system or with both. Detection of such perturbations at middle latitude stations gives no indication of the proximity of the auroral electrojets. However, as can be seen from Figure 3 the equatorward edge of the auroral electrojets and regions immediately equatorward of that boundary is marked by a significant Z-component perturbation (positive equatorward of an eastward electrojet and negative equatorward of a westward electrojet). The presence of a Z-component perturbation at a middle latitude station should be a sure sign of the proximity of an auroral electrojet and we shall use it as our proxy for the location of the auroral electrojets. One might think that it would be better to use the total perturbation magnetic field, as for situations in which an electrojet has its center directly above an observing station, no Z-component perturbation would be detected. However, as mentioned earlier substorm current wedges have field-aligned currents which produce significant magnetic perturbations in the horizontal plane despite the fact that the ionospheric wedge currents may be far distant from the observing site. This, we believe, is more of a source of error than having an electrojet accidentally stay with its center over a given observing site. Hence we used purely the Z-component perturbation as our proxy for the proximity of the auroral electrojets. In our study we used data from three years (1989, 1990 and 1991) from the three low latitude stations of Fredericksburg, Boulder and Fresno and the middle latitude station of Newport about 500 km north of Fresno. Most of the work involved the use of the low latitude stations, excluding Newport, which was treated separately.

The two parameters we shall be correlating are Dst (as a proxy for the strength of the symmetric component of the ring current) and the perturbation in the Z-component of the magnetic field dZ (as a measure of the proximity and strength of the auroral electrojets). Both of these parameters are imperfect measures of what they purport to show, however it is possible to correct them to make them somewhat more acceptable. In the case of Dst, there are three potential problems:

1. Since there are nowadays only four stations at most used for the computation of Dst, it is not possible to obtain a measure of the symmetric ring current without some contamination from the asymmetric ring current associated with substorm activity. Thus, there is always some "noise level" associated with contributions from non-ring current systems.

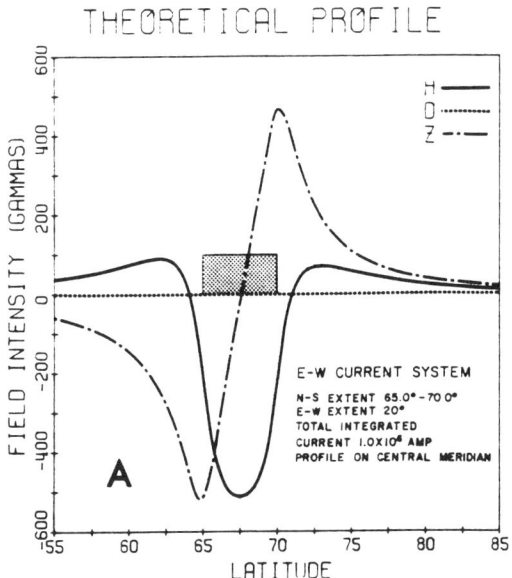 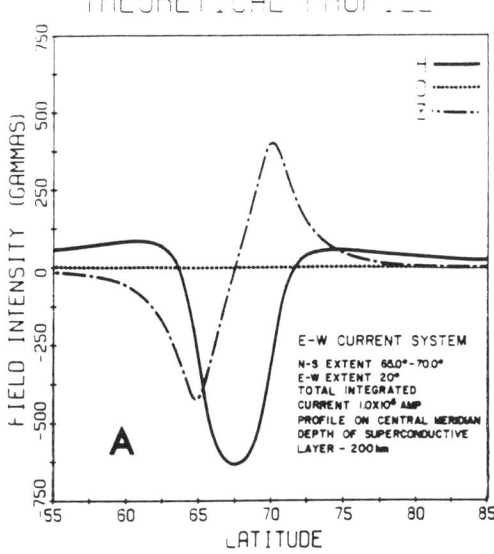

Fig. 3. Model latitude profiles from a current wedge involving a westward electrojet (after Kisabeth, 1972). The left hand panel shows the magnetic perturbations due to the current system not including effects of induced currents in the ground. The right hand panel shows the magnetic perturbations for the same source current system but including the effects of induced currents where the Earth is simulated by a superconductor starting 200 km below the surface. From profiles such as these one can establish the positions of the borders of the electrojet (which is uniformly distributed over 5 degrees of latitude between 65-70 degrees in this figure) and establish correction factors taking induction into account at each latitude.

2. Earth induction effects are not subtracted from the observed magnetic field perturbations before they are merged to compute the Dst index. Thus the magnitudes given for Dst contain an induction contribution and do not reflect the strength of the source (ring) current distribution alone.

3. Changes in solar wind dynamic pressure change the value of Dst.

For the first two problem areas, there is little that can be done at the present time. Fortunately, the second problem area identified does not have a severe impact except in cases where one of the Dst stations is sited near a subsurface conductivity anomaly. (This is the case for Honolulu, and should be a matter of concern for any researcher who looks for diurnal trends in Dst.) It is possible, however, to correct for solar wind dynamic effects using a relationship suggested by *Burton et al.* [1975] and developed further by *Gonzalez et al.* [1989], viz.

$$Dst_{corr} = Dst - (0.02 \, v \, n^{1/2} - 20 \, nT)$$

where v is the solar wind speed in km/s, n is the number density in particles per cc and 20 nT is a correction factor related to the effect of magnetopause currents for average solar wind conditions. The Dst data we shall present will be corrected in the above fashion.

In dealing with the perturbation dZ in the Z-component (which is the disturbed field value less the quiet time field value) there are three major concerns:

1. Finding a quiet time field value is not always very easy. It is possible to define a quiet day through use of indices such as Kp, however, even the quietest days defined in this fashion often feature some perturbations. Values of dZ calculated under such conditions will be slightly in error.

2. Subsurface induction effects due to anomalous conductivity distributions can sometimes lead to erroneous results.

3. The ring current actually contributes to the Z-component of the perturbation magnetic field at all latitudes except at the equator.

While there is little that can be done regarding the first two sources of error listed above, it is possible to correct the values of dZ for ring current contributions. To do this, one first corrects Dst for earth induction effects using the technique of *Kisabeth* [1972] to find the perturbation due to the source ring current system. One then finds the component of this perturbation field normal to the earth's surface at the latitude of the observing site. This vertical component is then corrected for induction effects to produce the term which must be added or subtracted from the observed Z-component perturbation to remove the ring current effect. The correction actually involves adding the ring current contribution to the (positive) Z-component perturbations in the afternoon hemisphere due

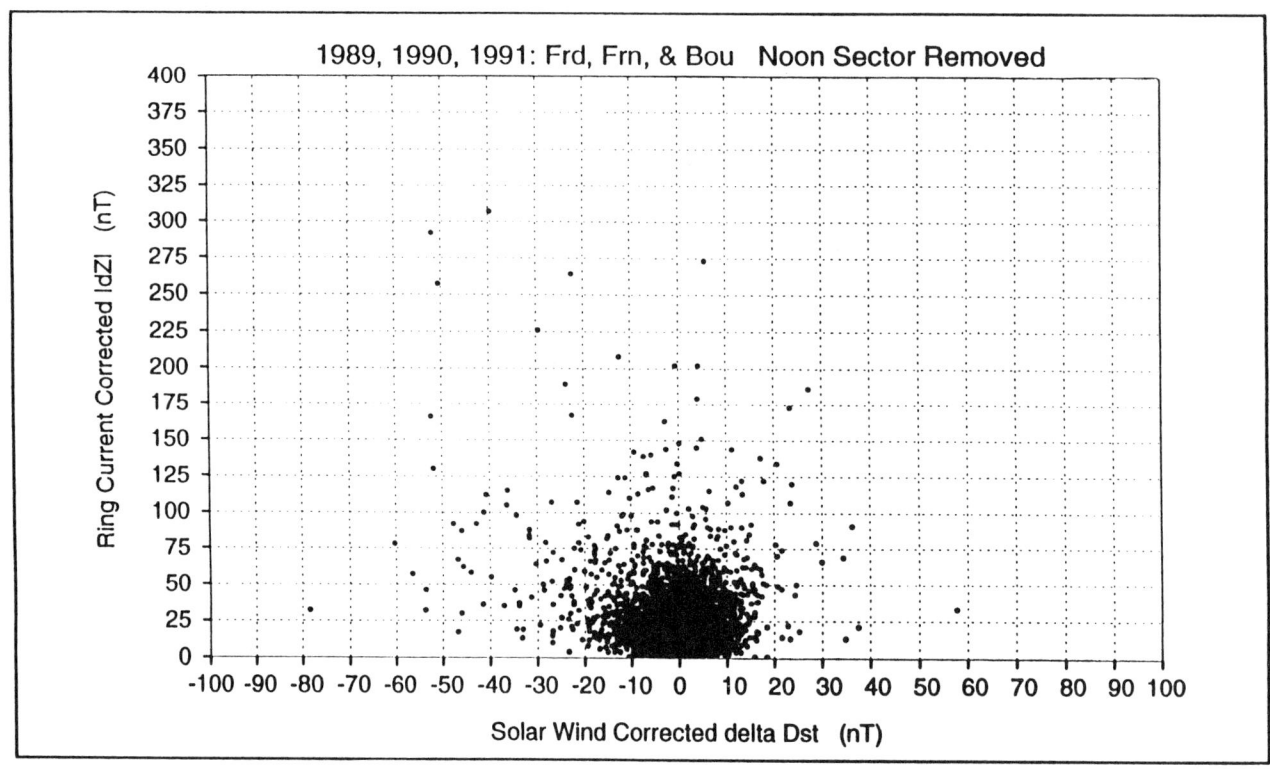

Fig. 4. Plot of the change in Dst from one hour to the next shown as a function of the dZ perturbation due to the edge effects of the auroral electrojets. Each point represents the maximum dZ perturbation at one of the three middle latitude stations of Fredericksburg, Boulder and Fresno for the hour in which the change in Dst would be taking place. The data shown are for the years 1989-1991 inclusive. No data from two hours on either side of local noon at Boulder are considered, as the electrojets in the noon sector are extremely weak in that local time interval and the absence of a significant disturbance dZ would not necessarily rule out the presence of the auroral electrojets at low latitude at other local times.

to the eastward electrojet while subtracting the ring current contribution from the (negative) Z-component in the morning hemisphere.

Figure 4 shows a plot of the change in Dst (dDst) from one hour to the next as a function of dZ in the hourly interval in which the Dst is in process of reaching its new value. Although there is a considerable amount of scatter in the points, there is clearly a trend for larger changes in Dst (i.e. changes making Dst more negative) to be associated with larger values of dZ suggesting that the auroral electrojets move to lower latitudes during periods of Dst growth. This pattern is more apparent when the data are averaged in 10 nT bins for dZ as shown in Figure 5; the data points here reflect cases where there are at least 10 data points in each bin. An interesting facet of Figure 5 is the association of positive increases in Dst with increased dZ. This is attributed to the effect of large substorm expansive phases affecting a significant longitudinal extent, so that the positive H-component within the longitudinal confines of the wedge actually contributes to Dst due to the inability of the small number of Dst stations to adequately identify the asymmetry in the disturbance field. Finally, in Figure 6 we present a set of contour plots reflecting the changes in Dst and dZ for values of |Dst| changes less than 10 nT. One would expect that a zero change in Dst would be associated with very small values of dZ. Clearly such is not the case, and this figure illustrates the sizes of errors one should anticipate in defining the magnitudes of changes in ground disturbance fields at low latitudes due to difficulties in treatment of the data discussed earlier in this section. From Figure 6, it is clear that one should not assign much meaning to changes in Dst of ~5nT or less or to changes in middle latitude ground magnetic perturbations of ~10 nT or less.

From the data shown in this section, we conclude that growth in Dst is indeed associated with equatorward motion of the directly driven auroral electrojets, and hence with the earthward motion of the inner edge of the crosstail current sheet. However, this does not yet answer the question of whether or not substorm expansive phases play a significant role in the energization of ring current particles. Our results thus far suggest merely that ring current particles will be energized only if the inner edge of the crosstail current sheet (as

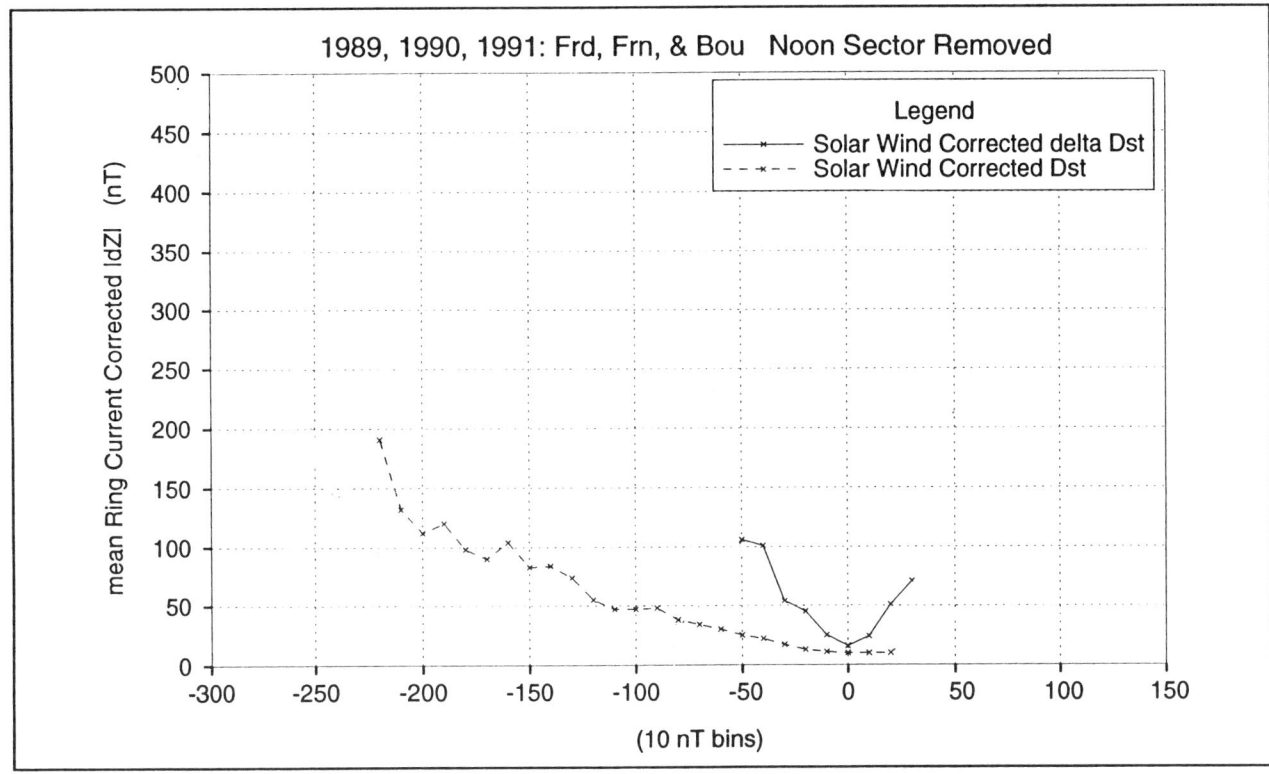

Fig. 5. The same data as shown in Figure 4 except averaged in 10 nT bins for dZ. Clearly the change in Dst from hour to hour (solid line) correlates with Z-component perturbation increases which can be attributed to the proximity of the auroral electrojets. An interesting feature is that positive changes in Dst also seem to be correlated with dZ. Also shown is the average value of Dst in the hour for which dZ is evaluated (dashed line); again the data suggest that increased ring current is associated with auroral electrojet expansion to low latitude.

identified by the proxy measurements of the locale of the equatorward border of the auroral electrojets) moves sufficiently close to the Earth. In terms of the expansive phase itself playing an important energization role, one would have the expectation that a large expansive phase occurring when the auroral oval has expanded far equatorward will have a visible impact on Dst. Figure 7 shows an example of an interval of time in which a ring current has clearly become established and large scale expansive phase activity is in progress. The key event in this sequence occurs in the hourly interval 0600-0800 UT, in which one sees a strong expansive phase onset at ~0620 UT. The large dZ perturbation associated with this onset attests to the fact that the auroral electrojets have moved far equatorward during the course of the main phase development. The value of Dst does not become more negative over the interval 0600-0800 UT suggesting that this strong expansive phase had little effect on ring current strength. One might argue that a large substorm expansive phase might feature a positive H-component disturbance large enough to balance or overpower the negative H-component disturbance expected from an enhanced ring current. However, in this case such an argument does not succeed because it is clear that, in the hour following the expansive phase (0700-0800 UT), Dst still did not become more negative. Since one expects the ring current lifetime to be significantly larger than that of substorm related currents, the fact that Dst did not become more negative in the hour following the expansive phase strongly suggests that the expansive phase occurring in the interval 0600-0700 UT did not cause a ring current enhancement as measured by Dst. This case is very suggestive that individual expansive phases do not have a significant impact on Dst regardless of how far earthward the inner edge of the crosstail current sheet has penetrated. A similar conclusion to this has been reached recently by Iyemori and Rao [1996], although they did not investigate the possible dependence on the latitude at which expansive phases are initiated. Based on the observations described above, we now outline what we believe to be the physics of storm time ring current development.

156 PHYSICS OF MAGNETIC STORMS

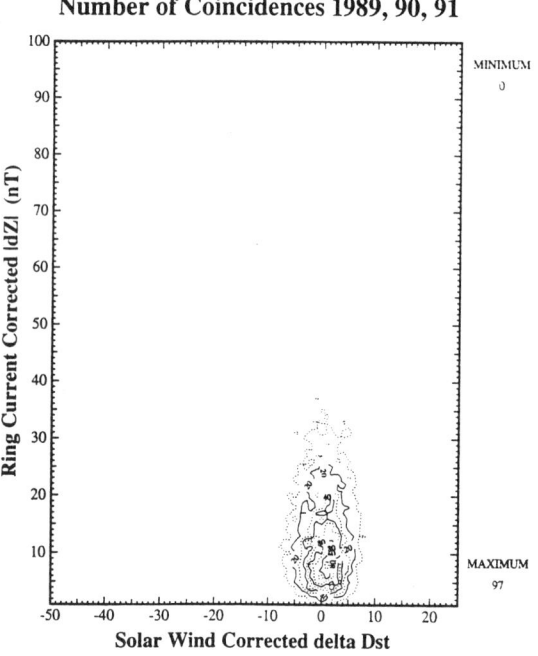

Fig. 6. Contour plots showing the level of uncertainty in correlations between changes in Dst and dZ. This figure suggests that hour to hour changes in Dst which are less than ~5 nT should not be considered as being significant, the same being true for changes in dZ of less than ~10 nT.

4. THE PHYSICS OF RING CURRENT DEVELOPMENT

We couch our description of how substorms affect ring current development in the framework of the boundary layer dynamics (BLD) model of substorms as recently presented by *Rostoker* [1994b, 1996b]. Here we shall concentrate on what happens relatively close to the Earth and not on the high latitude substorm perturbations which likely have little influence on ring current formation. Figure 8 shows the space charge distribution associated with the magnetosphere based on the assumptions of MHD in which space charge density can be expressed by

$$\rho = \varepsilon_0 \, \text{div} \, \mathbf{E} = -\varepsilon_0 \, \mathbf{B} \cdot \text{curl} \, \mathbf{v} + (1/c^2) \, \mathbf{v} \cdot \mathbf{J}_\perp \quad (1)$$

where **v** is the convection velocity and $\mathbf{J}_\perp$ is the component of current flow transverse to the magnetic field with **E** and **B** being the electric and magnetic fields respectively. The first term on the right hand side of Eq. (1) can be associated with vorticity in plasma flow while the right hand term is non-zero when there is a component of transverse current flow in the direction parallel or anti-parallel to the direction of the convective flow of plasma. The space charge associated with the second term on the right hand side of Eq. (1) reflects the

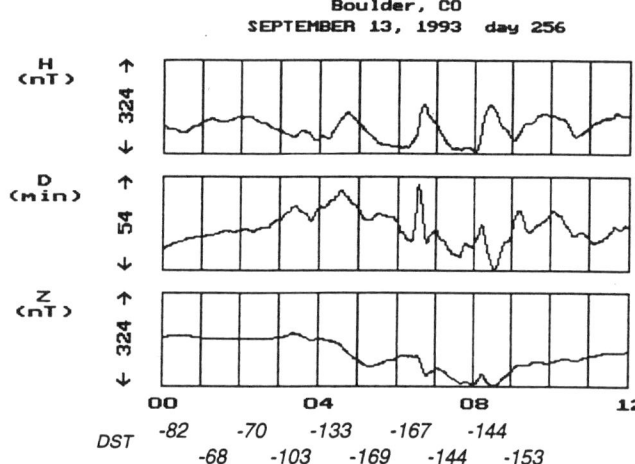

Fig. 7. Magnetogram from the middle latitude station of Boulder showing a period during which the Dst index increased to >-150 nT. The large expansive phase with onset at ~0620 UT does not appear to have any influence in terms of further expansive phase growth despite the fact that the Z-component indicates that the electrojets are displaced rather far equatorward.

shielding of the inner magnetosphere from the primary convection electric field imposed across the plasma sheet through the interaction of the solar wind with the magnetospheric boundary layers. We argue that the conditions for ring current growth begin with the enhancement of the primary convection electric field due to an increase in energy input from the solar wind to the magnetosphere. This increase in energy input could be thought of as being subdivided into two components. Part of the energy is deposited in the ionosphere through directly driven activity while the balance (less any energy deposited back into the solar wind) is stored in the magnetotail. That storage itself can be broken into three components. Part of the energy is stored in the tail magnetic field and the growth of this stored energy is consistent with the strengthening of the crosstail current in the plasma sheet. (This is equivalent to storage of magnetic field in an inductor.) A further portion of the stored energy is found in the convective motion of the plasma as it drifts earthward. (This is equivalent to the storage of electric energy in a capacitor.) The balance of the energy is stored in the gyrational, bouncing and gradient/curvature drift motion of particles acquired through Fermi and betatron acceleration of particles as they drift into the increasingly stronger and more dipolar magnetic field geometry during the development of the substorm growth phase.

The substorm expansive phase reflects a process in which the energy stored in the magnetic field and convective drift motion is converted into gyrational, bouncing and drift motion. Some of the particles so energized precipitate with the energy being lost to the upper atmosphere as heat. Others of the particles are trapped in stable drift orbits around the Earth, and these particles reflect storage of energy as kinetic

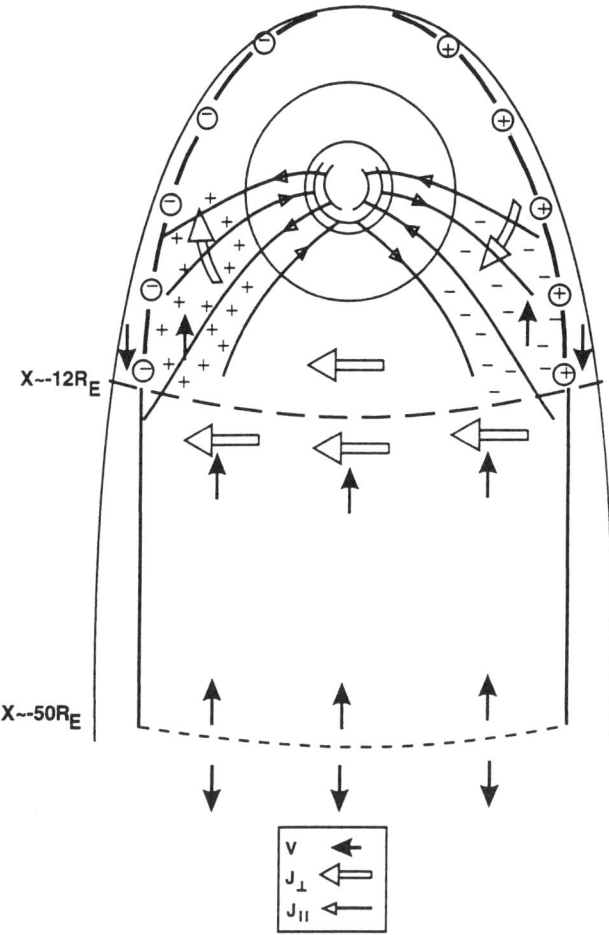

Fig. 8. Cartoon showing the projection of the magnetospheric space charge distribution, convective drift velocity and transverse component of magnetospheric current flow together with the Birkeland current flow shown in perspective (modified after Rostoker, 1994b). The space charge away from the flanks of the magnetosphere near the inner edge of the plasma sheet produces the shielding electric field. Substorm expansive phases are associated with sudden reductions in this shielding electric field which permit magnetotail plasma to penetrate closer to the Earth.

energy of motion and as magnetic energy associated with the flow of current associated with the particle motions. The current we speak of here is the ring current and the stored magnetic energy involves a magnetic field which identifies the ring current and which is represented by the Dst index.

One is then motivated to ask how much energy is available for the energization of particles in the substorm process and in what way this energy is transferred to those particles. If one examines the primary convection electric field, one might conclude that the maximum energy the particles might acquire is that associated with the crosstail electric field. This particular amount of energy would correlate well with the dawn-to-dusk component of the interplanetary electric field (viz., $v B_z$). Since the crosstail electric field can reach strengths in excess of ~100 keV during magnetospheric storm activity, a considerable portion of the energy of ring current particles should be attributable simply to energization by the convection electric field. This would explain the excellent correlation between ring current intensity (as quantified by Dst) and the dawn-to-dusk component of the interplanetary electric field reported by *Burton et al.* [1975].

However, there are other sources of energy which must be reckoned with. One such source is that associated with large amplitude waves. This concept is inherent in the thermal catastrophe model of *Goertz and Smith* [1989]. It may also explain the non-adiabatic acceleration events in the plasma sheet identified by *Huang et al.* [1992] at times of substorm expansive phase onset. Another possible source lies in heating of plasma sheet particles through reconfiguration of the near-Earth tail magnetic field. *Liu and Rostoker* [1995] have presented a magnetic pumping model for energization of plasma sheet particles through cyclical stretching (viz., growth phase) and collapsing (viz., dipolarization in an expansive phase) of magnetic field lines in the near-Earth plasma sheet. Figure 9 shows the changes in field geometry associated with the Liu and Rostoker proposal. The idea is that as the tail field stretches, ions with zero (90°) pitch angle are scattered to 90° (zero) pitch angles and when the magnetic field dipolarizes, these ions experience some energization. The energization is a second order process, and the ion distribution functions acquire the distinctive characteristics of a Kappa distribution of the type identified by *Christon et al.* [1988] in the near-Earth plasma sheet. We believe that such processes are capable of imparting energies of up to a few tens of keV to plasma sheet particles, bringing them up to sufficiently high "seed" energies that the primary convection electric field, during storm time is able to provide sufficient additional energy to make the ions capable of generating a significant ring current magnetic field.

Based on the above discussion, we consider the development of the ring current to proceed as follows. The enhancement in the primary convection electric field leads to the adiabatic energization of plasma sheet ions (cf. Hines, 1963) as they drift earthward. Near the inner edge of the plasma sheet, the crosstail current can become significantly enhanced [*Kauffman,* 1987] allowing the second term on the right hand side of Eq. (1) to become significant. This term represents the space charge associated with the shielding electric field and a buildup of this shielding charge prevents plasma from convecting closer to the Earth. If the plasma could not convect closer to earth, this would limit the energies to which it could be accelerated and hence restrict the ability of the ring current to grow. To counter this potential difficulty, the substorm expansive phase plays a very important role. As suggested by *Rostoker* [1994b], the fact that the substorm expansive phase involves the sudden reduction of the crosstail current near the inner edge of the plasma sheet is consistent with a sudden decrease in the space charge due to the second term in Eq. (1).

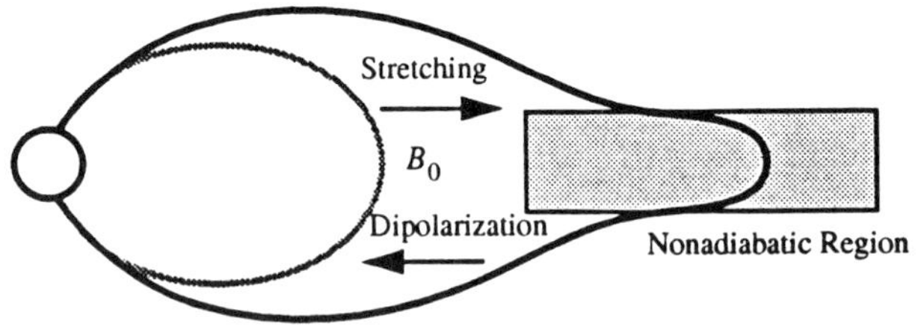

Fig. 9. Changes in field line form associated with the magnetic pumping concept proposed by W. W. Liu and G. Rostoker to energize near-Earth plasma sheet ions (after Liu and Rostoker, 1995). The cyclical stretching and dipolarization of the magnetic field together with pitch angle scattering of the ions leads the significant energization which provides a hot seed population for ultimate ring current growth.

The reduction of this space charge is equivalent to a reduction in the shielding electric field which permits plasma to convect further earthward under the influence of the primary convection electric field. A series of expansive phases under conditions of strong southward IMF for an extended period of time would permit a sequence of buildups and reductions in the space charge associated with shielding, each episode involving processes occurring closer and closer to the Earth. The currents involved in the reduction of shielding charge are, in fact, the Region 2 field-aligned currents. These currents thread the equatorward portion of the auroral oval and hence the earthward movement of the shielding currents would be reflected in the equatorward motion of the equatorward edge of the directly driven auroral electrojets.

We conclude by noting that a sequence of substorm expansive phases such as discussed above, would lead to rapid rises and falls in the electric field experienced by particles in the near-Earth plasma sheet. This type of variable electric field is characteristic of the conditions specified by *Chen et al.* [1993] for effective development of a storm time ring current.

5. DISCUSSION AND CONCLUSIONS

In this paper we have tried to outline the circumstances under which a storm time ring current develops and to acquire some physical insight as to how substorm expansive phases influence ring current growth. Based on our arguments presented above, we conclude that:

1. Ring current particles are energized primarily through the action of the convection electric field as the particles drift closer to the Earth.
2. A non-trivial amount of energy is acquired by the particles through acceleration due to stretching and subsequent dipolarization of the near-Earth tail magnetic field. The action of magnetic pumping is one means by which the particles acquire this energy. When one looks at ring current particles with energies around 150 keV or thereabouts, one should view only a very few tens of keV as being the consequence of dipolarization with the majority of the energy being attributable to the action of the convection electric field.
3. The primary role of the substorm expansive phases during storm main phase development is to break down the shielding electric field making it possible for particles near the inner edge of the plasma sheet to move further earthward and hence to be further energized through adiabatic processes.

It is worth pointing out at this point that the shielding effect that we have been discussing in this paper is unrelated to the shielding associated with the hot particle population responsible for the ring current (cf. Southwood, 1977). The transverse magnetospheric current in that case circles the Earth and thus the current carriers spend a significant portion of their lifetime on field lines threading the high conductivity dayside ionosphere. In our case, the Kaufmann current closes primarily on the near-Earth magnetopause, which results in shielding that differs significantly from that associated with the ring current in that the effect manifests itself outside the ring current regime and the breakdown of this shielding ultimately leads to ring current creation closer to the Earth. This, by no means, suggests that some shielding is not effected by the physical processes described by *Vasyliunas* [1972], *Jaggi and Wolf* [1973], and *Southwood* [1977]. We simply identify another source of shielding associated with the growth of the Kaufmann current and we suggest that the growth and breakdown of this component of shielding plays a key role in the substorm and storm processes.

While one may no longer point to substorm expansive phases as being a primary accelerator of ring current particles, the energy acquired through a sequence of growth and expansive phase cycles may turn out to be of some importance. Figure 10 shows the lifetimes of ring current particles calculated by *Smith et al.* [1981] for a plasma made up of H^+, He^+, He^{++} and O^+ with energies from 10-100 keV. This calculation, relevant for distances from the Earth corresponding to L=5, shows that the lifetime of H^+ depends very strongly on energy. For energies lower than ~40 keV, O^+ has a lifetime greater than that of H+. However, as the energy of the particles increases, the lifetime of the H^+ in the ring current increases significantly. From this diagram, it can be seen that a very few tens of keV can make an enormous difference in the lifetime of the ring current carried by those particles. For this reason, while it is clear that most of the energization of ring current particles is effected by the primary convection electric field, just how large the main phase ring current becomes and how long it persists after the end of the main phase growth may depend significantly on how much energy can be acquired by the ions through acceleration processes such as magnetic pumping.

We should note, in conclusion, that our discussion of substorms in this paper has not involved a treatment of the role of the substorm current wedge (cf. Baumjohann, 1983). The currents that flow as a consequence of reductions of the shielding space charge can be understood in the context of the directly driven system. (In fact, the resultant current flow is into the ionosphere in the evening sector and out of the iono-

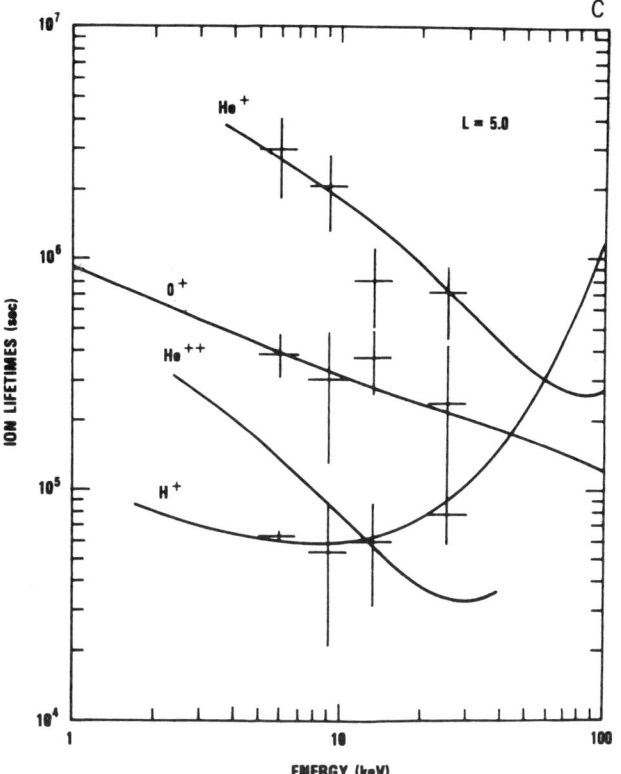

Fig. 10. Lifetimes of ring current ions as a function of their energy for a radial distance from the Earth of L=5 Re (after Smith et al., 1981). Note the steep increase in lifetime of hydrogen ions for energies approaching 100 keV. Clearly the longevity of the ring current is very sensitive to the energization achieved during the development of the storm main phase.

sphere in the morning sector characteristic of the Region 2 currents which are part of directly driven activity.) At this time we do not know if the substorm wedge currents reflect diversion of near-Earth crosstail current into the ionosphere (e.g. Lui, 1996) or whether the mechanism producing the field-aligned currents and associated westward ionospheric electrojet is located further back in the tail. In any event, the impact of the wedge field-aligned currents closer to the Earth is likely to be small when one considers ring current particle acceleration processes so we have chosen not to address the question of the origin of the substorm current wedge in this paper.

Acknowledgements. This research was supported by the Natural Sciences and Engineering Research Council of Canada through a grant to G.R.

REFERENCES

Akasofu, S.-I., The development of the auroral substorm, *Planet. Space Sci.,12*, 273, 1964.

Akasofu, S.-I., The development of geomagnetic storms without a preceding enhancement of the solar plasma pressure, *Planet. Space Sci., 13*, 297, 1965.

Akasofu, S.-I., *Polar and Magnetospheric Substorms*, D. Reidel Publ. Co., Dordrecht, Holland, 1968.

Akasofu, S.-I., What is a magnetospheric substorm?. in *Dynamics of the Magnetosphere*, edited by S.-I. Akasofu, p. 447, D. Reidel Publ. Co., Dordrecht, Holland, 1979.

Akasofu, S.-I. and S. Chapman, The ring current, geomagnetic disturbance and the Van Allen radiation belts, *J. Geophys. Res., 66*, 1321, 1961.

Baumjohann, W., Ionospheric and field-aligned current systems in the auroral zone: a concise review, in *Advances in Space Physics, 2*, 55, 1983.

Burton, R. K., R. L. McPherron and C. T. Russell, An empirical relationship between interplanetary conditions and Dst, *J. Geophys. Res., 80*, 4204, 1975.

Caan, M. N., R. L. McPherron, and C. T. Russell, Characteristics of the association between the interplanetary magnetic field and substorms, *J. Geophys. Res., 82*, 4837, 1977.

Chen, M. W., M. Schulz, L. R. Lyons and D. J. Gorney, Stormtime transport of ring current and radiation belt ions, *J. Geophys. Res., 98*, 3835, 1993.

Christon, S. P., D. G. Mitchell, D. J. Williams, L. A. Frank, C. Y. Huang, and T. E. Eastman, Energy spectra of plasma sheet ions and electrons from 50 eV/e to ~1 MeV during plasma sheet transitions, *J. Geophys. Res., 93*, 2562, 1988.

Goertz, C. K. and R. A. Smith, The thermal catastrophe model of substorms, *J. Geophys. Res., 94*, 6581, 1989.

Gonzalez, W. D., B. T. Tsurutani, A. L. C. Gonzalez, E. J. Smith, F. Tang, and S.-I. Akasofu, Solar-wind magnetosphere coupling during intense magnetic storms (1978-1979), *J. Geophys Res., 94*, 8835, 1989.

Hines, C. O., The energization of plasma in the magnetosphere: hydrodynamic and particle drift approaches, *Planet. Space. Sci., 10*, 239, 1963.

Huang, C.Y., L.A. Frank, G. Rostoker, J. Fennell and D.G. Mitchell, Nonadiabatic heating of the central plasma sheet at substorm onset, *J. Geophys. Res, 97*, 1481, 1992

Iijima, T. and T. A. Potemra, The amplitude distribution of field-aligned currents at northern high latitudes observed by Triad, *J. Geophys. Res., 81*, 2165, 1976.

Iyemori, T., and D. R. K. Rao, Decay of the Dst field of geomagnetic disturbance after substorm onset and its implication to storm-substorm relation, *Ann. Geophysicae, 14*, 608, 1996.

Jaggi, J. R., a nd R. A. Wolf, Self-consistent calculation of a sheet of ions in the magnetosphere, *J. Geophys. Res., 78*, 2852, 1973.

Joselyn, J. A., and B. T. Tsurutani, Geomagnetic sudden impulses and storm sudden commencements, *Eos, 71*, 1808-1809, 1990.

Kamide, Y., Relationship between substorms and storms, in *Dynamics of the Magnetosphere*, edited by S.-I. Akasofu, D. Reidel Publ. Co. Dordrecht, Holland, 1979.

Kamide, Y., Is substorm occurrence a necessary condition for a magnetic storm?, *J. Geomag. Geoelectr., 44*, 109, 1992.

Kauffman, R. L., Substorm currents: growth phase and onset, *J. Geophys. Res., 92*, 7471, 1987.

Kisabeth, J. L., The dynamical development of the polar electrojets, Ph.D. Thesis, University of Alberta, Fall, 1972.

Kisabeth, J. L. and G. Rostoker, The expansive phase of magnetospheric substorms 1. Development of the auroral electrojets and auroral arc configuration during a substorm, *J. Geophys. Res., 79*, 972, 1974.

Lennartsson, W., and E. G. Shelley, Survey of 0.1- to 16-keV/e plasma sheet ion composition, *J. Geophys. Res., 91,* 3061-3076, 1986.

Liu, W. W. and G. Rostoker, Energetic ring current particles generated by recurring substorm cycles, *J. Geophys. Res., 100,* 21,897, 1995.

Lui, A.T.Y., Current disruption in the earth's magnetopshere: observations and models, *J. Geophys. Res., 101,* 13,067, 1996.

Lyons, L. R., O. de la Beaujardière, G. Rostoker, J. S. Murphree and E. Friis-Christensen, Analysis of substorm expansion and surge development, *J. Geophys. Res., 95,* 10,575, 1990.

Rostoker, G., Geomagnetic indices, *Rev. Geophys. Space Phys., 10,* 935-950, 1972.

Rostoker, G., Triggering of expansive phase intensifications of magnetospheric substorms by northward trunings of the interplanetary magnetic field, *J. Geophys. Res., 88,* 6981, 1983.

Rostoker, G., Auroral signatures of magnetospheric substorms and constraints which they provide for substorm theories, *J. Geomag. Geoelectr., 43,* Suppl. 233, 1991.

Rostoker, G., The role of substorms in the development of the storm time ring current, in *Proceedings of the International Conference on Magnetic Storms,* edited by Y. Kamide, p. 109, Solar-Terrestrial Environment Laboratory, Toyokawa, Japan, 1994a.

Rostoker, G., A renovated boundary layer dynamics model for magnetospheric substorms, in *Proceedings of the Second International Conference on Substorms,* edited by J. R. Kan, J. D. Craven and S.-I. Akasofu, p. 189, Univ. of Alaska, Fairbanks, 1994b.

Rostoker, G., The role of substorms in the formation of the ring current, in *Workshop on the Earth's Trapped Particle Environment,* edited by G. D. Reeves, p. 33, American Institute of Physics, Woodbury, NY, 1996a.

Rostoker, G., The phenomenology and physics of magnetospheric substorms, *J. Geophys. Res., 101,* 12,955, 1996b.

Rostoker, G., S.-I. Akasofu, W. Baumjohann, Y. Kamide and R. L. McPherron, The roles of direct input of energy from the solar wind and unloading of stored magnetotail energy in driving magnetospheric substorms, *Space Sci. Rev. 46,* 93, 1987.

Rostoker, G., T. D. Phan and F. Pascal, Inference of magnetospheric and ionospheric electrical properties from the decay of geomagnetic activity, *Can. J. Phys., 69,* 921, 1991.

Samson, J. C., D. D. Wallis, T. J. Hughes, F. Creutzberg, J. M. Ruohoniemi and R. A. Greenwald, Substorm intensifications and field line resonances in the nightside magnetosphere, *J. Geophys. Res., 97,* 8495, 1992.

Smith, P. H., N. K. Bewtra and R. A. Hoffman, Inference of the ring current ion composition by means of charge exchange decay, *J. Geophys. Res., 86,* 3470, 1981.

Southwood, D. J., The role of hot plasma in magnetospheric convection, *J. Geophys. Res., 82,* 5512-5520, 1977.

Sugiura, M., Hourly values of the equatorial Dst for IGY, in *Ann. Int. Geophys. Year, 35,* 945, Pergamon Press, Oxford, 1964.

Vasyliunas, V. M., Mathematical models of magnetospheric convection and its coupling to the ionosphere, in *Particles and Fields in the Magnetosphere,* edited by B. M. McCormac, p. 60, D. Reidel, Norwood, MA, 1972.

Zmuda, A. J. and J. C. Armstrong, The diurnal flow of field-aligned currents, *J. Geophys. Res., 79,* 4611, 1974.

M. Dobbs, E. Friedrich, and G. Rostoker, Physics Department, University of Alberta, Edmonton, Alberta, Canada, T6G 2J1.

Modeling Convection Effects in Magnetic Storms

R. A. Wolf, J. W. Freeman, Jr., B. A. Hausman, and R. W. Spiro

Space Physics and Astronomy Dept., Rice University, Houston, Texas

R. V. Hilmer

Institute for Scientific Research, Boston College, Newton, Massachusetts

R. L. Lambour

U. S. Air Force Phillips Lab, Hanscom AFB, Massachusetts

Over the last twenty years, many quantitative models have been developed to represent the idea that the storm-time ring current consists primarily of particles from the plasma sheet that are injected into the inner magnetosphere by strong westward electric fields. These models, which have gradually become more sophisticated and realistic, have achieved rough quantitative agreement with observed ring-current fluxes, but the tests are still imprecise. The models are still not sophisticated enough to provide accurate and reliable theoretical estimates of *Dst*. Controversy still surrounds the question of whether the injection of the storm-time ring current mainly results from induction electric fields associated with magnetospheric substorms or from potential electric fields associated with periods of strong convection. A series of computer experiments has been carried out with the Magnetospheric Specification and Forecast Model in an attempt to illuminate this key question, as well as several other cause-and-effect relationships. The MSFM runs indicate that potential convection electric fields play a far more important role in ring current injection than do substorm-associated induction fields. The computer experiments also demonstrate the importance of fluctuations in the convection field: periods of very strong convection, separated by periods of weak flow, inject particles deeper into the magnetosphere than a long period of moderately strong convection. A run carried out with strong convection limited to one hour – an imitation of a very strong isolated substorm – shows the injection of only a weak ring current. Run results also show reduced populations of ~ 50 keV ions in regions that are in the shadow of the magnetopause; these depleted regions are particularly prominent when the magnetosphere is highly compressed.

1. INTRODUCTORY COMMENTS

The main pressure-bearing particles in the Earth's magnetosphere execute a combination of E×B, gradient, and curvature drift. The electric field consists primarily of a convection field that is directed roughly from dawn to dusk, an earthward field that results from the rotation of the Earth, and induction fields that result from changes in the magnetic-field configuration. The convection electric field is generally strong during the main phase of a magnetic storm, when the ring current is strengthening. The primary large-scale magnetic field changes that seem relevant to the creation of the storm-time ring current are compressions and expansions of the magnetosphere, due mainly to changes in solar-wind ram pressure, and collapse of the tail magnetic field near local midnight in the expansion phase of a substorm. The storm-time ring current seems to consist primarily of particles from the inner plasma sheet that are injected deeper into the magnetosphere by strong westward electric fields. Some of the particles that form the outer ring current before the storm are transported inward and energized by the storm-associated electric fields, and those particles also contribute to the storm-time current and particle energy [*Lyons and Williams*, 1980].

One purpose of this paper is to summarize very briefly the various attempts that have been made over the last twenty years to quantitatively model the ring-current injection process (Section 2). The modeling efforts have become increasingly sophisticated, as observational knowledge and computer capabilities have increased, but we still do not have a full theoretical model of ring-current injection that solves all of the basic equations. Section 2 also briefly discusses the status of specific modeling problems, particularly the theoretical calculation of *Dst*. Section 3 displays the results of a set of computer experiments that have been done with the Magnetospheric Specification and Forecast Model (MSFM), in an attempt to illuminate several cause-and-effect relations, while Section 4 offers a brief summary.

2. TYPES OF THEORETICAL MODELS

Near the Earth, drift motions of magnetospheric particles are dominated by gradient drift, curvature drift, and **E**×**B** drift in the radially inward field due to the rotation of the planet; the relative importance of these drifts depends on the energy and pitch angle of the particle, but they are all mainly zonal in direction. Out in the plasma sheet, particle drift is dominated by **E**×**B**-drift in the dawn-to-dusk convection electric field, which produces a sunward motion. For a simple case where the electric and magnetic fields are all constant in time, Figure 1 shows sample drift trajectories. A key feature is a separatrix between drift paths that connect to the tail and drift paths that circle the Earth. That boundary, which separates particles with different histories and therefore presumably different flux levels, was defined by *Kavanagh et al.* [1968] to be the "Alfvén layer." The

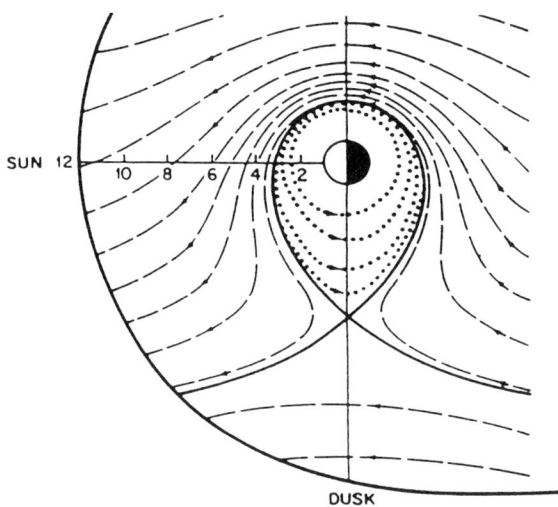

Fig. 1. Trajectories of particles with zero magnetic moment and zero second invariant, E×B–drifting in a uniform dawn-dusk electric field and an earthward corotation field. The dots represent motion in equal time intervals. From *Kavanagh et al.* [1968].

geocentric distance of the Alfvén layer depends on particle energy and decreases with increasing convection electric field. (See, e.g., *Wolf* [1995].) If the electric and magnetic fields are independent of time, then the inner edge of the plasma sheet for particles with given adiabatic invariants – and the outer edge of the ring current for the same particles – should coincide with the Alfvén layer for that type of particle. The Alfvén-layer picture of inner-magnetospheric particle populations, which is based on the idea that steep gradients in particle populations occur at separatrices calculated for assumed steady electric and magnetic-fields, has been successful in explaining several observational features. (See the review by *Kivelson et al.* [1979] or the commentary by *Kivelson et al.* [1987].) Of course, it is theoretically secure only in situations where the magnetospheric electric (and magnetic) fields have been steady for at least a particle drift time. Theoretical models based on the Alfvén-layer idea have typically assumed dipole fields and simple models for the cross-tail electric field, specifically either a uniform field with equatorial potential

$$\Phi = -E_o\, r \sin(\phi) \quad (1)$$

or a Stern-Volland potential (e.g., *Volland* [1978])

$$\Phi = -K\, r^2 \sin(\phi) \quad (2)$$

where E_o and K are constants, r is geocentric distance, and ϕ is a local-time angle (0 at local noon, $\pi/2$ at dusk, π at midnight).

A formalism that is more nearly adequate for treatment of ring-current injection is one in which the convection electric field is allowed to vary in time, but the magnetic field is still assumed time-independent (usually a dipole). Application of this type of model to explain features of observed ring-current injections is discussed in the accompanying paper by *Chen et al.* [1996], and we do not dwell on it here. However, one overall conclusion of the extensive comparisons between these models and observations is that reasonable time-dependent electric fields can inject plasma-sheet ions to approximately the energies, locations, and flux levels that are typical of the storm-time ring current.

The next level of model complexity entails varying both electric and magnetic fields in time. The Magnetospheric Specification and Forecast Model (as described in Section 3) is such a model. These models – more complex but less thoroughly tested and tuned than the ones described in the last paragraph – also exhibit rough overall agreement with the main features of observed magnetic storms.

An additional level of complexity is introduced in models that, rather than assuming an analytic form for the potential electric field, compute it by solving the fundamental equation of ionosphere-magnetosphere coupling:

$$\nabla_h \cdot \left(-\Sigma \cdot \nabla_h \Phi \right) = J_\| \sin(I) = -\hat{n} \cdot \nabla p \times \nabla \left(\int ds/B \right) \quad (3)$$

Here ∇_h is the horizontal gradient on a spherical shell representing the ionosphere, Σ is a 2×2 tensor that includes the field-line integrals of Hall and Pedersen conductivities, $J_\|$ is the density of Birkeland current down into the ionosphere (both hemispheres together), I is magnetic dip angle, $\hat{n}$ is a unit vector down into the northern ionosphere, p is pressure in magnetospheric particles (assumed isotropic), and $\int ds/B$ is the volume of a flux tube of unit magnetic flux [*Vasyliunas*, 1970]. Models employing (3) are described, for example, by *Vasyliunas* [1972], *Senior and Blanc* [1984], *Harel et al.* [1981], and *Erickson et al.* [1991]. These models have demonstrated that the inner edge of the plasma sheet tends to shield the region earthward of it from the convection electric field, and the currents $J_\|$ corresponding to this shielding are the region-2 Birkeland currents. They have also demonstrated that a sudden increase in convection causes a penetration of the dawn-dusk electric field into the inner magnetosphere, while a sudden decrease in convection causes a temporary dusk-dawn **E** in the same region – general features that have been confirmed by ground-radar observations (*Fejer and Schlerliess* [1995] and references therein).

Ring-current particle populations are, of course, affected by sources and losses. Two loss processes, namely charge exchange and Coulomb collisions, are well enough understood that the associated lifetimes can be calculated with reasonable accuracy (see, e.g., *Fok et al.* [1991, 1995]). Wave-induced pitch-angle scattering processes are harder to treat theoretically, partly because of the difficulty of estimating the wave spectrum. We generally do not have a reliable way to calculate those loss rates in large-scale models. A significant source for ring-current ions is upflow from the ionosphere, an effect that still awaits inclusion in global magnetospheric models. Most of the ring-current/inner-plasma-sheet models reviewed above have outer boundaries at 8 - 12 R_E in the equatorial plane, so that effects of ion upflow at large L can be included as a boundary condition. However, upflow may be important even for inner-magnetospheric flux tubes that just circle the Earth [e.g., *Daglis*, 1996], and the effect needs to be included in large-scale models.

No one has ever attempted a full theoretical calculation of ring-current injection, with self-consistently calculated electric and magnetic fields. The Rice Convection Model, for example, calculates the potential electric field self-consistently and has been used to model magnetic storms [*Wolf et al.*, 1982; *Spiro and Wolf*, 1984], but it does not use a self-consistent magnetic field. Magnetic-field models have been constructed to be consistent with given particle distributions (e.g., *Sckopke* [1972], *Cheng* [1995]), but those calculations do not include electric fields or the physics of ring-current injection. We are just now assembling the computational machinery required for fully self-consistent simulations of the inner magnetosphere [*Toffoletto et al.*, 1996], and the first such calculations of ring-current injections can be expected in the next few years. One possible obstacle to the full self-consistent calculation of ring-current injection lies in the intractability of the substorm problem. It may prove easier to do self-consistent calculations first with an outer calculational boundary at $L \sim 10$, thus excluding much of the complicated dynamics of the substorm process from the detailed calculation and, instead, representing the substorm effect on ring-current injection by observation-based boundary conditions.

An obvious test of theoretical ring-current-injection models would be to compare the model-predicted *Dst* curve for a given event with the observed *Dst* for the same event. However, such tests turn out to be difficult in practice, partly because of ambiguities in the physical meaning of the index. (See, e.g., *Campbell* [1996].) The easy way to estimate *Dst* theoretically is usually to calculate the total energy in ring-current particles and use the Dessler-Parker-Sckopke (DPS) relation [*Dessler and Parker*, 1959; *Sckopke*, 1966]; however, that relation was derived and is valid only for a dipole magnetic field. Much of the energy in the storm-time ring current resides in regions where β ~ 1, i.e., where the dipole field is significantly inflated; the DPS relation can be in error by a significant factor for such configurations. Further confusion results from the fact that a significant fraction of the total particle energy in the magnetosphere resides in the plasma sheet, which extends far out in the tail; in using the DPS relation with results from a realistic global model, it isn't clear at what radial distance one should stop counting particle energy. One way to compute *Dst* theoretically without using DPS would be to calculate a Biot-Savart integral over the whole three-dimensional magnetosphere-ionosphere current distribution: however, that calculation is long and tedious, and, in most cases, appropriate corrections need to be made for currents outside the modeling region. A more fundamental difficulty is that none of the ring-current-injection calculations carried out so far utilizes a model **B** whose curl agrees with the gradient/curvature/magnetization currents implied by the model-computed particle distributions. Though one can, in principle, do a Biot-Savart integration based on the gradient/curvature/magnetization current carried by the model-computed particle distribution in the model magnetic field, the calculated currents will be significantly in error if the assumed magnetic fields are wrong. In summary, we cannot yet reliably calculate *Dst* theoretically closer than ~ a factor of two, even though we occasionally get much closer agreement than that [e.g., *Wolf et al.*, 1982]. More accurate theoretical calculation of *Dst* awaits the development of fully self-consistent inner-magnetosphere models.

3. STORM SIMULATIONS WITH THE MAGNETOSPHERIC SPECIFICATION AND FORECAST MODEL

Description of the Magnetospheric Specification and Forecast Model

The Magnetospheric Specification and Forecast Model (MSFM) was designed for operational use by the U. S. Air Force, and it is consequently designed to run from input parameters that are available in real time. Here we present only a brief sketch of how the model works: for more detailed information on the model, see *Bales et al.* [1993], *Lambour* [1994], and *Freeman et al.* [1996]. The modeling region extends almost to the magnetopause at local noon; at local midnight, it extends to about twice the subsolar magnetopause standoff distance. The central equation of the MSFM is the particle-conservation equation

$$\left(\frac{\partial}{\partial t} + \mathbf{v}_d \cdot \nabla\right)\eta_s = -\frac{\eta_s}{\tau_s} \qquad (4)$$

where τ_s is the loss lifetime, η_s is the number of particles of type s per unit magnetic flux, being defined by a given chemical species (mass m_s, charge q_s) and given invariant energy λ_s, as defined by

$$\lambda_s = W_s \left(\int \frac{ds}{B}\right)^{2/3} \qquad (5)$$

where W_s is the particle energy (in gyrational and bounce motion), and $\int ds/B$ is the volume of a tube containing one unit of magnetic flux. The pitch-angle distribution is assumed to be isotropic, so that the bounce-averaged drift velocity $\mathbf{v}_d$ is given by

$$\mathbf{v}_d = \frac{\mathbf{E} \times \mathbf{B}}{B^2} + \frac{\lambda_s}{q_s} \frac{\mathbf{B} \times \nabla\left(\int ds/B\right)^{-2/3}}{B^2} \qquad (6)$$

Derivations of equations (4)-(6) are given, for example, by *Wolf* [1983]. Note that (4) implies that η_s is constant along the drift path of a particle, except for the effects of loss.

In the MSFM, the only sources of particles are the boundary conditions and the initial conditions. Fluxes are specified on those portions of the boundary where particles are drifting into the modeling region. Initial and boundary values of η_s are taken from an empirical model of particle fluxes as a function of species, energy, *L*, and *Kp*. However, the fluxes are reduced on some portions of the model boundary, to reflect the fact that the dayside magnetopause is a weak source of particles with energies of tens of kilovolts, for example,.

The MSFM uses magnetic- and electric-field models that are driven by the available real-time data. The crucial input data for the magnetic-field model [*Hilmer and Voigt*, 1995] are the magnetopause standoff distance; *Dst*; the auroral boundary index (ABI), which is the midnight equatorward boundary of the diffuse aurora at local midnight [*Gussenhoven et al.*, 1983; *Madden and Gussenhoven*, 1990]; and the collapse parameter, which indicates the collapse of model field lines near local midnight during the expansion phase of a substorm (= 1 when the field is uncollapsed, 0 when collapsed).

The electric-field model utilizes a scaled version of the Heppner-Maynard empirical model [*Heppner and Maynard*, 1987; *Rich and Maynard*, 1989] in the polar cap and analytical formulae based on experience with the Rice Convection Model at lower latitudes [*Lambour*, 1994; *Freeman et al.*, 1996]. The input parameters are the polar-cap potential drop, the polar-cap pattern type (A, BC, or DE Heppner-Maynard models), and the ABI. A novel feature of the MSFM's electric field model is that the penetration of the electric field equatorward of the auroral zone is estimated from the rate of change of the ABI: when the auroral zone is expanding, a westward electric field penetrates to low latitudes on the night side; when the auroral zone is contracting, the penetration field is eastward on the night side.

Event Simulation

In order to use the MSFM to examine cause-and-effect relationships, we have run the code through an idealized magnetic-storm main phase. MSFM input parameters for the idealized storm are shown in Figure 2. We assume a storm sudden commencement at 0600 UT, in which the magnetopause standoff distance decreases from 10 R_E to 6 R_E. We assume weak convection before the storm – the polar-cap potential drop is 14.6 kV, the auroral boundary index is 67.8°, and *Dst* and *Kp* are both set equal to zero. The first of two intervals of very strong convection starts at 0600 UT, coincident with the sudden commencement, and lasts until about 0900 UT; during these intervals, the potential drop increases to 150.6 kV, and the ABI decreases to 51.2°; *Kp* rises to 8. *Dst* is ramped down to −62 nT during the first interval and remains at that level for the rest of the simulation. The collapse parameter decreases starting at 0630 UT, causing the *B*-field model to collapse the tail field in the midnight sector; the tail-field collapse gradually recovers from 0700 to 0800 UT. Convection is weak again between 0900 and 1200 UT, with all of the indices back at their prestorm values except for the standoff distance and *Dst*. Convection is strong again from 1200 to 1500 UT, then weak again from 1500 to 1800. The simulation ends at 1800 UT: in this study, we have not attempted to represent the recovery phase of the storm.

Although the input parameters shown in Figure 2 are obviously highly idealized, we have tried to inject some elements of realism. The assumed changes in convection-associated quantities were intended to correspond to southward turnings of the interplanetary magnetic field at about 0600 and 1200 UT, and to northward turnings at about 0900 and 1500. The polar-cap potential drop was assumed to react to the changes over a period of 30 min, the auroral

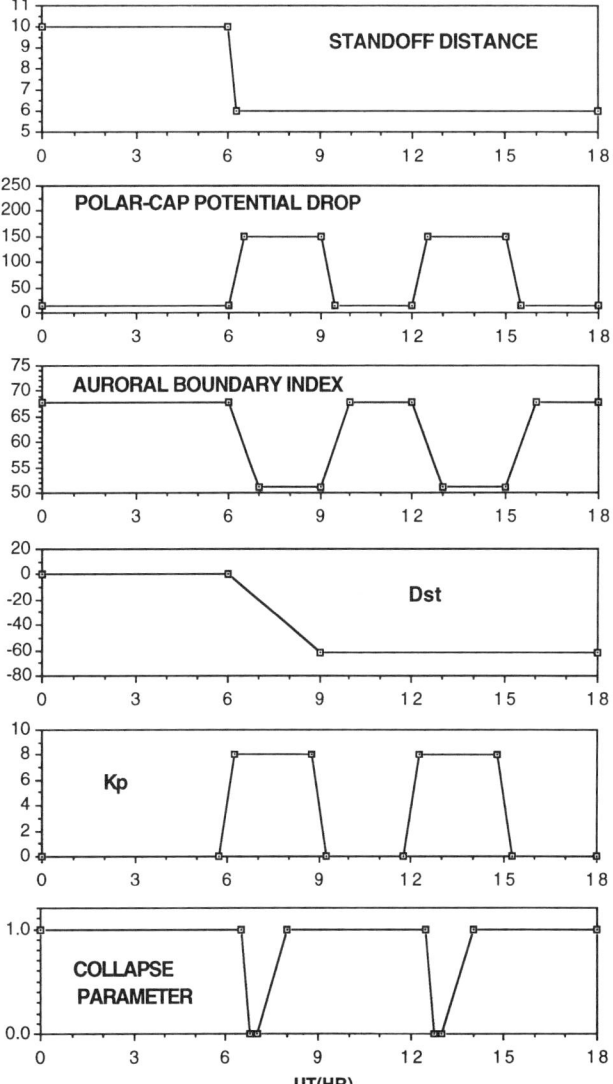

Fig. 2. MSFM input parameters for the idealized magnetic storm. Assumed magnetopause standoff distance (in R_E), polar-cap potential drop (in kV), auroral boundary index (in degrees latitude), *Dst* (in nanoTeslas), *Kp*, and a collapse parameter are plotted vs. universal time for the 18-hour period of the simulation.

boundary index over 1 hr, and *Dst* over 3 hr. (The MSFM's particle computation machinery ingests input parameters only at 15-min intervals.) A clean, strong substorm expansion phase – as signalled by the collapse of the field in the midnight sector – was assumed to begin 30 min after each southward turning. The expansion phase was assumed to last 45 min, followed by an hour-long recovery phase. Of course, three hours of strong convection might be expected to generate additional

166 CONVECTION EFFECTS IN MAGNETIC STORMS

complicated substorm-like activity, but we did not attempt to represent that in terms of a collapse parameter. In the input-data stream, the maximum and minimum values for potential drop, auroral boundary index, and Dst correspond roughly to average values for $Kp = 0$ and $Kp = 8$.

Figure 3 shows MSFM-computed values of η_s (particles per unit magnetic flux) for H^+ ions that have energy invariant λ_s equal to 1228 eV $(nT/R_E)^{2/3}$, which corresponds to about 8.6 keV at synchronous orbit, 33 keV at $L = 4$ and 70 keV at $L = 3$. According to (4), η_s is conserved along the drift path of a particle, except for the effects of loss. For the invariant energies for which results are displayed, charge-exchange lifetimes are longer than the simulation for $L > 3$, and we can therefore roughly associate colors with populations. The red and orange regions of the plots can be identified with particles of plasma-sheet origin. It should be noted that numerical diffusion in the code tends to blur fine features.

From 00 to 06 UT, the particle population moves toward equilibrium with the weak applied electric field. By 06 UT, the inner edge of the plasma sheet is at about 8 R_E at local midnight. By 07 UT, after the sudden compression and an hour of strong convection, the inner edge of the plasma sheet has come within about 4 R_E of the Earth on the night side. Populations have decreased on the day side, as some of the original plasma sheet and outer trapped particles have exited the magnetosphere, because of enhanced sunward convection of the plasma and the earthward motion of the magnetopause. By 09 UT, the minimum geocentric distance of the inner edge has reached about 3 R_E, and the ions have drifted around the dusk side almost to local noon, while part of the void that was created on the day side has drifted westward around the night side. From 09 until 12 UT, convection is weak again, and particle drift is dominated by westward gradient/curvature drift. By 12 UT, a tongue of enhanced particle population can be seen to extend westward almost all the way around the Earth. The particles near the tip of the tongue were part of the outer-belt trapped population before the storm, while the base of the tongue consists mainly of fresh plasma-sheet particles. The tongues of newly injected ions have spiral shapes, with geocentric distance decreasing westward, because, for given energy invariant λ_s, the angular velocity of ions about the Earth due to gradient/curvature-drift increases earthward like $L^{-5/3}$, so that the particles that penetrate deepest drift westward fastest. By 15 UT, at the end of the second period of very strong convection, the orange population (particles fresh from the plasma sheet) looks much as it did at 09 UT, after the first strong-convection period, but the pre-noon sector now contains higher populations, as a result of the first injection. The high-η region is also close to the Earth over a wider range of local time. By 18 UT, the end of another three-hour period of weak convection, a new high-population tongue of particles, primarily convected in from the tail during the modeled storm, has almost encircled the Earth.

In an effort to determine how different input parameters affect ring-current injection, we did a total of six MSFM runs through the eighteen-hour period. Table 1 summarizes the input parameters used in the various runs. The final configurations in the different cases are compared in Figure 4 for H^+ ions with $\lambda_s = 1228$ (same as Figure 3). Figure 5 shows a similar comparison for more energetic H^+ ions: $\lambda_s = 4000$ (corresponding to energies of 28 keV at $L = 6.6$, 106 keV at $L = 4$). Run 1 was a "control run", with all of the parameters held at the initial quiet values for the entire eighteen hours. Run 2 had no magnetospheric compression; convection parameters were held constant after 06 UT, at a level that was half way between the strong and weak convection levels shown in Figure 2. Run 3 assumed no compression and weak convection except between 06 and 07 UT, when convection rose to its maximum levels; it was designed to represent an extraordinarily strong isolated substorm. Run 4 used the input parameters shown in Figure 2, except that the magnetosphere was not compressed, and the midnight-region field was not collapsed; thus run 4 included no induction electric fields. Run 5 used the input parameters of Figure 2, except that the midnight region field was not collapsed. Run 6 used the full set of input parameters shown in Figure 2. (Figure 3 shows results from Run 6.)

The run results illustrate several possible cause-and-effect relationships:

1. Comparison of Run 2 and Run 4 shows the difference between assuming that the storm consists of alternating periods of very strong and very weak convection (Run 4) and constant fairly-strong convection (Run 2). It is clear that alternating periods of strong and weak convection are much more effective in injecting fresh particles deep into the magnetosphere. This result is not surprising, of course: in the limit of small fluctuations, it is the Fourier component of the electric field that resonates with the drift period of the particles that causes lasting radial transport, and previous investigations have focussed on the time variations of the electric field (e.g., *Chen et al.*, 1992, *Riley and Wolf*, 1992].

2. Run 3, representing an extremely strong isolated substorm, resulted in injection of fresh particles into the outer-ring-current region, but the new particles didn't penetrate further earthward than approximately $L = 4.5$.

3. Comparing Runs 4 and 5 suggests that the effects of magnetospheric compression on ring-current injection are

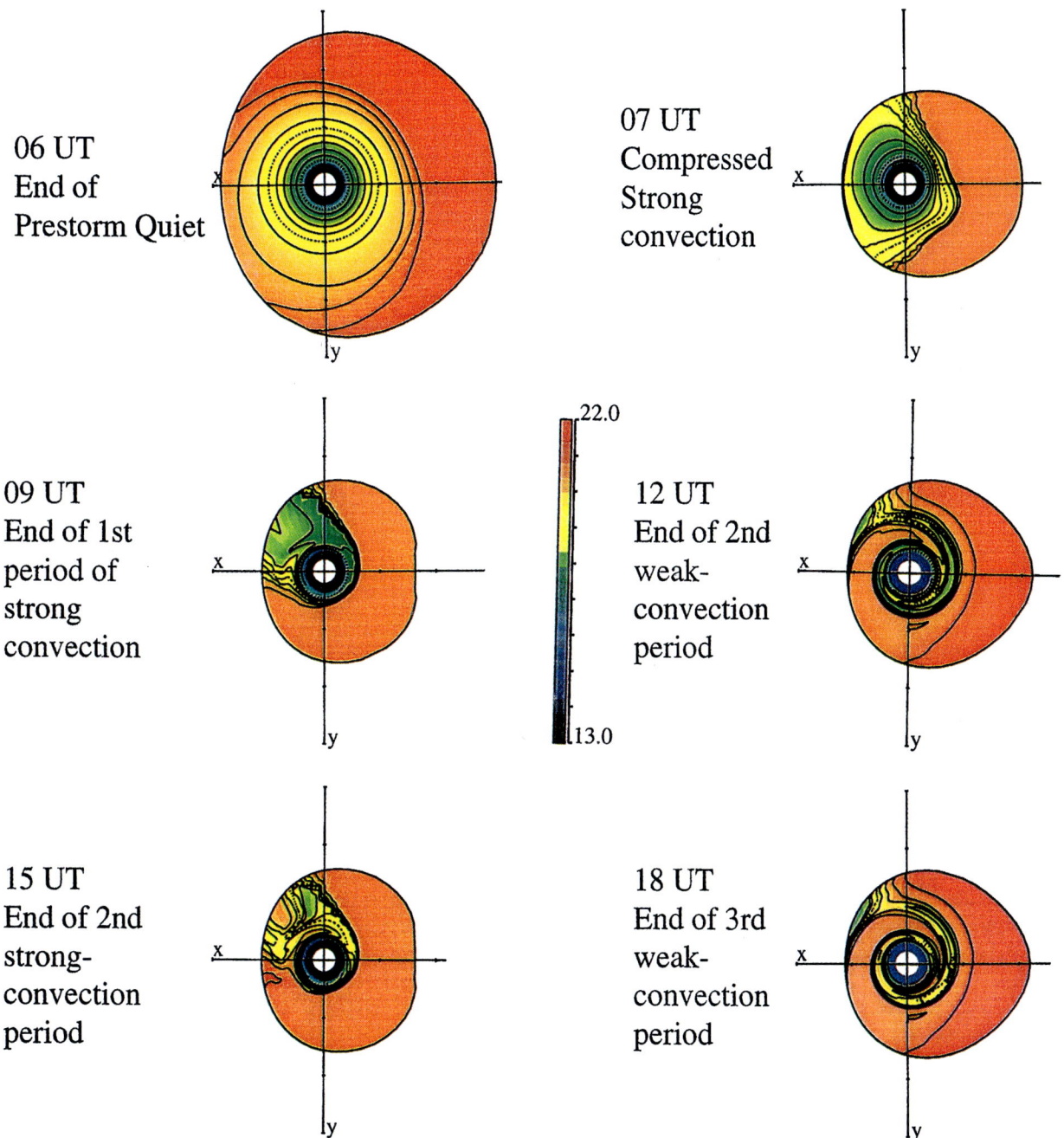

Fig. 3. Equatorial values of $\log_{10}(\eta_s)$ for six times during Run 6, which used the input parameters shown in Figure 2. The particles are H^+ ions with 8.6 keV at geosynchronous orbit. The Sun is to the left. Tic marks are at 5 R_E intervals. The boundary of the simulation is set in the ionosphere and mapped to the equatorial plane; variations in boundary shape result from mapping subtleties and are not important.

TABLE 1. Run Parameters

Parameter	Run 1	Run 2	Run 3	Run 4	Run 5	Run 6
Standoff Distance	10 for (0-18)*	10 for (0-18)	10 for (0-18)	10 for (0-18)	10 for (0-6), 6 for (6-18)	10 for (0-6), 6 for (6-18)
Polar-Cap Potential Drop (kV)	14.6 for (0-18)	14.6 for (0-6), 82.6 for (6.5-18)	14.6 for (0-6, 7.5-18), 150.6 for (6.5-7)	14.6 for (0-6, 9.5-12, 15.5-18), 150.6 for (6.5-9, 12.5-15)	14.6 for (0-6, 9.5-12, 15.5-18), 150.6 for (6.5-9, 12.5-15)	14.6 for (0-6, 9.5-12, 15.5-18), 150.6 for (6.5-9, 12.5-15)
Auroral Boundary Index (degrees)	67.8 for (0-18)	67.8 for (0-6), 59.5 for (6.5-18)	67.8 for (0-6, 8-18), 51.2 for UT=7	67.8 for (0-6, 10-12, 16-18), 51.2 for (7-9, 13-15)	67.8 for (0-6, 10-12, 16-18), 51.2 for (7-9, 13-15)	67.8 for (0-6, 10-12, 16-18), 51.2 for (7-9, 13-15)
Dst (nanoteslas)	0 for (0-18)	0 for (0-6), −20 for (12-18)	0 for (0-6, 13-18), −20.7 for (7-12)	0 for (0-6), −62 for (9-18)	0 for (0-6), −62 for (9-18)	0 for (0-6), −62 for (9-18)
Kp	0 for (0-18)	0 for (0-6), 4 for (6-18)	0 for (0-6), 2.67 for (6-9)	0 for (0-6, 9-12, 15-18), 8 for (6-9, 12-15)	0 for (0-6, 9-12, 15-18), 8 for (6-9, 12-15)	0 for (0-6, 9-12, 15-18), 8 for (6-9, 12-15)
Midnight Field Collapse Parameter	1 for (0,18)	1 for (0,18)	1 for (0,18)	1 for (0,18)	1 for (0,18)	1 for (0-6.5, 8-12.5, 4-18), 0 for (6.75-7, 12.75-13)

* Figures in parentheses give UT in hours. Parameters vary linearly with time in transition periods.

noticeable but not dramatic. The basic structure, with a westward-stretching tongue earthward of a low-density region, exists independent of the compression.

4. Comparison of Runs 5 and 6 indicates that midnight-region collapse of the tail field has very little effect on injection of the storm-time ring current.

5. Comparison of Figures 4 and 5 indicates that the more energetic ions do not penetrate as deeply into the magnetosphere as their less energetic colleagues. The deepest penetration of plasma-sheet ions in Figure 5 is about $L = 4$, where the particle energy is ~ 100 keV. This is consistent with many previous model calculations.

6. Results for Runs 5 and 6 show that an extensive low-flux region appears on the dawn side, beyond about $L = 5$. Particle trajectories in that low-flux region trace back to the dayside magnetopause. Of course, the magnetosheath is a weak source of ~ 30 keV ions, and that fact is built into the MSFM's particle boundary condition. In times of weak convection, which includes the last three hours of Runs 4-6, convection into the magnetopause is slower than gradient/curvature drift in through the same boundary, for particles with $\lambda_s = 4000$, leading to the formation of a distinct "shadow of the magnetopause" feature. Of course, that shadow penetrates much deeper into the magnetosphere when the magnetosphere is compressed, as it was in Runs 5-6, than when it was an average size, as in Run 4. The earthward edge of the magnetopause shadow is closely related to the boundary of the stable trapping region, in radiation-belt terminology. However, that boundary is usually calculated assuming just gradient and curvature drift. In the present case, we are dealing with particle energies such that the E×B drift is non-negligible and, indeed, can be dominant in times of strong convection.

The computer experiments presented here have verified again that reasonable time-varying convection electric fields can inject fresh particles deep into the magnetosphere to form a reasonable storm-time ring-current plasma population. However, we have not come close to exploring in detail the entire input-parameter space. In principle, we should consider a full range of different strengths of convection, of different durations of the convection surges, of repeated compressions and rarefactions, etc., *ad nauseum*. There are many possibilities. Thus the firmness of our conclusions is subject to "theoretical sampling errors." The strength of conclusion 4 is also limited by the fact that the tail-field collapse in the *Hilmer-Voigt* [1995] magnetic-field model does not correspond to the conventional substorm current loop. In our B-field model, the eastward

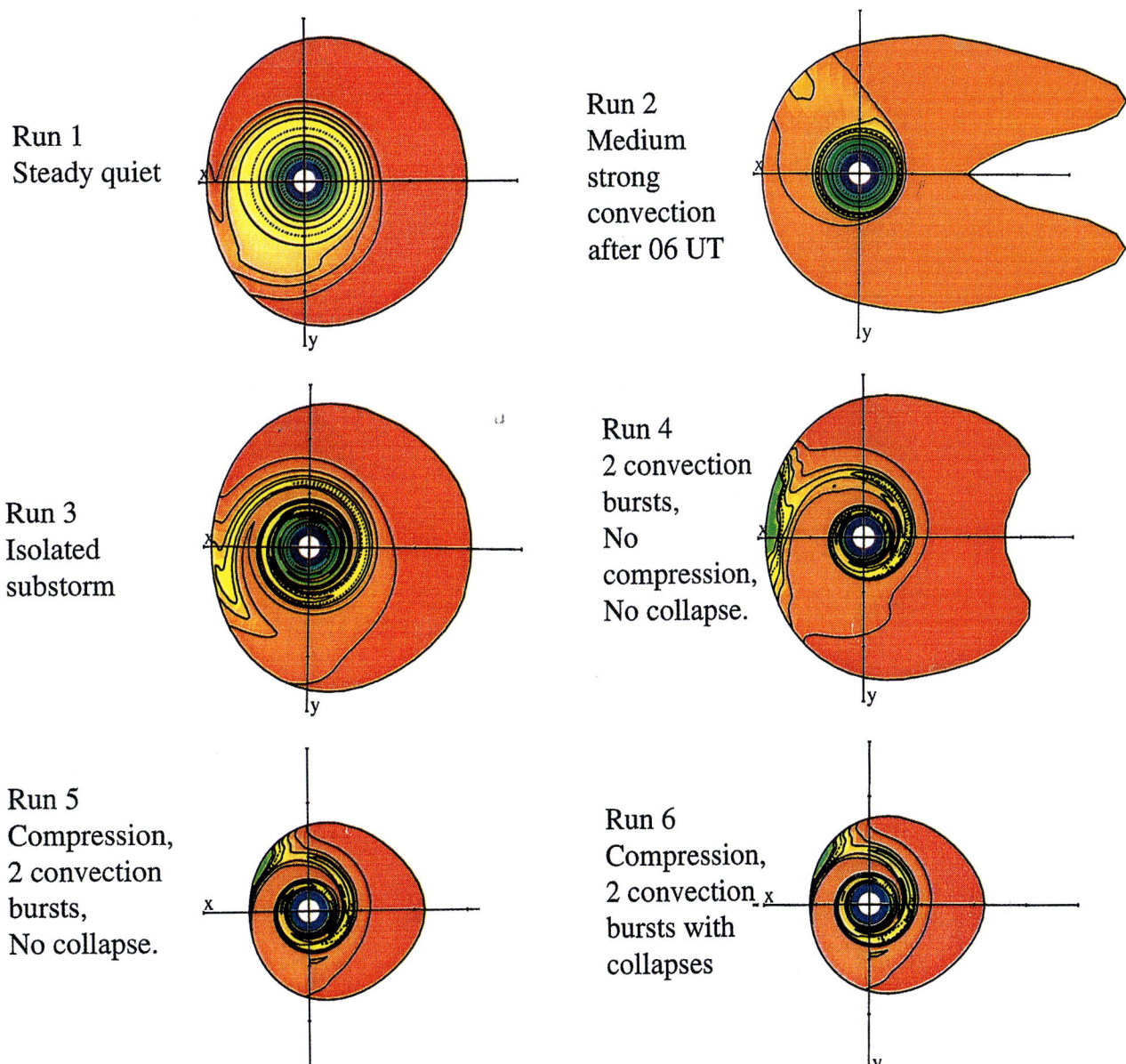

Fig. 4. Equatorial values of $\log_{10}(\eta_s)$ at the ends of the six runs described in the text. The format is the same as Figure 3, and the particles are H^+ ions with 8.6 keV at geosynchronous orbit.

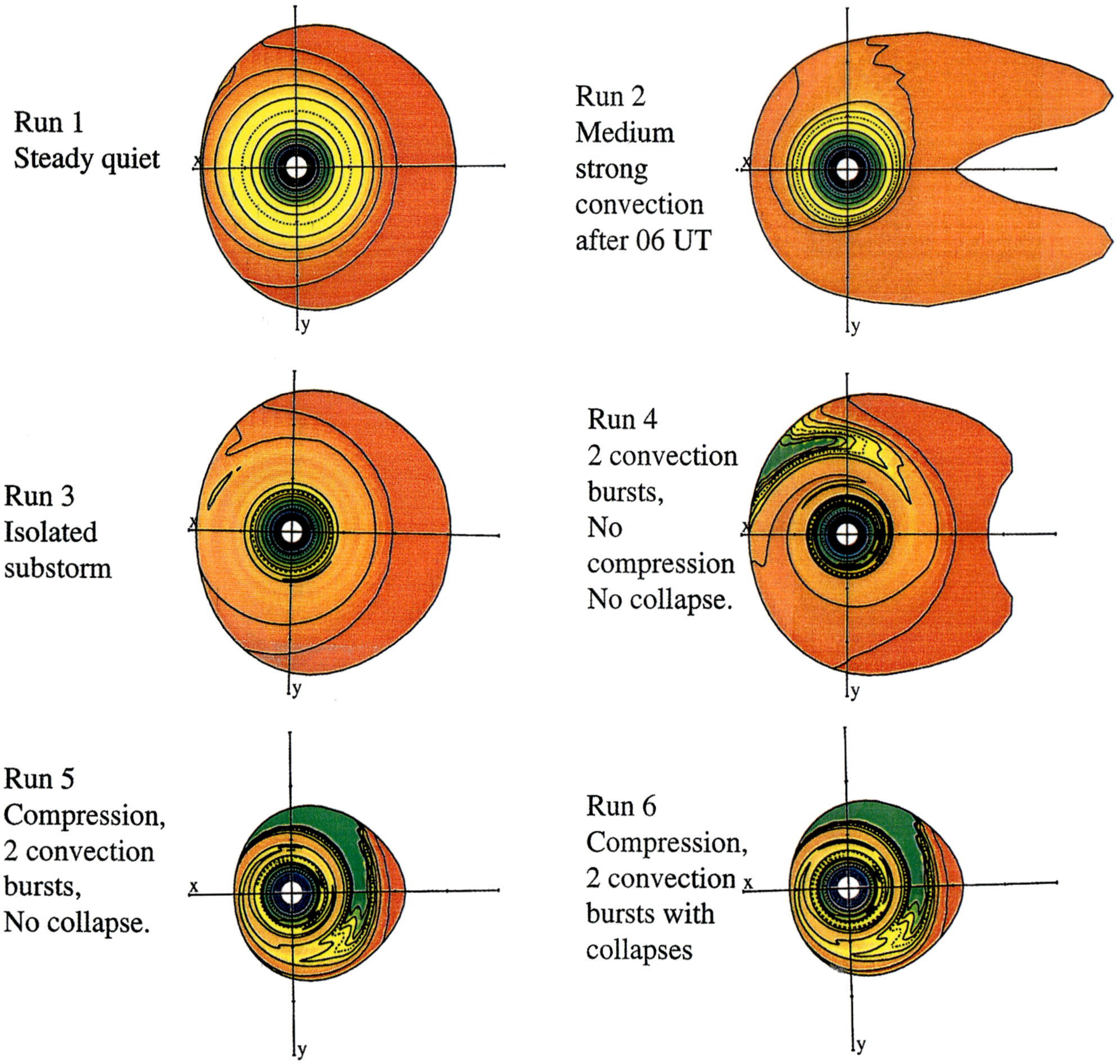

170 CONVECTION EFFECTS IN MAGNETIC STORMS

Run 1
Steady quiet

Run 2
Medium strong convection after 06 UT

Run 3
Isolated substorm

Run 4
2 convection bursts,
No compression
No collapse.

Run 5
Compression,
2 convection bursts,
No collapse.

Run 6
Compression,
2 convection bursts with collapses

Fig. 5. Equatorial values of $\log_{10}(\eta_s)$ at the ends of the six runs described in the text. The format is the same as Figure 3, and the particles are H^+ ions with 28 keV at geosynchronous orbit.

perturbation in the cross-tail current near local midnight is completed by currents up through the tail lobe rather than by currents flowing along field lines to the ionosphere, as in the usual picture; consequently, the substorm-associated perturbation ΔB's very close to the Earth are probably unrealistically weak in the Hilmer-Voigt model. However, both model-field collapses (near 0700 and 1300 UT) were clearly non-negligible in the crucial region. The collapse-associated increase in the northward magnetic field in the equatorial plane at local midnight was about 20 nT at 6 R_E geocentric, about 40 nT at 4 R_E. The field lines that crossed the equatorial plane at 6 R_E after the collapses extended to 9-10 R_E before collapse.

Our MSFM-based conclusion that the expansion-phase collapse of midnight-region field lines has little effect on ring-current injection seems qualitatively consistent with new observational evidence presented recently by *Iyemori et al.* [1996] and *McPherron* [1996], who showed that substorm expansions tend to have a positive (rather than a negative) effect on the *Dst* index. *Siscoe and Petschek* [1996] have proposed a theoretical interpretation of this observational result in terms of very general energy theorems.

4. CONCLUDING COMMENTS

The standard picture of a magnetic storm, with most of the storm-time ring current coming from plasma-sheet particles that are injected into the inner magnetosphere in a time of strong convection, seems to be roughly consistent with observations, on the whole. Although the quantitative comparisons with observations are still imprecise, present model calculations can usefully illustrate how the process works. However, dramatic progress in theoretical understanding of the ring current awaits models that self-consistently calculate electric fields, magnetic fields, and particle populations, as well as including an ionospheric particle source. Such models should appear within the next few years, given enlightened research-funding policies and selections.

Acknowledgments. The authors are grateful to Anthony Chan, Frank Toffoletto and George Siscoe for helpful conversations. This research was supported by the Atmospheric Sciences Section of the National Science Foundation under grants ATM-9613824 and ATM-9122310, by the U. S. Air Force under contract F19628-94-K-0002, and by the Space Physics Division of the National Aeronautics and Space Administration under grant NAGW-2826.

REFERENCES

Bales, B., J. Freeman, B. Hausman, R. Hilmer, R. Lambour, A. Nagai, R. Spiro, G.-H. Voigt, R. Wolf, W. F. Denig, D. Hardy, M. Heinemann, N. Maynard, F. Rich, R. D. Belian, and T. Cayton, Status of the development of the Magnetospheric Specification and Forecast Model, in *Solar-Terrestrial Predictions-IV: Proceedings of a Workshop at Ottawa, Canada, May 18-22, 1992*, ed. J. Hruska, M. A. Shea, D. F. Smart and G. Heckman, NOAA, Environmental Res. Labs, Boulder, 467-478, 1993.

Campbell, W. H., Geomagnetic storms, the Dst ring-current myth and lognormal distributions, *J. Atmos. Terr. Phys.*, *58*, 1171-1188, 1996.

Chen, M. W., M. Schulz, L. R. Lyons, and D. J. Gorney, Ion radial diffusion in an electrostatic impulse model for stormtime ring current formation, *Geophys. Res. Lett.*, *19*, 621-624, 1992.

Chen, M., M. Schulz, and L. R. Lyons, Modeling of ring current formation and decay: A review, this volume, 1996.

Cheng, C. Z., Three-dimensional magnetospheric equilibrium with isotropic pressure, *Geophys. Res. Lett.*, *22*, 2401-2404, 1995.

Daglis, I. A., The role of magnetosphere-ionosphere coupling in magnetic storm dynamics, this volume, 1996.

Dessler, A. J., and E. N. Parker, Hydromagnetic theory of geomagnetic storms, *J. Geophys. Res.*, *64*, 2239-2252, 1959.

Erickson, G. M., R. W. Spiro, and R. A. Wolf, The physics of the Harang discontinuity, *J. Geophys. Res.*, *96*, 1633-1645, 1991.

Fejer, B. G., and L. Scherliess, Time-dependent response of equatorial ionospheric electric fields to magnetospheric disturbances, *Geophys. Res. Lett.*, *22*, 851-854, 1995.

Fok, M.-C., J. U. Kozyra, A. F. Nagy, and T. E. Cravens, Lifetime of ring current particles due to Coulomb collisions in the plasmasphere, *J. Geophys. Res.*, *96*, 7861-7867, 1991.

Fok, M.-C., T. E. Moore, J. U. Kozyra, G. C. Ho, and D. C. Hamilton, Three-dimensional ring current decay model, *J. Geophys. Res.*, *100*, 9619-9632, 1995.

Freeman, J., R. Wolf, R. Spiro, B. Hausman, B. Bales, D. Brown, K. Costello, R. Hilmer, R. Lambour, and A. Nagai, The Magnetospheric Specification Model, in preparation, 1996.

Gussenhoven, M. S., D. A. Hardy, and N. Heinemann, Systematics of the equatorward diffuse auroral boundary, *J. Geophys. Res.*, *88*, 5692-5708, 1983.

Harel, M., R. A. Wolf, P. H. Reiff, R. W. Spiro, W. J. Burke, F. J. Rich, and M. Smiddy, Quantitative simulation of a magnetospheric substorm 1, Model logic and overview, *J. Geophys. Res.*, *86*, 2217-2241, 1981.

Heppner, J. P., and N. C. Maynard, Empirical high-latitude electric field models, *J. Geophys. Res.*, *92*, 4467-4489, 1987.

Hilmer, R. V., and G.-H. Voigt, A magnetospheric magnetic field model with flexible current systems driven by independent physical parameters, *J. Geophys. Res.*, *100*, 5613-5626, 1995.

Iyemori, T., and D. R. K. Rao, Decay of the Dst field of geomagnetic disturbance after substorm onset and its implication to storm-substorm relation, *Ann. Geophys.*, *14*, 608-618, 1996.

Kavanagh, L. D., Jr., J. W. Freeman, Jr., and A. J. Chen, Plasma flow in the magnetosphere, *J. Geophys. Res.*, *73*, 5511, 1968.

Kivelson, M. G., S. M. Kaye, and D. J. Southwood, The physics of plasma injection events, in *Dynamics of the Magnetosphere*, ed. S.-I. Akasofu, D. Reidel, Dordrecht-Holland, pp. 385-405, 1979.

Kivelson, M. G., J. Feynman, B. H. Mauk, and R. A. Wolf, Dialog on injection-boundary versus Alfvén-layer models, in *Magnetotail Physics*, ed. A. T. Y. Lui, The Johns Hopkins University Press, Baltimore, 403-407, 1987.

Lambour, R. L., Calibration of the Rice Magnetospheric Specification and Forecast Model for the Inner Magnetosphere, Ph. D. thesis, Rice University, 1994.

Lyons, L. R., and D. J. Williams, A source for the geomagnetic storm main phase ring current, *J. Geophys. Res.*, *85*, 523, 1980.

Madden, D., and M. S. Gussenhoven, Auroral boundary index from 1983 to 1990, Report GL-TR-90-0358, Geophysics Lab, Space Physics Division, USAF.

McPherron, R. L., The role of substorms in the generation of magnetic storms, this volume, 1996.

Rich, F. J., and N. C. Maynard, Consequences of using simple analytic functions for the high-latitude convection electric field, *J. Geophys. Res.*, *94*, 3687-3701, 1989.

Riley, P., and R. A. Wolf, Comparison of diffusion and particle drift descriptions of radial transport in the earth's inner magnetosphere, *J. Geophys. Res.*, *97*, 16865-16876, 1992.

Sckopke, N., A general relation between the energy of trapped particles and the disturbance field near the Earth, *J. Geophys. Res.*, *71*, 3125, 1966.

Sckopke, N., A study of self-consistent ring current models, 3, 330-348, 1972.

Senior, C., and M. Blanc, On the control of magnetospheric convection by the spatial distribution of ionospheric conductivities, *J. Geophys. Res.*, *89*, 261, 1984.

Siscoe, G. L., and H. E. Petschek, On storm weakening during substorm expansion phase, submitted to *Ann. Geophys.*, 1996.

Spiro, R. W. and R. A. Wolf, Electrodynamics of convection in the inner magnetosphere, *Magnetospheric Currents*, edited by T. A. Potemra, AGU, Washington, DC, p. 247, 1984.

Toffoletto, F. R., R. W. Spiro, R. A. Wolf, M. Hesse, and J. Birn, Self-consistent modeling of inner magnetospheric convection, submitted for publication in the proceedings of the ICS-3 substorm meeting, 1996.

Vasyliunas, V. M., Mathematical models of magnetospheric convection and its coupling to the ionosphere, in *Particles and Fields in the Magnetosphere*, ed. B. M. McCormac, D. Reidel, Dordrecht, Holland, pp. 60-71, 1970.

Vasyliunas, V. M., The interrelationship of magnetospheric processes, in *The Earth's Magnetospheric Processes*, ed. B. M. McCormac, D. Reidel, Dordrecht, Holland, pp. 29-38, 1972.

Volland, H., A model of the magnetospheric electric convection field, *J. Geophys. Res.*, *83*, 2695, 1978.

Wolf, R. A., The quasi-static (slow-flow) region of the magnetosphere, in *Solar Terrestrial Physics*, ed. R. L. Carovillano and J. M. Forbes, D. Reidel, Hingham, MA, 303-368, 1983.

Wolf, R. A., Magnetospheric configuration, in *Introduction to Space Physics*, ed. M. G. Kivelson and C. T. Russell, Cambridge University Press, Cambridge, England, pp. 288-329, 1995.

Wolf, R. A., M. Harel, R. W. Spiro, G.-H. Voigt, P. H. Reiff, and C. K. Chen, Computer simulation of inner magnetospheric dynamics for the magnetic storm of July 29, 1977, *J. Geophys. Res.*, *87*, 5949-5962, 1982.

R. A. Wolf, J. W. Freeman, Jr., B. A. Hausman, and R. W. Spiro, Space Physics and Astronomy Dept., Rice University MS#108, 6100 S. Main St., Houston, TX 77005.

R. V. Hilmer, Institute for Scientific Research, Boston College, Newton MA 02159.

R. L. Lambour, MIT Lincoln Laboratory, 244 Wood St., Lexington, MA 02173.

Modeling of Ring Current Formation and Decay: A Review

Margaret W. Chen

*Space and Environment Technology Center, The Aerospace Corporation,
Los Angeles, California*

Michael Schulz

*Advanced Technology Center,
Lockheed Martin Missiles & Space, Palo Alto, California*

Larry R. Lyons

*Space and Environment Technology Center, The Aerospace Corporation,
Los Angeles, California*

The development of a ring current is the defining characteristic of a magnetic storm. Ring current dynamics involve particle access, particle energization, particle loss, large-scale electric- and magnetic-field variations, and wave generation. Numerical modeling has led to qualitative and quantitative progress toward the understanding of these processes. This is a review of major advances in ring current modeling, with emphasis on understanding the essential transport and loss processes. Significant advances have been made by simulating the access of charged particles to the ring-current region and relating the associated transport to limiting idealizations (convective access and radial diffusion) under both quiescent and stormtime conditions. Stormtime (main phase) depressions in D_{st} can be understood quantitatively as the consequence of enhanced access of plasma sheet particles to the ring current region, along with the particle energization that accompanies such access. Reduced radial transport leads to the recovery phase (during which D_{st} decays back upward toward zero) by increasing the relative importance of loss processes such as charge-exchange, Coulomb drag, and wave particle interactions.

1. INTRODUCTION

This is a review of modeling work on ring current formation and decay. The intention is to provide an overview of such work with emphasis on more recent contributions.

The ring current consists of geomagnetically trapped 10–200 keV ions and electrons [e.g., *Frank*, 1967; *Williams*, 1981a]. It flows azimuthally but extends radially from $L \sim 7$ to as low as $L \sim 2$. The global intensity of the ring current is commonly measured by the geomagnetic index Dst. *Dessler and Parker* [1959], *Sckopke* [1966], and *Carovillano and Maguire* [1966] showed that the magnitude of Dst should be proportional to the total energy content of the trapped-particle population. A good ring current model must, therefore, account for the access and energization of charged particles during the main phase of a storm (typically ~ 3–12 hours), as well as for the largely collisional processes that cause the ring current to subsequently decay to its quiescent level on a typical time scale ~ 2–3 days.

2. FORMATION OF STORMTIME RING CURRENT

Ring current formation is a consequence of the stormtime transport of charged particles into and within the magnetosphere. It is thought that storm-associated enhancements in the convection electric field are mainly responsible for this transport. Early studies of charged-particle motion in the magnetosphere [*Kavanagh et al.*, 1968; *Chen*, 1970; *Ejiri*, 1978] investigated the effects of a steady convection electric field. For example, *Chen* [1970] described the adiabatic drift paths of equatorially mirroring protons with magnetic moments $\mu \leq 4$ MeV/G in a dipolar magnetic field, on which he superimposed a corotation electric field and a uniform steady-state convection electric field.

Such steady-state descriptions are useful for explaining gross features of particle transport. For example, the separatrix between open and closed drift trajectories, also called the Alfvén layer [e.g., *Kavanagh et al.*, 1968], is the innermost drift shell to which charged particles can convect inward from the night side along open drift trajectories. However, dynamic variations in the convection electric field play an important role in both injection and trapping of particles during storms. *Roederer and Hones* [1974] showed that charged particles ≤ 10 keV in the plasma sheet can be transported earthward during times of storm-associated enhanced convection electric fields and become trapped on closed drift paths when the enhanced electric field decays away. *Smith et al.* [1979] illustrated this effect by tracing the drift trajectories of equatorially mirroring 2-keV singly charged ions that had been injected at $L = 10$ on the night side in a dipolar field model. (This would correspond to $\mu = 6.4$ MeV/G.) They applied a Volland-Stern convection electric field [*Volland*, 1973; *Stern*, 1973] with a magnitude that varied with (interpolated) one-hour values of the Kp index for a real storm, as described by *Grewbowsky and Chen* [1975]. Figure 1 [*Smith et al.*, 1979] shows a time sequence of representative equatorial drift trajectories of singly charged ions. The Kp index and (thus) the modeled convection electric field magnitude increased between times $t = 4.5$ h and $t = 8.5$ h in the simulation (see Figures 1a and 1b). This can be seen to have led to an earthward penetration of ions. At times $t = 16.5$ h and $t = 20.5$ h, when Kp was smaller (see Figures 1c and 1d), some of the ions remained at lower L values and became trapped on closed drift trajectories. Similar drift tracings have been performed in the more realistic non-axisymmetric Mead-Fairfield and Tsyganenko magnetic field models [*Takahashi and Iyemori*, 1989; *Takahashi et al.*, 1990], which illustrate consequences of the day-night magnetic asymmetry in the magnetosphere.

The type of convective access to the inner magnetosphere described above applies mainly to low-μ particles ($\mu \leq 5$ MeV/G, which corresponds to $E \leq 50$ keV at $L = 3$). Such particles have azimuthal drift periods much longer than the main phase of a typical storm and thus remain at roughly a fixed magnetic longitude during the main phase. An enhanced convection electric field would transport such particles almost directly inward in L along nightside trajectories [cf. *Lyons and Williams*, 1980]. On the other hand, higher-μ (≥ 10 MeV/G, hence $E \geq 100$ keV at $L = 3$) particles have drift periods shorter than the typical storm's main phase, and thus drift several times around the Earth during the time of a storm-associated electric-field enhancement. For such particles it is important to consider storm-associated fluctuations in the convection electric field, as these can lead to a quasi-diffusive transport of the higher-μ particles. *Lyons and Schulz* [1989] treated the inward transport of ≥ 50-keV ring current protons by using radial diffusion theory as a first approximation. Using the random-impulse model of *Cornwall* [1968] to generate the fluctuation spectrum required in the resonant-particle diffusion theory of *Fälthammar* [1965], they found that storm-associated electric-field fluctuations having root-mean-square amplitudes typical of those observed (~ 1.0–1.5 mV/m) could account for the increased trapped-ion fluxes observed at $2 \leq L \leq 4$ during a 3-hour main phase. However, the validity of diffusion theory breaks down (for the reason noted above) for particles having $\mu \leq 10$ MeV/G.

Stormtime enhancements of ring current particle fluxes are observed at energies ~ 10–200 keV [e.g., *Frank*, 1967; *Williams*, 1981a], and much of the ring current energy content lies in the intermediate range (~ 50–100 keV at $L = 3$, hence μ ~ 5–10 MeV/G) for which particle drift periods are comparable to the duration of a typical storm's main phase. As the stormtime transport of these particles would be transitional between convective and diffusive, guiding-center simulations have been particularly helpful for understanding their transport and its consequences for stormtime ring current formation. In recent work [*Chen et al.*, 1992, 1993, 1994], we have traced the guiding-center drifts of singly charged equatorially mirroring ions to L ~ 3–5 in a magnetic field model constructed by adding a uniform southward field to the geomagnetic dipole field. We modeled a storm by adding to our quiet-time Volland-Stern electric field distribution a superposition of almost randomly occurring impulses (representing the constituent substorms) in a convection electric field that is uniform in L [e.g., *Brice*, 1967; *Nishida*, 1966]. Figure 2 shows how the cross-tail potential drop varied for prototypical storms with main phases of 3 hours, 6 hours, and 12 hours [*Chen et al.*, 1994]. (The average enhancement in the cross-tail potential drop in Figure 2 is 180 kV and thus corresponds to a very large storm.) *Chen et al.* [1993] found, in agreement

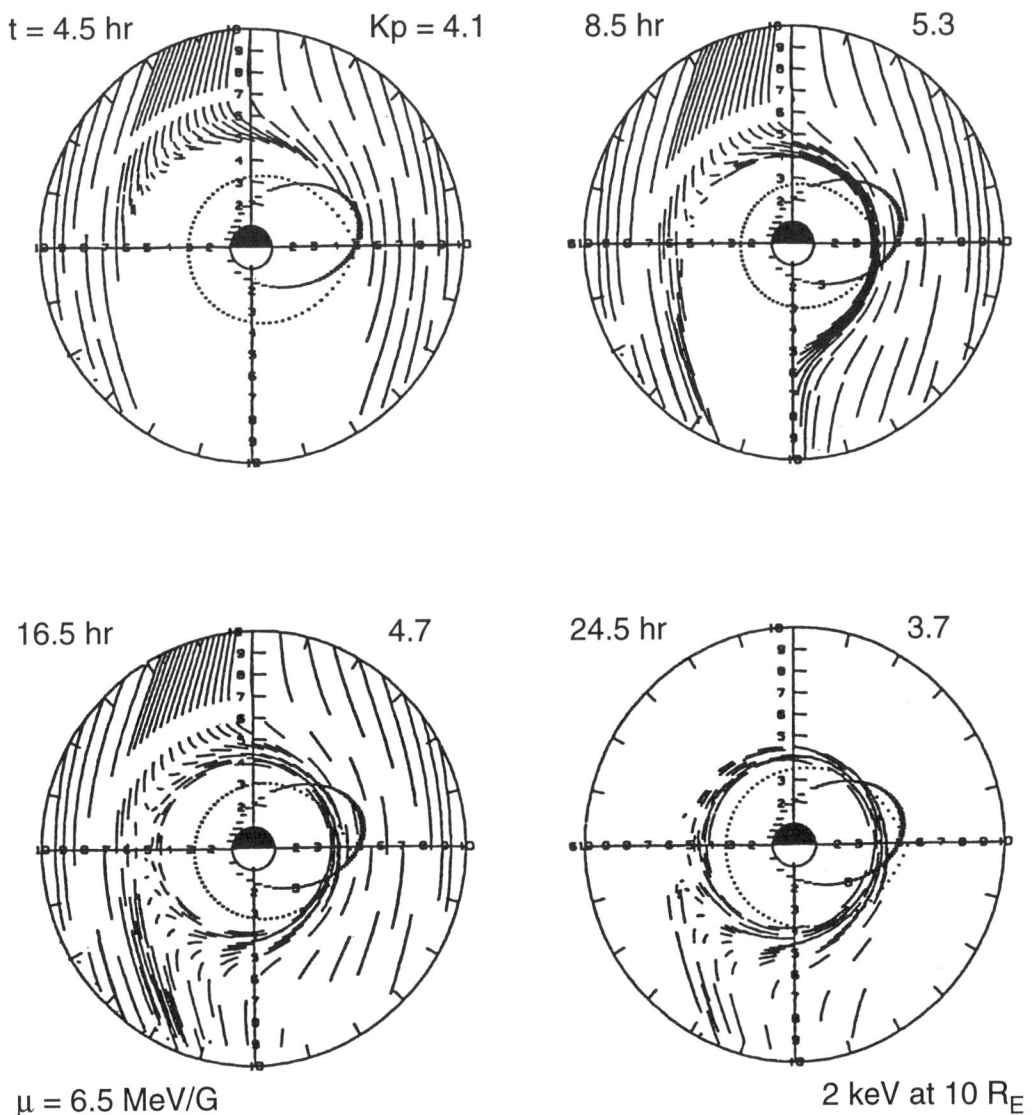

Figure 1. Plots of singly charged ion drift trajectories in the equatorial plane at (a) t = 4.5 h, (b) t = 8.5 h, (c) t = 16.5 h, and (d) t = 24.5 h in the simulation of Smith et al. [1979]. The magnitude of the Volland-Stern convection electric field was varied with Kp in the simulation corresponding to the storm of February 24, 1972.

with *Lyons and Williams* [1980], that the access of $\mu \sim 1–7$ MeV/G ions to $L \sim 3$ (where they would attain energies $\sim$ 10–70 keV) was largely a consequence of the enhanced mean value (rather than the impulsive character) of the convection electric field. Most such ions were transported to $L \sim 3$ along open drift paths from the nightside plasma sheet. Conversely, the transport of higher-μ ions ($\mu \gtrsim 13$ MeV/G, hence energies $E \gtrsim 150$ keV at $L = 3$) was found to be appropriately describable as radial diffusion [cf. *Lyons and Schulz*, 1989] across closed drift paths [*Chen et al.*, 1992].

In addition to being useful for investigating particle transport, guiding-center simulations can also be used to map stormtime phase-space distributions of ring current particles so as to account quantitatively for stormtime ring current formation. *Kistler et al.* [1989] used simulated guiding-center drift trajectories to perform point-to-point mappings of stormtime equatorial ion phase-space distributions from quiescent ion distributions obtained from AMPTE/CCE data. *Chen et al.* [1994] computed drift-averaged stormtime proton phase-space distributions $\bar{f}$ by mapping their simulation results from a quiescent (pre-

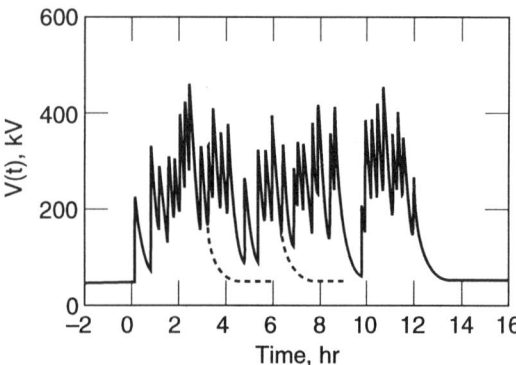

Figure 2. Cross-tail potential imposed by *Chen et al.* [1994], consisting of a quiescent value (50 kV) plus a superposition of exponentially decaying impulses (decay time t = 20 min). Impulses start at times distributed almost randomly over a main phase of 3 hours (terminating with the first dashed-curve extension), 6 hours (terminating with the second dashed-curve extension), and 12 hours (solid curve).

storm) distribution based on steady-state transport. On closed L shells, the quiescent distribution was obtained by balancing quiet-time radial diffusion against charge exchange, the dominant proton loss process. On open drift shells, the phase-space density was defined by the distribution at the boundary f^* where we had specified an exponential function of μ. This distribution assumes that particle losses can be neglected along open drift trajectories. Since we specify the boundary particle distribution, our model does not distinguish whether the particle originated from the solar wind or the ionosphere. The quiet-time distribution obtained for a cross-tail potential drop of 50 kV is shown by the dotted curves in Figure 3a. At the higher μ values, for which radial diffusion dominates charge exchange, the spectrum closely resembles the exponential spectrum on open drift trajectories. A "valley" appears in the distribution at the lower μ and R values ($R \equiv$ geocentric radial distance normalized by R_E) because the charge exchange rate increases more rapidly with decreasing μ and R than the radial-diffusion coefficient does. This result reproduces features similar to those found in proton phase-space distributions obtained by *Williams* [1981a] from ISEE 1 data.

Significantly enhanced ion fluxes at energies $E \sim$ 10–200 keV at $L \sim$ 2.5–7 (hence at $\mu \sim$ 1–20 MeV/G) are observed late in the main phase of a storm [e.g., *Frank*, 1967; *Williams*, 1981a]. Mapped proton phase-space distributions after applying our impulsively enhanced electric field for 3 hours to simulate a moderate storm are shown by the solid curves in Figure 3a. Stormtime enhancements in $\bar{f}$ are largest at the smaller R and μ values, where the pre-storm distribution had been depleted by charge exchange. The enhancements occur at energies $\sim$ 30–150 keV over the entire spatial range shown ($L \sim$ 2–5) and are due mainly to transport along open nightside drift paths. The enhance-

ments compare favorably to observations of *Frank* [1967] and *Williams* [1981a]. Stormtime access of particles to $L \sim$ 3 requires a transport time $\sim$ 1–2 hours. During storms with longer main phases, the stormtime ring current is augmented also by diffusive transport of higher-μ (and thus higher-energy) ions from initially closed drift paths (see dashed curves in Figure 3b). This can be seen by comparing the simulated distributions after 12 hours of enhanced fluctuating electric fields with those (solid curves in Figure 3b) after 3 hours of such electric fields. Diffusive transport is thus quite important for the longer (and typically more intense) storms [*Chen et al.*, 1994].

The phase-space description of ring current particle populations provides a framework for deriving useful physical quantities such as particle energy density. Both electrons (~20%) and ions (~80%) contribute to the energy density of charged particles in the stormtime ring current [*Frank*, 1967; *Williams*, 1981b], but our simulations so far have involved ions only. *Chen et al.* [1994] used simulated phase-space distributions to obtain radial profiles of normalized proton energy density. Results corresponding to the pre-storm ring current and to stormtime ring currents generated by the 3-hour and 12-hour model main phases are shown in Figure 4. The quiet-time ring current energy density peaks at $L \approx 4$, and the position of peak energy density moves progressively inward (from $L \approx 3$ to $L \approx 2.5$) with increasing storm duration, as one might have expected. To learn whether the impulsive character of our storm-associated enhancements in the convection electric field actually had important effects, we considered model storms in which an equally enhanced electric field was held constant at a level corresponding to the average value in Figure 2. The results for 3-hour and 12-hour model storms with a constant electric field are shown as X's and open triangles (respectively) in Figure 4. We found that both the 3-hour and the 12-hour storms with a constant electric field produced virtually the same energy-density profiles as a 3-hour impulsive storm. However, the energy density over $L \sim$ 2–4 is significantly larger for a 12-hour impulsive model storm than for 3- or 12-hour storms with a constant electric field. Thus, the impulsive character of enhancements in the convection electric field must have made a significant contribution to the ring current energy content only in model storms longer than 3 hours, which is approximately the stormtime convective access time to $L = 2.5$.

Chen et al. [1993] estimated the stormtime enhancement in proton energy content by integrating the energy density curves in Figure 4. We found that the overall proton-energy content increased by factors of 3 and 5 (respectively) during our 3-hour and 12-hour impulsive model storms. *Dessler and Parker* [1959], *Sckopke* [1966], and *Carovillano and Maguire* [1966] showed that the magnitude of D_{st} in a dipole field should be proportional to the total energy content of the trapped particle population. If the quiescent ring current contributes about 10–20 nT to the baseline of $-D_{st}$, then the stormtime D_{st} would be ~30–60

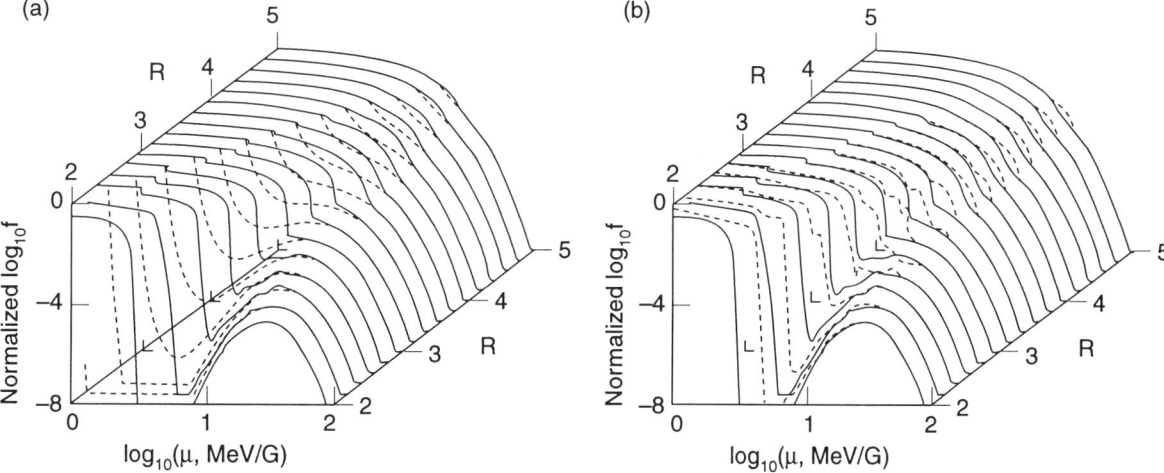

Figure 3. Isometric plots of normalized drift-averaged proton phase-space density, $\bar{f}$, versus $\log_{10}(\mu)$ versus R ($\equiv r/R_E$) for (a) pre-storm conditions (dotted curves) and after the main phase of a 3-hour storm (solid curves); and (b) after the main phase of a 3-hour storm (solid curves) and of 12-hour storm (dashed curves).

nT and ~50–100 nT for our respective 3-hour and 12-hour model storms. It seems that transport alone cannot account for the entire increase in $|D_{st}|$ characteristic of a large storm ($\gtrsim 80$ nT). Though the ring current is probably the dominant current at low- and mid-altitudes during a storm, D_{st} includes the effect of all magnetospheric currents. *Siscoe* [1970] used the virial theorem to obtain an expression for the field that includes the additional effect of magnetospheric boundary currents though the boundary terms are hard to estimate. Recently *Campbell* [1996] has reminded the community that Birkeland and cross-tail current may contribute to the stormtime D_{st} and their effects, even if relatively small, should be accounted for.

AMPTE/CCE observations of *Hamilton et al.* [1988], however, have revealed stormtime increases in the ion energy density at $L \sim 8$ (see Figure 5a) and even at the magnetopause. Accordingly, we have considered applying stormtime enhancements in the phase-space density at the outer boundary of our model. Such increased phase-space densities in the tail plasma sheet are a plausible consequence of enhanced energization of ions in the cross-tail current sheet [*Lyons and Speiser*, 1982]. The model profiles in Figure 5a reflect an enhancement in boundary phase-space density by a factor of 2 for the 3-hour storm and by a factor of 4 for the 6-hour and 12-hour storms. These profiles give an overall increase in proton-energy content by factors of 6, 13, and 16 during our 3-, 6-, and 12-hour model storms, respectively. These results suggest that stormtime transport in the presence of observationally supported [*Hamilton et al.*, 1988] stormtime increases in the boundary value of the phase-space density can account for the stormtime ring current of any intensity observed.

The profiles in Figure 5b qualitatively resemble the ion energy-density profiles obtained by *Hamilton et al.* [1988] from AMPTE/CCE data for storms of various intensities (shown in Figure 5a). For example, the quiescent ring current ion energy density peaked at $L \approx 4$, and this peak moved inward with increasing storm intensity (as measured by $|D_{st}|$). *Hamilton et al.* [1988] found that the ion energy content inferred from the AMPTE data could account for about 50%–70% of the $|D_{st}|$ encountered in several representative storms. *Roeder et al.* [1996] have found that the ion energy content inferred from CRRES data can account for about 50% of $|D_{st}|$ for a more recent storm. However, the cited studies did not include electrons, which can contribute a further ~25% to the stormtime ring current energy content [*Frank*, 1967]. They also did not take account of Earth induction, which should further increase the surface magnetic field depression ΔB by a factor of ≈ 1.4 [e.g., *Søraas and Davis*, 1968]. These additional effects might be sufficient to account fully for the depressions in D_{st} observed during most storms.

Hoffman and Bracken [1967] prescribed a parabolic radial profile for the plasma pressure in order to compute ring current magnetic-field profiles (see Figure 6a) that resemble observed $\Delta B(R)$ profiles [e.g., *Cahill* 1966]. However, it is no longer necessary to postulate the plasma pressure profile arbitrarily. The energy density profiles in Figures 4 and 5, derived either from phase-space mappings along simulated trajectories [*Chen et al.*, 1994] or from direct observations of particle distributions [*Hamilton et al.*, 1988], are tantamount to plasma pressure profiles. *Chen et al.* [1994] computed radial profiles of the ring current magnetic field $\Delta B(R)$ from energy density profiles similar to those in Figure 5b. Profiles of the resulting $\Delta B(R)$, normalized by the corresponding value of D_{st}, are shown in Figure 6b. Unnormalized depressions in the magnetic field grow progressively stronger (essentially in

178 MODELING OF RING CURRENT FORMATION AND DECAY: A REVIEW

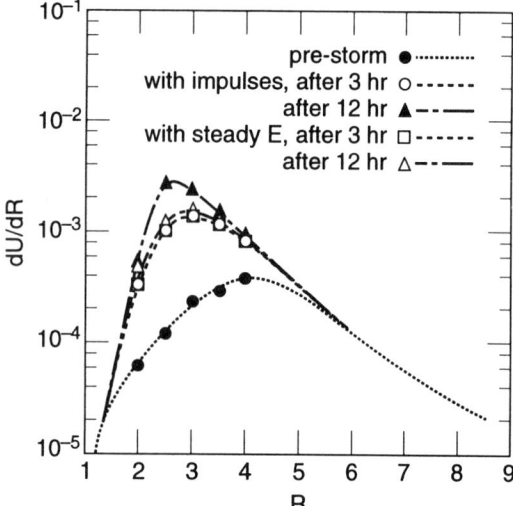

Figure 4. Profiles of normalized ring current energy density before the storm (dotted curve corresponding to filled circles) and after the main phases of our model storms: dashed curves corresponding to 3-hour storms and dash-dotted curves corresponding to 12-hour storms. Open circles and filled triangles correspond to 3-hour and 12-hour storms modeled by impulsive enhancements in the convection electric field. The ×'s and open triangles correspond to 3-hour and 12-hour storms modeled by steady enhancements in the convection electric field.

proportion to D_{st}) with storm duration, but the stormtime minima become sharper and move further inward with increasing main-phase duration. The profiles of $-\Delta B(R)/D_{st}$ thus obtained from simulated storms resemble some $\Delta B(R)$ profiles inferred by *Cahill* [1966] from Explorer 26 data (see Figures 6c and 6d). Indeed, the characteristic profile shapes and locations of the minima in $\Delta B(R)$ during the main phase and recovery phase (to be compared with stormtime and quiescent simulation results, respectively) show good qualitative agreement.

Energy density and magnetic field profiles of the ring current discussed so far have pertained to equatorially mirroring ions. (The quantity dU/dR in Figures 4 and 5a actually represents energy content per unit R in the equatorial plane.) Numerical models have recently begun to include the transport of particles that mirror off the equator so as to describe the latitudinal distribution of the ring current. *Peroomian and Ashour-Abdalla* [1995] followed the full bounce and drift motions of a large number of ions, launched from the auroral ionosphere into a steady-state electric field, to determine the contribution of ionospheric ions to the quiescent plasma sheet and outer ring current. A bounce-averaged formulation would have been inappropriate in that context since the guiding-center approximation fails near the equator at the boundary between closed and open field lines. *Jordanava et al.* [1994] and *Fok et al.* [1995] have developed ring current models by solving bounce-averaged time-dependent kinetic equations that

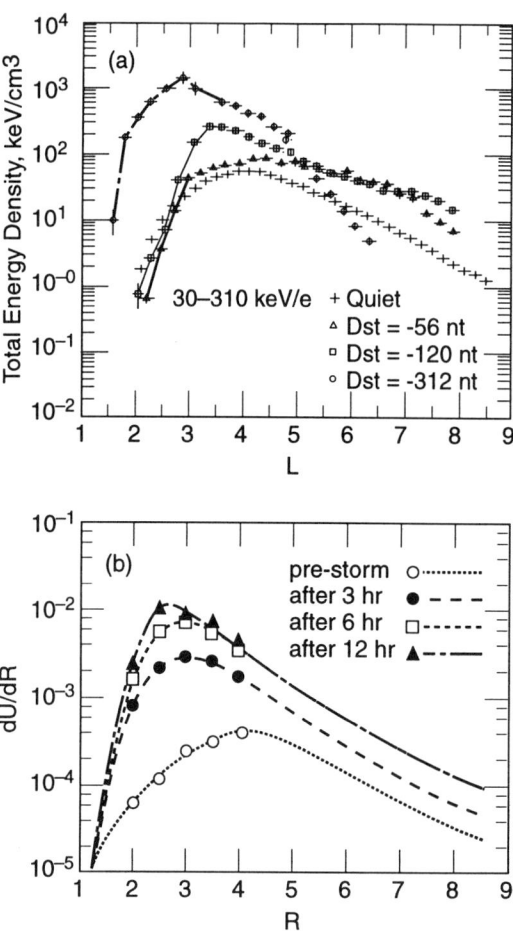

Figure 5. (a) Radial profiles of the total ion energy density compiled from AMPTE/CCE data [*Hamilton et al.*, 1988b]; and (b) normalized proton energy-density profiles taken from simulations of *Chen et al.* [1994].

include Kp-dependent ion drifts but not radial diffusion explicitly.

For the purpose of computing stormtime ring current proton pitch-angle distributions, we have recently extended our simulation model to three dimensions by including calculations of bounce-averaged ion drifts derived from a Hamiltonian approach in which the first two adiabatic invariants are held fixed [*Schulz and Chen*, 1995]. We made ten separate phase-space mappings by using the ten representative isotropic ISEE 1 plasma-sheet proton spectra (called A–J) of *Christon et al.* [1989] as boundary conditions on our model. Figure 7 shows pitch-angle distributions at selected energies (55 keV, 110 keV, and 167 keV) at $R = 3$ during quiescent times and at the end of a 3-hour model main phase, with phase-space mappings made from plasma-sheet proton spectrum A of *Christon et al.* [1989] as the boundary condition. The average value of the cross-tail potential drop for this model storm is 125 keV. Since a

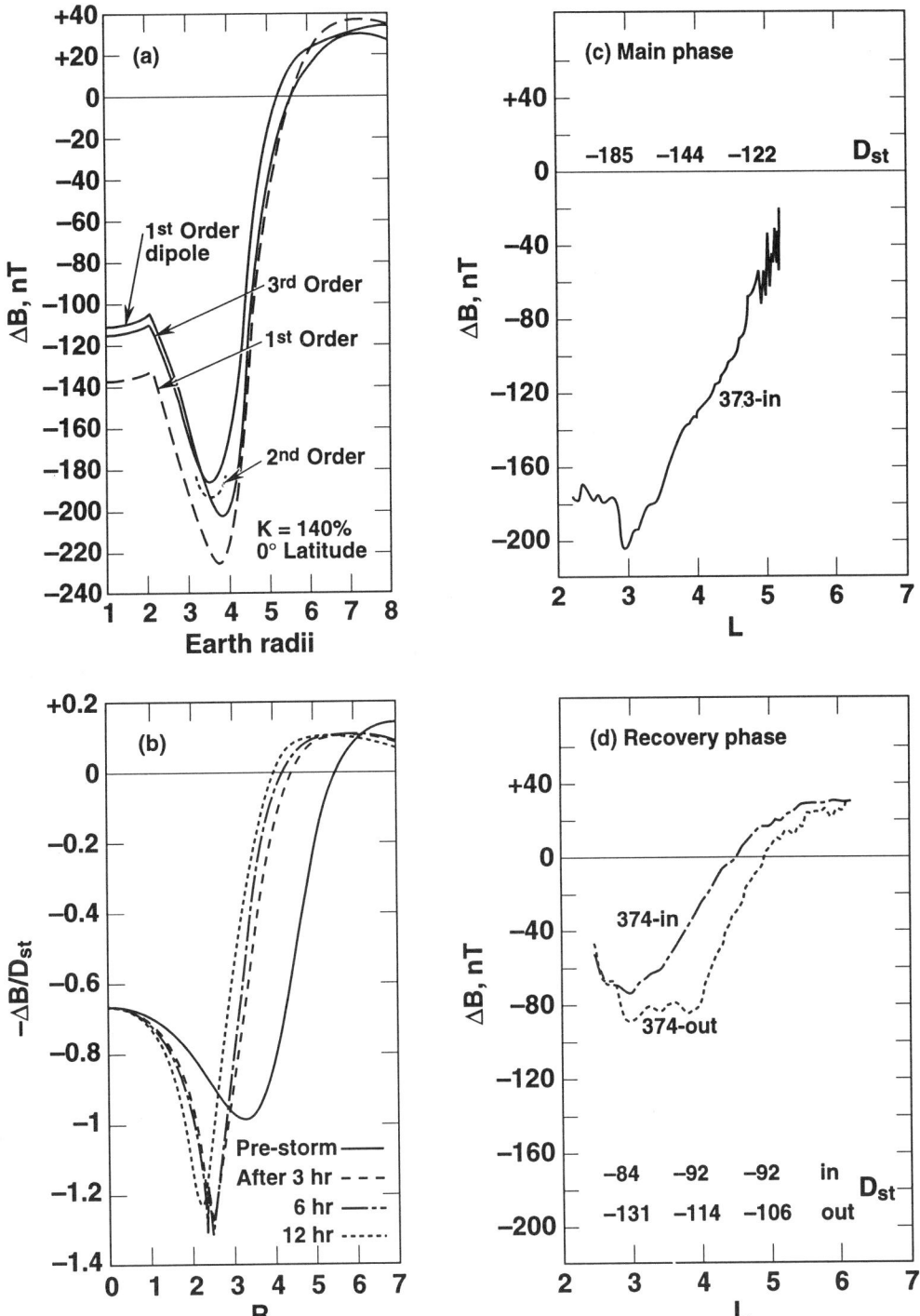

Figure 6. Profiles of ring current magnetic field obtained from (a) model of *Hoffman and Bracken* [1967], (b) simulation results of *Chen et al.* [1994], and (c, d) Explorer 45 data analyzed by *Cahill* [1966] during the main phase (c) and recovery phase (d) of the storm of April 17, 1965.

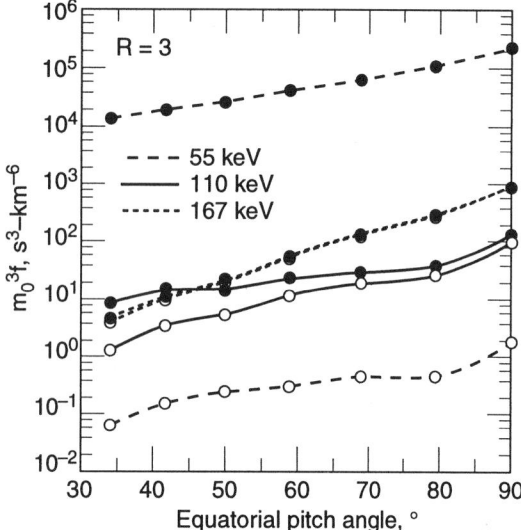

Figure 7. Pitch-angle distributions of protons at $R = 3$ having energies of 55 keV (dashed curve), 100 keV (solid curve), and 167 keV (dotted curve). Open and filled circles correspond to the pre-storm and stormtime (or post-main phase) conditions. The anisotropy $A \equiv (\langle f \sin\alpha \rangle / 2 \langle f \cos\alpha \rangle) - 1$ for the pre-storm/(stormtime pitch-angle distributions: 2.7 (2.7), 4.4 (2.2), and 7.7 (7.5) for 55 keV, 110 keV, and 167 keV protons at $R = 3$, respectively.

few people in the community had pointed out that our previously used average cross-tail potential drop of 180 kV was higher than what is typically observed (see e. g., [Reiff and Luhmann, 1986]), we now use a more realistic value of the cross-tail potential drop. At lower energies (e.g., 55 keV) the stormtime pitch-angle anisotropy can be accounted for largely from the mapping of the boundary (or plasma sheet) spectrum. This is because most protons of this energy have had direct access to the ring current from the plasma sheet along open drift paths. At high energies (e.g., 167 keV) there is barely any difference between the pre-storm and stormtime pitch-angle distributions since three hours of quasi-diffusive transport can produce little change in the phase-space density at any equatorial pitch angle. At intermediate energies (e.g., 110 keV), however, differences in the stormtime transport (e.g., along open versus closed drift paths) for different equatorial pitch angles can lead to stormtime anisotropies that differ significantly from the quiet-time ring current anisotropy. When we imposed the ten different ISEE 1 plasma-sheet spectra as boundary conditions on our mappings, we found the pitch-angle anisotropy of ring current protons to depend strongly on the plasma-sheet energy spectrum at $E \sim 0.1$–7 keV. *Nakada et al.* [1965] had found a qualitatively similar result for radiation-belt protons (though at much higher energies both in the source spectrum and in the inner magnetosphere).

Thus far we have focused on the development of the stormtime ring current via storm-associated enhancements in the convection electric field. Another process that has been believed (e.g., *Mauk and Meng* [1986]) to inject particles into the stormtime ring current are induced electric fields associated with magnetic field dipolarizations that occur during the expansion phase of substorms. Recently, *Lyons* [1996] estimated the effectiveness of substorm-associated magnetic field dipolarizations in injecting particles into the $L < 4$ region using a very simple calculation of the induced electric field. He assumed a change in the equatorial parallel magnetic field of 50 nT and only an azimuthal component of the induced electric field. He found that substorm dipolarizations can readily inject particles from much further out to near-synchronous orbit where substorm injections are often observed. However, the dipolarizations cannot displace particles at significant distances below $L = 4$ where the bulk of the stormtime ring current lies. *Wolf* [1996] simulated the effect of substorm dipolarizations in the Magnetospheric Specification Model and found that they have little effect on development of the symmetric ring current. These recent studies indicate that substorm expansions do not play an important role in the injection of stormtime ring current particles. (See *Lyons* [1996] for a brief discussion of observational evidence supporting this.)

In our own work, we have so far considered the dynamics of ring-current protons, while recognizing that electrons and (especially during major storms) oxygen ions are also significant constituents. The inclusion of ring-current electrons and oxygen ions in our simulations would require a few changes in the method of modeling (e.g., to let oxygen ions make transitions among various charge states and perhaps to allow relativistic kinematics for electrons). Since we introduce new particles at or near the neutral line in our model, it does not matter whether such particles have gained access to the plasma sheet from the ionosphere via auroral ion beams or from the solar wind via the magnetosheath.

B. T. Tsurutani (personal communication, 1996) has suggested that additional oxygen ions might achieve direct access to the ring-current region along field lines from the midlatitude ionosphere. Such access could be modeled by guiding-center simulations (as in the plasmasphere-replenishment problem) or by simulation of the full motion (as in the work of *Peroomian and Ashour-Abdalla* [1996] on auroral ion beams). Collisions and/or wave-particle interactions would likely be needed [e.g., *Schulz and Koons*, 1972] in order to let the corresponding ions become trapped. The main question would be how oxygen ions (or even protons) with direct access from the ionosphere to the ring-current region might have gained enough kinetic energy to contribute much to the ring current itself without having been exposed to the large-scale electric fields that produce auroral ion beams, neutral-sheet energization, or inward radial transport.

The simple time-varying Volland-Stern electric field model that we have been using in our studies captures the essence of particle transport leading to the formation of the stormtime ring current and has allowed us to make comparisons with theoretical approximations of particle transport. In order to simulate real storms, it is necessary to model the global electric field as realistically as possible. We are currently modeling real storms based on polar cap potential drop measurements obtained from DMSP drift meter data made available from the University of Texas, Dallas. More sophisticated electric field models that include transient low-latitude electric field responses to changes in the high-latitude convection have been used by *Spiro et al.* [1988] in the Rice Convection Model [*Wolf et al.*, 1986]. As monitoring of global magnetospheric electric fields improves, so will the modeling of ring current development.

3. RING CURRENT DECAY

Ring current particles can be removed by various collisional processes. These loss processes are usually neglected in comparison with enhanced radial transport during the main phase of a storm. However, the reduced radial transport rate characteristic of the recovery phase of a storm allows loss processes to become dominant so that the ring current decays (typically with a lifetime ~ 2 days) to its quiet-time level.

Ring current ions are lost via charge exchange with neutrals, Coulomb collisions with thermal plasma, and interactions with various magnetospheric waves. Of these three processes, wave-particle interactions are the least understood in terms of their role in causing the ring current to decay. Since the wave spectrum is usually not known, it is usual to simulate the effect of wave-particle interactions either by postulating a lifetime $\tau_w(E, L)$ against pitch-angle diffusion or to calculate $\tau_w(E, L)$ from an assumed wave spectrum (as *Lyons et al.* [1971] and *Lyons and Thorne* [1972] did for radiation-belt electrons). The drift-averaged ion distribution $\bar{f}$ should then satisfy the transport equation,

$$\frac{\partial \bar{f}}{\partial t} = L^2 \frac{\partial}{\partial L}\left[\frac{D_{LL}}{L^2}\frac{\partial \bar{f}}{\partial t}\right] - \mu^{-1/2} \frac{\partial}{\partial \mu}\left[\mu^{1/2}\left(\frac{d\mu}{dt}\right)_v \bar{f}\right]_{\mu, J, L} - \frac{\bar{f}}{\tau_q} - \frac{\bar{f}}{\tau_w} \quad (1)$$

whose RHS includes terms describing radial diffusion, Coulomb drag, charge exchange, and wave-particle interactions, respectively, where D_{LL} is a diffusion coefficient, v is a subscript connoting frictional drag, J is the second invariant, and τ_q is the lifetime against charge exchange.

Neglecting wave-particle interactions, *Spjeldvik* [1977] solved the steady-state transport equation [$\partial/\partial t \to 0$ in Eq. (1)] with the standard diffusion coefficient [*Cornwall*, 1968] to find quiescent radiation-belt proton distributions. Later, *Spjeldvik and Fritz* solved for quiescent radiation-belt helium [1978a] and oxygen [1978b] ion distributions. (The treatment of such heavier ions is more difficult since heavy ions can charge exchange among multiple charge states.) In the interest of obtaining quiescent ring current distributions, *Sheldon and Hamilton* [1993] extended this type of analysis to $\mu = 1-100$ MeV/G and $L = 2.5-7$.

Sheldon [1994] compared quiescent H^+, He^+, and O^+ distributions obtained with the standard diffusion coefficient and from a modified D_{LL} (enhanced at $L \sim 3$) against data compiled from AMPTE/CCE. Agreement between data and the standard diffusion model was not good, particularly at $L \lesssim 3$; but agreement with a suitably modified diffusion model was greatly improved. From the L-dependence of his optimally modified diffusion coefficient, *Sheldon* [1994] inferred that electric field fluctuations from the ionosphere [cf. *Brice and McDonough*, 1973], rather than from the solar wind, may be important for the transport that leads to quiet-time ring current formation. Since all charged particles must respond to electric field fluctuations, this hypothesis could be tested by examining whether Sheldon's modified diffusion coefficient is consistent with measured quiescent electron ring current and radiation-belt distributions.

Recent work [*Fok et al.*, 1991, 1993] has improved the understanding of the role of Coulomb drag on ring current decay. *Fok et al.* [1991] derived bounce-averaged H^+, He^+, and O^+ Coulomb-drag coefficients and deduced lifetimes from these by taking account of the corresponding energy spectra. Figure 8, taken from their paper, shows these lifetimes as functions of ion energy (solid curve) for representative plasmaspheric densities. For comparison, the dashed curves represent charge-exchange lifetimes at representative L values. Coulomb decay lifetimes thus deduced are longer than charge-exchange lifetimes for protons with energies between 5 and 200 keV. *Sheldon and Hamilton* [1993] have independently found that charge exchange dominates Coulomb drag except at very high energies ($\gtrsim 500$ keV). However, Coulomb decay lifetimes are comparable to or shorter than charge-exchange lifetimes for He^+ and O^+ (see Figures 8b and 8c) below about 30 keV. This suggests that Coulomb decay may be an important ring current loss process for heavy ions in the 10–30 keV energy range. To investigate the role of Coulomb collisions on ring current decay, *Fok et al.* [1993] calculated ion distributions from a time-dependent kinetic equation with charge exchange and Coulomb drag losses during the recovery phase of a storm. Their model includes Kp-dependent ion drift (but not radial diffusion explicitly)

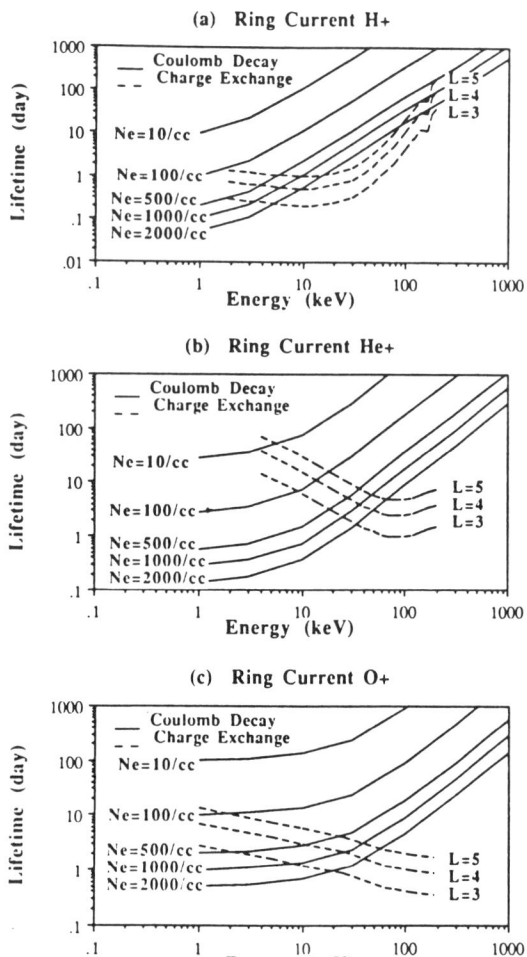

Figure 8. Comparison of Coulomb decay lifetime (solid curve) and charge exchange lifetime (dashed curve) of (a) H$^+$, (b) O$^+$, and (c) He$^+$ ions [*Fok et al.*, 1991]. Coulomb decay rates were calculated for different electron densities N_e.

in a dipolar **B** field model with Volland-Stern convection superimposed. They found that energy degradation of higher-energy ring current ions by Coulomb collisions within the plasmasphere leads to a build-up of the lower-energy population at $L \sim 2.5$. The time-scale for this build-up and eventual decay is longer for O$^+$ ions than H$^+$ ions since H$^+$ ions have shorter Coulomb lifetimes than O$^+$ in this energy range.

Recently, *Fok et al.* [1995] made direct comparisons of ion fluxes obtained from AMPTE measurements during the recovery phase of a real storm with ion fluxes predicted by their model, using initial fluxes taken directly from AMPTE data. Figures 9a and 9b show ion fluxes of H$^+$, He$^+$, and O$^+$ ions from the model (solid curves) and from the data (symbols) 16 hours after the storm main phase at $L = 3.25$ and $L = 6.25$, respectively. The model agrees well with the data for He$^+$ and O$^+$ ions but has underestimated the decay of proton ring current intensities at $E \leq 100$ keV.

Our model similarly underestimates the decay rate for ring current protons, as we learned while studying the time evolution of the anisotropies of proton pitch-angle distributions in our simulation model. Figure 10 shows time histories of the pitch-angle anisotropy

$$A \equiv (\langle f \sin\alpha \rangle / 2 \langle f \cos\alpha \rangle) - 1, \qquad (2)$$

where α is equatorial pitch angle) of protons having 55 keV (dashed curve), 100 keV (solid curve), and 167 keV (dotted curve) at $R = 3$ during the recovery phase of our 3-hour model storm. The pitch-angle anisotropy at any energy varies during the recovery phase because the charge exchange and quiet-time radial diffusion rates vary with different equatorial pitch angle. At first the pitch-angle, anisotropy of 55-keV protons increases because bounce-averaged charge exchange lifetimes are shorter for protons that mirror farther from the equator [cf. *Cornwall*, 1966; *Jordanava et al.*, 1994]. Later, the anisotropy decreases back to the pre-storm value as quiet-time charge exchange becomes balanced by radial diffusion. It takes about 21 days for the anisotropy and intensity of the pitch-angle distribution to relax to its quiescent state in our model, whereas the observed recovery of the proton ring current is typically ~2–3 days.

However, pitch-angle anisotropy can lead to an electromagnetic ion cyclotron wave (EMICW) instability, which has been neglected in the above-cited studies. Trapped particles may be pitch-angle scattered by the resulting waves and become lost into the atmosphere. Ion cyclotron wave generation has been analyzed in some detail [e.g., *Cornwall*, 1965, 1966; *Kennel and Petschek*, 1966; *Perraut et al.*, 1967; *Gendrin*, 1981; *Kozyra et al.*, 1984; *Horne and Thorne*, 1993] in the magnetosphere. *Kaye et al.* [1979] calculated ICW growth rates from anisotropic proton distributions that had evolved from isotropic source distributions under quiescent adiabatic convection. The resulting growth rates ($\gamma \leq 0.002\Omega_p$, where Ω_p is the proton gyrofrequency) were typically small. Increasing the source energy did not necessarily increase the wave growth rate in the inner magnetosphere since the more energetic ions could not drift to the inner L shells. More modeling work is needed to understand the role that wave-particle interactions play in the decay of the ring current. *Li et al.* [1993] investigated the effects of a prescribed wave distribution on particle loss by following test particle trajectories in the ring current region. Transport studies that include the pitch-angle distributions could form the basis for a study of wave generation processes in the ring current region that could include the effects of waves on particle loss.

4. UNRESOLVED PROBLEMS

Further modeling work can help to address many fundamental ring current issues that remain unresolved. For

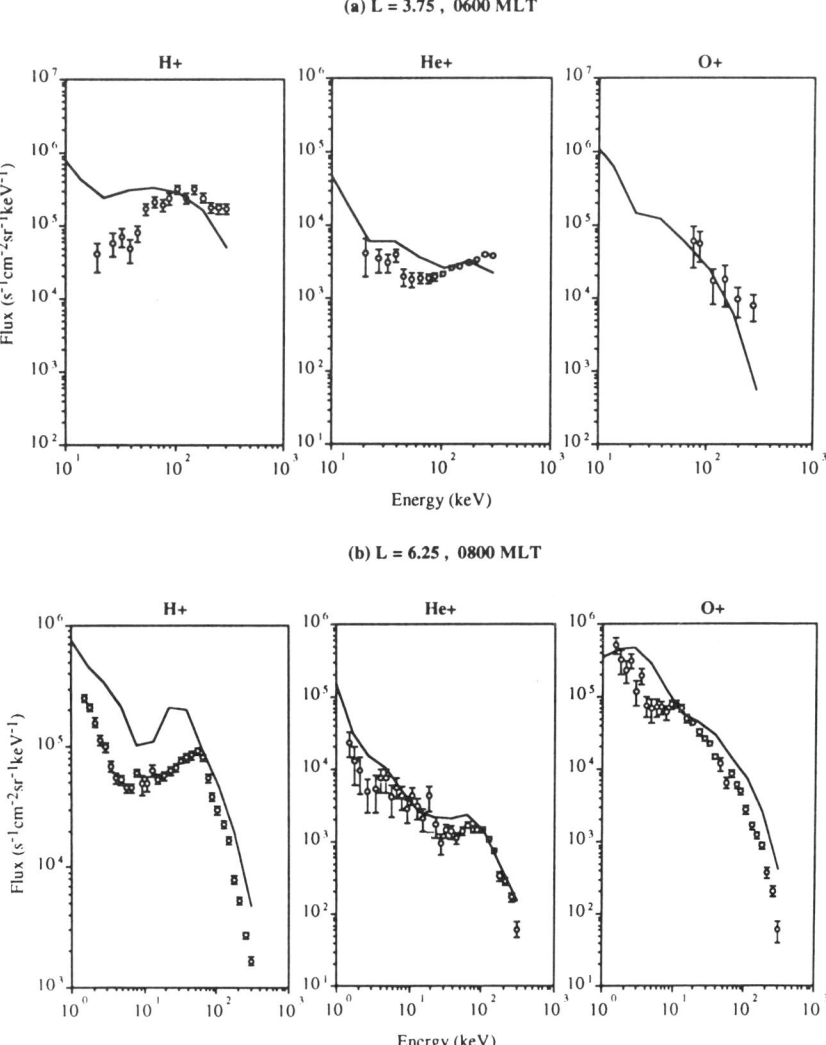

Figure 9. Ion spectra from AMPTE/CCE measurement (symbols) and as calculated (solid lines) for epoch $t = 16$ h into recovery phase at selected L values [*Fok et al.*, 1995].

example, the effect of electric fields induced as the magnetic field responds to the formation and decay of the ring current needs to be quantitatively investigated. This would require a self-consistent treatment of large-scale electric and magnetic fields. Another problem that modeling can help to address is the relative contribution of ionospheric and solar wind ions to the ring current. Observations indicate that O^+ ions can be a dominant ring current constituent during the main phases of very large storms [*Hamilton et al.*, 1988; *Roeder et al.*, 1996]. Modeling of heavy ions is complicated by the fact that they have multiple charge states. A refined understanding of the connection between the ring current and D_{st} is needed. More quantitative assessments of the relative contribution of the stormtime ring current to the depression in D_{st} would help resolve this issue. Also, an explanation (perhaps involving wave-particle interactions) for the rapid decay of the proton ring current is needed. In general, more work needs to be done on the origins and effects of wave-particle interactions in ring-current dynamics. Though very difficult, a self-consistent treatment of particle transport and wave instabilities would be very revealing.

Acknowledgments. We are grateful to D. C. Hamilton for providing Figure 5a of this paper. The work of M. W. Chen and L. R. Lyons in preparing this review was supported by the NSF grant NSF-ATM-9522288 and by the Aerospace Sponsored Research Program. The work of M. Schulz was supported by the Independ-

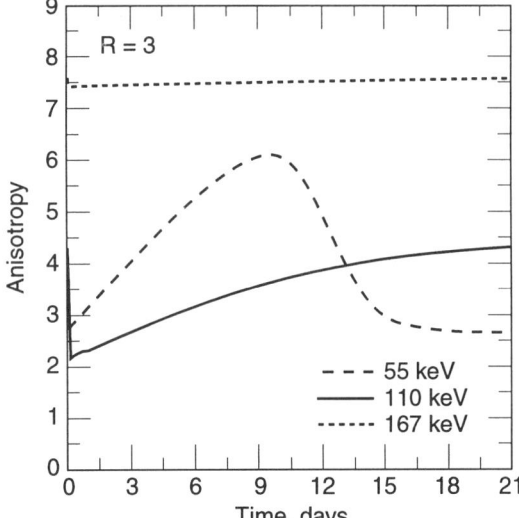

Figure 10. Pitch-angle anisotropy versus time at $R = 3$, as computed from our 3-hour storm simulation, for protons having energies there.

ent Research Program of Lockheed Martin Missiles and Space. Computing resources for the research work of M. W. Chen, M. Schulz, and L. R. Lyons were provided by the San Diego Supercomputer Center.

REFERENCES

Brice, N. M., Bulk motion of the magnetosphere, *J. Geophys. Res.*, 72, 1246–1256, 1967.

Brice, N. M., and T. R. McDonough, Jupiter's radiation belts, *Icarus*, 18, 206–219, 1973.

Cahill, L. J., Inflation of the inner magnetosphere during a magnetic storm, *J. Geophys. Res.*, 71, 4505–4519, 1966.

Campbell, W. H., Geomagnetic storms, the Dst ring-current myth and lognormal distributions, *J. Atm. Terr. Phys.*, 58, 1171–1188, 1996.

Carovillano, R. L., and J. J. Maguire, The energy of confinement of a shielded magnetic dipole field, *Geophys. J. R. Astr Soc.*, 12, 23–28, 1966.

Chen, A. J., Penetration of low-energy protons deep into the magnetosphere, *J. Geophys. Res.*, 75, 2458–2467, 1970.

Chen, M. W., M. Schulz, L. R. Lyons, and D. J. Gorney, Ion radial diffusion in an electrostatic impulse model for stormtime ring current formation, *Geophys. Res. Lett.*, 19, 621–624, 1992.

Chen, M. W., M. Schulz, L. R. Lyons, and D. J. Gorney, Stormtime transport of ring current and radiation belt ions, *J. Geophys. Res.*, 98, 3835–3849, 1993.

Chen, M. W., M. Schulz, and L. R. Lyons, Simulations of phase space distributions of stormtime proton ring current, *J. Geophys. Res.*, 99, 5745–5759, 1994.

Christon, S. P., D. J. Williams, D. G. Mitchell, L. A. Frank, C. Y. Huang, Spectral characteristics of plasma sheet ion and electron populations during undisturbed geomagnetic conditions, *J. Geophys. Res.*, 94, 13,409–13,424, 1989.

Cornwall, J. M., Cyclotron instabilities and electromagnetic emission in the ultra low frequency and very low frequency ranges, *J. Geophys. Res.*, 70, 61–69, 1965.

Cornwall, J. M., Micropulsations and the outer radiation zone, *J. Geophys. Res.*, 71, 2185–2199, 1966.

Cornwall, J. M., Diffusion processes influenced by conjugate-point wave phenomena, *Radio Sci.*, 3, 740–744, 1968.

Dessler, A. J. and E. N. Parker, Hydrodynamic theory of geomagnetic storms, *J. Geophys. Res.*, 64, 2239–2252, 1959.

Ejiri, M., Trajectory traces of charged particles in the magnetosphere, *J. Geophys. Res.*, 83, 4798–4810, 1978.

Fälthammar, C.-G., Effects of time dependent electric fields on geomagnetically trapped radiation, *J. Geophys. Res.*, 70, 2503–2516, 1965.

Fok, M.-C., J. U. Kozyra, A. F. Nagy, and T. E. Cravens, Lifetime of ring current particles due to Coulomb collisions in the plasmasphere, *J. Geophys. Res.*, 96, 7861–7867, 1991.

Fok, M.-C., J. U. Kozyra, A. F. Nagy, C. E. Rasmussen, and G. V. Khazanov, Decay of equatorial ring current ions and associated aeronomical consequences, *J. Geophys. Res.*, 98, 19,381–19,393, 1993.

Fok, M.-C., T. E. Moore, J. U. Kozyra, G. C. Ho, and D. C. Hamilton, Three-dimensional ring current decay model, *J. Geophys. Res.*, 100, 9619–9632, 1995.

Frank, L. A., On the extraterrestrial ring current during geomagnetic storms, *J. Geophys. Res.*, 72, 3753–3768, 1967.

Gendrin, R., General relationships between wave amplification and particle diffusion in a magnetoplasma, *Rev. Geophys. Space Phys.*, 19, 171, 1981.

Grewbowsky, J. M., and A. J. Chen, Effects of convection electric field on distribution of ring current type protons, *Planet. Space Sci.*, 23, 1045–1052, 1975.

Hamilton, D. C., G. Gloeckler, F. M. Ipavich, W. Stüdemann, B. Wilken, and G. Kremser, Ring current development during the great geomagnetic storm of February 1986, *J. Geophys. Res.*, 93, 14,343–14,355, 1988.

Hoffman, R. A., and P. A. Bracken, Higher-order ring currents and particle energy storage in the magnetosphere, *J. Geophys. Res.*, 72, 6039–6049, 1967.

Horne, R. B., and R. M. Thorne, On the preferred source location for the convective amplification of ion cyclotron waves, *J. Geophys. Res.*, 98, 9233–9247, 1993.

Jordanava, U. K., J. U. Kozyra, G. V. Khazanov, A. F. Nagy, C. E. Rasmussen, and M.-C. Fok, A bounce-averaged kinetic model of the ring current ion population, *Geophys. Res. Lett.*, 21, 2785–2788, 1994.

Kavanagh, L. D., J. W. Freeman, Jr., and A. J. Chen, Plasma flow in the magnetosphere, *J. Geophys. Res.*, 73, 5511–5519, 1968.

Kaye, S. M., M. G. Kivelson, and D. J. Southwood, Evolution of ion cyclotron instability in the plasma convection system of the magnetosphere, *J. Geophys. Res.*, 84, 6397–6406, 1979.

Kennel, C. F., and H. E. Petschek, Limit on stably trapped particle fluxes, *J. Geophys. Res.*, *71*, 1-28, 1966.

Kistler, L. M., F. M. Ipavich, D. C. Hamilton, G. Gloeckler, B. Wilken, G. Kremser, and W. Stüdemann, Energy spectra of the major ion species in the ring current during geomagnetic storms, *J. Geophys. Res.*, *94*, 3579-3599, 1989.

Kozyra, J. U., T. E. Cravens, A. F. Nagy, E. G. Fontheim, R. S. B. Ong, Effects of energetic heavy ions on electromagnetic ion cyclotron wave generation in the plasmapause region, *J. Geophys. Res.*, *89*, 2217-2233, 1984.

Li, X., M. Hudson, A. Chan, and I. Roth, Loss of ring current O^+ ions due to interaction with Pc 5 Waves, *J. Geophys. Res.*, *98*, 215-231, 1993.

Lyons, L. R., Ring current formation processes: A summary of current understanding and outstanding questions based on the 1996 Chapman Conference on Magnetic Storms, *Magnetic Storms*, ed. by B. T. Tsurutani and Y. Kamide, in this volume, Am. Geophys. Un., Washington, D. C., 1996.

Lyons, L. R., and M. Schulz, Access of energetic particles to stormtime ring current through enhanced radial "diffusion", *J. Geophys. Res.*, *94*, 5491-5496, 1989.

Lyons, L. R., and T. W. Speiser, Evidence for current sheet acceleration in the geomagnetic tail, *J. Geophys. Res.*, *87*, 2276, 1982.

Lyons, L. R., R. M. Thorne, and C. F. Kennel, Electron pitch angle diffusion driven by oblique whistler-mode turbulence, *J. Plasma Phys.*, *6*, 589, 1971.

Lyons, L. R., R. M. Thorne, and C. F. Kennel, Pitch angle diffusion of radiation belt electrons within the plasmasphere, *J. Geophys. Res.*, *77*, 3455-3474, 1972.

Lyons, L. R., and D. J. Williams, A source for geomagnetic storm main phase ring current, *J. Geophys. Res.*, *85*, 523-530, 1980.

Mauk, B. H., and C.-I. Meng, Macroscopic ion acceleration associated with the formation of the ring current in the Earth's magnetosphere, *Ion Acceleration in the Magnetosphere and Ionosphere, Geophys. Monogr. Ser.*, vol. 38, ed. by T. Chang, 351-361, AGU, Washington, D. C., 1986.

Nakada, M. P., J. W. Dungey, and W. N. Hess, On the origin of outer-belt protons, 1, *J. Geophys. Res.*, *70*, 3529-3532, 1965.

Nishida, A., Formation of a plasmapause, or magnetospheric plasma knee by combined action of magnetospheric convection and plasma escape from the tail, *J. Geophys. Res.*, *71*, 5669-5679, 1966.

Peroomian, V., and M. Ashour-Abdalla, Relative contributions of the solar wind and the auroral zone to near-earth plasmas, in *Cross-Scale Coupling in Space Plasmas, Geophys. Monogr. Ser.*, Vol. 93, edited by J. L. Horwitz, 213-217, AGU, Washington, DC, 1995.

Peroomian, V., and M. Ashour-Abdalla, Population of the near-Earth magnetotail from the auroral zone, *J. Geophys. Res.*, *101*, 15,387-15,401, 1996.

Perraut, S., R. Gendrin, and A. Roux, Amplification of ion cyclotron waves for various typical radial profiles of magnetospheric parameters, *J. Atmos. Terr. Phys.*, *38*, 1191, 1967.

Reiff, P. H., and J. G. Luhmann, Solar wind control of the polar-cap voltage, in *Solar-Wind Magnetosphere Coupling*, ed. by Y. Kamide and J. A. Slavin, 453-479, 1986.

Roeder, J. L., J. F. Fennell, M. W. Chen, M. Schulz, M. Grande, and S. Livi, CRRES observations of the composition of the ring-current ion population, *Adv. Space Res.*, *17*, 17-24, 1996.

Roederer, J. G., and E. W. Hones, Jr., Motion of magnetospheric particle clouds in a time-dependent electric field model, *J. Geophys. Res.*, *79*, 1432-1438, 1974.

Schulz, M., and M. W. Chen, Bounce-averaged Hamiltonian for charged particles in an axisymmetric but nondipolar model magnetosphere, *J. Geophys. Res.*, *100*, 5627-5635, 1995.

Schulz, M., and H. C. Koons, Thermalization of colliding ion streams beyond the plasmapause, *J. Geophys. Res.*, *77*, 248-254, 1972.

Sckopke, N., A general relation between the energy of trapped particles and the disturbance field near the earth *J. Geophys. Res.*, *71*, 3125-3130, 1966.

Sheldon, R. B., and D. C. Hamilton, Ion transport and loss in the Earth's quiet ring current, 1. Data and standard model, *J. Geophys. Res.*, *98*, 13,491-13,508, 1993.

Sheldon, R. B., Ion transport and loss in the Earth's quiet ring current, 2. Diffusion and magnetosphere ionosphere coupling, *J. Geophys. Res.*, *99*, 5705-5720, 1994.

Siscoe, G. I., The virial theorem applied to magnetospheric dynamics, *J. Geophys. Res.*, *75*, 5340-5350, 1970.

Smith, P. H., N. K. Bewtra, and R. A. Hoffman, Motions of charged particles in the magnetosphere under the influence of a time-varying large scale convection electric field, in *Quantitative Modeling of Magnetospheric Processes, Geophys. Monogr. Ser.*, Vol. 21, edited by W. P. Olson, pp. 513-535, AGU, Washington, DC, 1979.

Søraas, F., and L. R. Davis, Temporal variations of the 100 keV to 1700 keV trapped protons observed on satellite Explorer 26 during first half of 1965, *GSFC Report X-612-68-328*, Greenbelt, Md., 1968.

Spiro, R. W., R. A. Wolf, and B. G. Fejer, Penetration of high-latitude electric-field effects to low latitudes during SUNDIAL 1984, *Ann. Geophys.*, *6*, 39-50, 1988.

Spjeldvik, W. N., Equilibrium structure of equatorially mirroring radiation belt protons, *J. Geophys. Res.*, *82*, 2801-2808, 1977.

Spjeldvik, W. N., and T. A. Fritz, Energetic ionized helium in the quiet time radiation belts: Theory and comparison with observation, *J. Geophys. Res.*, *83*, 654-662, 1978a.

Spjeldvik, W. N., and T. A. Fritz, Theory for charge states of energetic oxygen ions in the Earth's radiation belts, *J. Geophys. Res.*, *83*, 1583-1584, 1978b.

Stern, D. P., A study of the electric field in an open magnetospheric model, *J. Geophys. Res.*, *78*, 7292-7305, 1973.

Takahashi, S., and T. Iyemori, Three-dimensional tracing of charged particle trajectories in a realistic magnetospheric model, *J. Geophys. Res.*, *94*, 5505-5509, 1989.

Takahashi, S., and T. Iyemori, and M. Takedo, A simulation of the storm-time ring current, *Planet. Space Sci.*, *38*, 1133-1141, 1990.

Volland, H., A semiempirical model of large-scale magnetospheric electric fields, *J. Geophys. Res.*, 78, 171–180, 1973.

Williams, D. J., Phase space variations of near equatorially mirroring ring current ions, *J. Geophys. Res.*, 86, 189–194, 1981a.

Williams, D. J., Ring current composition and sources: An update, *Planet. Space Sci.*, 29, 1195–1203, 1981b.

Wolf, R. A., R. W. Spiro, and G. A. Mantjoukis, Theoretical comments on the nature of the plasmapause, *Adv. Space Res.*, 6, 177–186, 1986.

Wolf, R. A., J. W. Freeman Jr., B. A. Hausman, R. W. Spiro, R. B. Hilmer, and R. L. Lambour, Modeling convection effects in magnetic storms, Magnetic Storms, ed. by B. T. Tsurutani and Y. Kamide, in this volume, Am. Geophys. Un., Washington, D. C., 1996.

M. W. Chen, and L. R. Lyons, Space and Environment Technology Center, M2-260, The Aerospace Corporation, P. O. Box 92957, Los Angeles, CA 90009-2957. (e-mail: chen@dirac2.dnet.nasa.gov; lyons@dirac2.dnet.nasa.gov)

M. Schulz, Advanced Technology Center, O/91-20, B/252, Lockheed Martin Missiles and Space, 3251 Hanover Street, Palo Alto, CA 94304. (e-mail: schulz@agena.space.lockheed.com)

Modeling of the Contribution of Electromagnetic Ion Cyclotron (EMIC) Waves to Stormtime Ring Current Erosion

J. U. Kozyra[1], V. K. Jordanova[2], R. B. Horne[3], and R. M. Thorne[4]

The impact of losses due to wave scattering on the global ring current evolution is an important unresolved question. The present study attempts to address this question by incorporating wave scattering (specifically due to ring current proton resonance with ion cyclotron waves) into the Ring Current - Atmosphere Interaction Model (RAM). RAM is a drift-loss model that follows the evolution of three major ring current ion species (H^+, He^+, and O^+) considering adiabatic drift motions, collisional interactions with the hydrogen geocorona and with a time-dependent plasmasphere, and pitch-angle scattering of protons in the fields of ion cyclotron waves. A time-dependent global model of convective wave gain was constructed to be consistent with velocity space distributions in the RAM model and with simulations of wave propagation and amplification in the HOTRAY warm plasma ray tracing code. The plasmapause density gradient was found to have a major impact on the path-integrated gain. A simple algorithm to connect wave gain to wave amplitudes was devised, based upon observed amplitudes of Pc1 waves in the inner magnetosphere. A region of strong ion cyclotron wave activity forms just inside and along the plasmapause. The integrated energy loss from the ring current during the one-hour simulation interval, due to the scattering of protons into the loss cone, caused an additional ~8 nT recovery in the Dst index and thus was found to be important to the global energy balance of the ring current. The corresponding globally-averaged energy loss time scale for H^+ due to wave scattering is ~11 hours, somewhat longer than globally-averaged collisional loss time scales for this storm. A careful consideration of feedback of the waves on the ion distribution function is required before the full temporal history of the wave scattering losses can be estimated. The result depends sensitively on the assumed waved amplitudes.

[1]Space Physics Research Lab, University of Michigan, Ann Arbor, MI

[2]Space Science Center, University of New Hampshire, Durham, NH

[3]British Antarctic Survey, Natural Environment Research Council, Cambridge, England

[4]Department of Atmospheric Sciences, University of California, Los Angeles, CA

1. INTRODUCTION

The integrated effect on the dissipation of the storm-time ring current of losses due to the scattering of ions by plasma waves is one of the major unresolved questions in ring current dynamics. Isolated single-point observations of plasma waves and/or changes in ion pitch angle distributions attributed to plasma waves have been made on spacecraft which document the presence and impact of these wave modes on the ring current distribution in restricted local time intervals [*Perraut* 1982; *Fraser and McPherron*, 1982; *Studemann*, et al., 1987; *LaBelle et al.*, 1988]. On the other hand, observations [c.f. *Anderson et al.*, 1992] and recent theoretical models [c.f., *Thorne and Horne*, 1992] indicate that the occurrence of particular plasma wave modes may be limited in time and/or confined to localized regions (such as the plasmapause gradient) in the inner magnetosphere, bringing into question the ability of wave scattering processes to affect the global energy balance of the ring current.

One particularly promising and well studied interaction, which is the focus of the present modeling effort, is between ring current ions and ion cyclotron waves in the inner magnetosphere, first suggested by *Cornwall et al.*, [1970] as an important ring current loss process. Time scales for scattering of ions into the loss cone during amplification of ion cyclotron waves can be quite rapid [*Lyons and Thorne*, 1972]. This is attractive since studies of the ring current energy balance [c.f., *Gonzalez et al.*, 1989; *Prigancova and Feldstein*, 1992] suggest that energy loss time scales during the main phases of intense-to-great geomagnetic storms may reach values as low as 0.5-1.0 hrs, far too rapid to be the result of charge exchange or Coulomb collision processes [see *Fok et al.*, 1991 for a discussion of the relative magnitudes of charge exchange and Coulomb loss time scales]. *Feldstein et al.*, [1994] report decay times for the asymmetric component of the ring current with values of the order of an hour in the dusk to noon MLT sector. In addition, distributions unstable to the amplification of ion cyclotron waves are produced naturally in the inner magnetosphere through the betatron acceleration of ions moving along adiabatic drift paths [c.f., *Cornwall et al.*, 1970]. The enhanced charge exchange loss of ring current ions with small pitch angles [*Cornwall*, 1977] deepens the loss cone and increases the anisotropy of the drifting ion distributions making them even more unstable to the generation of plasma waves.

Recently developed large-scale drift-loss models of the ring current, including major sources for the generation of anisotropy in the ion distribution functions [*Fok et al.*, 1995, 1996] provide further motivation for considering the possible influence of waves on the ring current distribution. *Fok et al.*, [1995] used a bounce-averaged drift-loss model of the ring current, which includes charge exchange and Coulomb losses, to describe the evolution of the ring current during a major magnetic storm interval in early February 1986. The calculated ion fluxes agreed reasonably well with observed values except that the model consistently overestimated proton fluxes at tens of keV. One candidate process suggested by the authors to provide the necessary additional loss is pitch-angle scattering of ions into the loss cone. Comparisons between modeled and observed pitch angle distributions during a storm period in May 1986, using an improved version of the same ring current drift-loss model, revealed a persistent disagreement at high ring current energies (>100 keV) [*Fok et al.*, 1996]. The observed pitch-angle distributions were much more rounded at moderate to small pitch angles than the ring current model predicted. The model ring current flux distributions were very flat with respect to pitch angle with the exception of a deep atmospheric loss cone at the lowest pitch angles. The inclusion of weak scattering (corresponding to wave amplitudes of ~0.1nT) throughout the inner magnetosphere was sufficient to bring the model results into agreement with observations by diffusing particles from intermediate pitch angles into the atmospheric loss cone.

The central problem in modeling wave-particle interactions on a global scale is the need to represent wave growth and particle diffusion self-consistently and to relate wave convective gains to resultant wave amplitudes. The source of the wave amplification is the free-energy in the ring current ion distribution. In addition, wave growth rates depend sensitively on the composition of the thermal and energetic plasma [c.f., *Kozyra et al.*, 1984] and on gradients in the thermal plasma which effect the wave normal angle and wave propagation paths [*Thorne and Horne*, 1992]. As a result, waves are amplified over limited spatial regions that vary in location and spatial extent as the ring current evolves during the magnetic storm interval. The wave-induced diffusion, which also varies with the plasma composition [c.f., *Jordanova et al.*, 1996b], drives the ring current ions toward marginal stability, limiting the time interval during which the waves exist. Injections of fresh particles from the near-Earth plasma sheet into the inner magnetosphere provide additional free energy to sustain wave instability. The wave amplitude, that results from a given wave gain, depends on the background fluctuation level that is being amplified. This background level may be a result of thermal fluctuations or embryonic wave populations. The instability is assumed to be limited by convection of the waves out of the growth region.

An initial attempt to model the global impact of ion cyclotron waves on the ring current was made by *Jordanova et al.*, [1996c] using a ring current drift-loss model (RAM) which follows the evolution of three major ring current ion species (H^+, He^+, and O^+) considering adiabatic drift motions, collisional interactions with the hydrogen geocorona and with a time-dependent plasmasphere, and pitch-angle scattering of protons in the fields of ion cyclotron waves, during a moderate magnetic storm interval. The authors used a simplified model to construct a global pattern of wave convective gain. Equatorial field-aligned growth rates [*Kozyra et al.*, 1984] were calculated throughout the equatorial plane based on the thermal and energetic ion distributions from the RAM simulation. These convective growth rates were integrated along ray paths assumed to be field-aligned and to extend between ±10° magnetic latitude. Wave gain exceeding 1 e-folding was taken to define spatial regions of instability. Within these regions, wave amplitudes of 1 nT were adopted based on observed amplitudes at frequencies between the oxygen and helium gyrofrequencies in the inner magnetosphere [*Anderson et al.*, 1992]. For this initial study, O^+ ring current distributions were used in the calculation of the convective growth rates but were assumed to be unaffected by the waves during the simulation. Only H^+ was allowed to diffuse in the wave fields. Changes in the H^+ distribution were calculated using quasi-linear theory and considering only pitch angle diffusion which dominates energy diffusion in this interaction. Quasi-linear diffusion coefficients, adopted for the simulation, were derived to take into account the multi-component thermal ion plasma environment that exists in the inner magnetosphere [*Jordanova et al.*, 1996b]. The changing H^+ distribution was used in the calculation of wave growth rates and thus provided some self-consistency in the calculation of unstable wave regions. The model produced order-of-magnitude enhancements in the ion precipitation as a result of diffusion in the ion cyclotron wave fields within the unstable regions, which were limited initially to the post-noon sector around L~5. The precipitation zones moved to lower L value, the maximally-unstable frequency of the waves in the equatorial plane increased with time, and the zones disappeared as the recovery phase progressed. For this case, which represented a relatively short-lived moderate magnetic storm condition, no significant impact of the wave losses was seen in the global energy balance though the waves reduced the anisotropy in the proton pitch angle distributions locally. This is consistent with the much longer time scales for ring current loss during modest magnetic activity levels that can generally be explained by collisional loss processes without the necessity of invoking wave scattering as an energetically important global process.

The present study continues and improves upon this earlier work by considering wave activity during a major magnetic storm (minimum Dst ~-120 nT) which occurred during November 2-6, 1993. Attention was restricted to characteristic wave frequencies covered in a statistical study of the inner magnetosphere by *Anderson et al.*, [1992]. In this statistical study, waves in the dusk sector were mainly observed in the frequency range between the oxygen and helium gyrofrequency. The treatment of oxygen cyclotron wave growth is deferred to later studies. A description of this magnetic storm interval and the associated ring current simulation is presented in Section 2. An improved scheme for estimating the global distribution and amplitude of ion cyclotron wave activity, which uses a warm plasma ray tracing program (HOTRAY) [*Horne*, 1989; *Horne and Thorne*, 1993] to identify the most unstable waves and determine the path integrated gain, is described in Section 3. Section 4 contains the results of the ring current simulation including wave scattering. The impact of the scattering losses on the global energy balance of the ring current is discussed in Section 5, followed by summary and conclusions in Section 6.

2. THE NOVEMBER 2-6, 1993 STORM AND RING CURRENT SIMULATION

The present study uses the results of a recently completed simulation of the November 2-6, 1993 magnetic storm period presented in *Kozyra et al.*, [1995]. The November 2-10, 1993 geomagnetic storm interval was selected for intensive study by NSF's Global Environmental Modeling (GEM) program. As a result, a large body of observational information has been accumulated over this interval related to solar wind conditions, energetic plasma properties at geosynchronous orbit, ion precipitation, plasmaspheric structure and related quantities that provide valuable input to models as well as important observational "ground-truth". The availability of these data sets was a primary motivation for selected this particular storm interval.

The stormtime buildup of the terrestrial ring current during this interval was simulated using a version of the RAM model recently modified to include pitch angle scattering processes. This model is described in *Jordanova et al.*, [1996 a, c]. In general, RAM is a bounce-averaged drift-loss model of the ring current ions which includes adiabatic drifts, charge-exchange and Coulomb drag losses, and pitch angle scattering due to Coulomb collisions and interactions with ion cyclotron waves. RAM incorporates a time-dependent plasmasphere model [*Rasmussen et al.*, 1993], a hydrogen geocoronal model from *Rairden et al.*, [1986], a Volland-Stern convection potential [*Volland*, 1973; *Stern*, 1975], and a dipolar magnetic field configuration throughout the magnetospheric region $2 \leq L \leq 6.5$. The plasmasphere model is a total density model and therefore further assumptions are required to specify the composition. A typical plasmaspheric composition of 77% H^+, 20% He^+ and 3% O^+ is adopted.

The RAM simulation begins in the quiet period preceding the storm (November 2 at 0400 UT). The statistical quiet time proton ring current model of *Sheldon and Hamilton* [1993] was adopted as an initial condition on the proton ring current fluxes. Initial temperature anisotropy of these quiet-time distributions was estimated from *Garcia and Spjeldvik* [1985]. Since the quiet time ring current consists predominantly of protons, the initial O^+ composition was assumed to be zero. Time-dependent nightside outer boundary conditions on the ion flux, temperature and temperature anisotropy were supplied by the MPA and SOPA instruments on the Los Alamos geosynchronous satellites 046 and 095 during the entire storm interval. Details of the MPA instruments are given in *McComas et al.*, [1993] and for the SOPA instrument in *Belian et al.*, [1992]. Since compositional information is not available from these satellites, the ratio of O^+ to H^+ in the number density of ions at geosynchronous orbit was specified through its dependence on Kp and F10.7 using the statistical model of *Young et al.*, [1982]. This results in a ring current composed of ~30% O^+ within the L value range 2 to 6.5 near the time of minimum Dst.

RAM was run for 48 hrs with this quiet time distribution, fixed outer boundary conditions specified by geosynchronous plasma observations on November 2, and constant Kp of 2, to allow open drift paths to become populated in the inner magnetosphere. By the end of the 48 hrs, the model had achieved a quasi-steady state. The storm simulation was started using this quasi-steady state as the initial quiet-time condition of the inner magnetosphere and model elapsed time (referred to as t_m) was reset to zero. The start time of the model corresponds to 0400 UT on November 2 (the quiet day preceding storm commencement). Geosynchronous observations on the nightside were used to drive the model; while comparisons to dayside geosynchronous measurements provided a "ground-truth" to check model results. Reasonable agreement between predicted and observed dayside fluxes was achieved throughout the four days of the simulation. Of importance for the ring current simulation, a superdense plasmasheet was observed by the geosynchronous satellites beginning at ~ 1800 UT on November 3 and persisting for about 12 hours. This case and others, along with possible source mechanisms, are discussed in *Borovsky et al.*, [1996]. Since the near-Earth plasma sheet is a major source of plasma for the storm-time ring current, the effects on the ring current formation and evolution were watched with great interest during the simulation.

The November 2-6, 1993 magnetic storm interval began with a sudden commencement at 1720 UT on November 2. A comparison between the modeled and observed ring current contribution to the Dst is reproduced in Figure 1. The dashed line is the observed Dst index. The Dst was corrected to remove the contribution of the conducting Earth (dotted line) using estimates provided by *Dessler and Parker* [1959]. Observations of the solar wind density and speed by IMP-8 (provided courtesy of the *MIT Space Plasma Physics Group*) prior to November 4 0100 UT and by GEOTAIL in the magnetosheath after this time *(provided courtesy of T. Mukai, M. Nakamura, and S. Kokubun,* 1996) were used to estimate the contribution of magnetopause currents (open circle) given by

$$B_{mp}(nT) = 0.02 v_{sw}(km/s)\sqrt{n_{sw}(cm^{-3})}$$

where B_{mp} is the contribution of the magnetopause currents to the Dst index, n_{sw} is the density and v_{sw} is the speed of

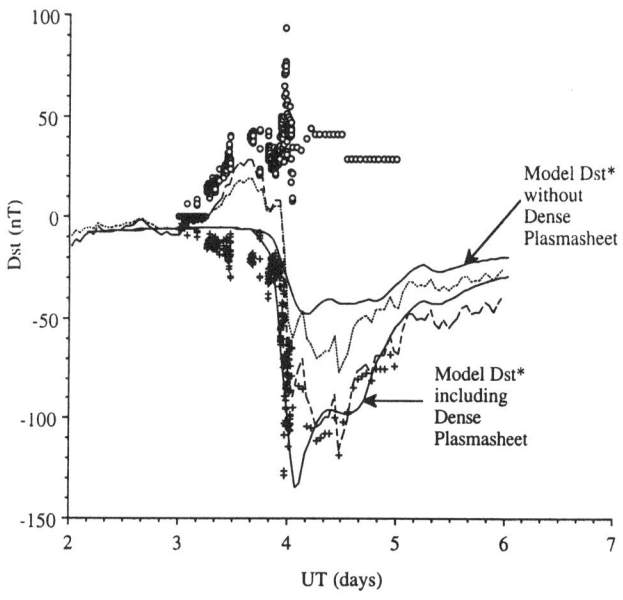

Fig. 1. Comparison of observed and modeled Dst* (ring current contribution to the Dst index). Shown are the observed Dst (dashed line), the observed Dst corrected for the effects of a conducting Earth (dotted line) and the Dst corrected for both the conducting Earth and the presence of magnetopause currents, called Dst*(+). The contribution of magnetopause currents (o) was estimated using IMP 8 observations prior to November 4 0100 UT [*courtesy of the MIT Plasma Physics Group*] and GEOTAIL observations after this time *(provided courtesy of T. Mukai, M. Nakamura, and S. Kokubun,* 1996)]. The modeled Dst* is indicated with and without the presence of the observed dense plasmasheet.

the solar wind. The contribution of the ring current to the Dst index (indicated by + in the figure), called Dst*, was obtained by removing the contributions from both the conducting earth and the magnetopause currents. The solid lines are the model predictions of Dst* for two different outer boundary conditions. During one simulation the density of the inner plasmasheet was held fixed at its prestorm value to observe the energization due solely to the enhanced cross-tail electric field. In the second simulation, the density was allowed to vary as indicated by the observations at geosynchronous orbit. This is a dramatic illustration of the importance of the superdense plasma sheet, observed during the storm main phase, in producing the ring current. The need for enhanced plasma sheet densities to produce observed ring current intensities was discussed recently by *Chen et al.*, [1994]. The remarkable agreement between the predicted and observed Dst* indicates that the ring current distributions are reasonably well represented in the RAM model for this storm interval. These distributions along with the plasmaspheric populations are key elements in constructing a global ion cyclotron wave model. Selected results from the RAM simulation will be presented in later sections as needed to explain wave-scattering results. The role of the superdense plasma sheet in promoting ion cyclotron wave growth is an interesting question but one which cannot be addressed by the current preliminary study. The wave growth rates depend sensitively on the magnitude of the ring current distribution function as well as its shape in velocity space. How these attributes vary between ring currents generated in the presence and absence of a superdense plasmasheet is not clear at present. The answer awaits further studies comparing growth rates during a variety of storm conditions.

3. IMPROVED GLOBAL ION CYCLOTRON WAVE MODEL

Recent observational studies of ion cyclotron waves in the inner magnetosphere indicate that these waves exist over limited spatial regions that vary with time [*Anderson et al.*, 1992]. Studies with Viking [*Erlandson et al.*, 1990] indicate the plasmapause as a preferred location for wave excitation. *Thorne and Horne* [1992] examined the propagation and amplification of ion cyclotron waves in the plasmasphere. Using the HOTRAY warm plasma ray-tracing code, they found that the plasmapause density gradient plays an important role in the net path-integrated gain of the waves. It allows the waves to maintain small wave normal angles over a larger portion of their ray paths. In some cases the waves were able to make multiple passes through the equatorial growth region with small wave normal angles. In such cases, the wave gain can be enhanced significantly.

The focus of the present model is on ion cyclotron waves with frequencies between the oxygen and helium gyrofrequencies. For purposes of constructing a time-dependent model of the ion cyclotron wave convective gain, several key elements must be specified. The energetic ring current ions are the source of the free-energy used in the wave amplification. Bi-Maxwellian fits to the output from the RAM model are used to determine the density, temperature and temperature anisotropy of two major ring current ions, H^+ and O^+ and the evolution of these quantities from the quiet period preceding the storm (starting 0400 UT on November 2, 1993) into the late recovery phase of the storm (0000 UT on November 6, 1993). The thermal plasma density influences the ion resonant energies and, related to this, the most unstable wave frequency. In addition, the large-scale structure of the thermal plasma is fundamental in determining the ray path and path integrated gain of the amplified waves. The time-dependent plasmaspheric density and structure is calculated in RAM throughout the storm interval using the plasmasphere model of *Rasmussen et al.*, [1993]. A first approximation to the global pattern of wave convective gain is constructed using thermal and energetic plasma populations from a version of RAM which includes only collisional losses (charge exchange, Coulomb drag and Coulomb scattering). Global patterns of wave gain are obtained by (1) calculating field-aligned convective growth rates according to [*Kozyra et al.*, 1984] and (2) integrating these along field-aligned ray paths which extend to ±5° magnetic latitude.

Having identified regions of enhanced wave gain by this method, the HOTRAY code is employed within these regions to identify the most unstable waves and to determine the path integrated gain at selected locations. The aim is to incorporate the HOTRAY results, including modification of the wave gain in the vicinity of the plasmapause, into the global wave model, thereby significantly improving the representation of the waves. In fact, the reduction in the length of the wave growth region from the ±10° magnetic latitude used in the *Jordanova et al.*, [1996c] study to the ± 5° magnetic latitude in the present model is a result of preliminary HOTRAY studies using the RAM ring current and thermal plasma from the November 1993 storm simulation.

Figure 2a is a radial cut through the equatorial plane in the inner magnetosphere from the RAM simulation at 1700 MLT, 0400 UT on November 4 (t_m = 48 hrs). The open circles are values of the thermal electron density from

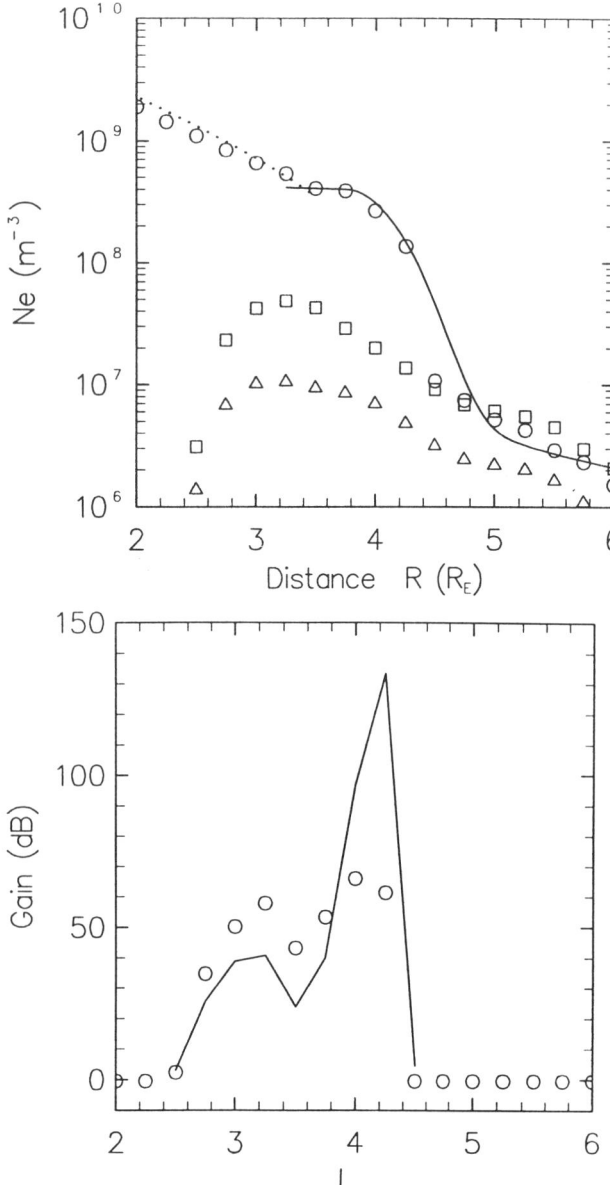

Fig. 2. (a) Radial cut through the equatorial plane in the inner magnetosphere from the RAM simulation at 0400 UT on November 4 (t_m = 48 hrs), 1700 MLT. The open circles are values of the thermal electron density from RAM. The solid (dotted) line is the fit to the thermal electron density (Ne) radial profile for $L \geq 3.5$ ($L \leq 3.25$) that was used in the HOTRAY model. The ring current proton (oxygen ion) density from RAM is given by the open squares (triangles). (b) Resultant wave gain (open circles) from the simple convective amplification model used in RAM, which assumes a ±5° magnetic latitude path length and field-aligned convective growth rates. The solid curve is the wave gain calculated by the HOTRAY code.

RAM. The solid (dotted) line is the fit to the thermal electron density (Ne) radial profile for $L \geq 3.5$ ($L \leq 3.25$) that was used in the HOTRAY model. The ring current proton (oxygen ion) density from RAM is given by the open squares (triangles). Figure 2b shows the resultant wave gain (open circles) from the simple convective amplification model used in RAM, which assumes a ±5° magnetic latitude path length and field-aligned convective growth rates. The solid curve is the wave gain calculated by the HOTRAY code. At each L shell, the ray with the maximum path-integrated gain was selected after performing a series of runs with HOTRAY at different frequencies spanning the unstable frequency band. The effects of the plasmapause density gradient on the wave gain are immediately obvious. The wave gain at this location is increased from ~70 dB to over 130 dB.

Comparisons between HOTRAY results and the simple convective amplification model were made throughout the unstable region. Based on these comparisons, the following algorithm was devised to take into account the important effects of the plasmapause gradient on the path-integrated wave gain. At locations well inside the plasmasphere, propagation in the heterogeneous medium leads to a rapid increase in the wave normal angle. Wave amplification is therefore confined to a limited region close to the equator. An acceptable simulation of the net path-integrated gain can be obtained using the field-aligned equatorial convective growth rate over a path with end points at ±5° magnetic latitude centered on the equator. Near the steep density gradient associated with the plasmapause, the wave normal direction is guided and remains roughly field-aligned over an extended region (±10° magnetic latitude). This leads to a substantial enhancement in path-integrated gain. This enhancement is applied in the global simulation of gain whenever the thermal density drops by a factor of 10 or more over a distance of 0.25 R_E. This criterion allows the plasmapause region to be selected in an automated fashion during the storm simulation. Smaller enhancements which result from a ray path extending between ±7.5° magnetic latitude are applied over the regions 0.25 R_E wide on either side of the plasmapause gradient. An example of the plasmaspheric electron density along with the location of the plasmapause gradient, as specified by this algorithm, are shown in Figure 3 for November 4 at 0400 UT. The wave gain calculated by RAM was multiplied by a factor of 2 at the plasmapause, and by a factor of 1.5 at 0.25 R_E distance on both sides of the plasmapause to obtain agreement with HOTRAY results. A comparison between the wave gain calculated by RAM using the simple convective amplification model, and the gain modified to take into account the guiding of the waves along the

Model output on November 4, 0400 UT

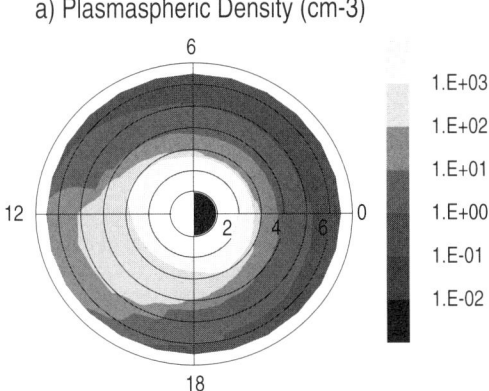
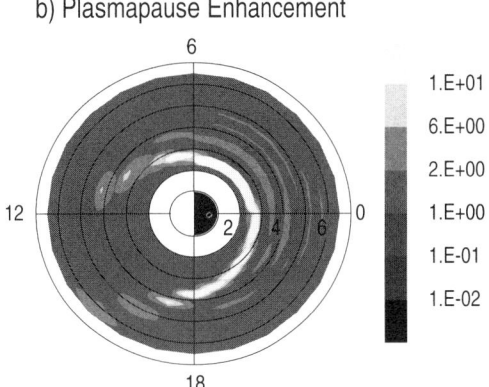
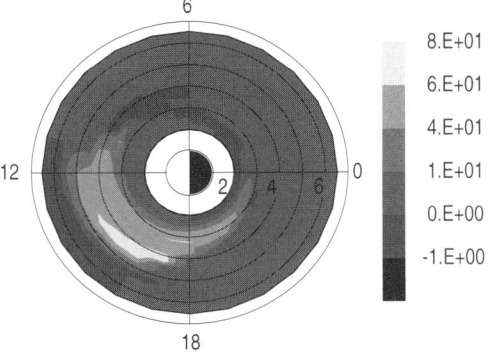
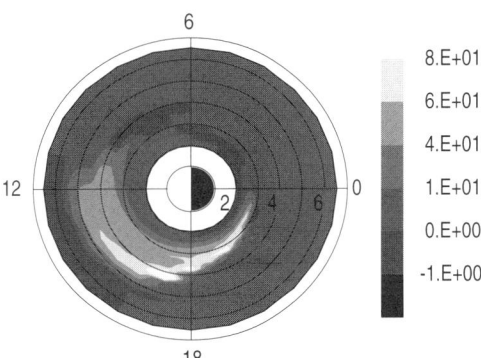

Fig. 3. The plasmaspheric electron density (upper left panel) for November 4 at 0400 UT is shown along with the location of the plasmapause as identified by the automated procedure described in the text (upper right panel). A comparison between the wave gain calculated by using the simple convective amplification model (lower left panel), and the gain modified to take into account the guiding of the waves along the plasmapause gradient (lower right panel) is given.

plasmapause gradient is shown in Figure 3. The enhancement near the plasmapause in the postnoon sector is clearly seen.

Figure 4 presents an overview of the initial calculation of unstable regions using output from the RAM simulation. A region of instability develops shortly after the sudden commencement at 17:20 UT (elapsed model time, $t_m = 37$ hrs 20 mins). At this time the unstable region extends in local time from just before midnight, around the duskside, to just past noon. The time of minimum Dst* is ~0200 UT on November 4 ($t_m = 46$ hrs). Continuing injections occur until ~0000 UT on November 5 ($t_m \sim 68$ hrs). After this time the ring current begins to relax toward prestorm conditions. Just following the time of minimum

194 ION CYCLOTRON WAVE DIFFUSION AND RING CURRENT EROSION

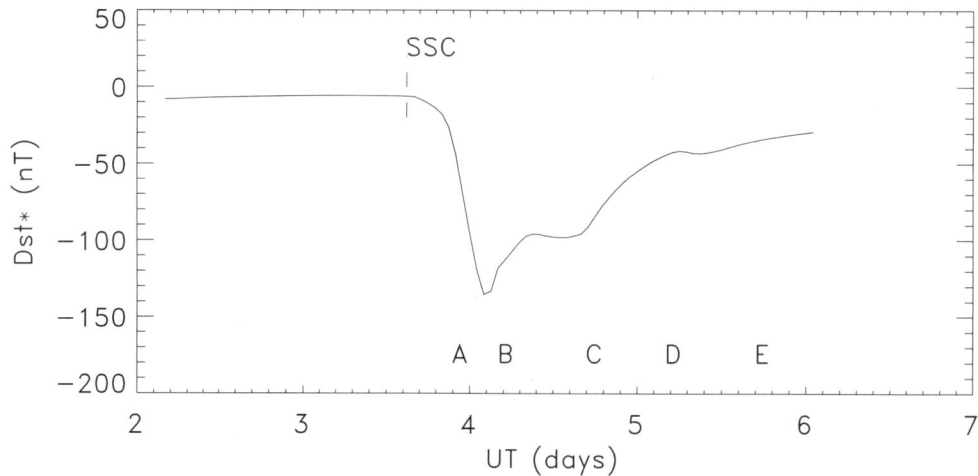

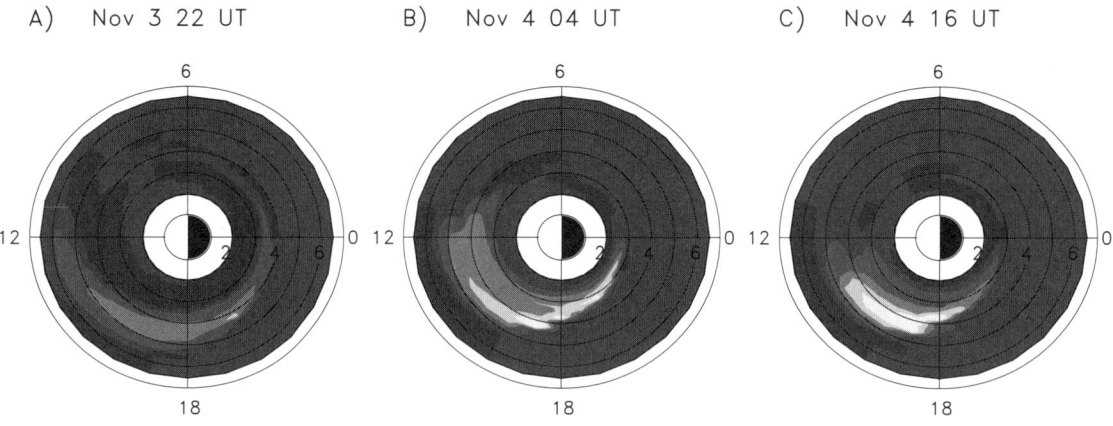

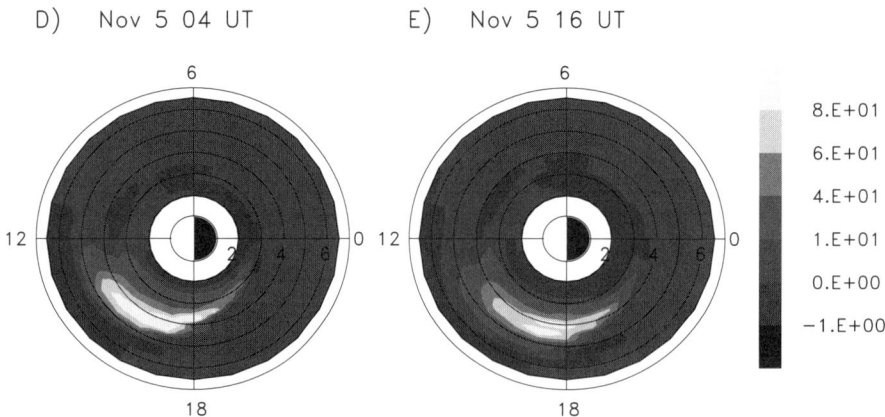

Fig. 4. Dial plots of the wave gain calculated using the RAM outputs and the algorithm for enhancing the gain at the plasmapause gradient derived from HOTRAY results.

Dst*, the unstable region extends to lower L value and shrinks in local time toward the post-noon local time sector. The strongest region of wave growth is collocated with the extension of the plasmasphere to higher L values in this local time sector.

The questions arises as to what characteristics of the thermal and energetic plasma populations combine to produce significant wave gain in the post noon and dusk sectors confined to a restricted interval in the radial direction. Figure 5 shows the ring current H$^+$ density, parallel and perpendicular temperature, temperature anisotropy and the average energy of the H$^+$ normalized by $B^2/8\pi N_e$ at 0400 UT on November 4. The thermal plasma density is also shown for the same time. The normalized proton energy reaches a maximum in a region collocated in local time with the wave gain but this maximum in the normalized proton energy spans a much wider interval in the radial direction. In the same local time interval, the energetic proton density drops off dramatically toward larger and smaller radial distances. The rapid decrease in the proton density with radial distance confines the region of significant wave gain to L values less than ~5. The plasmapause density gradient, evident in the thermal density plot, amplifies the wave gain along the duskward boundary of the unstable region.

The wave gain cannot be converted directly into absolute wave power because information on the background embryonic noise level (which is amplified to produce the unstable waves) is unavailable. Observations indicate that wide band wave amplitudes at frequencies between the helium and oxygen gyrofrequency during storms reach values as high as 10 nT but more commonly 1-3 nT in the inner magnetosphere in the vicinity of the plasmapause (Anderson, private communication, 1996). The following wave model is adopted for the simulation based on these observations.

Gain, g (dB)	Wave Amplitude (nT)
> 70	$B_{sat} = 10$
30 – 70	$B_{sat} \times 10^{\{(g-70)/20\}}$
< 30	< 0.1 nT (neglect in model)

These estimates do not take into account any focusing (defocusing) effects which would increase (decrease) the wave power. However, since these waves are guided by the magnetic field within ten degrees or so of the equator, we would expect this effect to be small. Using this model to associate wave gain with absolute wave amplitude, the wave gains in Figure 4 are converted to wave amplitudes and displayed in Figure 6. In the upper portion of the figure, the selected times for the wave amplitude dial plots are indicated with respect to the Dst* development during the storm. For the simple wave amplitude model employed here, wave gains of >70 dB throughout the interval imply the continuous presence of ion cyclotron waves with amplitudes of 10 nT in a region collocated with the duskside plasmapause.

In the present study, the feedback on the location and duration of the unstable wave regions, that results from wave-induced diffusive changes in the proton distribution, is not taken into consideration. The investigation is limited to the first hour following the initiation of wave activity in the model. This time interval was selected to be of the same order of magnitude as the diffusion time scales shown in Figure 7 and thus short enough to allow neglect of feedback effects while still providing a first look at the importance of wave-induced scattering to ring current erosion. To most clearly see the possible impact of waves on the ring current evolution, the unstable region that appears on November 4 at 0400 UT is selected for study. Prior to this time, the presence of waves is ignored in RAM. Following this time, the waves are allowed to interact with the proton distribution for one hour and the impacts are assessed.

Throughout the simulation, the ring current O$^+$ distribution is assumed to be unchanged by the waves for this initial study. This restriction will be removed in future versions of the RAM. Since the O$^+$ distribution has an important impact on the proton cyclotron convective growth rates, it is taken into account in deriving the instability regions. The potential implications of this global wave model for ring current evolution are examined in the next section.

4. RING CURRENT SIMULATION INCLUDING WAVE SCATTERING

Quasi-linear theory is employed to solve for the effects of the assumed global wave distribution on the ring current proton population. The impacts on the other ring current ion species are left for future improvements of the model. Wave particle interactions are incorporated into RAM through the solution of a bounce-averaged pitch angle diffusion equation. The details of this equation and its use in RAM are given in *Jordanova et al.*, [1996c].

The physics of the interaction between the ring current protons and the ion cyclotron waves as well as the wave characteristics are incorporated into the diffusion equation through the use of bounce-averaged diffusion coefficients. Diffusion coefficients that take into account the important

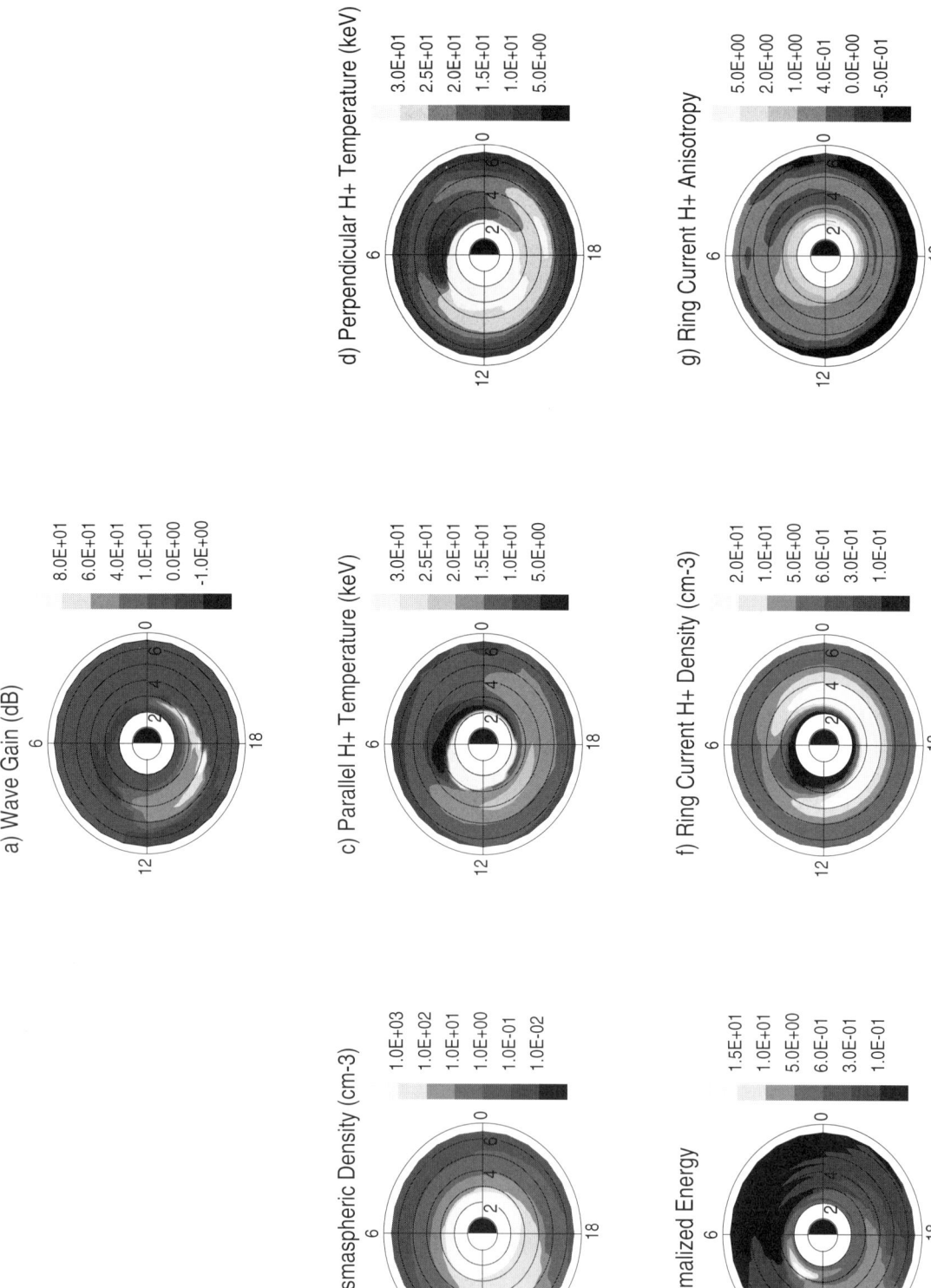

Fig. 5. Dial plots of wave gain at 0400 UT on November 4 along with associated values of ring current H+ parallel and perpendicular temperature, anisotropy, density and the parallel energy of the H+ normalized to $B^2/(8\pi N_e)$. Also shown is the thermal plasmaspheric density.

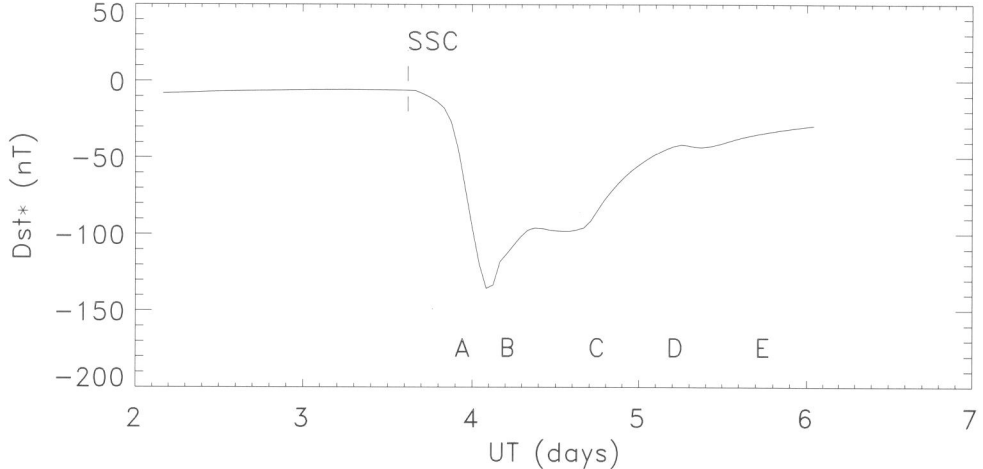
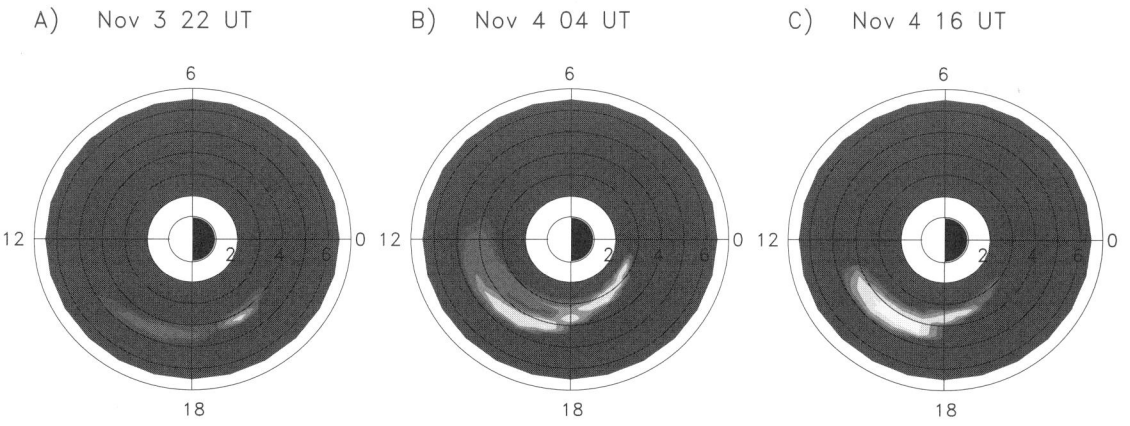
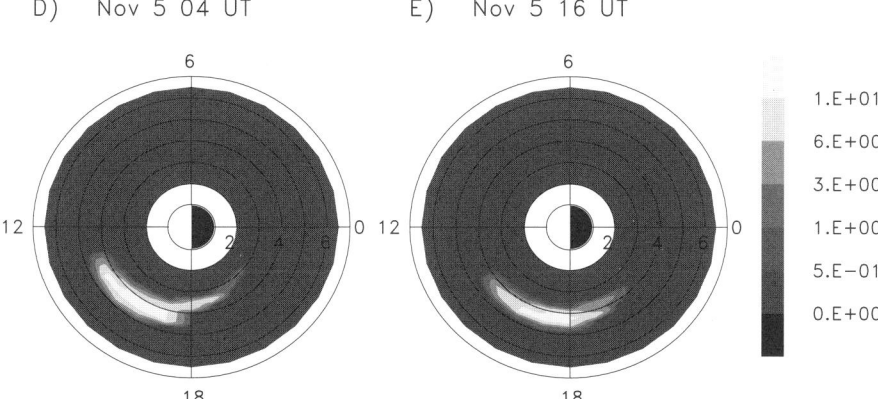

Fig. 6. Wave amplitudes associated with the wave gains given in Figure 4. See text for a description of the algorithm relating these two quantities.

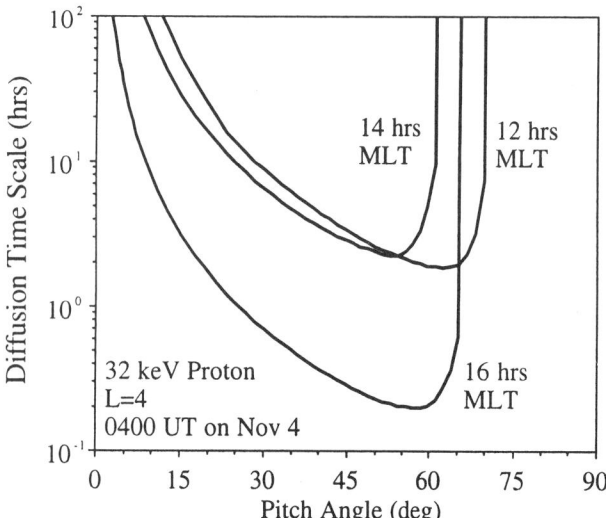

Fig. 7. Characteristic diffusion time scales (defined as the inverse of the diffusion coefficients), taken from RAM at 0400 UT on November 4 (t_m= 48 hrs) at L=4, and MLT = 12 hrs, 14 hrs and 16 hrs.

heavy ion components of the plasmaspheric plasma were derived by *Jordanova et al.*, [1996b]. An example of characteristic diffusion time scales (defined as the inverse of the diffusion coefficients), taken from RAM at 0400 UT on November 4 (t_m= 48 hrs) for 32 keV protons at L=4, and selected magnetic local times, are presented in Figure 7. The diffusion time scales are sensitive to the assumed wave amplitudes which are given in Figure 6 for this time interval. Time scales for diffusion in pitch angle are typically of the order of hours but can reach values less than a half hour in restricted local time intervals. These times are still long compared to the bounce period of a 32 keV proton.

Proton pitch angle distributions at L=4 and 16 hrs MLT at 0400 UT and 0500 UT on November 4 (t_m=48 and 49 hrs) are given in Figure 8 for 20 keV, 30 keV and 40 keV proton energies with and without the inclusion of wave-induced diffusion. The loss cone boundary at this location is indicated by a vertical dashed line. The proton pitch angle distributions, that result after diffusion in the ion cyclotron wave fields, are no longer well represented by a Bi-Maxwellian. A sharp discontinuity in the distribution

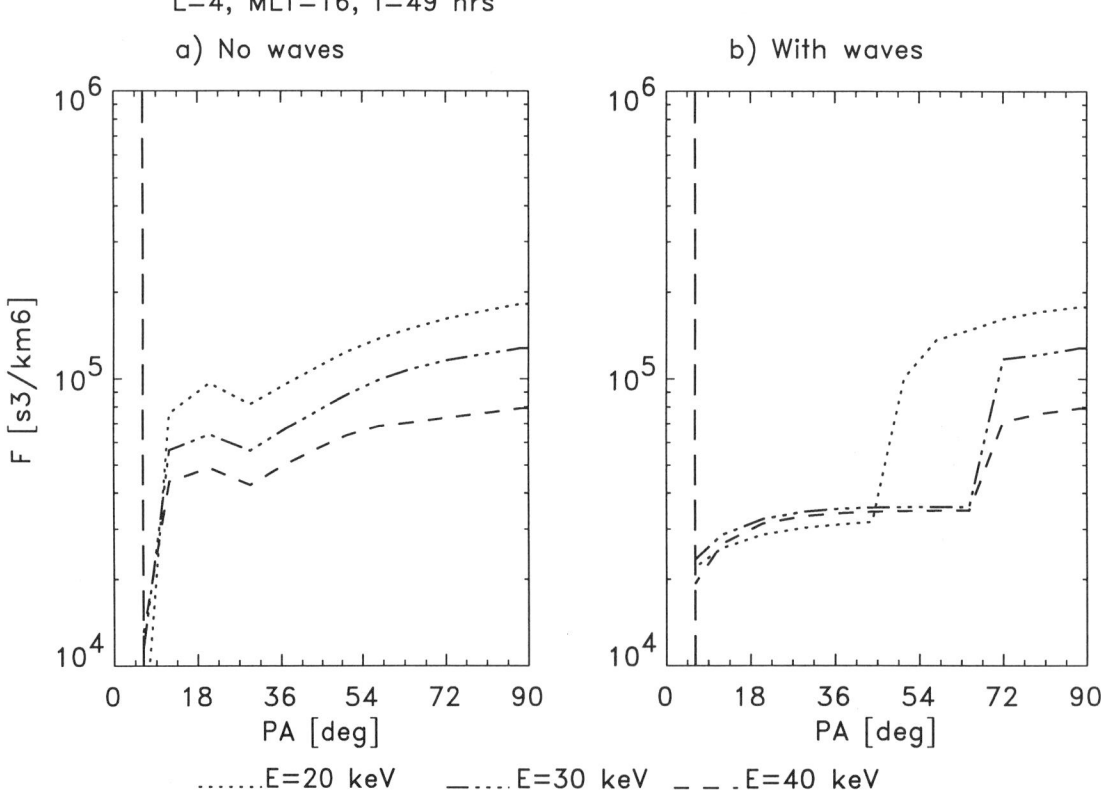

Fig. 8. Proton pitch angle distributions at L=4 and 16 hrs MLT at 0500 UT on November 4 (after 1 hour of wave-induced diffusion) are given for 20 keV, 30 keV and 40 keV proton energies. The corresponding proton distributions in the absence of waves are also shown. The loss cone boundary at this location is indicated by a vertical dashed line

functions of the protons develops at intermediate pitch angles. At pitch angles above this discontinuity, the proton's parallel energy falls below that required for cyclotron resonance with the specified wave distribution. The discontinuity in the distribution shifts to higher pitch angles as the proton energy increases. For the highest energy protons, the parallel energy, even though it decreases with increasing pitch angle, still remains in the range necessary for resonance at the largest pitch angles. These types of proton distributions were observed in the ring current by *Williams and Lyons*, [1974 a, b]. At small pitch angles the distributions tend to become more isotropic.

The pitch angle scattering of ions into the atmospheric loss cone results in significantly enhanced ion precipitation over a localized region in the post noon sector during the storm. Figure 9 gives dial plots of the global pattern of total precipitating ion flux at 200 km at 0500 UT on November 4 (t_m = 49 hrs) with and without wave scattering. To construct this plot, equatorial ion fluxes in the loss cone are averaged over pitch angle and integrated over selected energy ranges (0.15 - 1 keV, 1 - 40 keV and 40-325 keV). The characteristic zones of ion precipitation that appear in the absence of wave scattering were discussed in detail in *Jordanova et al.*, [1996a]. These are due predominantly to the convection of ions earthward on open drift paths. The pitch angle of the ions increases due to the conservation of the first adiabatic invariant as the ions move earthward. For small pitch angle ions, the increase in their pitch angle does not occur fast enough to avoid being transported into the widening atmospheric loss cone and they precipitate. The zones move to higher L values and the precipitating flux decreases as the convection slows and the boundary between open and closed drift paths moves to higher L values during later stages of the storm.

When pitch angle scattering due to interactions with the specified global ion cyclotron wave distribution are in-

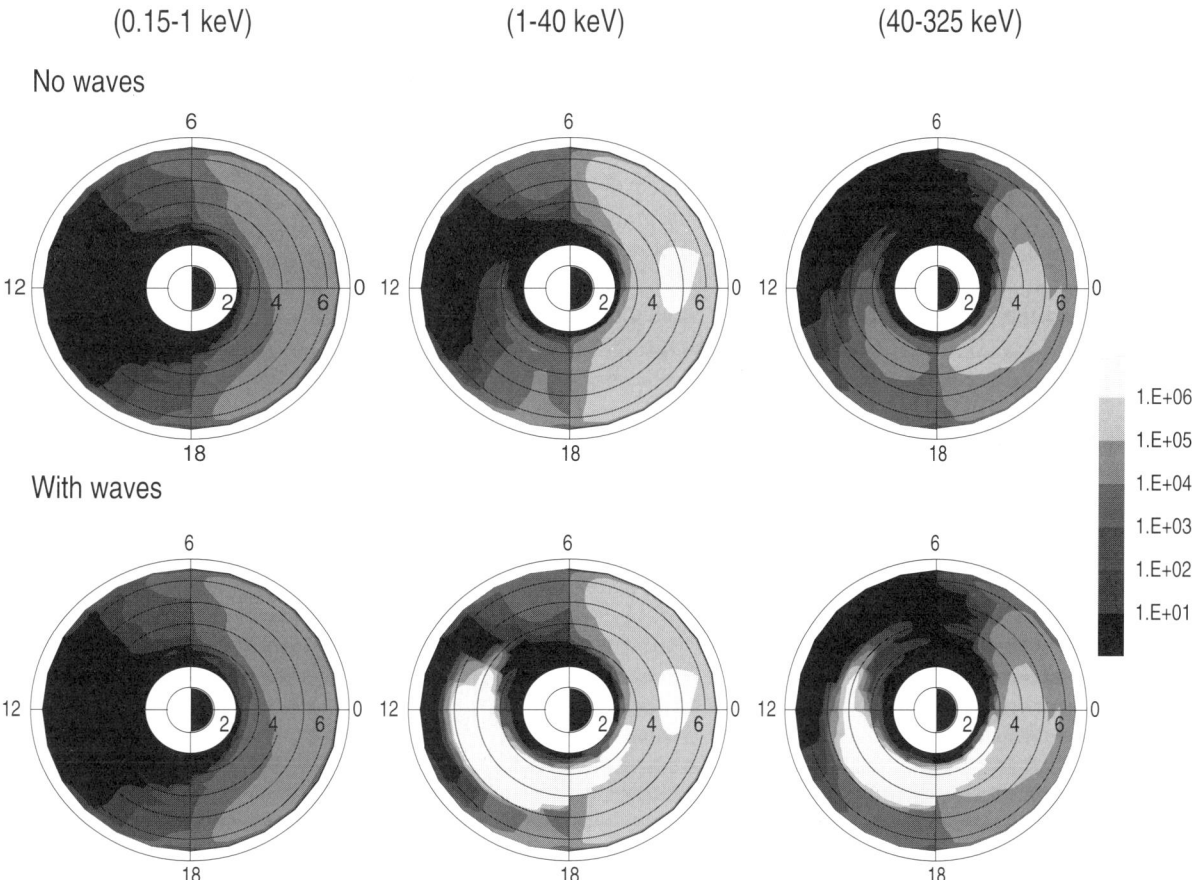

Fig. 9. Dial plots of the global pattern of total precipitating ion flux at 200 km are given at 0500 UT on November 4 (t_m=49 hrs) with and without wave scattering. Equatorial ion fluxes in the loss cone are averaged over pitch angle and integrated over selected energy ranges (0.15 - 1 keV, 1 - 40 keV and 40-325 keV).

cluded in RAM, an enhanced region of proton precipitation develops that extends from the post-noon sector into the dusk sector in the vicinity of the plasmapause. Proton fluxes in this region exceed 10^8 cm^{-2} s^{-1}, two orders of magnitude larger than maximum nightside precipitating fluxes in the absence of the waves. As seen previously, protons at the lowest energies are not in cyclotron resonance with the waves. Precipitation at these energies is unchanged in the presence of the waves.

5. IMPACT OF WAVE SCATTERING ON THE GLOBAL ENERGY BALANCE

The question remains of the importance of wave scattering in this case to the global ring current decay. One way to address this issue is to look for the impact of the waves on the predicted Dst* which is a function of the ring current total energy [c.f., *Dessler and Parker*, 1959]. Figure 10 is a plot of the Dst* from the two simulations, one with and one without wave scattering, all else being identical. The effects of the wave-induced losses can be clearly seen in the figure as an additional 8 nT recovery in the Dst* during the one-hour interval of wave scattering. This corresponds to a loss timescale for the H$^+$ component of the ring current due to wave scattering of roughly 11 hrs which is of the same order as collisional loss timescales but far in excess of the 0.5 to 1.0 hrs loss timescales estimated to occur by global energy balance models during the main phase of major magnetic storms. Charge exchange lifetimes for O$^+$ and H$^+$ at this time are ~20.8 hrs and 14.5 hrs, respectively and the time scale for their combined charge exchange loss is 15.5 hrs. The H$^+$ loss lifetime for combined H$^+$ charge exchange and wave scattering losses is

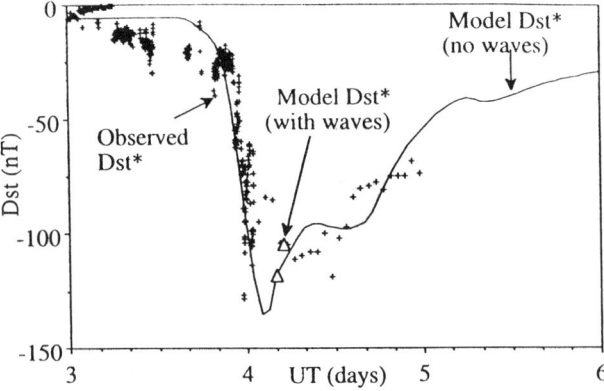

Fig. 10. Plot of the Dst* that results from the RAM simulation in the absence of wave scattering (-) and the observed Dst* (+) are reproduced from Figure 1. Plotted for comparison is the Dst* that results before and after one hour of wave-induced scattering (open triangles).

~6.4 hrs. Global loss lifetime, combining H$^+$ and O$^+$ and considering all loss processes, are of the order of 8-9 hrs during the one hour simulation period. The magnitude of the wave scattering contribution to the ring current losses, of course, depends on the assumed wave amplitudes which are a major uncertainty in the present work. A planned comparison between predicted and observed ion precipitation levels during this storm period will allow a refinement in the assumed wave amplitude model and a more realistic simulation of the November 1993 storm period.

6. SUMMARY AND CONCLUSIONS

The present work is an attempt to model the possible effects of wave scattering (specifically due to electromagnetic ion cyclotron waves) on the global evolution of the ring current during a well-studied magnetic storm period, November 2-6, 1993. The role of waves in the ring current decay is an important, unresolved issue that requires a close interplay between global models and observations for its resolution.

A drift-loss model of the ring current evolution (RAM), which represents adiabatic drifts, charge exchange, Coulomb drag, Coulomb scattering and proton diffusion due to wave scattering, is used to simulate the temporal development of energetic ion populations from the quiet time preceding the storm through the beginning of the late recovery phase. RAM also incorporates a plasmasphere model [*Rasmussen et al.*, 1993] enabling a time-dependent representation of the thermal plasma density and structure during the storm interval critical to calculating wave gain. A time-dependent global model of convective wave gain was constructed in two stages. RAM energetic and thermal plasma outputs were used to calculate field-aligned proton cyclotron wave convective growth rates at frequencies between the oxygen and helium gyrofrequency throughout the equatorial plane. These convective growth rates were integrating along field-aligned paths extending between ± 5° magnetic latitude to obtain wave gain. In the second stage, the HOTRAY warm plasma ray tracing model was used to simulate the wave propagation and amplification at selected locations within these predicted unstable regions. The plasmapause density gradient was found to have a major impact on the path-integrated gain. A scheme was devised to identify the plasmapause gradients in the RAM output and increase the wave gain at these locations in keeping with the HOTRAY results. In this initial modeling attempt, the unstable wave regions were not updated during the simulation to take into account modifications in their location and intensity due to changes in the energetic proton distribution (the source of the wave energy) as protons diffuse in the ion cyclotron wave fields. Because of this, the simulation of wave-induced scattering effects was limited to a one-hour time interval, reasonably short compared to diffusion time scales so that feedback could be ignored but long enough to allow the potential effects of the scattering on the ring current erosion to be evaluated. For this particular simulation, ring current O^+ was assumed to be unaffected by the presence of the waves.

A simple algorithm to connect wave gain to wave amplitudes was devised based upon observed amplitudes of Pc1 waves, at frequencies between the oxygen and helium gyrofrequency in the inner magnetosphere (*B. J. Anderson, private communication*, 1996). As a consequence of this model, the largest observed wave amplitudes, ~10 nT were assigned to regions where the wave gain exceeded 70 dB. Smallest wave amplitudes of 0.1 nT were assigned wherever wave gains reached 30 dB. For gains below 30 dB, waves were assumed to reach insignificant amplitudes and were ignored for the purposes of this simulation.

According to this model, a region of strong proton cyclotron wave activity forms just inside and along the plasmapause. During the one-hour simulation of wave-induced scattering, a strong enhancement in the proton precipitation occurs in association with the region of wave activity. The integrated energy loss from the ring current during this hour due to the scattering of protons into the loss cone during interaction with the prescribed ion cyclotron waves is important to the global energy balance of the ring current. With the quoted parameters, the wave losses cause an additional ~8 nT recovery of the Dst* in one hour, corresponding to a H^+ loss lifetime of ~ 11 hrs, which is shorter than collisional loss lifetimes during this time interval. The global loss lifetime (including collisional and wave scattering induced losses of all species) reduces to ~8-9 hrs with the inclusion of wave scattering during the one hour interval of the simulation.

Several improvements to the simulation are being pursued. These are (1) the proper inclusion of the feedback of wave-induced changes in the proton distribution on the location and magnitude of unstable wave regions, (2) the treatment of wave-induced diffusion of the O^+ distribution, and (3) a more realistic assessment of wave amplification in the non-Maxwellian distributions that evolve during the simulation. A major uncertainty in the present simulation is in the estimation of wave amplitudes from the convective wave gains. A planned comparison between predicted and observed ion precipitation during the November 2-6, 1993 storm period will provide a critical test of the assumptions that were used in constructing the global wave model.

Acknowledgments. Computational resources for the HOTRAY program were provided by the Natural Environmental Research Council. This work was supported by NASA grant NAGW 4611, NSF grants ATM 93 13158 and ATM 94 12363 and NATO grant CRG 930237.

REFERENCES

Anderson, B.J., R. E. Erlandson, and L. J. Zanetti, A statistical study of Pc1-2 magnetic pulsations in the equatorial magnetosphere, 1. Equatorial occurrence distributions, *J. Geophys. Res., 97,* 3075, 1992

Belian, R. D., G. R. Gisler, T. Cayton and R. Christensen, High Z energetic particles are geosynchronous orbit during the great solar proton event of October 1989, *J. Geophys. Res., 97,* 16892, 1992.

Borovsky, J. E., M. F. Thomsen, and D. J. McComas, The superdense plasma sheet: Plasmaspheric origin, solar-wind origin, or ionospheric origin?, *J. Geophys. Res.,* in press, 1996.

Chen, M. W., L. R. Lyons and M. Schulz, Simulations of phase space distributions of storm time proton ring current, *J. Geophys. Res., 99,* 5745, 1994.

Cornwall, J. M., F. V. Coroniti, and R. M. Thorne, Turbulent loss of ring current protons, *J. Geophys. Res., 75,* 4699, 1970.

Cornwall, J. M., On the role of charge exchange in generating unstable waves in the ring current, *J. Geophys. Res., 82,* 1188, 1977.

Dessler, A. J., and E. N. Parker, Hydromagnetic theory of geomagnetic storms, *J. Geophys. Res., 64,* 2239, 1959.

Erlandson, R. E., L. J. Zanetti, T. A. Potemra, L. P. Block, and G. Holmgren, Viking magnetic and electric field observations of Pc1 waves at high latitude, *J. Geophys. Res., 95,* 5941, 1990.

Feldstein, Y.I., A. E. Levitin, S. A. Golyshev, L. A, Dremukhina, U. B. Vestchezerova, T. E. Valchuk, and A. Grafe, Ring current and auroral electrojets in connection with inerplanetary medium parameters during magnetic storms, *Ann. Geophys., 12,* 602, 1994.

Fok, M.-C., J. U. Kozyra, A. F. Nagy, and T. E. Cravens, Lifetimes of ring current particles due to Coulomb collisions in the plasmasphere, *J. Geophys. Res., 96,* 7861, 1991.

Fok, M.-C., T. E. Moore, J. U. Kozyra, G. C. Ho and D. C. Hamilton, Three-dimensional ring current decay model, *J. Geophys. Res., 100,* 9619, 1995.

Fok, M.-C., T. E. Moore, M. E. Greenspan, Ring current development during storm main phase, *J. Geophys. Res., 101,* 15311, 1996.

Fraser, B. J., and R. L. McPherron, Pc1-2 magnetic pulsation spectra and heavy ion effects at synchronous orbit: ATS 6 results, *J. Geophys. Res., 87,* 4560, 1982.

Garcia, H. A., and W. N. Spjeldvik, Anisotropy characteristics of geomagnetically trapped ions, *J. Geophys. Res., 90,* 347, 1985

Gonzalez, W. D., B. T. Tsurutani, A. L. C. Gonzalez, E. J. Smith, F. Tang, and S.-I. Akasofu, Solar wind-magnetosphere coupling during intense magnetic storms, *J. Geophys. Res., 94,* 8835, 1989.

Horne, R. B., Path-integrated growth of electrostatic waves: The generation of terrestrial myriametric radiation, *J. Geophys. Res., 94,* 8895, 1989.

Horne, R. B. and R. M. Thorne, On the preferred source location for the convective amplification of ion cyclotron waves, *J. Geophys. Res., 98,* 9233, 1993.

Jordanova, V. K., L. M. Kistler, J. U. Kozyra, G. V. Khazanov, and A. F. Nagy, Collisional losses of ring current ions, *J. Geophys. Res., 101,* 111, 1996a.

Jordanova, V. K., J. U. Kozyra and A. F. Nagy, Effects of heavy ions on the quasi-linear diffusion coefficients from resonant interactions with EMIC waves, *J. Geophys. Res., 101,* 19771, 1996b.

Jordanova, V. K., J. U. Kozyra, A. F. Nagy, G. V. Khazanov, Kinetic model of the ring current-atmosphere interaction, in press, *J. Geophys. Res.,* 1996c.

Kozyra, J.U., T. E. Cravens, A. F. Nagy, E. G. Fontheim and R. S. B. Ong, Effects of energetic ions on electromagnetic ion cyclotron wave generation in the plasmapause region, *J. Geophys. Res., 89,* 2217, 1984.

Kozyra, J.U., V. K. Jordanova, A. F. Nagy, J. E. Borovsky, M. F. Thomsen, T. E. Cayton and D. J. McComas, Simulation of the ring current formation and loss during the November 4-6, 1993 magnetic storm period, EOS, Transactions of the American Geophysical Union, 76, F501, 1995.

LaBelle, J., R. A. Treumann, W. Baumjohann, G. Haerendel, N. Sckopke, G. Paschmann, and H. Luhr, The duskside plasmapause/ring current interface: Convection and plasma wave observations, *J. Geophys. Res., 93,* 2573, 1988.

Lyons, L.R., and R. M. Thorne, Parasitic pitch angle diffusion of radiation belt particles by ion cyclotron waves, *J. Geophys. Res., 77,* 5608, 1972

McComas, D. J., S. J. Bame, B. L. Barraclough, J. R. Donart, R. C. Elphic, J. T. Gosling, M. B. Moldwin, K. R. Moore, and M. F. Thomsen, Magnetospheric plasma analyzer: initial three-spacecraft observations from geosynchronous orbit, *J. Geophys. Res., 98,* 13453, 1993

Perraut, S., Wave-particle interactions in the ULF range: GEOS-1 and -2 results, *Planet. Space Sci., 30,* 1219, 1982.

Prigancova, A., and Ya. I. Feldstein, Magnetospheric storm dynamics in terms of energy output rate, *Planet. Space Sci., 40,* 581, 1992.

Rairden, R. L., L. A. Frank and J. D. Craven, Geocoronal imaging with Dynamics Explorer, *J. Geophys. Res., 91,* 13613, 1986.

Rasmussen, C. E., S. M. Guiter, and S. G. Thomas, Two-dimensional model of the plasmasphere: refilling time constants, Planet. *Space Sci., 41,* 35, 1993.

Sheldon, R. B. and D. C. Hamilton, Ion transport and loss in the earth's quiet ring current 1. Data and standard model, *J. Geophys. Res., 98,* 13491, 1993.

Stern, D. P., The motion of a proton in the equatorial magnetosphere, *J. Geophys Res., 80,* 595, 1975.

Studemann, W. et al., The May 2-3, 1986 magnetic storm: First energetic ion composition observations with the MICS instrument on Viking, *Geophys Res Lett, 14,* 455, 1987.

Thorne, R. M. and R. B. Horne, The contribution of ion-cyclotron waves to electron heating and SAR-Arc excitation near the storm-time plasmapause, *Geophys. Res. Lett., 19,* 417, 1992.

Volland, H., A semiempirical model of large-scale magnetospheric electric fields, *J. Geophys. Res., 78,* 171, 1973.

Williams, D. J. and L. R. Lyons, The proton ring current and its interactions with the plasmapause: Storm recovery phase, *J. Geophys. Res., 79,* 4195, 1974a.

Williams, D. J. and L. R. Lyons, Further aspects of the proton ring current interaction with the plasmapause: Main and recovery phases, *J. Geophys. Res., 79,* 4791, 1974b.

Young, D. T., H. Balsiger, and J. Geiss, Correlations of magnetospheric ion composition with geomagnetic and solar activity, *J. Geophys. Res., 87,* 9077, 1982.

Richard B. Horne, British Antarctic Survey, Natural Environment Research Council, High Cross Madingley Road, Cambridge, England CB3 0ET.

Vania K. Jordanova, Space Science Center, University of New Hampshire, Durham, NH 03824-3525.

Janet U. Kozyra, Space Physics Research Laboratory, 2455 Hayward Street, University of Michigan, Ann Arbor, MI 49109-2143.

Richard M. Thorne, Department of Atmospheric Sciences, 405 Hilgard Avenue. University of California, Los Angeles, CA 90024.

How Does the Thermosphere and Ionosphere React to a Geomagnetic Storm?

T.J. Fuller-Rowell and M.V. Codrescu

CIRES, University of Colorado and NOAA, Space Environment Center, Boulder, Colorado

R.G. Roble and A.D. Richmond

High Altitude Observatory, National Center for Atmospheric Research, Boulder, Colorado

Unraveling the ionospheric and thermospheric response to a geomagnetic storm has been a challenge for many decades, due largely to the complex interactions between the plasma and neutral species. Geomagnetic storm sources to the upper atmosphere are caused by an increase in the convective electric field and auroral precipitation, that give rise to Joule heating, the primary driver of global atmospheric change. Driven by the impulsive energy input, wave surges propagate and interact globally, and are dependent on Universal Time (UT) and the time history of the source. There is a strong preference for surges to maximize on the nightside and in the longitude sector adjacent to the magnetic pole. Equatorward wind surges drive the plasma upwards and can initiate a positive ionospheric change. The divergent nature of the wind field causes upwelling and changes to the neutral composition that can be transported by the storm and background wind fields. Negative ionospheric phases result from increased molecular species. Ionosondes have recorded the apparently chaotic ionospheric response for more than 50 years, but it is only recently that local time (LT) and seasonal dependencies have been quantified. Numerical models have shed light on the physical processes; the LT response is caused by the diurnal wind field migration of the composition "bulge," and the seasonal dependence is controlled through the transport of the bulge by the summer-to-winter prevailing circulation. Neutral density changes and satellite airglow observations support this basic concept. At low latitudes, electrodynamic changes are initiated by penetration of magnetospheric fields followed by rapid shielding. After shielding, the electrodynamics is forced by dynamo action of the disturbed neutral atmosphere, driving a sequence of equatorial plasma drifts for more than a day. The precise mechanisms responsible for this equatorial response have yet to be defined, but it is tempting to associate the time-scales with those of the global dynamical and composition response of the neutral atmosphere. Despite an increase in our understanding of the causes of the mid-latitude ionospheric response, simulation of a real storm has yet to confirm theory. We are limited by accurate knowledge of the source function, and by the lack of comprehensive data coverage of both the neutral and ionospheric parameters. Both are needed before theory and models can be thoroughly tested.

1. INTRODUCTION

A geomagnetic storm, from the perspective of Earth's upper atmosphere, is a period of intense energy input from the magnetosphere. The consequences can be visually spectacular, with brilliant aurora extending into mid-latitudes [*Allen et al., 1989*], and can also effect technological systems. These include the more rapid decay of low-Earth orbiting satellites, disruption of HF, VHF, and UHF communications, and induced currents in power system transformers that occasionally cause power outages. These consequences can be moderated if we can comprehend the ways the upper atmosphere responds to these periods of dramatic increase in energy input.

Geomagnetic storms have come to be defined by the low-latitude geomagnetic index, D_{st}, which is a measure of the strength of the magnetospheric ring current. The D_{st} is driven by the same magnetospheric processes that dump energy into the upper atmosphere. Although the ring current is not the driver of the upper atmosphere they have common sources; D_{st} often reflects the level of magnetospheric energy input to the upper atmosphere. It is a coincidence that the recovery time of the ionosphere is similar to the recovery time of the ring current and the D_{st} index. A clear example of this can be seen in the time sequence of the ratio of storm to quiet ionospheric density in response to a storm in December 1982 (Figure 1). The Australian sector shows a fairly rapid decrease in the ratio (a negative ionospheric storm) followed by a gradual recovery. The shape and time-scales involved are not unlike those seen for a typical storm-time D_{st} index. The reason for this similarity will be explained later.

This paper will describe the sequence of processes that result from the flow of energy from the magnetosphere to the upper atmosphere, and will illustrate the complex global response of the thermosphere and ionosphere to this energy input.

2. OBSERVED IONOSPHERIC RESPONSE

Much of the interest in understanding the response of the upper atmosphere to geomagnetic storms has stemmed from the need to predict the ionospheric response. The need arises for practical reasons; some communication systems require signals to pass from ground-to-ground using HF radio via the ionosphere, and from ground-to-satellite through the ionosphere at higher frequencies. The parameter that has received a great deal of attention is the peak F region electron density (NmF2), which is related to the maximum usable frequency (MUF) for oblique propagation of radio waves. The total electron content (TEC) is also strongly correlated with NmF2, and is significant for the phase delay of high frequency ground-to-satellite signals.

It has been known for several decades that the ionospheric response to geomagnetic storms varies with season. The summer-winter differences are shown in Figure 2 (taken from Prölss [*1980*]). Figure 2 shows the variation of NmF2 following a 12-hour burst of magnetic activity, as shown by the a_p index. Compared to the monthly median value, the noon NmF2 is less than half at the southern (summer) station and has doubled at the northern (winter) station. A negative (F-layer) storm occurs in the summer hemisphere and a positive storm occurs in the winter hemisphere; both stations are at middle geomagnetic latitudes, -41° and 36°, respectively.

Wrenn et al. [*1987*] and Rodger et al. [*1989*] derived the average F2-layer storm response at three southern stations, as a function of local time, season, and a modified a_p index, using data from many storms that occurred during 1971-1981. Figure 3 shows the variation of ln(N/No) at Argentine Islands (65°S), where N and No are the storm (a_p>30) and quiet-day values of NmF2. On this scale, a variation of -0.5 units represents a reduction of NmF2 by 40 percent [exp(-0.5)=0.6].

Each month shows a similar local-time variation, with a minimum in the morning hours around 0600 LT and a maximum in the evening hours around 1800 LT. The local-time "AC" variation is superimposed on a "DC" shift of the mean level that varies with season, being most positive in winter (May-July) and most negative in summer (October-February). Figure 3 also shows the prevalence of positive storms in winter, and of negative storms in summer, when there is hardly any increase of NmF2 at any local time. It should be noted that individual storms show large deviations from the average behavior. The dependence of the storm-time ionosphere to both local-time and season can be explained as a response to neutral composition changes and their movement by the global wind field; this will be described later.

The information presented in Figure 3 was compiled from many active periods, and represents an average of storms commencing at different UT's and of varying duration. Figure 1, taken from Fuller-Rowell et al. [*1996a*], illustrates the ionospheric response to a particular storm in December 1982, and shows the dependence on longitude, or UT start-time, and the duration of a storm. The data is from the Ionospheric Digital Database CD-ROM compiled by NGDC, and has been grouped into six mid-latitude regional sectors, including Europe, N.E. Asia, and North America in the north, and, in similar respective longitude sectors, Africa, Australia, and South America in the southern Hemisphere. Ionosonde observations from each sector have been averaged, and the time series of the ratio of the storm-time NmF2 to the monthly median is plotted in the six panels (the scale runs from a ratio of 0.4 to 2.0). As in Figure 2 and 3 the seasonal dependence can be seen. The northern winter sectors are more variable and have no substantial negative phase, but the southern summer sectors all show a negative excursion, with the Australia sector experiencing the strongest decrease. Fuller-Rowell et al. [*1996a*] showed this variation in longitude to be a consequence of the UT start time and the duration of the storm.

3. MAGNETOSPHERIC ENERGY SOURCES

The magnetospheric sources deposit energy and momentum in three discrete ways. Precipitating auroral electrons

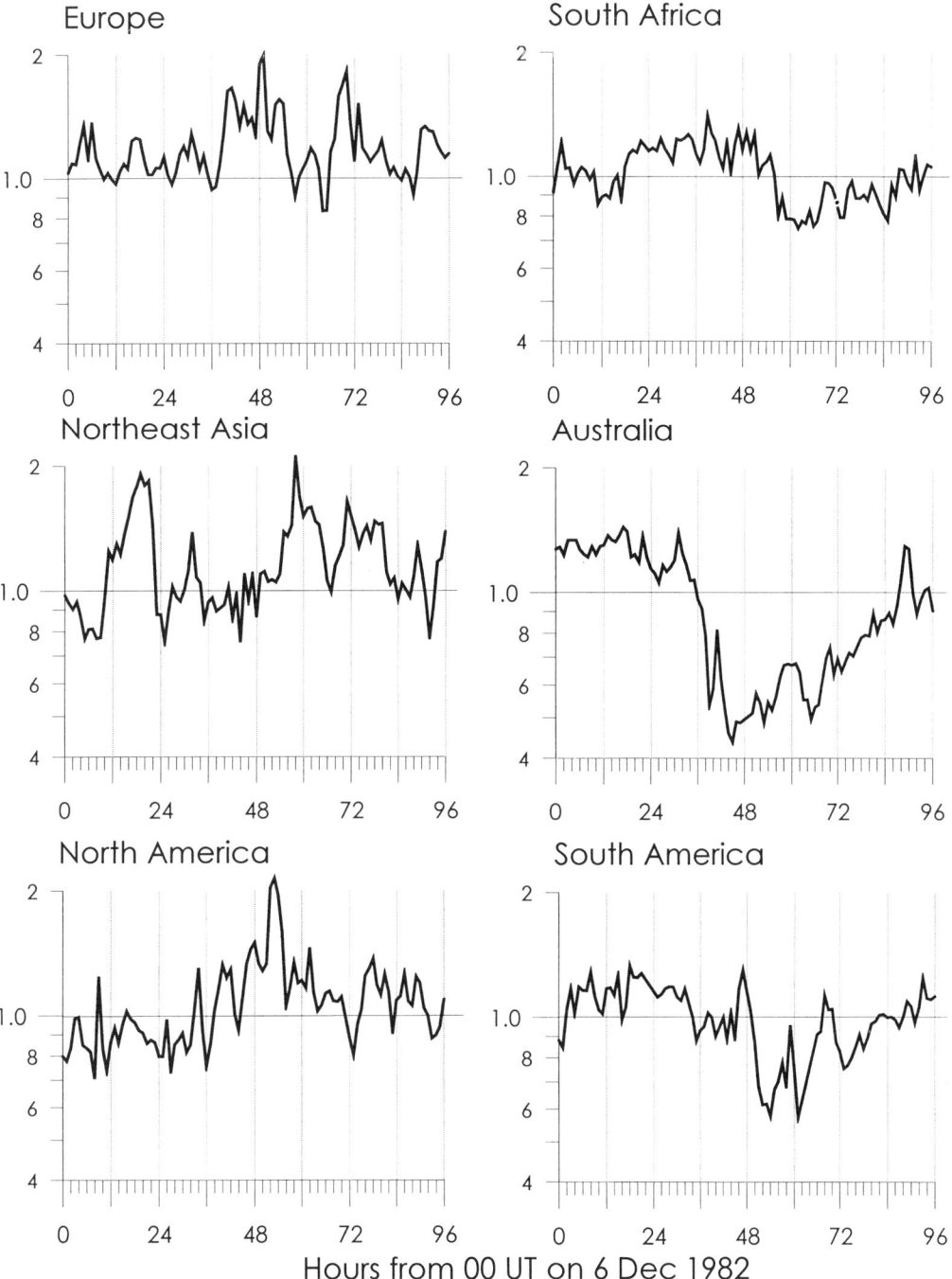

Figure 1. The response of the ratio of storm time NmF2 to the monthly median for six midlatitude longitude sectors for a storm in December 1982.

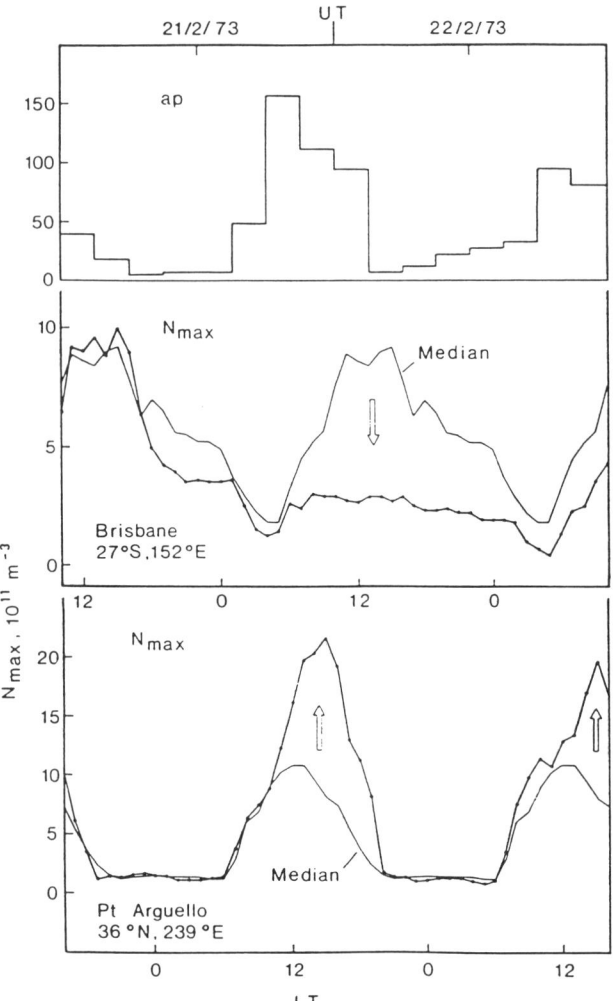

Figure 2. The ionospheric response to a storm in February 1973 at two midlatitude stations, taken from Prölss [*1980*]. The southern hemisphere station of Brisbane recorded a strong "negative storm"; the station at Pt. Arguello, in the North, recorded a "positive storm."

sphere can exceed a terawatt, possibly dumping thousands of terajoules of energy during the course of a storm. This energy is eventually lost, mostly by infrared radiation, over the day or so following the storm.

The third mechanism, ion drag, deposits kinetic energy into the atmosphere. The ionosphere responds to the convective electric field in the upper thermosphere by drifting in the E×B direction. An electric field of 50 mV/m would cause a drift of about 1 km/s, and the collision of ions with the atmosphere acts as a momentum source to the neutral gas, forcing the material to follow the ions. Other forces on the neutral gas, such as Coriolis and pressure gradients, prevent perfect matching of the ion and neutral velocity so that neutral drag and Joule heating continue to impose a load on the magnetosphere. At lower altitudes the collision frequency increases, and by an altitude of 150 km the ion velocity begins to rotate in the direction of the electric field. The kinetic energy imparted to the neutral gas is dissipated

and protons collide with the atmosphere and deposit heat directly into the neutral gas. Particle heating rates increase, from less than 10 to 20 gigawatts during quiet times to over 100 gigawatts during geomagnetic storms; this source of heat, however, contributes only a small part (20 - 30 percent) of the total storm energy input [*Lu et al., 1995; Evans et al., 1988*]. Auroral ionization increases the conductivity of the thermosphere, and the conductivity combined with the magnetospheric convection electric field produces Joule heating, which is the dominant atmospheric energy source during a storm. Joule heating in one hemisphere can increase from tens of gigawatts during quiet times, to hundreds of gigawatts during severe geomagnetic disturbances [*Lu et al., 1995; Evans et al., 1988*]. The combined globally-integrated rate of energy input from the magneto-

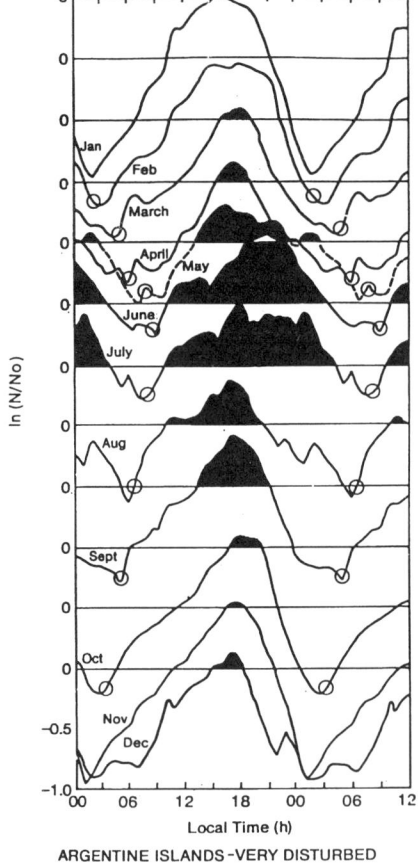

Figure 3. The average seasonal and local time variations in ln(N/No), where N/No is the observed storm/quiet ratio of peak F2 layer electron density, at Argentine Islands (65°S) for 1971-1981, taken from Rodger et al. [*1989*]. The zero level (N/No)=1 is shown for each month, and the shading denotes times of increased electron density where (N/No)>1.

by viscosity, heats the neutral gas, and is eventually radiated to space.

As well as increasing in magnitude, the magnetospheric sources expand spatially from their quiet time locations [*Foster et al., 1986; Evans et al., 1988; Knipp et al., 1989; Allen et al. 1989*]. Aurora and convection fields typically reach magnetic latitudes between 50° and 60° at least 10° to 15° more equatorward than normal. The net result is a large increase in the rate of energy input into the upper atmosphere, for a number of hours, at latitudes substantially closer to the equator than normal. This increase in energy input perturbs the neutral atmosphere and ionosphere globally; it is these changes that are described in detail in the following sections.

4. HIGH LATITUDE RESPONSE

The region of the atmosphere directly affected by the increase in magnetospheric energy responds in a variety of ways. The ionospheric E-region density increases in response to more intense auroral particle precipitation, and the expanded convective electric field redistributes high-latitude plasma [*Sojka and Schunk, 1983; Prölss et al., 1991*]. Tongues of ionization extend over the polar cap, and trough features move to lower latitudes. The response of the high latitude thermosphere can also be quite dramatic. During disturbed intervals, neutral winds strengthen at all thermospheric altitudes driven by enhanced ion drag, and speeds can exceed 1 km/s in the upper thermosphere. Upper thermospheric temperatures can rise by several hundred degrees Kelvin [*Maeda et al., 1989*], causing thermal expansion, changes in neutral composition on a height surface [*Prölss, 1980*], and cells of increased and decreased neutral density [*Crowley et al., 1996*].

The increase in temperature and neutral winds in the upper thermosphere, from a numerical simulation of a storm interval, is shown in Plate 1. The figure shows the relatively quiet thermosphere on the left, and disturbed conditions on the right, from 40°N latitude to the pole. The model is forced by the convective electric field and auroral precipitation tuned to the time history of the NOAA/TIROS auroral power index from a storm period on December 7-8, 1982. Full details of the storm forcing can be found in Fuller-Rowell et al. [*1996a*]. The temperature increases by more than 300 K at high latitudes; the temperature pattern is not an image of the Joule heating rate, but is a result of the integrated heat input over the previous few hours of the storm. The Joule heating region plows through the atmosphere [*Killeen and Roble, 1984*] during the storm, as the Earth rotates, heating the gas over a wide geographic area. The temperature distribution is also under the influence of transport by the wind field and losses by heat conduction and infrared radiation.

The winds increase by about 400 m/s in the dusk-sector auroral oval, 400 m/s over the polar cap, and by about 150 m/s in the dawn sector, due to ion drag. The asymmetric response of the dawn and dusk sectors has been previously explained [*Fuller-Rowell, 1985*] as a consequence of the balance between Coriolis and curvature forces.

The polar temperature increase creates large-scale pressure gradients that drive a global circulation.

5. GLOBAL THERMOSPHERIC WIND RESPONSE

Dynamics drive many of the ionospheric changes; if the dynamical response is understood, many of the subsequent mechanisms fall into place. Plasma moving along the magnetic field in response to the meridional wind raises or lowers the F2 peak into a region of different neutral composition, and the divergent global wind field drives upwelling that causes changes in neutral composition. The new chemical loss rates in both cases affect the plasma density. A possible chemical change, not driven by dynamics, is the production of vibrationally excited molecular nitrogen [*Richards and Torr, 1986*].

Plate 2 shows an example of the meridional wind change during a numerical simulation of a 12-hour geomagnetic storm from Fuller-Rowell et al. [*1994*]. The storm source in this case is a simple step function increase of electric field and auroral precipitation over a 12-hour period. The plots are snapshots of the response at storm times of 3, 6, 12, and 18 hours, at about 300 km altitude, and equatorward of 70° latitude in both hemispheres. Each snapshot is plotted with local time adjusted to put 0° longitude at the far left. The velocity scale covers winds from -100 m/s (northward) to +100 m/s (southward); the other scale for polar cap potential is not relevant here.

Surges of equatorward wind have propagated from both polar regions after 3 hours. The longitude sector with the strongest response is on the nightside. Notice that the variation in longitude, or local time, of the response in the north is weaker than in the south. For a storm commencing 12 hour later, the southern hemisphere would have the more uniform longitude response. Two effects are important for the strength of the wind surges: the location of the longitude sector of the geomagnetic pole, and the preference for wind and wave propagation on the nightside. If the two effects coincide, a strong response in that sector is generated. This can be seen, for example, in the southern hemisphere, near midnight local time and 120° east. The longitude sector of the magnetic pole is on the nightside, a big surge is seen in meridional wind, and a strong longitude dependence is generated. The longitude sector 180° away in those same hemispheres has a weak response because the distance to the magnetic pole is greater and because propagation on the dayside is inhibited. Restraint of dayside propagation may be a consequence of the prevailing poleward wind, but it is more likely due to higher ion density causing ion-drag to dissipate the surge more rapidly.

The leading edge of the disturbance is in the longitude sector of the geomagnetic pole. The surge behind the leading edge, however, is smaller when this sector is in daylight

208 THERMOSPHERE-IONOSPHERE STORMS

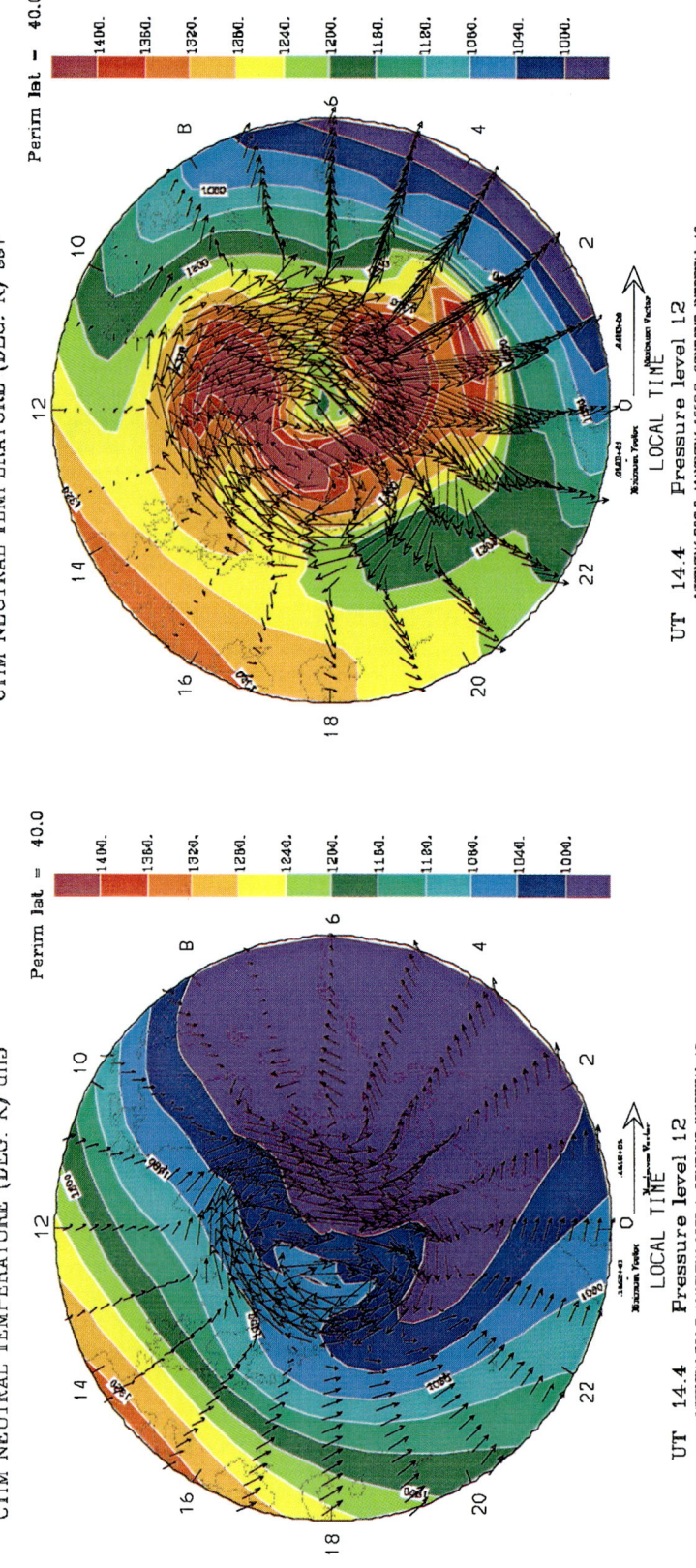

Plate 1. Example of the response of neutral temperature and winds in the upper thermosphere from a numerical simulation of a storm in December 1982. On the left is the quiet day, and on the right is the same UT a few hours into the storm interval.

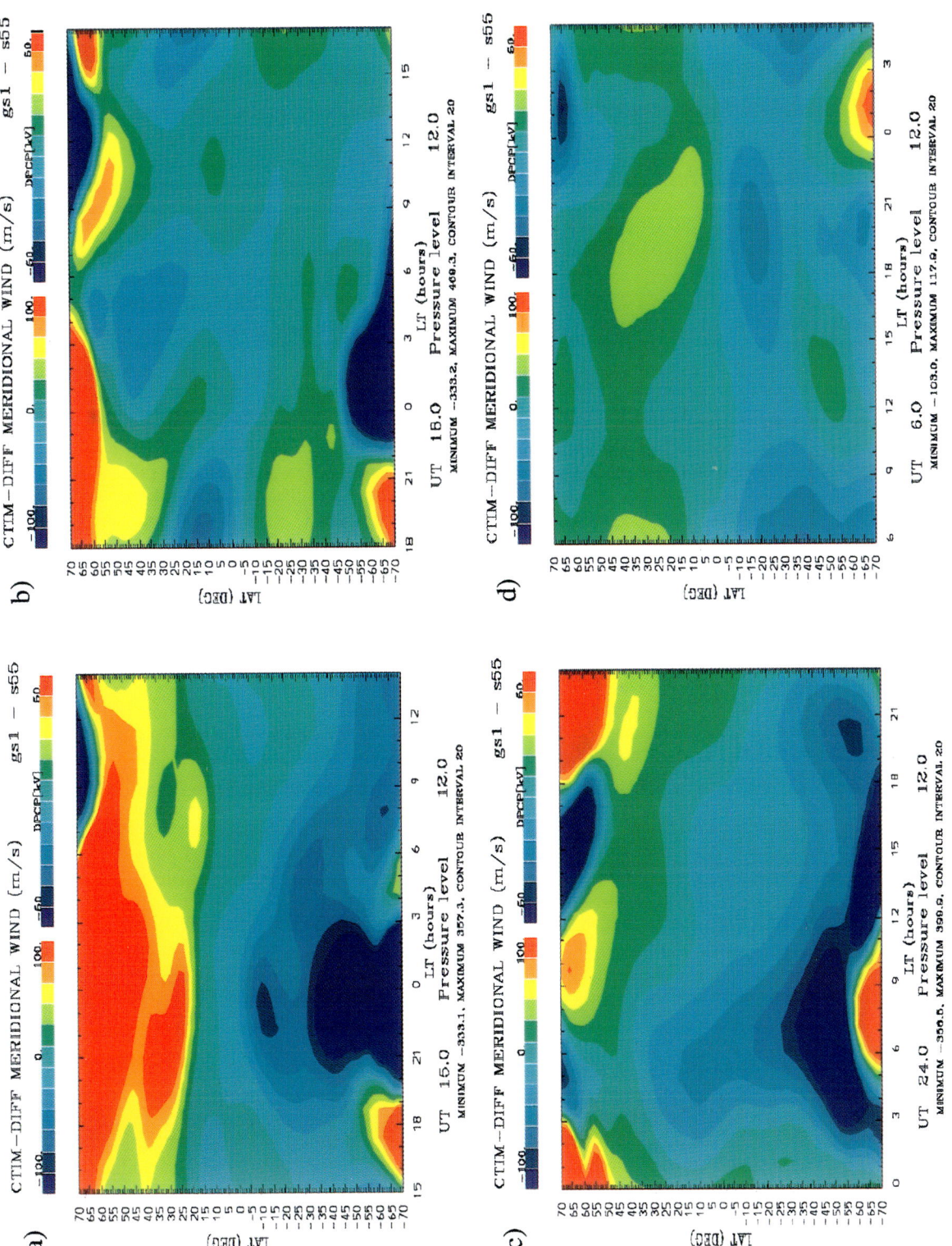

Plate 2. The simulation of the dynamics of the global upper thermosphere (near 300 km) in response to geomagnetic storm forcing. The change in the meridional wind from quiet to disturbed simulations is shown, from 70°S to 70°N latitude, at storm times of (a) 3 hours, (b) 6 hours, (c) 12 hours, and (d) 18 hours.

than when it is in the night sector. This indicates that the speed of propagation is not dominated by the prevailing poleward winds on the dayside inhibiting the wave, but that ion drag is effective in reducing the amplitude.

The rate of propagation of the storm surge is consistent with a large-scale gravity wave, with a phase speed of 600 to 700 m/s. The wave front appears to have a double structure, but it is important to note that the meridional circulation, exceeding 100 m/s, is established very soon after the wave front has passed a particular location. The global circulation change progresses not at the rate of the bulk velocity of the atmosphere, but closer to the phase speed of the wave front. The wave and circulation changes are manifestations of one surge.

Within another hour it is clear that the waves from each hemisphere will make contact, interact, and affect the dynamics of the thermosphere on a global scale. At a storm duration of 6 hours the two waves and circulation have interfered and partially canceled each other. Poleward winds, in fact, are created over a wide area at this time, and surges penetrating into the opposite hemisphere are a dominant feature of the response and form an integral part of the global wind system.

After a further 6 hours (12 hours into the storm) and near the end of the driven phase of the disturbance, a global circulation has re-established itself. The two hemispheres do not respond independently, however; winds from the north and from the south are drawn towards the sector of their respective geomagnetic pole, so they do not compete.

The meridional wind difference 6 hours after the end of the storm is seen in the last plate of Plate 2. The main wind system recovers fairly quickly after the storm, although a residual wave-like circulation persists for at least 12 hours. Richmond and Roble [*1979*] and Fujiwara et al. [*1996*] demonstrated a similar response to an impulsive energy input, and indicated that a global oscillation in the wind field can persist for at least two cycles.

The dynamic response of the thermosphere provides clues to many of the ionospheric changes. Plate 3 shows the response to the same simulated storm in the 0° longitude sector, equatorward of 70° latitude in both hemispheres. The scale runs from -100 m/s (northward) to +100 m/s (southward); on the right hand side is the change in polar cap potential (DPCP), shown immediately beneath the color scales.

Evidence for the interhemispheric penetration of the winds is again displayed. A clear signature of a surge is propagating from the north at the start of the storm, into the southern hemisphere 4 hours later. The equivalent wave from the other hemisphere is weaker, the waves interact at low latitudes as they cross, and the double feature of the response is displayed. The initial surge from each hemisphere is interrupted when it encounters the other, and a second development of the meridional wind follows while the storm input remains.

There is a clear indication that the circulation is again subsiding before the end of the storm input. In this case, however, it does not appear to be due to interference from the other hemisphere. If the storm input remains imposed indefinitely the global circulation reaches a new equilibrium. Plate 2c, which depicts the global circulation after 12 hours, indicates that a new equilibrium is not a uniform increase in equatorward flow at all local times or longitude sectors. The second meridional wind increase is part of the transient response to the storm input; its subsequent decrease is due to the local time dependence of the approaching new global equilibrium.

The global circulation is restricted by a number of basic physical processes; the first contribution arises from simple dynamics. If a global scale latitude-pressure gradient exists in a idealized friction-free atmosphere, a geostrophic balance will be established with zero meridional wind and a zonal jet will develop. In this situation the Coriolis force in the meridional direction, due to the zonal wind, balances the original pressure gradient. Zonal winds will eventually increase and limit the meridional circulation. The thermosphere does not satisfy pure geostrophic balance, but part of the curtailment of the second meridional surge is caused by the development of a zonal wind circulation. The buildup of a westward zonal wind can also be interpreted by conservation of angular momentum [*Blanc and Richmond, 1980*]. Equatorward winds transport air from high latitudes where the angular momentum associated with Earth's rotation is less, and as the air parcel arrives at middle latitudes a westward motion must develop, in the Earth's frame, to conserve angular momentum.

A further limit to meridional motion is due to a reduction of pressure gradients by the motion of air parcels which transport energy from high latitude. The meridional pressure gradient is reduced as this effect limits global circulation. The F region thermosphere is strongly influenced by ion drag which also limits the neutral wind.

The meridional winds do not simply recover after the storm ceases; they appear to oscillate for at least one cycle. As the atmosphere begins to cool, the latitude pressure gradient subsides, the zonal winds that developed are no longer balanced, and a poleward flow forms. This can be seen clearly in Plate 3 at 0200 UT. A further reversal then occurs at about 0600 UT.

All of the features described above from the numerical simulation have been observed. Incoherent scatter observations and Fabry-Perot interferometers [*Buonsanto et al. 1989; 1990*] have observed surges in equatorward wind (particularly on the nightside) in response to geomagnetic disturbances. Codrescu et al. [*1992*] showed the penetration of winds and ionospheric positive and negative phases in the Japanese longitude sector during a storm. At low latitudes observations from the Dynamics Explorer satellite have shown the sequence of wave surges propagating from both polar regions [*Burns and Killeen, 1992*]. Fesen et al.

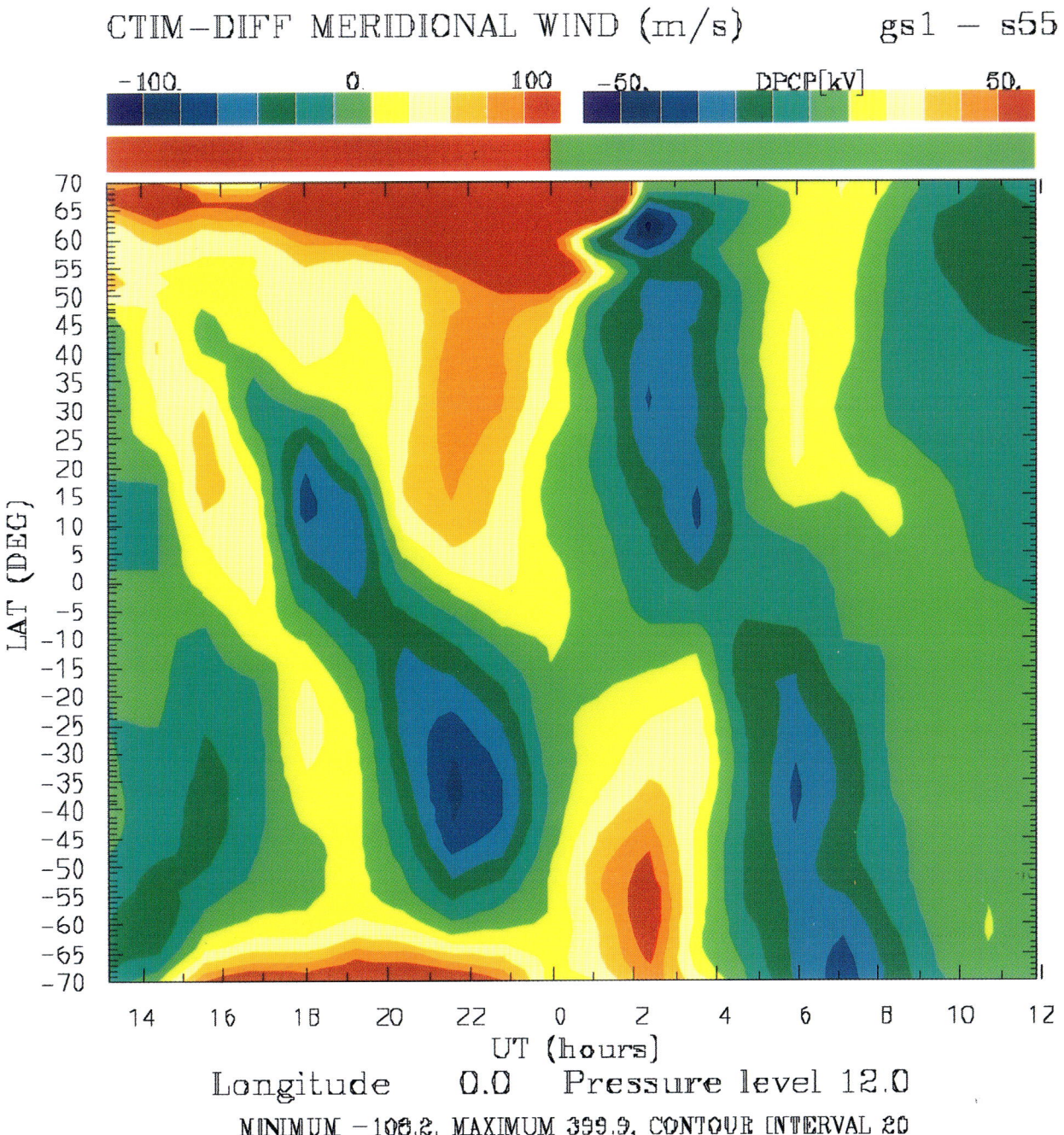

Plate 3. The dynamic response as seen from the 0° longitude sector. The change in the meridional wind from quiet to disturbed simualtions is shown, from 70°S to 70°N latitude, for the 24 hour period of the simulation.

[1989] demonstrated the ionospheric consequence of these wind surges at low latitudes, as the plasma oscillates from side to side along the nearly horizontal field lines. These observations confirm the basic global scale nature of the dynamic changes that occur during a geomagnetic storm.

6. THERMOSPHERIC COMPOSITION RESPONSE

The equatorward divergent flow from high latitudes causes upwelling through the pressure surfaces and increases of mean molecular mass [Rishbeth et al. 1987; Prölss, 1987; Burns et al. 1991]. The region of increased mean molecular mass has been termed a composition bulge, and it has been recently shown that the bulge can be transported by the background and storm-time wind field. Plate 4 shows the time history of the change in mean molecular mass (m) for a 12-hour storm at solstice, in the summer hemisphere. The six plots each show mean mass at a constant pressure near 300 km altitude, from 10° latitude to the north pole. The storm times depicted are 6, 12, 18, 24, 36, and 48 hours. Each plot has the same scale to illustrate the rise and fall in magnitude; the highest color range corresponds to values that increase by more than 2.4 atomic mass units (amu), and the lowest color range corresponds to a decrease by 0.6 amu. The first two images were taken during the driven phase of the storm and show the development of the composition bulge. Subsequent times show the period of recovery, when the region of disturbed composition is most clearly transported by the background wind field.

The maximum response occurs at the end of the driven phase of the storm, with areas exceeding 2.4 amu. The composition change is an integrated effect of upwelling over the entire storm period and, again, is modified by transport and diffusion that attempts to restore equilibrium. As a reference, if mean mass changes from 19 to 22 amu the proportion of molecular nitrogen in the atmosphere at that pressure level changes from 25 percent to 50 percent.

Soon after the end of the storms driven phase the divergent wind field ceases and upwelling stops. At this point the bulge of composition begins to subside as diffusion gradually restores the atmosphere to pre-storm conditions. This recovery process, however, is very slow, and continues for one or two days following the geomagnetic forcing. This slow recovery of neutral composition is the cause for the slow recovery of the negative ionospheric phase in the Australian sector of Figure 1. The cause of both the seasonal and local-time dependence of the ionospheric storm time response shown in Figures 2 and 3 can be understood by examining the evolution of the composition bulge in the recovery phase of the storm.

Once the bulge is created and the driven phase has ceased, the expectation is for the bulge to simply rotate with Earth; this is a scenario described by Prölss [1993]. A new perspective is obtained by examining the results of the numerical simulation during the recovery. The size of the bulge decreases, as would be expected, as diffusion continues to restore equilibrium. The location of the bulge, however, does not simply rotate with Earth. As the bulge moves into the dayside it moves poleward, and on entering the nightside it moves equatorward. Once it has completed one rotation the bulge does not return to its starting point, but it has migrated equatorward by more than 10° latitude.

These features can be studied by comparing the images in Plate 4. A few hours after the end of the driven phase the composition bulge lies in the midnight and early morning sectors over northeast Canada, the north Atlantic, and western Europe (see image 7c at 0600 UT). Over the next 6 hours the bulge is on the nightside and is subjected to equatorward winds. At 1200 UT (24-hours storm time and 12 hours after the storm input has ceased), the peak composition change has moved equatorward and westward; it is now over the eastern United States, approaching Florida (image 7d). At this time the bulge is moving onto the dayside and is subjected to the dayside poleward winds that oppose the prevailing seasonal circulation from summer to winter. Over the next 12 hours the bulge continues westward and also moves poleward, centering itself over western Canada. From 2400 UT to 1200 UT (images 7e and f) there is rapid equatorward movement. During this time the bulge moves through the nightside, the equatorward winds from the diurnal cycle reinforce the prevailing circulation, and the bulge moves from northern Canada to Mexico, by this time the magnitude of the bulge has decreased by more than 50 percent. The movements of the composition feature are consistent with the superposition of a diurnal wind field with an amplitude of 50 m/s, and a prevailing summer-to-winter wind of 25 m/s.

The winter-time and equinox response were also shown by Fuller-Rowell et al. [1996b]. The winter hemisphere, with prevailing winds flowing towards the polar region, was shown to have a much more spatially constrained region of enhanced mean mass. The explanation of the data presented in Figures 2 and 3 is now clear. It is well known that the F2-layer ion density depends on the neutral gas composition because the ion production rate depends on the atomic oxygen concentration and the loss rate depends on the molecular concentrations. Therefore an increase of mean molecular mass that is associated with an increase in the proportion of molecular gases causes a negative storm. A decrease in mean molecular mass conversely causes a positive storm [Rishbeth, 1989]. The local time variation (the so-called AC effect) found by Rodger et al. [1989] may be attributed to an oscillation in latitude of the composition bulge in response to the background wind field. The seasonal effect (the so-called DC offset in the mean level) is caused by the movement of the bulge by the prevailing summer-to-winter circulation. Support for the concept of transport of the bulge is presented by Skoblin and Forster [1993], who showed a case where steep gradients in thermospheric composition could be advected by the meridional wind.

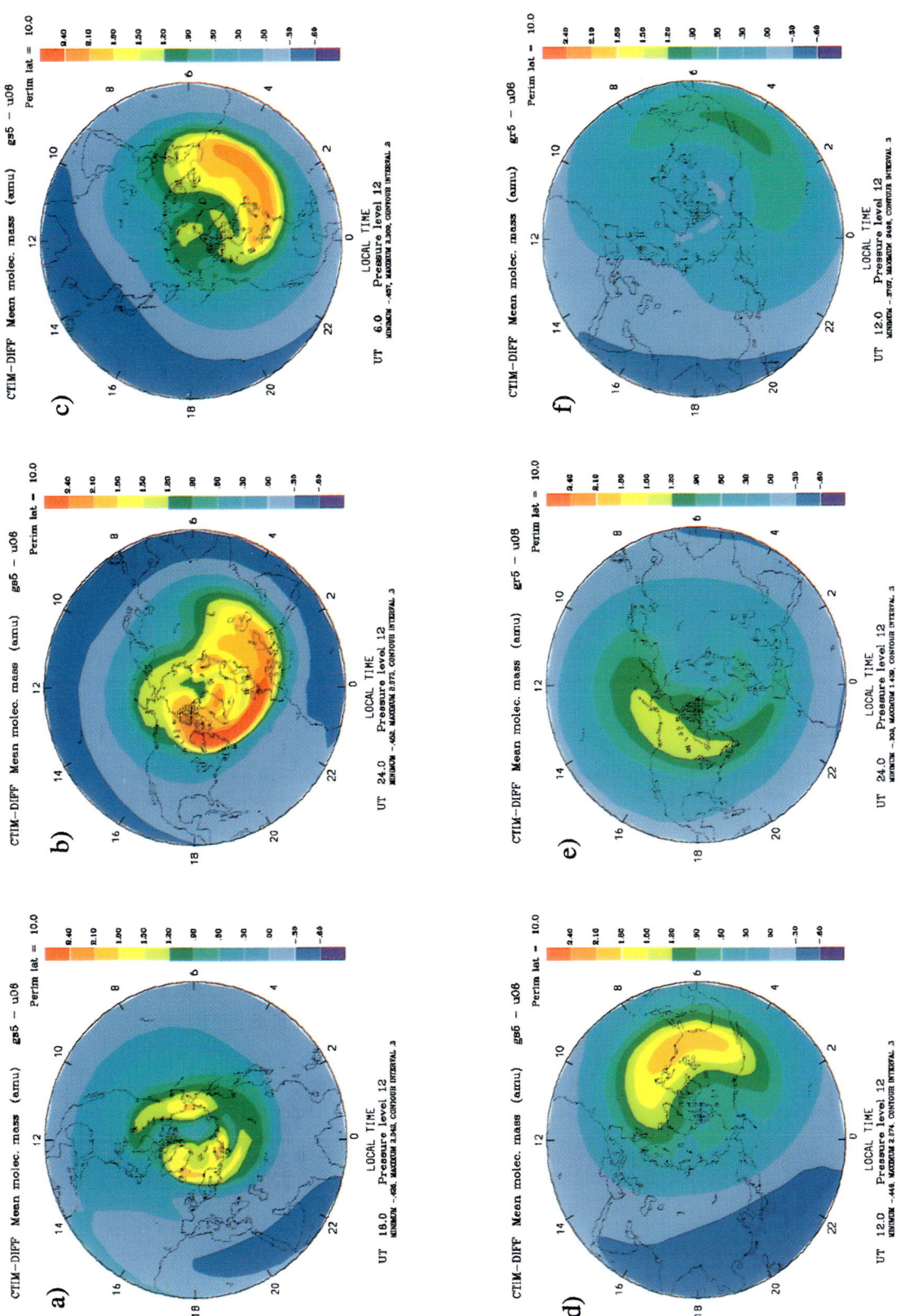

Plate 4. Series of six images tracking the evolution of a composition bulge during a modeled storm in June. The change in mean molecular mass (amu), on a constant pressure (near 300 km), is shown for the northern hemisphere from the pole to 10° latitude. The figure shows the storm times of (a) 6 hours, (b) 12 hours, (c) 18 hours, (d) 24 hours, (e) 36 hours, and (f) 48 hours.

Just as the numerical simulations shed light on the local time and seasonal dependence of the ionospheric response to storms seen in Figures 2 and 3, the model can also help us understand the UT or longitude effect seen in Figure 1. During a particular storm there will be a preferred location for the development of the composition bulge. For the storm in December 1982, shown in Figure 1, Fuller-Rowell et al. [1996a] demonstrated that the main driven phase of the storm occurred as the Australian sector moved through the midnight sector. This is entirely consistent with the ionospheric response seen in the data. This sector then experiences the maximum magnetospheric forcing and also the largest upwelling and composition change. A storm commencing at a later UT would generate a composition bulge in a different longitude sector.

The explanation of longitude dependence implies a response that is sensitive to the spatial distribution of the energy input; Plate 5 shows a comparison of the pattern of Joule heating from the statistical electric field model of Foster et al [1986] and at the peak phase of a substorm from the empirical model of Kamide et al. [1996]. The statistical model is an average over many time periods, so day-to-day distortions of the pattern have been averaged out. The empirical model, however, is specifically designed to capture the changing shape of the convection pattern during the different phases of a substorm; only the pattern associated with the peak phase is illustrated here. The pattern of energy input is quite different. Joule heating from the statistical model peaks in the dusk sector auroral oval, but the substorm pattern has a more localized, intense peak, in the midnight sector. The consequence for heating of the atmosphere, and subsequent upwelling and changes in composition, will also be quite different. The implication is that an accurate picture of the spatial distribution, and temporal variation, of the sources must be available in order to predict the thermosphere and ionosphere changes.

A number of papers have investigated the seasonal changes of storm-induced composition and ionization perturbations [e.g. Prölss, 1977; Prölss and von Zahn, 1977; Prölss, 1993]. A number of the observed features appear to be reproduced by the model—in particular the different extent and latitude shape of composition changes in winter and summer [Prölss, 1993]. There is also some evidence that a bulge in mean mass, detached from the high latitude region, has actually been observed by satellite [Prölss, 1980].

Two other examples of observations that support the idea of composition bulge transport will be examined. The first is an observation by the Dynamics Explorer I satellite in October 1981 [Cravens et al., 1994]. The Auroral Imager instrument [Frank et al., 1981] shows an extensive region at mid-latitude of reduced airglow at 130.4 nm; this is an emission feature from atomic oxygen. The interpretation for the reduced airglow is that the region is depleted of atomic oxygen and rich in molecular species [Meier et al, 1995]. The scenario suggested by the numerical simulations, that a region of molecular rich gas is transported to mid-latitude by the global circulation, is consistent with observations from the satellite.

The second example was recently published by Forbes et al. [1996]. They studied the evolution of atmospheric density at a constant height from a fixed local time sector. The data clearly showed the penetration of regions of increased neutral density from the summer polar regions, and the lack of such penetration in the winter hemisphere.

7. MODELLED IONOSPHERIC RESPONSE

Many of the global ionospheric changes during a geomagnetic storm arise from coupling with the neutral atmosphere. The most pronounced effect is in the negative ionospheric phase. These periods tend to be more long-lived, and coherent, over a sizable geographic region. The negative phase over Australia in Figure 1, for instance, is typical of the response seen over most of the continent and New Zealand during this storm period. Thermospheric composition changes have been suggested as the cause for many years, and this has been demonstrated clearly with satellite data [Prölss, 1996, this issue]. Numerical simulations confirm the basic physical coupling; they can predict similar magnitudes in the depth of the response, and capture some of the regional differences. Plate 6 illustrates the high correlation between the regions of enhanced molecular species in the neutral gas (on the left hand side), and the decrease in the ratio of the storm to quiet NmF2 (on the right hand side). The southern geographic frame is shown from -10° latitude to the pole, based on a numerical simulation of the December 7, 1982 geomagnetic storm. The region of depleted ionosphere off the west coast of Australia, reaching ratios of about 0.5, correlates with the large composition bulge. A comparison with Figure 1 reveals the cause of the large negative phase seen in the Australian sector for this particular storm. Matching of the model and observed data is far from perfect, with the model capturing about two-thirds of the magnitude seen in the data. The previous section suggests that this discrepancy could well arise from inaccurate information about the spatial distribution of the magnetospheric sources. Alternatively, it could be due to the neglect of vibrationally excited molecular nitrogen (N_2^*) in the simulation. Comparison of NmF2 at Millstone Hill during a major storm with the Field Line Interhemispheric Plasma (FLIP) model [Torr et al. 1990] indicated that the increase in (N_2^*) decreased NmF2 by a factor of 2 or more.

It is also possible for winds to cause a negative phase. Plates 2 and 3 demonstrate, and agree with observations, that the wind and wave propagation from one hemisphere can easily penetrate well into the opposite hemisphere. Poleward winds, even during the driven phase of the storm, are not unexpected, and they would drive plasma downwards. The plasma in a region of enhanced molecular species would decay faster, causing a negative phase. This may appear as a transient effect due to the more dynamic na-

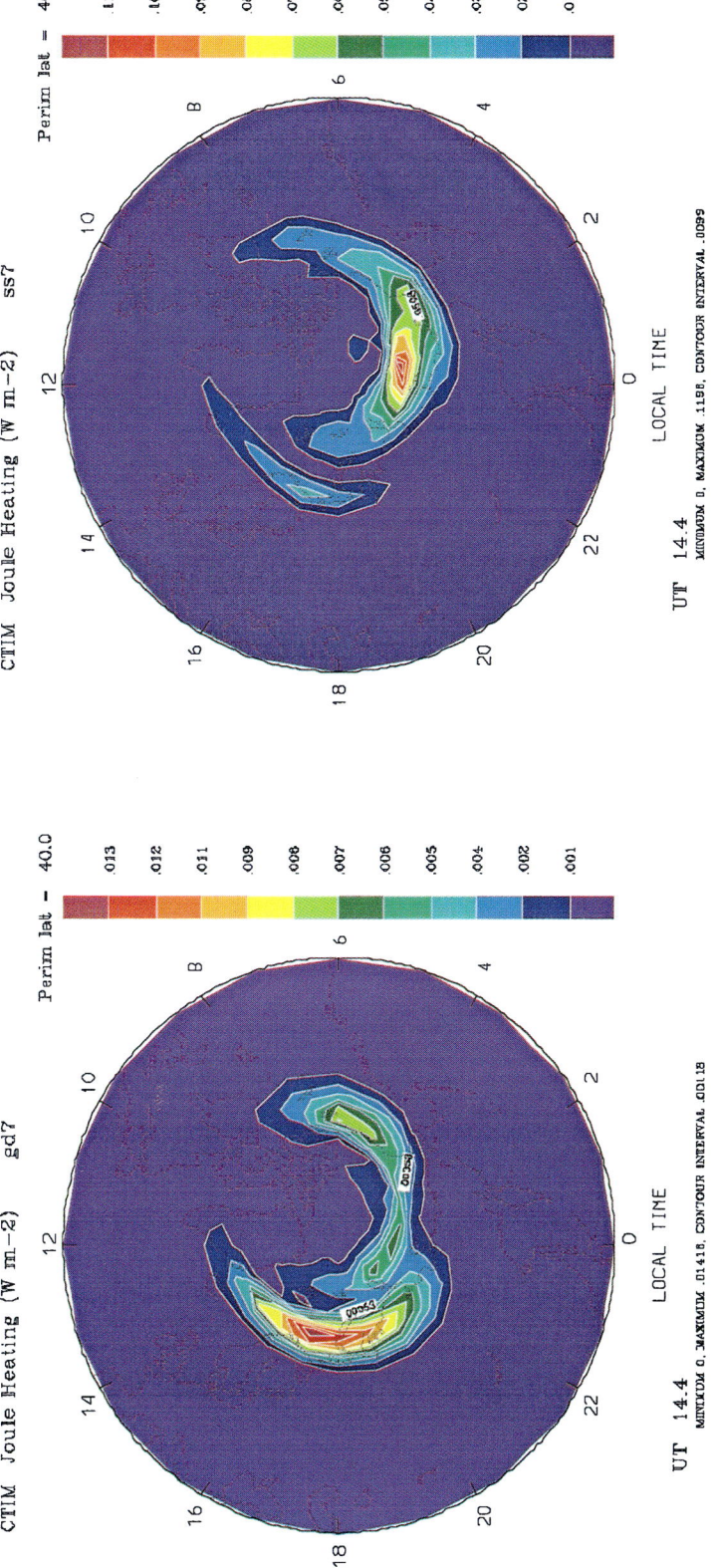

Plate 5. Comparison of the pattern of Joule heating from the statistical electric field model of Foster et al. [1986], on the left hand side, and the empirical model of Kamide et al. [1996], on the right.

216 THERMOSPHERE-IONOSPHERE STORMS

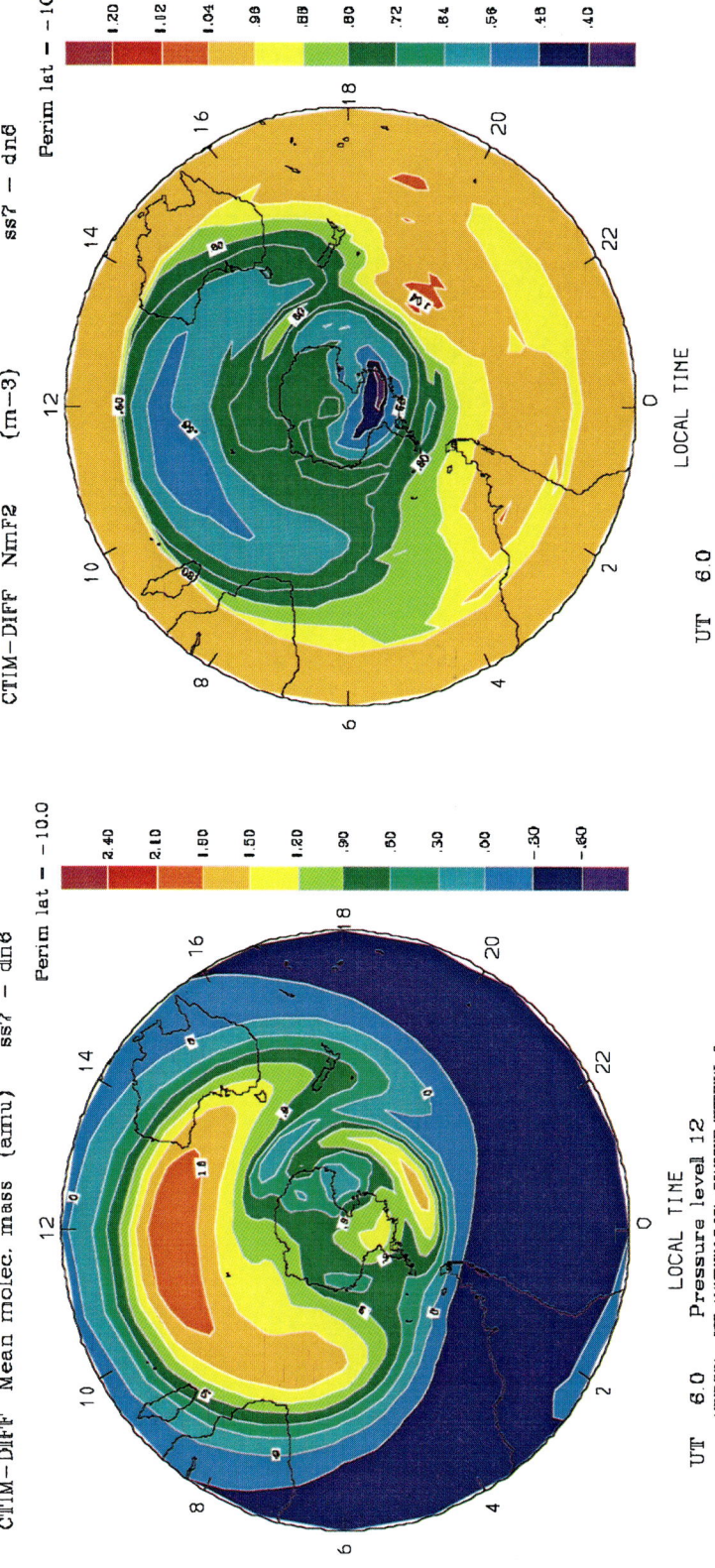

Plate 6. Illustration of the high correlation between the regions of enhanced molecular species in the neutral gas (on the left hand side) and the decrease in the ratio of the storm to quiet NmF2 (on the right hand side).

ture of the wind field, rather than the more consistent long-lived signature expected from actual composition changes.

Positive ionospheric changes are usually associated with equatorward meridional winds raising the F2 layer height and placing the peak density in a region of reduced molecular species. There are, however, two other mechanisms that could drive a positive phase. The first is illustrated in Plate 6, which shows that regions outside the magnetospheric sources are associated with downwelling. Notice that there are extensive regions, even in the summer hemisphere, that experience a decrease in mean mass. In winter mid-latitudes, the regions of decreased molecular species predominate and contribute significantly to positive ionospheric phases, particularly in the long-lived features seen during the recovery phase, like those in Figure 3.

The other cause of positive phases, the so called dusk effect based on Millstone Hill incoherent scatter data, has been described by Foster [1993]. At sub-auroral latitudes in the dusk sector at certain UTs the convection region moves equatorward, plowing though geographic latitude. This has the effect of picking up dayside solar-produced, high density plasma from lower latitudes advecting and accumulating the plasma in the dusk sector. Buonsanto [1995] recently suggested an alternative mechanism for the observations; he proposed that the dusk effect may be explained by a combination of mechanisms, including the one offered by Foster, together with convergence of plasma along the field line from neutral wind gradients, and from composition changes. The dusk effect has not yet been simulated by numerical models, and confirmation of the exact mechanisms still has to be demonstrated.

8. ZONAL MEAN RESPONSE

In addition to the response of the upper thermosphere to geomagnetic storms there are also major changes in the lower thermosphere. To illustrate the global response of the entire thermosphere to storm time forcings, difference fields of the zonal mean structure between two case studies are presented from simulations using a Thermosphere Ionosphere General Circulation Model (TIGCM). Differences are taken between a run with an auroral hemispheric power input of 16 GW and a cross-polar cap potential drop of 70 kV and a TIGCM run with an auroral hemispheric power input of 5 GW and a 30 kV potential drop, respectively. These represent comparisons between a prolonged period of moderate geomagnetic activity and a prolonged period of quiet geomagnetic activity. Both simulations are in steady-state and diurnally-reproducible, and were described in detail by Roble [1992]. The zonal mean neutral gas temperature difference field and the difference fields for O_2, N_2, O, NO, $N(^2D)$, $N(^4S)$ and Ne at 00:00 UT are shown in Figures 4 and 5 for an equinox simulation. The neutral temperature difference, shown in Figure 4a, is largest at high latitudes and becomes progressively smaller at the low latitudes. Maximum temperatures are 200 K in the upper thermosphere in both polar regions in response to the increase in auroral activity and Joule heating. Significant temperature differences on the order of 25-50 K occur down to the lower thermosphere. At the lower boundary of the TIGCM, 95 km, the temperature differences are artificially set to zero so the model cannot determine the depth into the Earth's atmosphere that the effects of auroral variability penetrate. However, recently Roble and Ridley [1994] extended the lower boundary of the model to 30 km altitude, allowing the study of this penetration into the mesosphere in the future.

The percent zonal mean difference fields for O_2, N_2 and O for equinox solar minimum conditions are shown in Figure 4b-d respectively. The differences in major constituent composition produced by auroral processes are almost entirely driven by neutral dynamics instead of atmospheric chemistry. Even though auroral particle precipitation can dissociate O_2 and N_2, the amount dissociated is small relative to the background reservoir of the species. The amount of additional O produced by auroral particle precipitation is also small compared with its background; the effects shown in the figure are caused by dynamics affecting the mass mixing ratios along constant pressure surfaces and not due to thermospheric chemistry, which is very slow in the middle and upper thermosphere. In the lower thermosphere, upward motion produced by auroral heating enhances the heavier molecular species and causes the O density to decrease. In the polar lower thermosphere, O decreases on the order of 40-50 percent and N_2 and O_2 increase on the order of 50-100 percent respectively, near 150 km. In the upper thermosphere, at a fixed height, the density of all species increases in response to the general expansion of the atmosphere from auroral heating. Even at low latitudes the general pole-to-equator mean meridional circulation causes the composition to change in response to the storm. Atomic oxygen densities increase by about 10 percent and molecular oxygen densities decrease by a similar amount. The above-described compositional changes occur in the TIGCM between two steady-state runs and represent equilibrium conditions.

The calculated zonal mean difference fields for NO, $N(^4S)$, $N(^2D)$ and Ne for the equinox conditions are shown in Figure 5a-d, respectively. Auroral particle precipitation dissociates N_2 at high latitudes and the resulting N reacts with O_2 to produce NO. As a result the NO densities increase greatly at high latitudes, and, by changing compositional structure and temperature, even at mid-latitudes. Several hundred percent increases occur in the lower thermosphere from the pole to 60 degrees latitude. NO increases at low latitudes are relatively small. The increase in NO in the polar lower thermosphere depletes $N(^4S)$ below about 250 km as shown in Figure 5b. Since NO has a larger density than $N(^4S)$ in the lower thermosphere, it destroys $N(^4S)$ through the "cannibalistic" chemical reaction $NO+N(^4S) \longrightarrow N_2+O$. Above about 250 km, the $N(^4S)$ num-

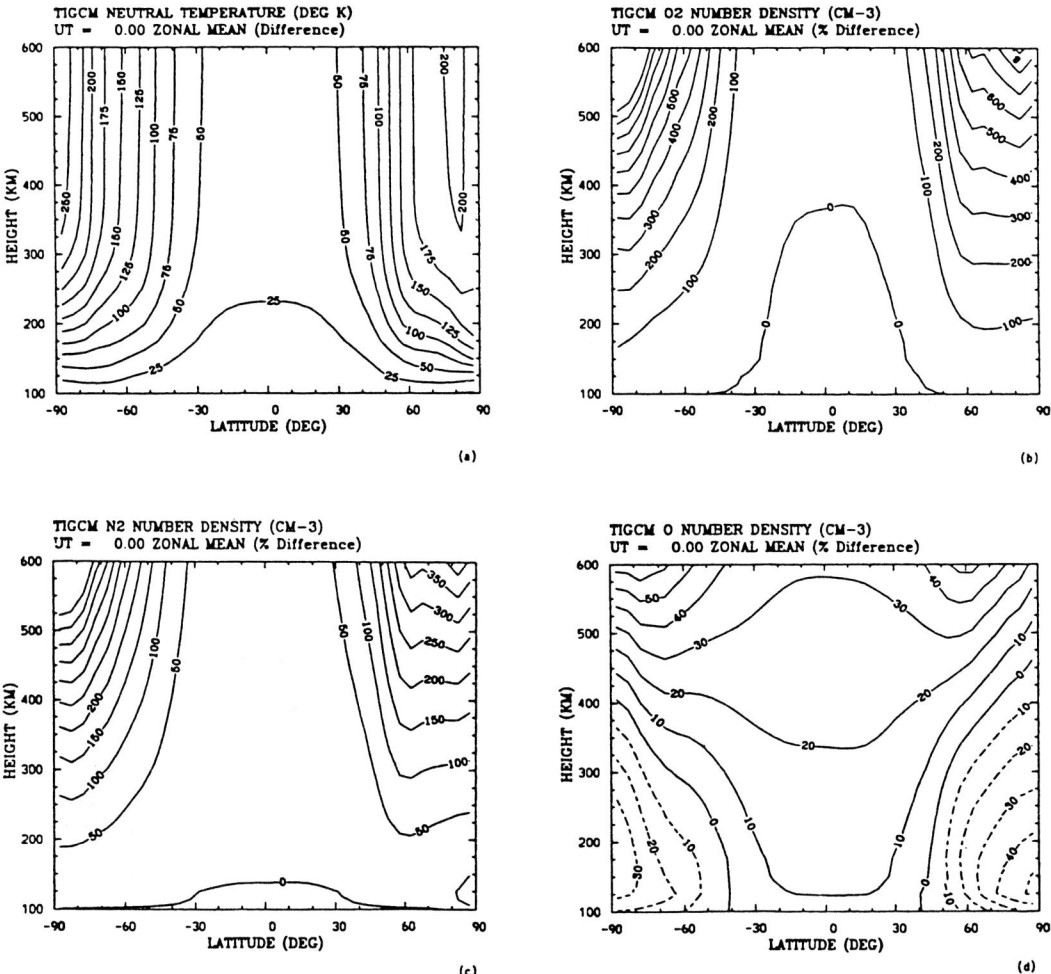

Figure 4. Zonally-averaged difference fields for the case of moderate geomagnetic activity minus quiet conditions, at equinox, (a) temperature (K); (b), (c), and (d) are percent difference fields for O_2, N_2, and O, respectively.

ber density increases at fixed altitudes due to the upward expansion of the atmosphere caused by auroral heating.

Figure 5c shows the increase in $N(^2D)$ caused by auroral activity and by compositional and temperature changes associated with thermospheric dynamics. In the lower thermosphere the large enhancements are caused primarily by particle precipitation, but some enhancement occurs because of O depletion and the consequent reduction of $N(^2D)$ quenching by O.

The zonal mean difference field of electron density, shown in Figure 5d, indicates enhancements from particle precipitation in the lower thermosphere, depletions in the upper thermosphere from enhanced recombination caused by increases in the N_2 and O_2 number densities, and horizontal plasma transport due to enhanced ion drifts. Enhancements in electron densities at low latitudes are caused by an increase in solar production from enhanced O_2, O and N_2 number densities at fixed altitudes. It follows that

the changes in the global distribution of electron density from dynamic and chemical effects are obviously quite complex. In these calculations the effects of altered magnetosphere/ionosphere plasma heat and mass exchange are not included. Additional electron density variations should therefore be superimposed upon these dynamic and chemical perturbations caused by magnetosphere/ ionosphere plasma exchange processes primarily in the upper ionosphere.

A similar atmosphere and ionosphere response occurs for solstice conditions as discussed by Roble [*1992*]; there is, however, a major summer-winter asymmetry. The solar driven mean meridional circulation is from the summer-to-winter hemisphere with an auroral heating superimposed mean circulation from the pole-to-low-latitudes. In the summer hemisphere the auroral heating reinforces the summer-to-winter circulation, where in the winter hemisphere it opposes the solar driven circulation. As a result, the storm

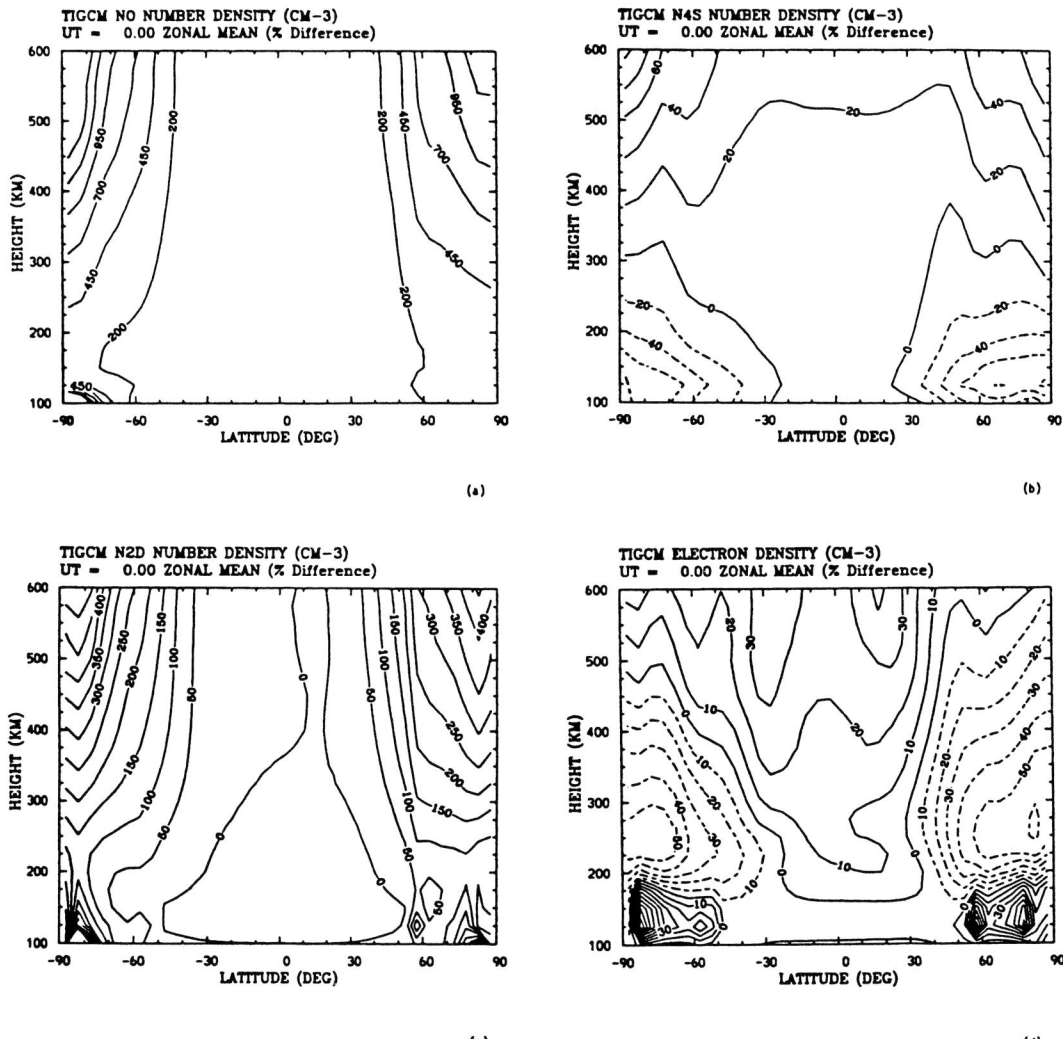

Figure 5. Zonally-averaged difference fields for the case of moderate geomagnetic activity minus quiet conditions, at equinox, (a), (b), (c), and (d) are percent difference fields for NO, N(^{4}S), N(^{2}D), and Ne, respectively.

effects are transported much further to lower latitudes in the summer hemisphere, but confined to the polar regions in the winter hemisphere. The species response discussed for equinox conditions are similar for solstice conditions but with a pronounced asymmetry.

9. DYNAMO EFFECTS OF DISTURBANCE WINDS

The changes in thermospheric circulation during a storm influence the global ionospheric electric fields and currents through dynamo action [*e.g., Axford and Hines, 1961*]; the winds move the electrically conducting ionospheric plasma through the geomagnetic field. This creates an electromotive force that drives currents and leads to polarization electric fields. The electric fields and currents produced by the disturbed winds have certain characteristic features.

Banks [*1972*] noted that the high-latitude winds set into motion by the rapidly convecting ions can generate an electric field that continues the plasma motion, if the source of magnetospheric field-aligned currents into the ionosphere is cut off. This effect was compared to a flywheel, since it would tend to maintain magnetospheric plasma convection in the aftermath of strong magnetospheric activity. Lyons et al. [*1985*], Deng et al. [*1991, 1993*], Thayer and Vickrey [*1992*], and Lu et al. [*1995*] examined this phenomenon from a different perspective. They calculated the current that would be driven by the winds alone, without an electric field. Effectively, this can be considered to correspond to a short circuit, with no electric field, as opposed to Banks' [*1972*] open circuit, with no field-aligned current.

Plate 7, adapted from Lu et al. [*1995*], illustrates the relationship between the currents driven by the magnetospheric

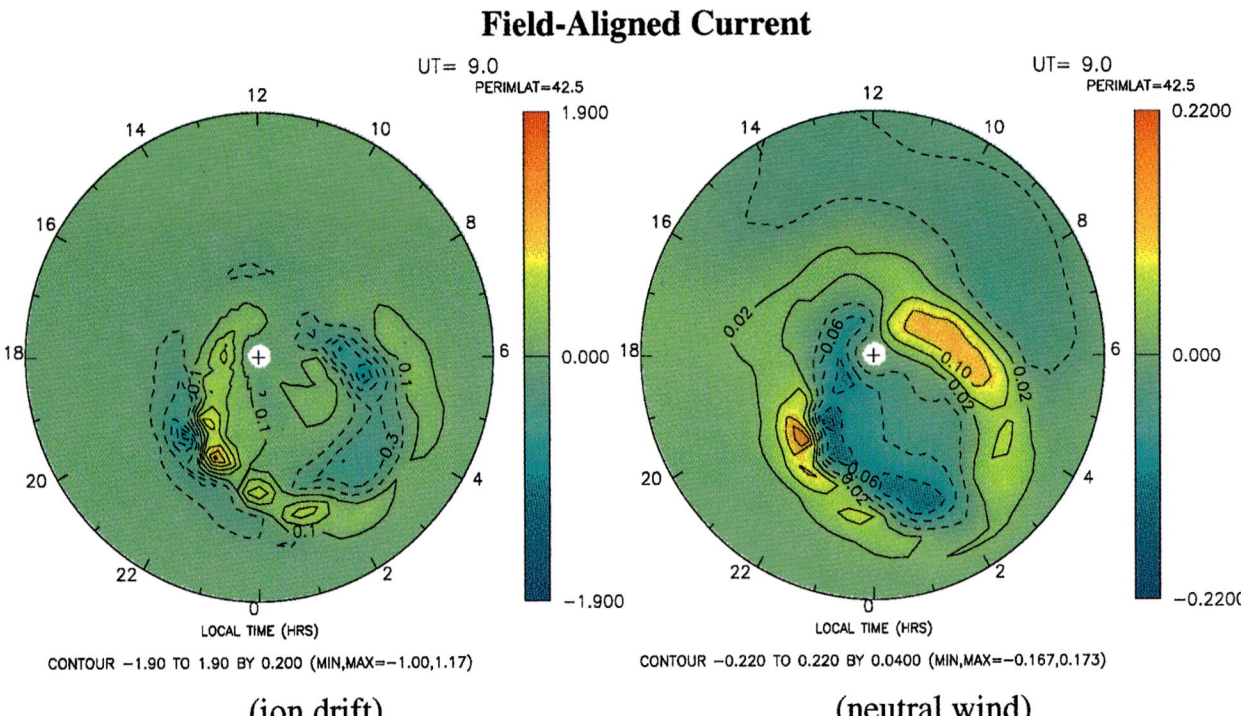

Plate 7. Illustration of the relationship between field-aligned currents driven by the magnetospheric electric field, on the left hand side, and those driven by the high-latitude neutral wind system, on the right.

electric field and those driven by the high-latitude winds, as modeled with the National Center for Atmospheric Research Thermosphere-Ionosphere-Electrodynamics General Circulation Model (NCAR TIE-GCM). The magnetospheric convection electric field, which is imposed on the model as an external input, drives height-integrated horizontal electric currents whose divergence corresponds to the vertical field-aligned current density shown on the left. On the poleward side of the auroral oval typical Region I field-aligned currents are seen, upward in the evening and downward in the morning, while on the equatorward side of the auroral oval typical Region II field-aligned currents are seen, downward in the evening and upward in the morning. The winds that have been spun up by the ion convection drive horizontal currents whose divergence gives the field-aligned currents shown on the right. The directions of the wind-driven and electric-field-driven current components are generally opposite, but the magnitude of the wind-driven current is only about 20 percent that of the electric-field-driven current. For a given electric field, then, the effect of the wind is to reduce the total current. If the magnetospheric source tries to maintain a given field-aligned current flow, however, then the effect of the wind will be to require development of a stronger electric field. If the current is cut off, a residual electric field will remain—the "flywheel" field.

Forbes and Harel [1989] examined how the flywheel effect influences the so-called shielding effect associated with energetic magnetospheric plasma. Under steady-state conditions this plasma tends to produce a quasi-equipotential layer at the inner edge of the ring current; ionospheric regions equatorward of this inner edge, as mapped along geomagnetic field lines to the ionosphere, then tend to be electrically shielded from higher latitudes [e.g., Wolf et al. 1986]. The time constant for shielding to become established varies inversely to the ionospheric conductance. Since the winds tend to reduce the current, their effect is somewhat similar to a reduction in the ionospheric conductivity, and therefore an enhancement of the shielding.

A variant of the flywheel concept is what has been called the fossil-wind effect [Spiro et al., 1988; Fejer, 1991]. During a magnetic storm, the auroral oval and the pattern of high-latitude convection expand, so that the wind system accelerated by ion drag extends down in latitude to upper mid-latitudes. As the storm calms, this wind system continues to exert a dynamo influence. The additional feature of the fossil-wind concept is the relation of this dynamo influence to the shielding layer. In the calming phase of a magnetic storm, the ionospheric projection of the ring-current inner edge retreats to higher magnetic latitudes, exposing the lower-latitude ionosphere to the full dynamo effects of the winds that were previously accelerated by ion convection. Some of the polarization electric field associated with this dynamo effect can spread all the way to the magnetic equator. Spiro et al. [1988] and Fejer et al. [1990] found that the effect could, potentially, help explain certain long-lived electric-field perturbations seen at the magnetic equator (see next section).

10. LOW LATITUDE ELECTRODYNAMICS

At mid-latitude many of the ionospheric changes result from wind or composition changes in the neutral atmosphere. The effects of electrodynamics may play some role but it is at low latitude, due to the geometry of the Earth's magnetic field, where the effects of electrodynamics are most pronounced.

Fejer and Scherliess [1995] used extensive radar measurements of F-region vertical plasma drifts from Jicamarca, and auroral electrojet indices to determine the storm time dependence of equatorial electric fields. The storm-time ion drifts at low latitudes result from a combination of prompt penetration of electric fields, and the dynamo action of disturbed winds that reach the equatorial ionosphere a few hours after the onset of magnetic activity. The signature of the equatorial disturbance depends on the relative strengths, and time histories, of these two components.

Figure 6 presents an example of the equatorial vertical drift for an idealized variation of the AE index, from Fejer and Scherliess [1995]. To allow a direct comparison with global convection models, an equivalent scale of polar cap potential drop is also shown, based on an empirical relationship with AE. The lower five panels show the drift at all local times during the disturbance shown in the top panel. At time t0, 30 minutes after onset, the initial penetration of the magnetospheric field has forced a downward plasma drift in the morning sector. One hour later, at time t1, plasma redistribution in the magnetosphere has effectively shielded the equatorial ionosphere. Four-and-a-half hours after onset of the disturbance an upward drift is predicted. This is about the time the first signature of the dynamo action of the storm-time winds are seen at low latitudes. Notice that this time is similar to the time for propagation of gravity waves from high latitudes (see Plate 3), and it therefore could be a local F-region dynamo effect. Following the decrease in AE after a simulated storm time of five hours, the reverse penetration effect combines positively with the existing dynamo field to produce the maximum upward drift velocities. One hour later, shielding has again removed the penetration field, leaving the long-lived dynamo fields to persist.

Fejer and Scherliess [1995] showed that penetration of magnetospheric electric fields in their empirical model has good agreement with that modeled by the RICE convection model. The dynamo fields predicted by their model was further shown to be in good agreement with that simulated by Blanc and Richmond [1980].

Figure 6 showed the short term response following the rise and fall of magnetospheric convection as indexed by AE. Evidence for an equatorial disturbance dynamo electric field, with time delays of 18-24 hours, was presented by Fejer [1996]. Using the relative efficiency of the storm-

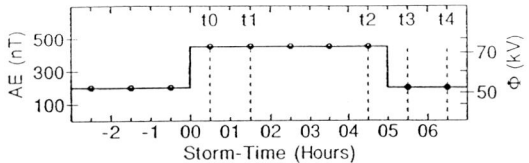

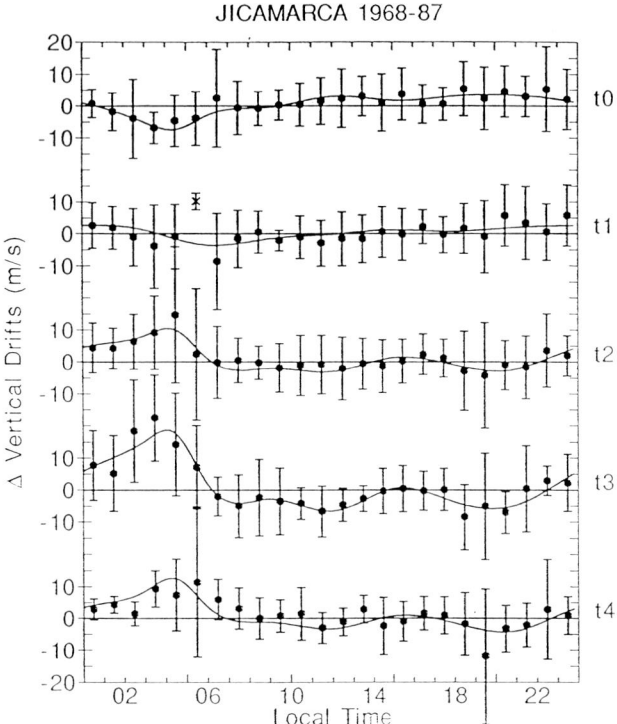

Figure 6. Idealized variation of the AE index and of the polar cap potential drop, and equatorial vertical plasma drift perturbation for the five storm times shown in the upper panel. The x denotes an average velocity from less than five samples; the scatter bars are also shown [after *Fejer and Scherliss, 1995*].

time ionospheric dynamo for generation of equatorial vertical plasma drift perturbations, he showed that following the first pulse in the early morning sector, a second increase occurred lasting several hours. He also indicated that the subsequent maximum in the efficiency arose 24 hours after the AE onset. It is tempting to associate the various time scales with phenomena in the neutral atmosphere. The initial response after 4 hours could be associated with gravity wave propagation and the F region dynamo. The second surge could be associated with the slower build up and penetration of the dynamo winds deeper into the thermosphere. The final signature, on the second day, may be associated with the transport of composition changes to the equator causing changes in conductivity.

Blanc and Richmond [*1980*] also showed that the effects of the disturbance dynamo from the globally altered wind system extends to low latitudes and can persist for a day or more beyond the storm. The model of Blanc and Richmond [*1980*] predicted that the disturbance-dynamo contribution to low-latitude east-west electric fields generally produces downward plasma drifts at day and upward drifts at night, opposite to the regular daily variation.

11. MIDDLE AND LOWER ATMOSPHERE RESPONSE

The atmospheric response to geomagnetic storms is most pronounced in the upper thermosphere where the heating per unit mass by auroral particle and Joule heating maximizes. Most of the geomagnetic storm energy is deposited in the lower thermosphere near 115 km, and it mainly affects the atmosphere above that altitude. In general, this energy does not directly perturb the atmosphere dynamically a few scale heights below the altitude of maximum heating. The large-scale geomagnetic storm generated electric fields can, however, penetrate downward to the ground and alter the vertical electric field at the ground that is maintained by world-wide thunderstorm activity as discussed by Roble [*1991*]. These affects are mainly confined to the polar cap where the storm electric field is enhanced.

As a general rule only 30 percent of the energy directly deposited by auroral precipitation ends up heating the atmosphere. Most of the particle energy is radiated back to space as either airglow and IR emissions, and the remainder of the energy is used to dissociate O_2. Since O can only recombine by a three-body recombination, any O dissociated in the upper thermosphere must be transported by diffusion to the lower thermosphere near 100 km where the recombination is sufficiently fast to reclaim the dissociation energy. Energy is then transported from the upper thermosphere to the lower thermosphere where it is radiated to space by CO_2 at 15 μm and NO at 5.3 μm. The energetics of the thermosphere and mesosphere has been discussed in detail by Roble [*1995*].

Auroral produced NO that is generated in the lower thermosphere can be transported downward by the wind system into the middle atmosphere, especially in the polar night region during solstice, where the lifetime of NO is long because of the absence of solar photodissociation. Indeed, if the enhanced NO can reach the vicinity of 40-50 km it can then catalytically destroy ozone. This mechanism has been observed and modeled for energetic solar proton events but it has not been investigated for geomagnetic storm produced NO.

Changes in the mean circulation in the lower thermosphere may also have an important effect in altering the transmission of waves propagating upward from the lower atmosphere, such as tide and planetary waves. Auroral generated gravity waves propagating away from the auroral zone may also interact with waves propagating upward from below. These wave-wave and wave-mean flow interactions have not been investigated, but they may be quite important for coupling solar-terrestrial energy between atmospheric regions.

The response of the upper atmosphere to geomagnetic storms has been studied in some detail and is quite complex. The geomagnetic storm response in the lower thermosphere and upper mesosphere is not well understood and there is an important need for measurements describing this response to improve our overall understanding of the effects of geomagnetic storms on the Earth's atmosphere.

12. CONCLUSIONS

Understanding the response of the upper atmosphere to geomagnetic storms has come long way, but there are still many challenges for the future. It is well known that the neutral atmosphere composition changes cause negative ionospheric phases, at mid-latitude for instance, but is it responsible for most of the change? This question cannot be answered yet, and to do so will require a comprehensive data set showing the spatial structure of changes in composition and plasma density, before, during, and after the driven phase of a storm. Models can be used to fill gaps in the data to try to answer the questions, but the problem then is that storm inputs are not known with enough accuracy. Knowledge of the spatial distribution, magnitude, and temporal variation of the input fields are essential before predictions of the global thermosphere and ionosphere changes during a storm can be modeled. Only when this information is available will it be possible to determine if additional mechanisms are required. Questions remain regarding the role of vibrationally excited molecular nitrogen in the recovery phase, and about plasmasphere loss and refilling at mid-latitudes. Understanding the fascinating electrodynamic changes observed at low latitudes is also a challenge for the future. There are also many questions relating to the impact of storms on the middle and lower atmosphere to be addressed in the future.

Acknowledgments. The first two authors acknowledge discussions over the years with Henry Rishbeth, whose insight into the workings of the upper atmosphere has led to many of the advances in understanding of the midlatitude response to geomagnetic storms, and to discussions with Bela Fejer on the nature of the dynamic and electrodynamic coupling at low latitude. Support of the first two authors for this work was by NASA grant NAGW-3530. The National Center for Atmospheric Research is sponsored by the National Science Foundation; this research was also supported by NASA Space Physics Theory Program.

REFERENCES

Allen, J., L. Frank, H. Sauer, and P. Reiff, Effects of the March 1989 solar activity, *EOS, 70, No. 46,* 1479-1488, 1989.

Axford, W. I., and C. O. Hines, A unifying theory of high-latitude geophysical phenomena and geomagnetic storms, *Can. J. Phys., 39,* 1433-1464, 1961.

Banks, P.M., Magnetospheric processes and the behavior of the neutral atmosphere, *Space Res., 12,* 1051-1067, 1972.

Blanc, M., and A.D. Richmond, The ionospheric disturbance dynamo, *J. Geophys. Res., 85,* 1669-1686, 1980.

Buonsanto, M.J., A case study of the ionospheric storm dusk effect, *J. Geophys. Res., 100,* 23857-23869, 1995.

Buonsanto, M., J.E. Salah, K.L. Miller, W.L. Oliver, R.G. Burnside, and P.G. Richards, Observations of neutral circulation at mid latitudes during the Equinox Transition Study, *J. Geophys. Res., 94,* 16987-16997, 1989.

Buonsanto, M., J.C. Foster, A.D. Galasso, D.P. Sipler, and J.M. Holt, Neutral winds and thermosphere/ionosphere coupling and energetics during the geomagnetic disturbances of March 6-10, 1989, *J. Geophys. Res., 95,* 21033-21050, 1990.

Burns, A.G., and T.L. Killeen, The equatorial neutral wind response to geomagnetic forcing, *Geophys. Res. Lett., 19,* 977-980, 1992.

Burns, A.G., T.L Killeen, and R.G. Roble, A theoretical study of thermospheric composition perturbations during an impulsive geomagnetic storm, *J. Geophys. Res., 96,* 14153-14167, 1991.

Codrescu, M.V., R.G. Roble, and J.M. Forbes, Interactive ionospheric modeling: A comparison between TIGCM and ionosonde data, *J. Geophys. Res., 97,* 8591-8600, 1992.

Cravens, J.D., A.C. Nicholas, L.A. Frank, and D.J. Strickland, Variations in FUV dayglow brightness following intense auroral activity, *Geophys. Res. Lett., 25,* 2793-2796, 1994.

Crowley, G., J. Schoendorf, R.G. Roble, and F.A. Marcos, Cellular structures in the high-latitude thermosphere, *J. Geophys. Res., 101,* 211-224, 1996.

Deng, W., T.L. Killeen, A.G. Burns, R.G. Roble, J.A. Slavin, and L.E. Wharton, The effects of neutral inertia on ionospheric currents in the high latitude thermosphere following a geomagnetic storm, *J. Geophys. Res., 98,* 7775-7790, 1993.

Deng, W., T.L. Killeen, A.G. Burns, and R.G. Roble, The flywheel effect: ionospheric currents after a geomagnetic storm, *Geophys. Res. Lett., 18,* 1845-1848, 1991.

Evans, D.S., T.J. Fuller-Rowell, S. Maeda, and J. Foster, Specification of the heat input to the thermosphere from magnetospheric processes using TIROS/NOAA auroral particle observations, *Adv. Astron. Sci., 65,* 1649-1667, 1988.

Fejer, B.G., The electrodynamics of the low-latitude ionosphere: Recent results and future challenges, *J. Atmos. Terr. Phys.,* in press, 1996.

Fejer, B.G., and L. Scherliess, Time dependent response of equatorial ionospheric electric fields to magnetospheric disturbances, *Geophys. Res. Lett., 22,* 851-854, 1995.

Fejer, B.G., Low latitude electrodynamic drifts: a review, *J. Atmos. Terr. Phys., 53,* 677-693, 1991.

Fejer, B. G., R. W. Spiro, R. A. Wolf, and J. C. Foster, Latitudinal variation of perturbation electric fields during magnetically disturbed periods: 1986 SUNDIAL observations and model results, *Ann. Geophysicae, 8,* 441-454, 1990.

Fesen, C.G., G. Crowley, and R.G. Roble, Ionospheric effects at low latitudes during the March 22, 1979, geomagnetic storm, *J. Geophys. Res., 94,* 5405-5417, 1989.

Forbes, J. M., and M. Harel, Magnetosphere-thermosphere coupling: an experiment in interactive modeling, *J. Geophys. Res.*, *94*, 2631-2644, 1989.

Forbes, J.M., R. Gonzalez, F.A. Marcos, D. Revelle, and H. Parish, Magnetic storm response of lower thermosphere density, *J. Geophys. Res.*, *101*, 2313-2320, 1996.

Foster, J.C., Storm-time plasma transport at middle and high latitude, *J. Geophys. Res.*, *98*, 1675-1687, 1993.

Foster, J.C., J.M. Holt, R.G.Musgrove, and D.S. Evans, Ionospheric convection associated with discrete levels of particle precipitation, *Geophys. Res. Lett.*, *13*, 656-659, 1986.

Frank, L.A., J.D. Craven, K.L. Ackerson, M.R. English, R.H. Eather, and R.L. Carovillano, Global auroral imaging instrumentation for the Dynamics Explorer mission, *Space Sci. Instrum.*, *5*, 369-393, 1981.

Fujiwara, H., S. Maeda, H. Fukinishi, T.J. Fuller-Rowell, and D.S. Evans, Global variations of thermospheric winds and temperature caused by substorm energy injections, *J. Geophys. Res.*, *101*, 225-240, 1996.

Fuller-Rowell, T.J., A two-dimensional, high-resolution, nested-grid, model of the thermosphere, 2. Response of the thermosphere to narrow and broad electrodynamic features, *J. Geophys. Res.*, *90*, 6567-6586, 1985.

Fuller-Rowell, T.J., M.V. Codrescu, R.J. Moffett, and S. Quegan, Response of the thermosphere and ionosphere to geomagnetic storms, *J. Geophys. Res.*, *99*, 3893-3914, 1994.

Fuller-Rowell, T.J., M.V. Codrescu, and I. Kutiev, Can modelling help us predict the ionospheric response to geomagnetic storms, *Proceedings of the 5th Solar Terrestrial Predictions Workshop*, Hitachi, Japan, January 23-27, 1996a.

Fuller-Rowell, T.J., M.V. Codrescu, R.J. Moffett, and S. Quegan, On the seasonal response of the thermosphere and ionosphere to geomagnetic storms, *J. Geophys. Res.*, *101*, 2343-2353, 1996b.

Kamide, Y., W. Sun, and S.-I. Akasofu, The average ionospheric electrodynamics for the different substorm phases, *J. Geophys. Res.*, *101*, 99-109, 1996.

Killeen, T.L., and R.G. Roble, An analysis of the high latitude thermospheric wind pattern calculated by a thermospheric general circulation model, 1. Momentum forcing, *J. Geophys. Res.*, *89*, 7509-7522, 1984.

Knipp, D.J. et al., Electrodynamic patterns for September 19, 1984, *J. Geophys. Res.*, *94*, 16913-16923, 1989.

Lu, G., A.D. Richmond, B.A. Emery, and R.G. Roble, Magnetospheric-ionospheric- thermospheric coupling: Effect of neutral winds on energy transfer and field-aligned currents, *J. Geophys. Res.*, *100*, 19643-19659, 1995.

Lyons, L. R., T. L. Killeen, and R. L. Walterscheid, The neutral wind flywheel as a source of quiet-time polar-cap currents, *Geophys. Res. Lett.*, *12*, 101-104, 1985.

Maeda, S., T.J. Fuller-Rowell, and D.S. Evans, Zonally averaged dynamical and compositional response of the thermosphere to auroral activity during Sept 18-24, 1984, *J. Geophys. Res.*, *94*, 16869-16883, 1989.

Meier, R.R., R. Cox, D.J. Strickland, J.D. Craven, and L.A. Frank, Interpretation of Dynamics Explorer far UV images of the quiet time thermosphere, *J. Geophys. Res.*, *100*, 5777-5794, 1995.

Prölss, G.W., Seasonal variation of atmospheric-ionospheric disturbances, *J. Geophys. Res.*, *82*, 1635-1640, 1977.

Prölss, G.W., and U. von Zahn, Seasonal variations in the latitude structure of atmospheric disturbances, *J. Geophys. Res.*, *82*, 5629-5632, 1977.

Prölss, G.W., Magnetic storm associated perturbations of the upper atmosphere: Recent results obtained by satellite-borne gas analyzers, *Reviews of Geophys. and Space Phys.*, *18*, 183-202, 1980.

Prölss, G.W., Storm-induced changes in the thermospheric composition at middle latitudes, *Planet. Space Sci.*, *35*, 807-811, 1987.

Prölss, G.W., L.H. Brace, H.G. Mayr, G.R. Carignan, T.L. Killeen, and J.A. Klobuchar, Ionospheric storm effects at subauroral latitudes: A case study, *J. Geophys. Res.*, *96*, 1275-1288, 1991.

Prölss, G.W., On explaining the local time variation of ionospheric storm effects, *Ann. Geophys.*, *11*, 1-9, 1993.

Prölss, G.W., Magnetic storm associated perturbations of the upper atmosphere, *AGU Geophysical Monograph*, this issue, 1996.

Richards, P.G., and D.G. Torr, A factor of 2 reduction in theoretical F2 peak electron density due to enhanced vibrationally excitation of N_2 in summer at solar maximum, *J. Geophys. Res.*, *91*, 11331-11336, 1986.

Richmond, A.D., and R.G. Roble, Dynamic effects of aurora-generated gravity waves on the mid-latitude ionosphere, *J. Atmos. Terr. Phys.*, *41*, 841-852, 1979.

Rishbeth, H., T.J. Fuller-Rowell, and A.D. Rodger, F-layer storms and thermospheric composition, *Physica Scripta*, *36*, 327-336, 1987.

Rishbeth, H., F-region storms and thermospheric composition, *Electromagnetic Coupling in the Polar Clefts and Cap*, 393-406, Eds. P.E. Sandholt and A. Egeland, 1989.

Roble, R.G., The Polar Lower Thermosphere, *Planet. Space Sci.*, *40*, 271-297,1992.

Roble, R. G., On modeling component processes in the Earth's global electric circuit, *J. Atmos. Terr. Phys.*, *53*, 831-847, 1991.

Roble, R. G., Energetics of the mesosphere and thermosphere, *The Upper Mesosphere and Lower Thermosphere*, Geophysical Monograph 87, 1-21, 1995.

Roble, R. G., and E. C. Ridley, A thermosphere-ionosphere-mesosphere electrodynamics general circulation model (TIME-GCM): Equinox solar cycle minimum simulations (30- 500 km), *Geophys. Res. Lett.*, *21*, 417-421, 1994.

Rodger, A.S., G.L. Wrenn, and H. Rishbeth, Geomagnetic storms in the Antarctic F- region. II. Physical interpretation, *J. Atmos. Terr. Phys.*, *51*, 851-866, 1989.

Skoblin, M.G., and M. Forster, An alternative explanation of ionization depletions in the winter night-time storm perturbed F2-layer, *Ann. Geophys.*, *11*, 1026-1032, 1993.

Sojka, J.J., and R.W. Schunk, A theoretical study of the high latitude F regions response to magnetospheric storm inputs, *J. Geophys. Res.*, *88*, 2112-2122, 1983.

Spiro, R. W., R. A. Wolf, and B. G. Fejer, Penetration of high-latitude-electric-field effects to low latitudes during SUNDIAL 1984, *Ann. Geophysicae*, *6*, 39-50, 1988.

Thayer, J. P., and J. f. Vickrey, On the contribution of the thermospheric neutral wind to high-latitude energetics, *Geophys. Res. Lett.*, *19*, 265-268, 1992.

Torr, M. R., D.G. Torr, P.G. Richards, and S.P. Yung, Mid-and low-latitude model of thermospheric emissions, 1, $O^+(2P)$ 7320 Å, and $N_2(2P)$ 3371 Å, *J. Geophys. Res.*, *95*, 21,147-21,168, 1990.

Wolf, R. A., G. A. Mantjoukis, and R. W. Spiro, Theoretical comments on the nature of the plasmapause, *Adv. Space Res.*, *6*, 177-186, 1986.

Wrenn, G.L., A.S. Rodger, and H. Rishbeth, Geomagnetic storms in the Antarctic F- region. I. Diurnal and seasonal patterns for main phase effects, *J. Atmos. Terr. Phys.*, *49*, 901-913, 1987.

T.J. Fuller-Rowell and M.V. Codrescu CIRES, University of Colorado and NOAA Space Environment Center, 325 Broadway, Boulder, CO 80303, USA.

R.G. Roble and A.D. Richmond High Altitude Observatory, National Center for Atmospheric Research, Boulder, CO 80308, USA.

Magnetic Storm Associated Perturbations of the Upper Atmosphere

Gerd W. Prölss

Institut für Astrophysik und Extraterrestrische Forschung, Universität Bonn, Germany

This review attempts to summarize what is presently known about the large-scale morphology and the physics of magnetic storm associated perturbations of the upper atmosphere. First the dissipation of electrical energy and the resulting basic atmospheric disturbance effects at higher latitudes are described. These include large plasma and neutral wind velocities, high plasma and neutral gas temperatures, and changes in the neutral gas composition. Parts of these perturbations are transported toward lower latitudes by traveling atmospheric disturbances and by large-scale wind circulation. Traveling atmospheric disturbances, for example, are thought to be responsible for the transient density perturbations observed at equatorial latitudes; large-scale winds are made responsible for the transport of composition perturbations toward middle latitudes. Local time and seasonal variations are attributed to the interaction of magnetospheric storm and solar radiation driven winds. Since charged particles are firmly embedded in their neutral gas environment, any perturbation of the neutral atmosphere will cause ionospheric disturbance effects. These include short-duration positive storms due to traveling atmospheric disturbances, long-duration positive storms due to changes in the large-scale wind circulation, and negative storm effects due to neutral composition changes. Even though significant progress has been made in understanding upper atmospheric storms, many open questions remain, and some of these are summarized at the end of this review.

1. INTRODUCTION

During magnetic storms a large amount of energy ($\simeq 10^{12}$ W) is dissipated in the polar regions, leading to profound changes in the global morphology of the upper atmosphere. Such perturbations form an important link in the complex chain of solar-terrestrial relations since their energy is ultimately supplied by the solar wind. They are also of great practical interest since they shorten the lifespans of satellites, degrade satellite ephemeris predictions, and disturb transionospheric radio communications.

Magnetic storm associated perturbations of the *ionized* component of the upper atmosphere were reported as far back as 1929 by *Hafstad and Tuve* [1929]. Even prior to this, radio engineers found that communication circuits involving wave propagation through the ionosphere were materially affected during magnetic storms. It was concluded that ionospheric disturbances — or "ionospheric storms", as they were to be called later on — must coincide with these events [e.g. *Espenschied et al.*, 1925; *Anderson*, 1928]. The first systematic investigations of this effect are credited to Appleton, Kirby, and Berkner and their co-workers [e.g. *Appleton et al.*, 1937; *Kirby et al.*, 1937; *Berkner and Seaton*, 1940]. Since then, more than 350 papers have been published on this subject, testifying to the complexity of this phenomenon. More recent reviews of this topic include those by *Matuura* [1972], *Danilov and Morozowa* [1985], *Abdu* [1991], and *Prölss* [1995].

At the time ionospheric storms were discovered, little was known about the properties and behavior of the *neutral* upper atmosphere. Early studies of auroral phenomena and of ionospheric perturbations suggested that the neutral atmosphere might be heated during magnetic storms [e.g. *Birkeland*, 1913; *Petersen*, 1927; *Appleton and Ingram*, 1935]. However, direct evidence that this is indeed the case was first obtained many years later when *Jacchia* [1959] detected magnetic storm associated changes in the decay rate of artificial satellites. Following this discovery, the "geomagnetic activity effect" on the neutral atmosphere has been studied extensively by using both satellite and ground-based observations. Even a cursory inspection of the review literature [e.g. *Roemer*, 1972; *Mayr et al.*, 1978; *Roble*, 1983; *Prölss et al.*, 1988] and of more recent publications [e.g. *Prölss*, 1993a; *Burns et al.*, 1995; *Fuller-Rowell et al.*, 1995] yields more than 100 references on this subject. In spite of this effort, many properties of neutral upper atmospheric perturbations (or "thermospheric storms") remain incompletely documented and insufficiently understood, indicating the need for additional work in this field.

The present review attempts to summarize what is presently known about the morphology and physics of these upper atmospheric perturbations. The material is organized in the following way. Section 2 describes the basic disturbance effects observed at higher latitudes. Their propagation to lower latitudes is discussed in Section 3. This section also summarizes some systematic variations of composition perturbations at middle latitudes. Ionospheric storms are the topic of Section 4. Finally, Section 5 lists some of the problem areas which need further study.

2. DISTURBANCE EFFECTS AT POLAR LATITUDES

It has long been suspected that auroral particle precipitation constitutes an important energy source for the polar upper atmosphere [e.g. *Birkeland*, 1913]. Later it was realized that electrodynamic (joule) heating should also significantly affect the energy balance of this region [*Cole*, 1962]. Measurements indeed confirmed the dominant role of these energy dissipation mechanisms [e.g. *Banks*, 1977]. They also indicated that globally, electrodynamic heating should be the more important energy source. For example, with the results obtained by *Spiro et al.* [1982], *Baumjohann and Kamide* [1984], *Ahn et al.* [1989], *Kosch and Nielsen* [1995], and *Lu et al.* [1995], a global electrodynamic to particle precipitation heat input ratio of 4 to 6 is derived.

The way in which the electrical energy is dissipated in the upper atmosphere is summarized in the block diagram of Fig. 1. In a first step, the ionospheric plasma is set into motion by the electric field. Without the neutral atmosphere an unimpeded $E \times B$ drift in the polar magnetic field would be established. In the presence of the neutral atmosphere, however, a signicant part of the drift energy is transferred to the neutral gas via collisions. This way the neutral gas is accelerated in the direction of the plasma drift, and strong neutral winds (sometimes in excess of 1200 m/s) are generated [e.g. *Fedder and Banks*, 1972; *Killeen et al.*, 1984]. Collisions also produce nondirectional motions, and this heat is evenly distributed among the colliding partners. Since there are much fewer plasma particles than neutral gas particles ($\leq 1\%$), the plasma temperature is increased by a much larger amount. This in turn leads to a heat exchange between both gas components which decreases the plasma temperature and increases the neutral gas temperature. It is this combination of heat exchange and direct frictional heating which is generally referred to as electrodynamic or joule heating [e.g. *St.-Maurice and Schunk*, 1981]. Model calculations indicate that more than 90% of the electrical energy dissipated is converted into heat [e.g. *Lu et al.*, 1995].

The block diagram also indicates that the neutral gas is in addition heated by gradients in the drift-generated

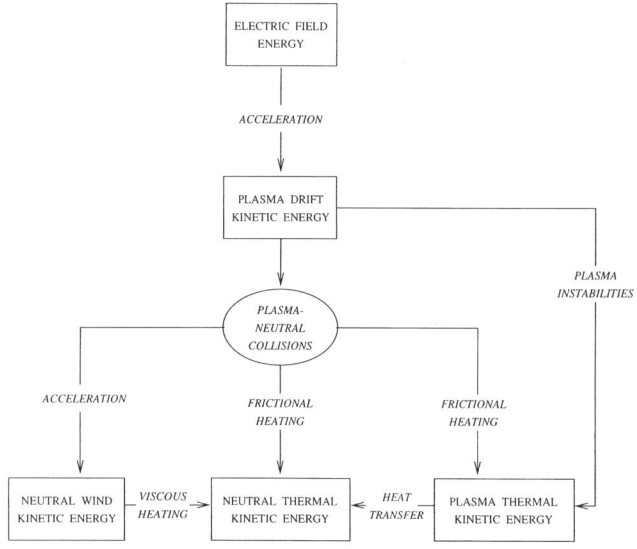

Fig. 1. Dissipation of electrical energy in the upper atmosphere [*Prölss et al.*, 1988].

winds (viscous heating) and by instabilities in the accelerated plasma component. Whereas viscous heating is probably less important [e.g. *St.-Maurice and Schunk*, 1981], plasma instabilities may well affect the heat balance of the lower thermosphere [e.g. *Schlegel and St.-Maurice*, 1981].

According to the processes summarized in the block diagram and schematically illustrated in Fig. 2, a satellite passing through a region of enhanced electrical energy dissipation should observe conspicuous changes in the upper atmosphere. This is indeed the case, and Fig. 3 shows the latitudinal variations of the ion drift and neutral wind velocities, of the ion and neutral gas temperatures, of the argon, nitrogen, oxygen and helium densities, and of the N_2/O density ratio as they are observed on a disturbed winter evening. The velocity and temperature measurements refer to a height range between 260 and 350 km and should only weakly depend on altitude. In contrast, the strongly height-dependent density data have been adjusted to a common altitude of 300 km and have been normalized to suitable reference values.

The most prominent feature of the measurements presented in Fig. 3 is the fairly abrupt transition from the more or less regular winter evening upper atmosphere to the disturbed polar atmosphere at about 53 degrees invariant latitude. This transition is marked by a large increase in the ion drift velocity which reaches nearly 2.6 km/s at 60 degrees latitude. The drifting ions have accelerated the neutral gas to peak wind velocities of more than 800 m/s. Due to frictional heating, a significant increase in the ion temperature and a moderate increase in the neutral gas temperature are observed. Since the ion temperature depends on the square of the difference between the ion and neutral gas velocities, a narrow spike is observed at 60 degrees latitude.

Another striking change concerns the density variations. Contrary to the general trend, an increase of the heavier gases argon and nitrogen and a decrease of the lighter gases oxygen and helium are observed. This surprising behavior may be explained as follows [e.g. *Shimazaki*, 1972; *Mayr and Volland*, 1972; *Burns et al.*, 1991]. Within the central heating region around 130 km altitude, the increase in temperature leads to an expansion and upwelling of density-rich gases, as illustrated in Fig. 2. This explains why the density of the major constituent N_2 increases at satellite altitudes inspite of the temperature enhancement. Continuous upwelling will eventually set up a vertical circulation cell of the Hadley type, as indicated in Fig. 4. Con-

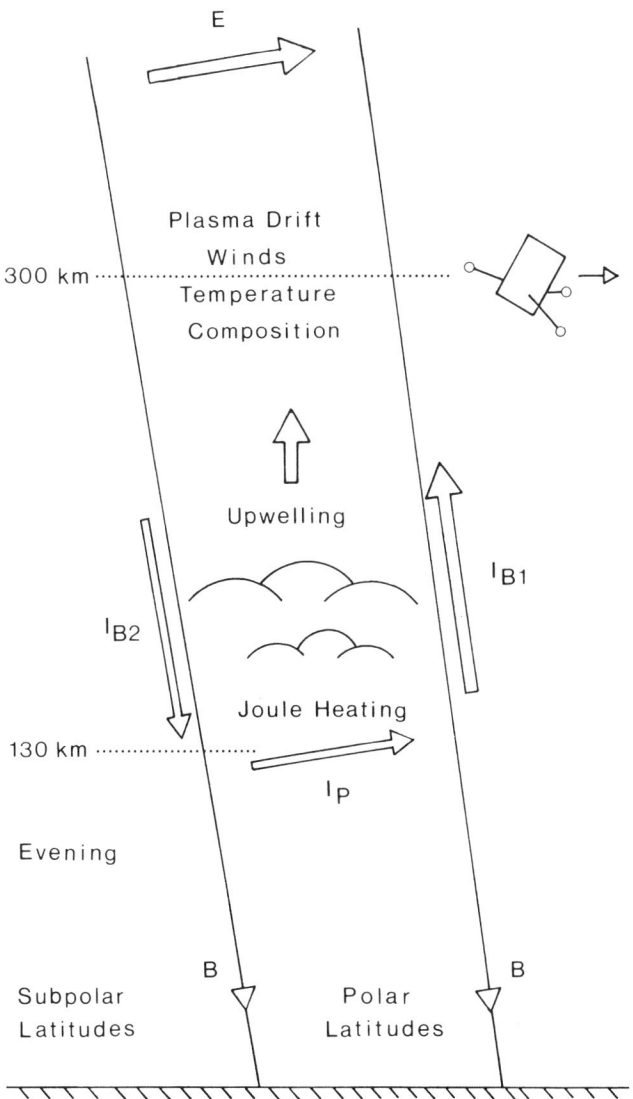

Fig. 2. Measurement situation of a satellite passing through a region of increased electrical energy dissipation. At the height of the satellite, the electric field E causes an ion drift perpendicular to both the electric and magnetic fields. This $E \times B$ drift accelerates the neutral gas in the direction of the plasma motion. At lower altitudes in the E-region of the ionosphere, the same electric field drives Pedersen currents (I_P) which heat the ambient atmosphere. The subsequent upwelling of the hot gases causes density and composition perturbations at the height of the satellite. The current circuit closes via region 1 (I_{B1}) and region 2 (I_{B2}) Birkeland currents in the magnetosphere. The situation shown refers to the evening sector of the northern hemisphere. Note that the distance between the field-aligned currents is of the order of a few degrees.

tinuity within this circulation cell requires that in the region of upwelling (where the horizontal divergence is assumed to be small), the particle flux of the major gas N_2 is approximately constant with height. Therefore the upward velocity has to increase in proportion to the decreasing density of the major gas. This increase in vertical velocity is much larger than would be required for a lighter gas like helium with its much larger scale height. Also it is much smaller than would be

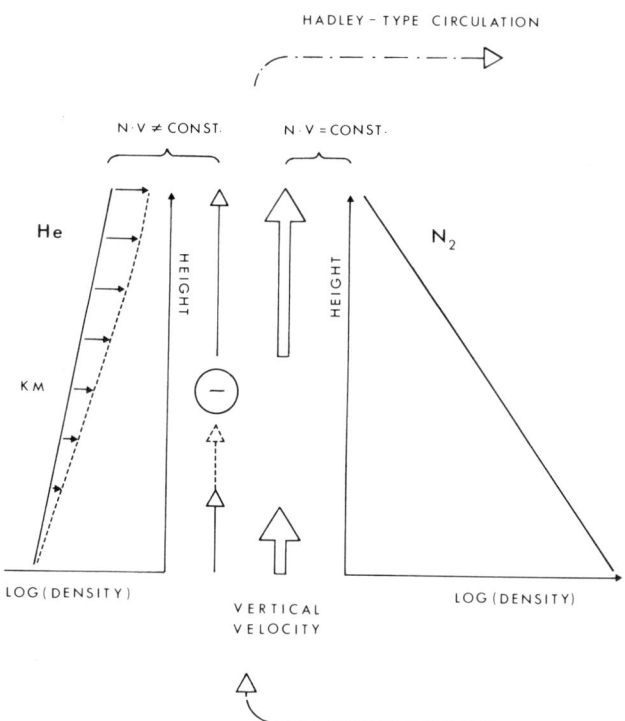

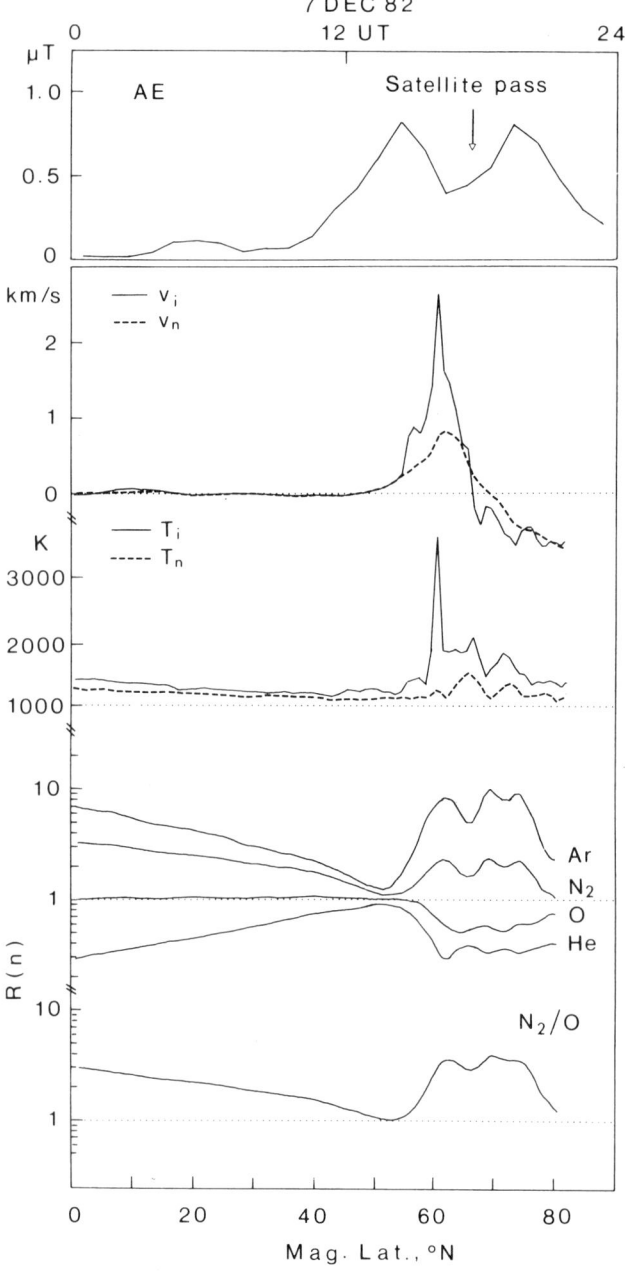

Fig. 4. Density perturbation of a minor gas (He) caused by wind-induced deviations from the diffusive equilibrium distribution; see text for details [Prölss et al., 1988].

required for the flux preservation of a heavier gas like argon. Thus if a circulation cell is established in the major gas and if this motion is impressed on the minor gas, this will lead to a depletion or enhancement of the

Fig. 3. Upper atmospheric perturbations in the evening sector of the polar region. The upper panel indicates the level of magnetic activity during the observations. The panel below shows the zonal ion drift velocity v_i and the zonal neutral wind velocity v_n; the ion temperature T_i and the neutral gas temperature T_n; the argon (Ar), molecular nitrogen (N_2), atomic oxygen (O), and helium (He) densities; and the N_2/O density ratio. These quantities are plotted as functions of magnetic (eccentric dipole) latitude. The drift and wind velocities are positive in the westward direction. Within the height range of the measurements (260 to 350 km), the velocities and the temperatures are assumed to be nearly independent of altitude. On the other hand, the gas densities and the N_2/O density ratio have been adjusted to a common height level of 300 km using standard hydrostatic techniques. These quantities have also been normalized to values obtained just outside the disturbance zone. Relative changes $R(n)$ with respect to these "undisturbed" values are then plotted. The same scale applies to all density variations, but the ordinate values apply only to the oxygen data. The measurements were obtained by the DE 2 satellite in the northern hemisphere at about 1600 UT on December 7, 1982, and refer to the 1800 LT sector.

minor gas. For example, more helium is transported away by the large velocities at higher altitudes than is supplied by the smaller velocities at lower altitudes. Indeed, if both gases were tightly coupled through collisional interactions, the depletion of the lighter gases and the simultaneous increase of the heavier gases would continue until both gases had the same scale heights. This is of course counteracted by diffusion, which tries to reestablish the original state of equilibrium. In the lower thermosphere, however, diffusive separation is too slow to fully compensate for the wind effects. Accordingly, lighter gases are depleted and heavier gases are enriched. This mechanism also predicts a reverse effect at places where downward-directed winds close the circulation cell.

The bottom part of Fig. 3 demonstrates that the molecular nitrogen to atomic oxygen density ratio is an excellent parameter to study the latitudinal structure of the composition disturbance. Whereas inside the perturbation zone this density ratio shows a very significant increase, indicating both the extent and magnitude of the density perturbations, the lower latitude regime is characterized by a lack of larger changes in this parameter. Later on, this density ratio will be used to document some systematic variations of the composition disturbance.

3. THERMOSPHERIC STORMS AT LOWER LATITUDES

Upper atmospheric perturbations are not restricted to the polar region. Rather, they may extend all the way to equatorial latitudes. This has led to speculation about an additional disturbance source at lower latitudes. In particular, heating by neutralized ring current particles has been suggested repeatedly in this context but has been shown to be inadequate to explain the observations [e.g. *Noël and Prölss*, 1993]. Now it is generally believed that most of the thermospheric density perturbations are *transported* from high to low latitudes. This transport is affected by traveling atmospheric disturbances and by large-scale wind circulation.

3.1. Traveling atmospheric disturbances

Thermospheric density perturbations observed at lower latitudes differ from those in the polar region. This is documented in Fig. 5, which shows magnetic storm associated changes in the argon, molecular nitrogen, atomic oxygen and helium densities as observed in

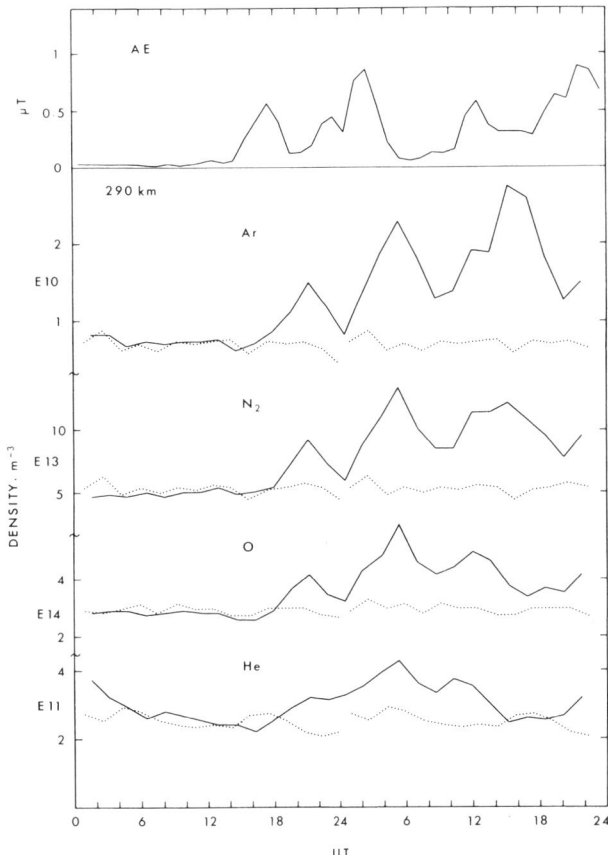

Fig. 5. Thermospheric storm effects at low latitudes. The upper panel shows the development of magnetic activity during the January 19/20, 1973 disturbance event. Associated changes in the argon (Ar), molecular nitrogen (N_2), atomic oxygen (O), and helium (He) densities are plotted in the lower panel. Measurements taken on January 18 serve as a quiet-time reference (dotted lines). The density values are averages which refer to the 5–10°S geographic latitude range. They have been adjusted to a common altitude of 290 km. Solar local time of observation is approximately 0045. Different linear scales have been used for the four constituents, with E10 meaning 10^{10} etc. [*Prölss*, 1982].

the equatorial region. All densities refer to a common altitude of 290 km, and measurements taken prior to the storm serve as a quiet-time reference.

In contrast to the composition changes at higher latitudes, all constituents are observed to increase. The increase is largest for argon, followed by that in nitrogen, oxygen, and helium. This suggests that part of the disturbance is caused by a thermal expansion of the atmosphere which carries density-rich air from lower to higher altitudes. Direct measurements confirm this increase in temperature at lower latitudes [e.g. *Blamont and Luton*, 1972; *Nisbet et al.*, 1977; *Biondi* and Meriwether, 1985; Burrage et al., 1992]. The enhancement of the helium density, on the other hand, is certainly a nonthermal effect. This is because such a light constituent is hardly affected by temperature changes. Therefore additional mechanisms like downwelling and/or compression must be invoked to explain the observations.

Figure 5 also indicates that the time delay between substorm (or convection) activity at higher latitudes and density perturbations at lower latitudes is relatively short. Thus a time lag of approximately 4 hours is derived for the first substorm period. If, with less confidence, the second peak in the density variation is associated with the larger increase in substorm activity around 0200 UT on the second day, a similar time delay is obtained. Assuming that the density perturbations originate at higher latitudes, average propagation velocities of many hundreds of meters per second are required to transport them to equatorial latitudes. Such large propagation velocities are typical for so-called *traveling atmospheric disturbances* (TADs). By this we mean pulse- or surge-like superpositions of atmospheric gravity waves. According to numerical modeling, such TADs are generated during substorm activity and propagate with high velocity from polar to equatorial latitudes [e.g. *Testud et al.*, 1975; *Richmond and Matsushita*, 1975; *Fuller-Rowell and Rees*, 1981; *Fesen et al.*, 1989; *Crowley et al.*, 1989; *Fuller-Rowell et al.*, 1994; *Fujiwara et al.*, 1996]. Therefore it seems plausible to attribute the initial density perturbations at equatorial latitudes to the energy dissipation of TADs launched in both polar regions. This disturbance scenario is illustrated schematically in Fig. 6. One may speculate that it is the equatorward-directed winds carried along by both TADs which cause a transient compression and heating of the low-latitude atmosphere.

3.2. Large-scale storm circulation

After the initial substorm bursts, activity may continue at a high level, as is the case during magnetic storms. This prolonged energy injection will lead to changes in the large-scale wind circulation. During daytime, for example, the high pressure area developing above the polar region will decrease the poleward-directed winds usually observed in this local time sector. This in turn will lower the energy drainage from this region, leading to a temperature enhancement at middle and lower latitudes [*Burns et al.*, 1992]. During nighttime, on the other hand, the same polar high pressure zone will reinforce the regular equatorward-directed winds and, together with ion drag, will pro-

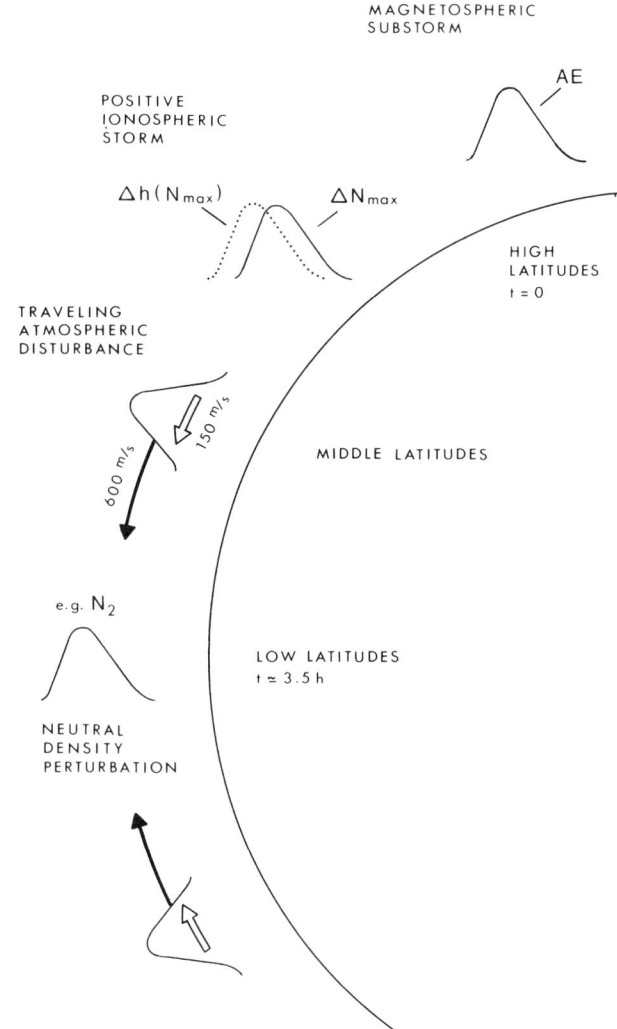

Fig. 6. Causal relationship between magnetospheric substorms (AE index), traveling atmospheric disturbances, short-duration positive ionospheric storms ($\Delta h(N_{max})$, ΔN_{max}), and density perturbations at low latitudes (N_2) [*Prölss*, 1993a].

duce storm surges. These high speed winds will advect air of disturbed composition out of the polar region toward middle latitudes, as is illustrated schematically in Fig. 7. In this manner an extended composition disturbance zone is generated like the one illustrated in Fig. 8.

Once generated, the composition perturbations at middle latitudes — while being moved around by winds [e.g. *Skoblin and Förster*, 1993; *Fuller-Rowell et al.*, 1994] — will corotate with the earth. This way they will also appear in the daytime sector. Figure 13 documents that extended composition perturbation zones

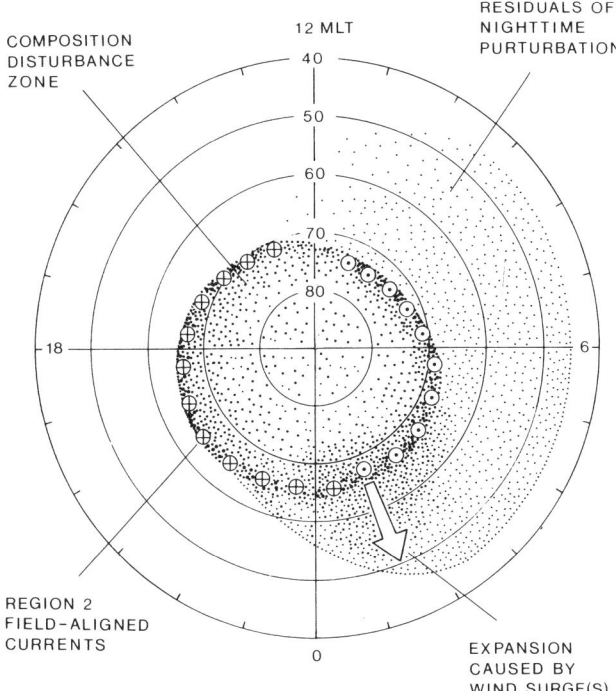

Fig. 7. Extent of the composition disturbance zone during moderately disturbed conditions. The polar coordinates are invariant magnetic latitude and magnetic local time. The dotted area indicates the distribution of the composition perturbation, and the circles mark the position of region 2 Birkeland currents flowing into and out of the ionosphere. An arrow indicates the expansion of the composition disturbance zone toward middle latitudes which occurs in the nighttime / early morning sector [Prölss, 1981].

larger in winter but extend to lower latitudes in summer (lower panel). Since seasonal winds blow from the summer to the winter hemisphere, they will inhibit the expansion of the composition disturbance in winter, but enhance it during summer.

Universal time / longitude variations are also to be expected since solar radiation effects are best ordered

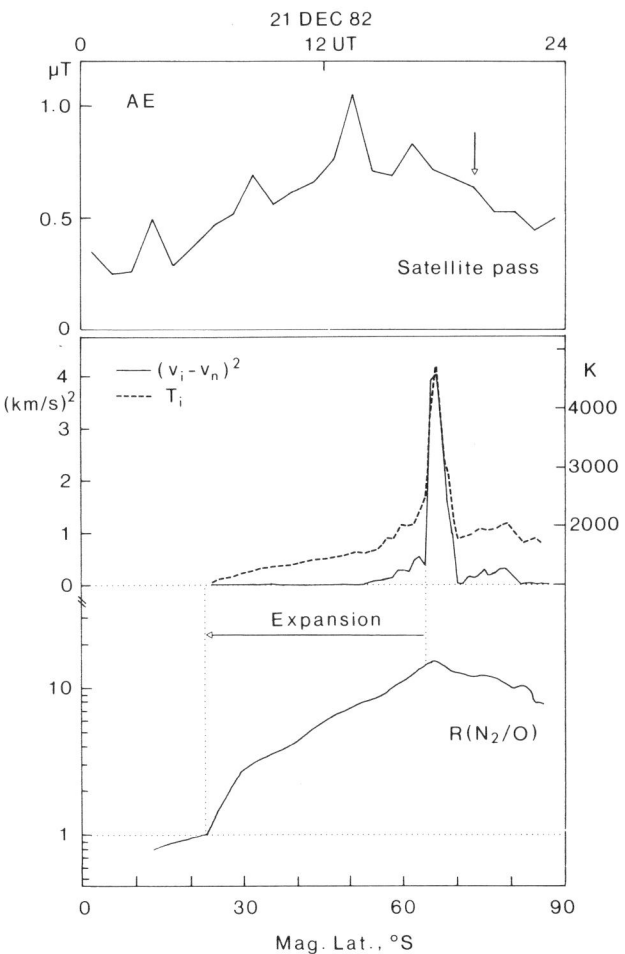

Fig. 8. Extent of the composition perturbation zone in the early morning sector. The upper panel describes the level of magnetic activity during the observation. The lower panel shows the latitudinal distribution of the joule heating and of the composition disturbance. The degree of joule heating is indicated by the square of the difference between the zonal ion (v_i) and neutral gas (v_n) velocities and by the ion temperature (T_i). Within the height range covered by the measurements (370 to 490 km), these parameters are assumed to be nearly independent of altitude. The composition perturbation is described by the N_2/O density ratio. This parameter has been adjusted to an altitude of 450 km and normalized to a value recorded outside the disturbance zone. The measurements were obtained by the DE 2 satellite in the southern hemisphere at about 1930 UT on December 21, 1982, and refer to the 0430 LT sector.

may be observed, for example, in the late morning sector.

3.3. Systematic variations of composition perturbations

Whereas composition perturbations like those illustrated in Figs. 3 and 8 exhibit a considerable degree of irregularity, certain systematic variations are also observed. These include changes with the intensity of magnetic activity, with local time, and with season [e.g. Prölss et al., 1988; Zuzic et al., 1996]. Such variations are documented in Fig. 9, where the N_2/O density ratio is used as an indicator.

Not surprisingly, the disturbance amplitude and extent grow larger with increasing magnetic activity (upper panel). Also, perturbations extend to much lower latitudes in the morning sector than, for example, in the evening sector (middle panel). This has already been attributed to the combined actions of regular and storm-induced winds. Finally, disturbance effects are

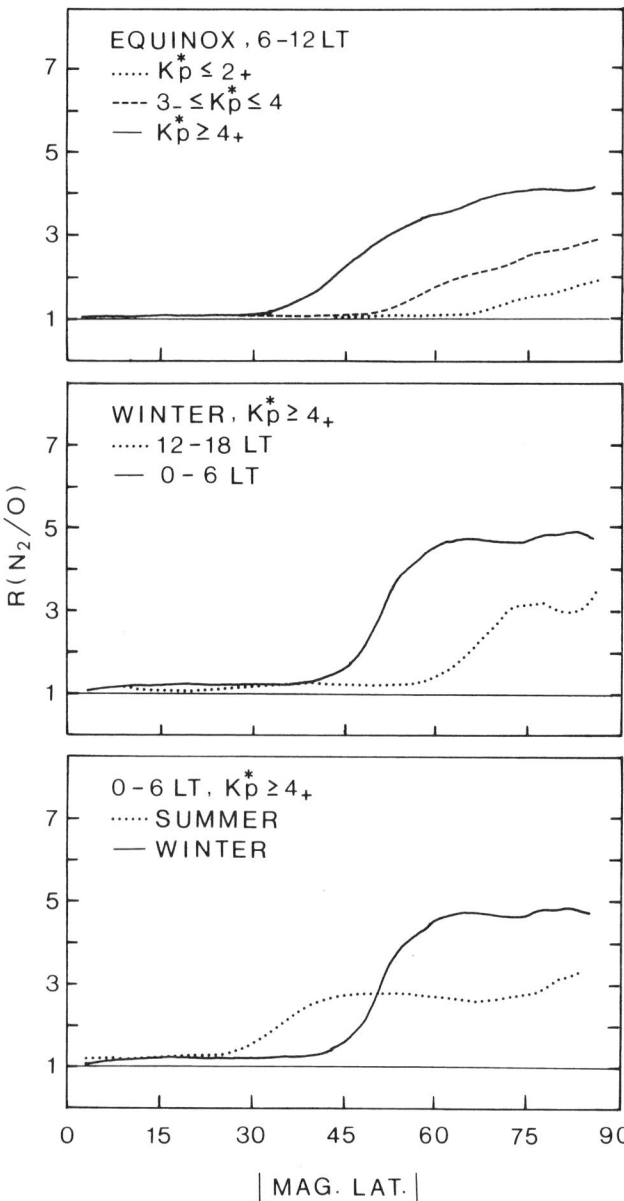

Fig. 9. Magnetic activity, local time, and seasonal variations in the magnitude and extent of neutral composition perturbations. Relative changes in the molecular nitrogen to atomic oxygen density ratio, $R(N_2/O)$, are plotted as a function of magnetic latitude. $R(N_2/O) = 1$ serves as a reference, meaning no change with respect to quiet conditions. All data refer to a common altitude of 280 km. The Kp* index is a modified Kp index which takes the development of the disturbance into account. Note that the curves represent mean latitudinal profiles which have been obtained by superimposing individual profiles using the equatorward boundary of the disturbance zone as a common reference location. This averaging procedure preserves the typical latitudinal structure of the disturbance [Zuzic et al., 1996].

in geographic coordinates, and magnetospheric effects in geomagnetic coordinates [e.g. Fuller-Rowell et al., 1994]. So far, however, such variations have been documented only for weakly disturbed conditions [e.g. Laux and von Zahn, 1979; Hedin and Carignan, 1985].

4. IONOSPHERIC STORMS

Charged particles are minor constituents of the upper atmosphere which are firmly embedded in their neutral gas environment. Therefore any perturbation of the neutral atmosphere will immediately entail corresponding perturbations of the ionosphere. The disturbance effects observed on these occasions are numerous, and all ionospheric parameters and regions are affected. Here we are interested in perturbations of the ionization density near the peak of the ionosphere. As illustrated in Fig. 10, this density can increase or decrease, and traditionally such deviations are denoted as *positive* and *negative ionospheric storms*, respectively. In what follows, three different kinds of disturbance effects will be discussed which can be attributed to perturbations of the neutral atmosphere. These are short-duration positive storm effects caused by traveling atmospheric disturbances; long-duration positive storm effects caused by changes in the large-scale wind circulation; and prolonged negative storm effects caused by changes in the neutral composition.

4.1. Short-duration positive storms

A very important feature of traveling atmospheric disturbances is that they carry along equatorward-directed winds of moderate magnitude. In Fig. 6 these winds are assumed to be of the order of 150 m/s. At middle latitudes such winds will drag the ionization up the inclined magnetic field lines, causing a transient increase in layer height ($\Delta h(Nmax)$ in Fig. 6). This uplifting of the ionization will lead to an increase in the ionization density ($\Delta Nmax$), which corresponds to a positive storm of limited duration.

Actual data supporting this disturbance scenario are presented in Fig. 11. Using the AE index as an indicator, the upper panel shows an isolated burst of substorm activity. In response to this energy injection and with a delay of about 2.5 hours, a positive storm develops at middle latitudes (bottom panel). This increase in ionization density is indeed preceded by a significant increase in layer height (middle panel).

That an increase in layer height will lead to positive storm effects is easily understood if the height dependences of the ionization production and loss rates are

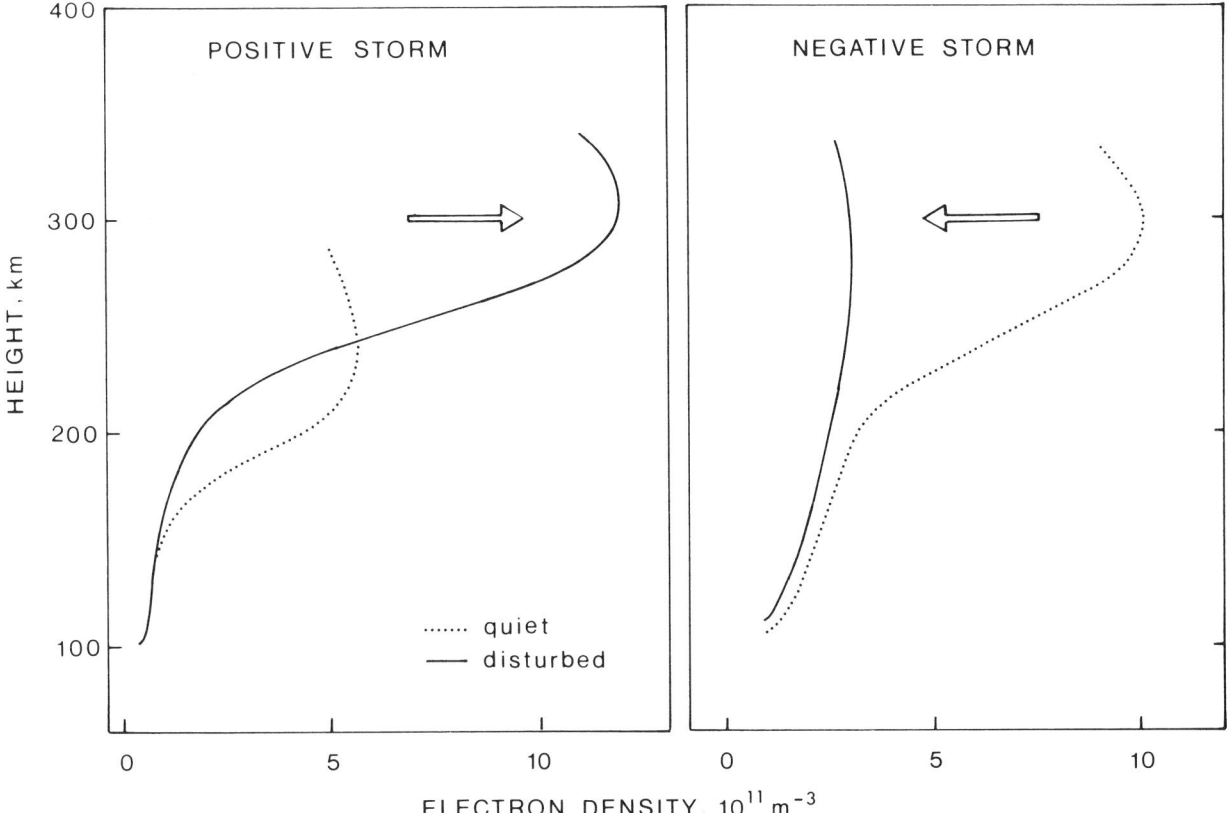

Fig. 10. Magnetic storm associated perturbations of the ionospheric electron density. The left panel presents a case when the ionization density has increased and a positive storm is observed. The height profiles of the electron density were recorded at Rome (42°N, 13°E) on March 4 (Ap = 4) and March 6 (Ap = 28), 1973. The local time of observation is 1700 h. The right panel illustrates a case when the ionization density is observed to decrease and a negative storm occurs. The density profiles were recorded at Mundaring (32°S, 116°E) on February 19 (Ap = 10) and on February 22 (Ap = 46), 1973. The local time of observation is 1100 h. The density values above the peak represent extrapolations.

considered. According to theory, the loss rate of ionization near the peak of the ionosphere is controlled by molecular nitrogen and molecular oxygen densities [e.g. *Rishbeth and Garriot*, 1969]. Therefore it decreases with height much faster than the production rate, the latter being proportional to the atomic oxygen density. An upward displacement of the ionization layer will thus lead to an effective increase in the ionization density [e.g. *Prölss*, 1995].

4.2. Long-duration positive storms

Besides the TAD-produced positive storms of short duration, longer lasting increases in the ionization density are observed. These long-duration events are also preceded by an increase in layer height, as is demonstrated in Fig. 12. In response to the continuing energy injection at polar latitudes (AE index), the layer height and the ionization density remain elevated for more than 6 hours.

Such prolonged increases in layer height are also attributed to meridional winds [e.g. *Prölss et al.*, 1991; *Codrescu et al.*, 1992]. However, this time changes in the large-scale wind circulation are required to maintain the upward drift of ionization for such a long duration. Here it suffices that the polar high pressure area reduces the intensity of the poleward-directed daytime winds, allowing a more "regular" build-up of the ionization density.

A special situation is encountered near the dip equator where the magnetic field is horizontal. Here meridional winds will not lead to a rise in the F-layer height and therefore cannot be directly responsible for positive storm effects. Nevertheless, storm-induced increases in the ionization density are observed and are attributed to a modification of the equatorial anomaly [e.g. *Prölss*, 1995]. For example, changes in the large-scale wind circulation will modify the dynamo electric field responsible for the fountain effect [e.g. *Blanc and Richmond*, 1980; *Fejer and Scherliess*, 1995]. Or equatorward-directed winds will oppose the poleward transport of ionization along the magnetic field lines and thus hinder the formation of the equatorial anomaly [*Burge et al.*, 1973].

4.3. Negative ionospheric storms

Neutral composition changes like those illustrated in Figs. 3 and 8 have important implications for the ionosphere. This is because a decrease of the oxygen density will decrease the production of ionization, and an increase in the molecular nitrogen density will increase the loss of ionization. Thus both changes combine to reduce the ionization density. Any station located below a composition disturbance zone should therefore observe negative ionospheric storm effects, and this is indeed the case.

Figure 13 shows storm-induced changes in the atomic oxygen and molecular nitrogen densities as observed in the late morning sector. In contrast to Figs. 3 and 8, a constant pressure coordinate system is used as is appropriate when correlating composition changes with ionospheric peak densities [*Rishbeth and Edwards*, 1989]. In addition, the locations of two ionosonde stations situated below the composition disturbance zone are indicated. The lower part of Fig. 13 shows the local time variation of the maximum electron density as observed at these stations. As is evident, both register very significant negative storm effects, which is in agreement with expectations.

Given the close connection between perturbations of the neutral composition and negative ionospheric storms, both exhibit the same kinds of systematic variations. These include changes with the intensity of the disturbance, with local time, and with season. As an example, Fig. 14 demonstrates that negative ionospheric storm effects extend to much lower latitudes during the summertime, in accord with corresponding asymmetries of the composition perturbation.

5. UNSOLVED PROBLEMS

While significant progress has been made toward understanding upper atmospheric storms, much remains

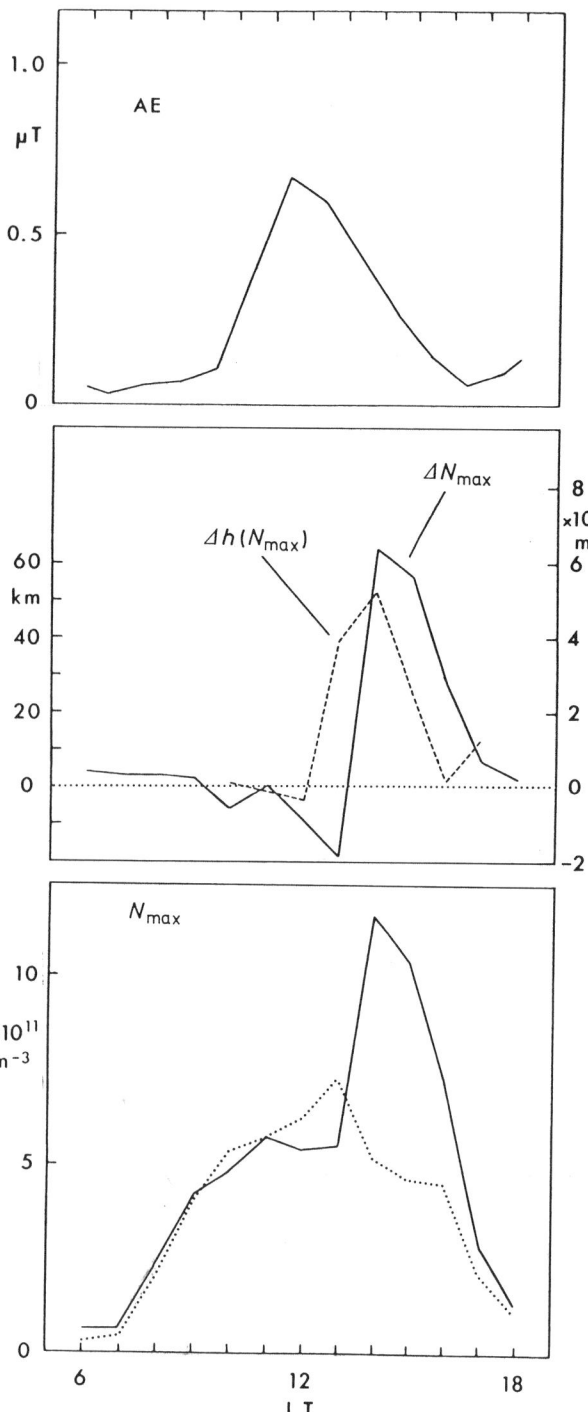

Fig. 11. Example of a short-duration positive ionospheric storm. In response to an isolated burst of substorm activity on January 23, 1973 (hourly averaged AE index, upper panel), the ionosonde at Slough (51°N, 359°E) observes a pulse-like increase in the maximum electron density (N_{max}, lower panel). Data obtained on January 22 serve as a quiet-time reference (dotted line). The middle panel demonstrates that this change in the maximum electron density (ΔN_{max}) is preceded by a significant - if transient - rise in the height of the maximum electron density ($\Delta h(N_{max})$) [*Prölss*, 1993b].

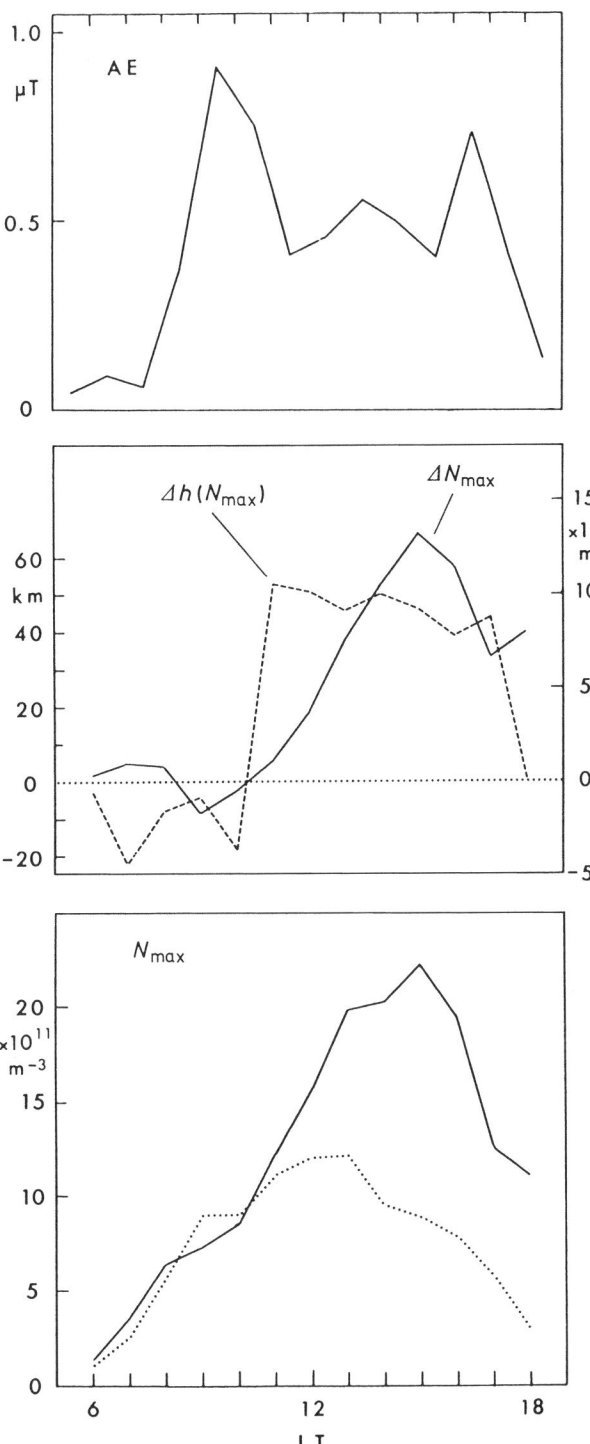

Fig. 12. Example of a long-duration positive ionospheric storm. The presentation corresponds to that of Fig. 11. The storm was observed at Pt. Arguello (36°N, 239°E) on February 21, 1973. Data recorded on February 18/19 serve as a quiet-time reference [Prölss, 1993b].

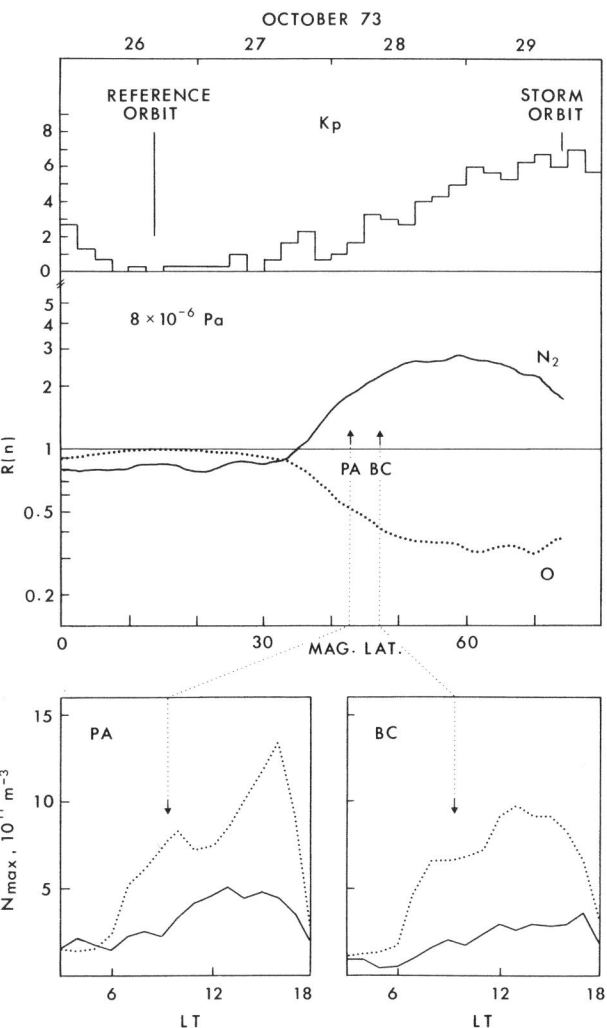

Fig. 13. Magnetic storm associated changes in the neutral gas composition and negative ionospheric storm effects. The upper panel shows the development of magnetic activity during a four-day interval in October 1973. It also indicates the times at which the storm data and the quiet-time reference data were recorded. The panel below presents storm-associated changes in both the molecular nitrogen and the atomic oxygen densities. Relative changes are plotted, and $R(n) = 1$ serves as a reference, meaning no change with respect to quiet times. The data refer to a common pressure level of 8×10^{-6} Pa ($\simeq 290$ km during quiet times). Local time and longitude of measurement are approximately 9 SLT and 245°E, respectively. The lower part of the figure shows the local time variations of the maximum electron density of the F2 layer as observed at two ionosonde stations: Boulder (40°N, 255°E) and Pt. Arguello (36°N, 239°E), whose relative positions with respect to the atmospheric disturbance zone are indicated by arrows. Stormtime data (October 29, solid lines) are compared to quiet-time reference data (October 26, dotted lines). The time of the satellite measurement is indicated by arrows [Prölss, 1991].

238 UPPER ATMOSPHERIC STORMS

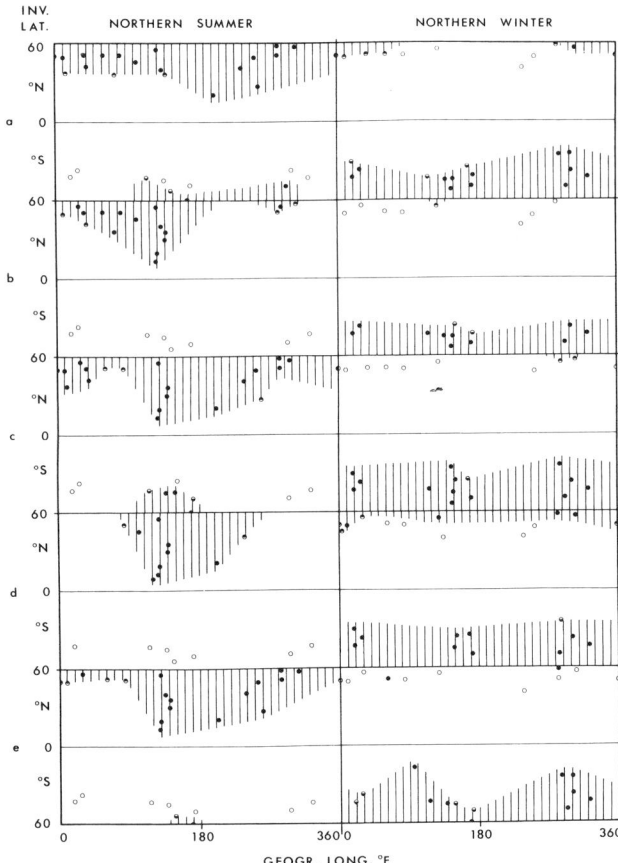

Fig. 14. Summer-winter asymmetry in the distribution of negative ionospheric storm effects. In an invariant latitude / geographic longitude coordinate system, the extent of the disturbance effects is described by the hatched area. The solid, semisolid, and open circles indicate that significant, smaller, and no negative storm effects, respectively, are observed at these locations. The data were obtained in 1973 during five northern summer (left side) and five northern winter (right side) storm events [Prölss, 1977].

to be learned about this intriguing phenomenon. This applies to both morphological and physical aspects. The following examples are just a small selection of problem areas.

1. Whereas neutral composition perturbations, like those illustrated in Fig. 3, are now generally attributed to vertical winds, this is by no means the only possible disturbance mechanism. Turbulent mixing, for example, has long been considered an alternate explanation [e.g. Chandra and Stubbe, 1971; Blum and Prölss, 1987], and so far the available data are not sufficient to discard this process. Therefore measurements obtained specifically in the disturbed lower thermosphere are urgently needed.

2. Also, the transport of composition perturbations from polar to middle latitudes has never been verified directly. Airglow measurements, like those obtained by the DE 1 satellite [Craven et al., 1994; see also Parish et al., 1994; Meier et al., 1995], should prove to be a very efficient means to monitor this propagation process.

3. Another open question is whether the composition perturbations observed at middle latitudes are really sufficient to explain the attendant negative ionospheric storm. It has been repeatedly suggested that vibrationally excited nitrogen is needed to explain some of the larger disturbance effects [e.g. Pavlov, 1994; Buonsanto, 1995].

4. Density perturbations at lower latitudes are another problem area. For example, is it correct that these changes are produced by the interaction of traveling atmospheric disturbances, as is assumed in this review? And can the equatorward-directed meridional winds carried along by TADs launched in opposite polar regions explain the transient compression and heating of the gases? Detailed theoretical simulations are needed to clarify this point.

5. Confusion also reigns over the question whether the density perturbations at lower latitudes include significant composition changes, and if they do, what are they? Satellite data like those shown in Fig. 5 indicate that the densities of all species increase and that larger changes in the N_2/O density ratio do not occur. However, Thermosphere Ionosphere General Circulation Models predict that such changes do occur and that they are large enough to explain long-duration positive ionospheric storms [e.g. Burns et al., 1995a,b; Fuller-Rowell et al., 1991, 1995; see also Mikhailov et al., 1995].

6. In the present review, long-duration positive ionospheric storms are attributed to changes in the large-scale meridional wind circulation. Unfortunately, experimental evidence in support of this view is difficult to obtain. For example, ground-based optical wind measurements are available only for nighttime conditions. A limited set of satellite data exists, but it is of a rather crude spatial and temporal resolution. Incoherent backscatter data are relatively scarce and are partially based on empirical models which may not be valid during the rapidly changing conditions during storms. Therefore, up to now no firm conclusions can be drawn on the basis of these measurements.

7. The interaction of TADs with the ionosphere is another problem area which requires further study. Since we are dealing with a highly dynamical situation, moving the ionization to higher altitudes may be only

partly responsible for positive ionospheric storms. In fact, compression (and dilution) of ionization by gradients in the vertical distribution of the meridional winds may well play an important role in explaining some of the disturbance features observed [e.g. *Burnside et al.*, 1991; *Millward et al.*, 1993; *Bauske and Prölss*, 1996).

8. In this context it is also very important to clarify the significance of storm-induced electric fields. Are these of secondary importance when it comes to explaining the average positive ionospheric storm, as has been tacitly assumed in this review? There are a number of authors who do not share this opinion [e.g. *Reddy and Nishida*, 1992; *Jakowski et al.*, 1992; *Pi et al.*, 1995]. Also, how important are convection electric fields in explaining positive storms at subauroral latitudes [e.g. *Foster*, 1993; *Buonsanto*, 1996]? To answer these questions, experimental and theoretical investigations of storm-induced upper atmospheric electric fields certainly need to be continued [e.g. *Fejer et al.*, 1990].

Admittedly, the points listed above are but a small selection of problem areas. However, solving only these problems will require considerable effort and concerted action from experimenters, data analysts, and theoreticians.

Acknowledgments. I would like to thank all those who participated in collecting the data used in this study. In particular, I am very grateful to G. Carignan, B. Hanson, R. Heelis, and N. Spencer, who provided the DE 2 data used in Figs. 3 and 8 and to U. von Zahn for providing the ESRO 4 data used in Figs. 5, 9 and 13. The ionospheric data were obtained with the help of World Data Centers A and C1.

REFERENCES

Abdu, M.A., J.H.A. Sobral, E.R. de Paula, and I.S. Batista, Magnetospheric disturbance effects on the Equatorial Ionization Anomaly (EIA): an overview, *J. Atmos. Terr. Phys.*, 53, 757–771, 1991

Ahn, B.-H., H.W. Kroehl, Y. Kamide, D.J. Gorney, S.-I. Akasofu, and J.R. Kan, The auroral energy deposition over the polar ionosphere during substorms, *Planet. Space Sci.*, 37, 239–252, 1989

Anderson, C.N., Correlation of long wave transatlantic radio transmission with other factors affected by solar activity, *Proc. Inst. Radio Eng.*, 16, 297–347, 1928

Appleton, E.V., and L.J. Ingram, Magnetic storms and upper-atmospheric ionisation, *Nature*, 136, 548–549, 1935

Appleton, E.V., R. Naismith, and L.J. Ingram, British radio observations during the second international polar year 1932-33, *Phil. Trans. Roy. Soc.*, A236, 191–259, 1937

Banks, P.M., Observations of joule and particle heating in the auroral zone, *J. Atmos. Terr. Phys.*, 39, 179–193, 1977

Baumjohann, W., and Y. Kamide, Hemispherical joule heating and the AE indices, *J. Geophys. Res.*, 89, 383–388, 1984

Bauske, R., and G.W. Prölss, Numerical simulation of positive ionospheric storm effects, *Adv. Space Res.* (in press), 1996

Berkner, L.V., and S.L. Seaton, Systematic ionospheric changes associated with geomagnetic activity, *Terr. Magn. Atmos. Elec.*, 45, 419–423, 1940

Biondi, M.A., and J.W. Meriwether, Jr., Measured response of the equatorial thermospheric temperature to geomagnetic activity and solar flux changes, *Geophys. Res. Lett.*, 12, 267–270, 1985

Birkeland, K., *The Norwegian aurora polaris expedition 1902-03*, Vol. 1, H. Aschehoug, Christiana, 1913

Blamont, J.E., and J.M. Luton, Geomagnetic effect on the neutral temperature of the F region during the magnetic storm of September 1969, *J. Geophys. Res.*, 77, 3534–3556, 1972

Blanc, M., and A.D. Richmond, The ionospheric disturbance dynamo, *J. Geophys. Res.*, 85, 1669–1686, 1980

Blum, P., and G.W. Prölss, Changes in thermospheric density caused by turbulence variations, *Adv. Space Res.*, 7, No. 10, 247–254, 1987

Burge, J.D., D. Eccles, J.W. King, and R. Rüster, The effects of thermospheric winds on the ionosphere at low and middle latitudes during magnetic disturbances, *J. Atmos. Terr. Phys.*, 35, 617–623, 1973

Burns, A.G., T.L. Killeen, and R.G. Roble, A theoretical study of thermospheric composition perturbations during an impulsive geomagnetic storm, *J. Geophys. Res.*, 96, 14153–14167, 1991

Burns, A.G., T.L. Killeen, and R.G. Roble, Thermospheric heating away from the auroral oval during geomagnetic storms, *Can. J. Phys.*, 70, 544–552, 1992

Burns, A.G., T.L. Killeen, W. Deng, G.R. Carignan, and R.G. Roble, Geomagnetic storm effects in the low-to middle-latitude upper thermosphere, *J. Geophys. Res.*, 100, 14673–14691, 1995a

Burns, A.G., T.L. Killeen, G.R. Carignan, and R.G. Roble, Large enhancements in the O/N$_2$ ratio in the evening sector of the winter hemisphere during geomagnetic storms, *J. Geophys. Res.*, 100, 14661–14671, 1995b

Buonsanto, M.J., Millstone Hill incoherent scatter F region observations during the disturbances of June 1991, *J. Geophys. Res.*, 100, 5743–5755, 1995

Burnside, R.G., C.A. Tepley, M.P. Sulzer, T.J. Fuller-Rowell, D.G. Torr, and R.G. Roble, The neutral thermosphere at Arecibo during geomagnetic storms, *J. Geophys. Res.*, 96, 1289–1301, 1991

Burrage, M.D., V.J. Abreu, N. Orsini, C.G. Fesen, and R.G. Roble, Geomagnetic activity effects on the equatorial neutral thermosphere, *J. Geophys. Res.*, 97, 4177–4187, 1992

Chandra, S., and P. Stubbe, Ion and neutral composition changes in the thermospheric region during magnetic storms, *Planet. Space Sci.*, 19, 491–502, 1971

Codrescu, M.V., R.G. Roble, and J.M. Forbes, Interactive ionospheric modeling: A comparison between TIGCM and ionosonde data, *J. Geophys. Res.*, 97, 8591–8600, 1992

Cole, K.D., Joule heating of the upper atmosphere, *Aust. J. Phys.*, 15, 223–235, 1962

Craven, J.D., A.C. Nicholas, L.A. Frank, D.J. Strickland, and T.J. Immel, Variations in the FUV dayglow after intense auroral activity, *Geophys. Res. Lett.*, 21, 2793–2796, 1994

Crowley, G., B.A. Emery, R.G. Roble, H.C. Carlson, Jr., and D.J. Knipp, Thermospheric dynamics during September 18–19, 1984. I. Model simulations, *J. Geophys. Res.*, 94, 16925–16944, 1989

Danilov, A.D., and L.D. Morozova, Ionospheric storms in the F region. Morphology and physics (Review), *Geomagn. Aeron.*, 25, 593–605, 1985

Espenschied, L., C.N. Anderson, and A. Bailey, Transatlantic radio telephone transmission, *Bell Syst. Techn. J.*, 4, 459–507, 1925

Fedder, J.A., and P.M. Banks, Convection electric fields and polar thermospheric winds, *J. Geophys. Res.*, 77, 2328–2340, 1972

Fejer, B.G., and L. Scherliess, Time dependent response of equatorial ionospheric electric fields to magnetospheric disturbances, *Geophys. Res. Lett.*, 22, 851–854, 1995

Fejer, B.G., R.W. Spiro, R.A. Wolf, and J.C. Foster, Latitudinal variations of perturbation electric fields during magnetically disturbed periods: 1986 SUNDIAL observations and model results, *Ann. Geophys.*, 8, 441–454, 1990

Fesen, C.G., G. Crowley, and R.G. Roble, Ionospheric effects at low latitudes during the March 22, 1979, geomagnetic storm, *J. Geophys. Res.*, 94, 5405–5417, 1989

Foster, J.C., Storm time plasma transport at middle and high latitudes, *J. Geophys. Res.*, 98, 1675–1689, 1993

Fujiwara, H., S. Maeda, H. Fukunishi, T.J. Fuller-Rowell, and D.S. Evans, Global variations of thermospheric winds and temperatures caused by substorm energy injection, *J. Geophys. Res.*, 101, 225–239, 1996

Fuller-Rowell, T.J., and D. Rees, A three-dimensional, time-dependent simulation of the global dynamical response of the thermosphere to a geomagnetic storm, *J. Atmos. Terr. Phys.*, 43, 701–721, 1981

Fuller-Rowell, T.J., D. Rees, H. Rishbeth, A.G. Burns, T.L. Killeen, and R.G. Roble, Modelling of composition changes during F-region storms: a reassessment, *J. Atmos. Terr. Phys.*, 53, 541–550, 1991

Fuller-Rowell, T.J., M.V. Codrescu, R.J. Moffett, and S. Quegan, Response of the thermosphere and ionosphere to geomagnetic storms, *J. Geophys. Res.*, 99, 3893–3914, 1994

Fuller-Rowell, T.J., M.V. Codrescu, H. Rishbeth, R.J. Moffett, and S. Quegan, On the seasonal response of the thermosphere and ionosphere to geomagnetic storms, *J. Geophys. Res.*, 101, 2343–2353, 1996

Hafstad, L.R., and M.A. Tuve, Note on Kennelly-Heaviside layer observations during a magnetic storm, *Terr. Magn. Atmos. Elec.*, 34, 39–43, 1929

Hedin, A.E., and G.R. Carignan, Morphology of thermospheric composition variations in the quiet polar thermosphere from Dynamics Explorer measurements, *J. Geophys. Res.*, 90, 5269–5277, 1985

Jacchia, L.G., Corpuscular radiation and the acceleration of artificial satellites, *Nature*, 183, 1662–1663, 1959

Jakowski, N., A. Jungstand, K. Schlegel, H. Kohl, and K. Rinnert, The ionospheric response to perturbation electric fields during the onset phase of geomagnetic storms, *Can. J. Phys.*, 70, 575–581, 1992

Killeen, T.L., P.B. Hays, G.R. Carignan, R.A. Heelis, W.B. Hanson, N.W. Spencer and L.H. Brace, Ion-neutral coupling in the high-latitude F region: Evaluation of ion heating terms from Dynamics Explorer 2, *J. Geophys. Res.*, 89, 7495–7508, 1984

Kirby, S.S., N. Smith, T.R. Gilliland, and S.E. Reymer, The ionosphere and magnetic storms, *Phys. Rev.*, 51, 992–993, 1937

Kosch, M.J., and E. Nielsen, Coherent radar estimates of average high-latitude ionospheric Joule heating, *J. Geophys. Res.*, 100, 12201–12215, 1995

Laux, U., and U. von Zahn, Longitudinal variations in thermospheric composition under geomagnetically quiet conditions, *J. Geophys. Res.*, 84, 1942–1946, 1979

Lu, G., A.D. Richmond, B.A. Emery, and R.G. Roble, Magnetosphere-ionosphere-thermosphere coupling: Effect of neutral winds on energy transfer and field-aligned current, *J. Geophys. Res.*, 100, 19643–19659, 1995

Matuura, N., Theoretical models of ionospheric storms, *Space Sci. Rev.*, 13, 124–189, 1972

Mayr, H.G., and H. Volland, Magnetic storm effects in the neutral composition, *Planet. Space Sci.*, 20, 379–393, 1972

Mayr, H.G., I. Harris, and N.W. Spencer, Some properties of upper atmospheric dynamics, *Rev. Geophys. Space Phys.*, 16, 539–565, 1978

Meier, R.R., R. Cox, D.J. Strickland, J.D. Craven, and L.A. Frank, Interpretation of Dynamics Explorer far UV images of the quiet time thermosphere, *J. Geophys. Res.*, 100, 5777–5794, 1995

Mikhailov, A.V., M.G. Skoblin, and M. Förster, Daytime F2-layer positive storm effect at middle and lower latitudes, *Ann. Geophys.*, 13, 532–540, 1995

Millward, G.H., S. Quegan, R.J. Moffett, T.J. Fuller-Rowell, and D. Rees, A modelling study of the coupled ionospheric and thermospheric response to an enhanced high-latitude electric field event, *Planet. Space Sci.*, 41, 45–56, 1993

Nisbet, J.S., B.J. Wydra, C.A. Reber, and J.M. Luton, Global exospheric temperatures and densities under active solar conditions, *Planet. Space Sci.*, 25, 59–69, 1977

Noël, S., and G.W. Prölss, Heating and radiation production by neutralized ring current particles, *J. Geophys. Res.*, 98, 17317–17325, 1993

Parish, H.F., G.R. Gladstone, and S. Chakrabarti, Interpretation of satellite airglow observations during the March 22, 1979, magnetic storm, using the coupled ionosphere-thermosphere model at University College London, *J. Geophys. Res.*, 99, 6155–6166, 1994

Pavlov, A.V., The role of vibrationally excited nitrogen in the formation of the mid-latitude negative ionospheric storms, *Ann. Geophys.*, 12, 554–564, 1994

Petersen, H., Über die Temperatur in den höheren Schichten der Atmosphäre, *Phys. Zeitschr.*, 28, 510–513, 1927

Pi, X., M. Mendillo, P. Spalla, and D.N. Anderson, Longitudinal effects of ionospheric responses to substorms at middle and lower latitudes: a case study, *Ann. Geophys.*, 13, 863–870, 1995

Prölss, G.W., Seasonal variations of atmospheric-ionospheric disturbances, *J. Geophys. Res.*, *82*, 1635–1640, 1977

Prölss, G.W., Latitudinal structure and extension of the polar atmospheric disturbance, *J. Geophys. Res.*, *86*, 2385–2396, 1981

Prölss, G.W., Perturbation of the low–latitude upper atmosphere during magnetic substorm activity, *J. Geophys. Res.*, *87*, 5260–5266, 1982

Prölss, G.W., Thermosphere-ionosphere coupling during disturbed conditions, *J. Geomag. Geoelectr.*, *43*, 537–549, 1991

Prölss, G.W., Common origin of positive ionospheric storms at middle latitudes and the geomagnetic activity effect at low latitudes, *J. Geophys. Res.*, *98*, 5981–5991, 1993a

Prölss, G.W., On explaining the local time variation of ionospheric storm effects, *Ann. Geophys.*, *11*, 1–9, 1993b

Prölss, G.W., Ionospheric F-region storms, in *Handbook of Atmospheric Electrodynamics*, Vol. 2 (ed. H. Volland), 195–248, CRC Press/Boca Raton, 1995

Prölss, G.W., M. Roemer, and J.W. Slowey, Dissipation of solar wind energy in the earth's upper atmosphere: The geomagnetic activity effect, CIRA 1986, *Adv. Space Res.*, *8*, No. 5, 215–261, 1988

Prölss, G.W., L.H. Brace, H.G. Mayr, G.R. Carignan, T.L. Killeen, and J.A. Klobuchar, Ionospheric storm effects at subauroral latitudes: A case study, *J. Geophys. Res.*, *96*, 1275–1288, 1991

Reddy, C.A., and A. Nishida, Magnetospheric substorms and nighttime height changes of the F2 region at middle and low latitudes, *J. Geophys. Res.*, *97*, 3039–3061, 1992

Richmond, A.D., and S. Matsushita, Thermospheric response to a magnetic substorm, *J. Geophys. Res.*, *80*, 2839–2850, 1975

Rishbeth, H., and R. Edwards, The isobaric F2-layer, *J. Atmos. Terr. Phys.*, *51*, 321–338, 1989

Rishbeth, H., and O.K. Garriot, Introduction to Ionospheric Physics, Academic Press, New York / London, 1969

Roble, R.G., Dynamics of the earth's thermosphere, *Rev. Geophys. Space Sci.*, *21*, 217–233 (U.S. national report to International Union of Geodesy and Geophysics 1979–1982), 1983

Roemer, M., Recent observational results on the thermosphere and exosphere, *CIRA 1972* (exec. ed. A.C. Stickland), Akademie-Verlag, Berlin, 341–396, 1972

Schlegel, K., and J.P. St.-Maurice, Anomalous heating of the polar E region by unstable plasma waves 1. Observations, *J. Geophys. Res.*, *86*, 1447–1452, 1981

Shimazaki, T., Effects of vertical mass motions on the composition structure in the thermosphere, *Space Res.*, *12*, 1039–1045, 1972

Skoblin, M.G., and M. Förster, An alternative explanation of ionization depletions in the winter night-time storm-perturbed F2-layer, *Ann. Geophys.*, *11*, 1026–1032, 1993

Spiro, R.W., P.H. Reiff, and L.J. Maher, Precipitating electron energy flux and auroral zone conductances - an empirical model, *J. Geophys. Res.*, *87*, 8215–8227, 1982

St.-Maurice, J.-P., and R.W. Schunk, Ion-neutral momentum coupling near discrete high-latitude ionospheric features, *J. Geophys. Res.*, *86*, 11299–11321, 1981

Testud, J., P. Amayenc, and M. Blanc, Middle and low latitude effects of auroral disturbances from incoherent-scatter, *J. Atmos. Terr. Phys.*, *37*, 989–1009, 1975

Zuzic, M., L. Scherliess, and G.W. Prölss, Latitudinal structure of thermospheric composition perturbations, *J. Atmos. Terr. Phys.* (in press), 1996

G. W. Prölss, Institut für Astrophysik und Extraterrestrische Forschung, Universität Bonn, 53121 Bonn, Germany

AI Techniques in Geomagnetic Storm Forecasting

Henrik Lundstedt

Swedish Institute of Space Physics, Solar-Terrestrial Physics Division, Box 43, S-221 00 Lund, Sweden

This review deals with how geomagnetic storms can be predicted with the use of Artificial Intelligence (AI) techniques. Today many different AI techniques have been developed, such as symbolic systems (expert and fuzzy systems) and connectionism systems (neural networks). Even integrations of AI techniques exist, so called Intelligent Hybrid Systems (IHS). These systems are capable of learning the mathematical functions underlying the operation of non-linear dynamic systems and also to explain the knowledge they have learned. Very few such powerful systems exist at present. Two such examples are the Magnetospheric Specification Forecast Model of Rice University and the Lund Space Weather Model of Lund University. Various attempts to predict geomagnetic storms on long to short-term are reviewed in this article. Predictions of a month to days ahead most often use solar data as input. The first SOHO data are now available. Due to the high temporal and spatial resolution new solar physics have been revealed. These SOHO data might lead to a breakthrough in these predictions. Predictions hours ahead and shorter rely on real-time solar wind data. WIND gives us real-time data for only part of the day. However, with the launch of the ACE spacecraft in 1997, real-time data during 24 hours will be available. That might lead to the second breakthrough for predictions of geomagnetic storms.

1. INTRODUCTION

The earth's magnetosphere acts as an obstacle to the solar plasma, unless the plasma (or distorted plasma ahead of the interplanetary structure) contains a magnetic field oppositely directed to the earth's field. If that is the case, then energy can enter, dissipate or be temporarily stored and then dissipated. The energy might be dissipated as e.g. the ring current and auroral substorms. The ring current is produced by a belt of energetic ions and electrons that circles the earth [Baumjohann and Haerendel, 1987]. The largest geomagnetic storms are mainly caused by large negative values of the southward directed interplanetary space magnetic field component resulting from compressing of different plasma streams (interaction regions) and clouds (CMEs) [Lindsay et al., 1995; Farrugia et al., 1993; Gosling et al., 1991; Tsurutani et al., 1988; Klein and Burlaga, 1982]. If the southward directed magnetic field in the solar phenomenon or surrounding interplanetary magnetic field $B_z < -10$ nT during more than 3 hours, then an intense geomagnetic storm $D_{st} < -100$ nT will occur, typically at an 80% confidence level [Gonzalez et al., 1994; Tsurutani et al., 1995]. It should also be emphasized that coronal mass ejections are not a necessary and sufficient condition to cause intense magnetic storms, but long duration, intense southward interplanetary magnetic fields are [Tsurutani and Gonzalez, 1992].

Gonzalez et al. [1994] discuss different definitions of a "geomagnetic storm" and suggest a new definition: "An interval of time when a sufficiently intense and long-lasting interplanetary convection electric field ($E_y = VB_z$) leads, through a substantial energization in the magnetosphere-ionosphere system, to an intensified ring current sufficiently strong to exceed some key threshold of the quantified storm-time D_{st} index". The variation of the geomagnetic storm index Dst, corrected due to the magnetopause currents, may be described by the following differential equation [Gonzalez et al., 1994]

$$\frac{dD_{st}(t)}{dt} = Q(t) - D_{st}(t)/\tau \qquad (1)$$

where $Q(t) = 2.5 \cdot 10^{21} U(t)$ in Gaussian units, U(t) the rate of energy into the ring current and τ the decay time. Q is most often set equal to E_y. E_y should also be rectified or gated with some function of angle. A nonlinear state appears as the decay time τ varies.

Intense geomagnetic storms can severely affect space-borne [Lundstedt et al., 1995] and ground-based technological systems [Lundstedt, 1992c; Boteler, 1993; Viljanen and Pirjola, 1994]. It is therefore of great economical and social importance, and also a great scientific challenge, to predict geomagnetic storms.

Many different methods [Baker, 1986] have been used in predicting geomagnetic storms such as ordinary statistical methods, linear, non-linear filtering and artificial intelligence (AI)-methods. Tom Detman covers the use of non-AI-methods for predictions in this monograph. I will focus on only the AI-methods. Most often AI is defined as the study of human intelligent behavior. However, non-human related definitions have also been suggested. AI has been defined as "the capability of a system to adapt its behavior to meet its goal in a range of environments". This definition has been used in evolutionary computation. Intelligence and learning are a part of the evolution.

AI can be coded in many different ways, which result in various methods such as expert/inductive systems, fuzzy systems, genetic algorithms, artificial life, artificial neural networks (ANN), and various combinations of these systems. In this review I will mostly discuss neural networks and the use of neural networks.

One of the outcomes of the workshop "AI Applications in Solar-Terrestrial Physics" held in Lund 1993, was the need for building bridges between symbolic (expert/fuzzy) systems and connectionism systems (ANN) [McPherron, 1993]. The weakness of one AI-method can then be removed by using the strength of another method. A major breakthrough is therefore foreseen by using so called "Hybrid Intelligent System". Thus far, only few hybrid systems have been developed or are under development [Freeman et al., 1993, Lundstedt, 1996].

2. AI-METHODS

As earlier mentioned, AI has been defined as either the study of human intelligence or of non-human-behavior. Here I will discuss only coding of human intelligent behavior, so called symbolic respectively connectionism systems. They may be looked upon as descriptions of human intelligence on different levels of abstraction. Expert systems (symbolic approach) are excellent on high-level reasoning (cognitive tasks) and neural networks (connectionism approach) are excellent on low-level reasoning (pattern recognition tasks).

An expert system [Jackson, 1986] describes intelligence with symbols. It consists of a knowledge base (rules and facts), an inference engine, an explanation facility and an interface. Many successful expert systems have been built in space engineering. However, they are expensive, have long development times, and they cope badly with brittleness.

Fuzzy systems [Kosko, 1992] also consist of rules. However, they can deal with fuzzy concepts. Fuzzy logic can describe degrees of truth ranging from 0 to 1. Membership functions translate symbolics to numerical values. Fuzzy systems can cope with brittleness and are good at both high and low-level reasoning. Rules and the membership functions must on the other hand be given to the fuzzy systems. A neural network could do this.

Genetic algorithms [Goldberg, 1989] are based on the biological principle of "survival of the fittest". The solution of a problem is improved for each successive generation. How the population should be changed is determined by genetic principles such as selection, cross-over and mutation. Genetic algorithms are very powerful tools for optimizing tasks. They have been used to optimize the learning parameters of neural networks.

An artificial neural network [Hertz et al., 1991] is essentially a group of interconnected computing elements (neurons). Typically a neuron computes the sum of its input and passes this sum through a nonlinear function (an activation function). Each neuron has only one output, but this output is multiplied by a weighting factor if it is used as the input to another neuron. The neural networks typically exhibit two types of behaviour. If no feedback (recurrent) loops connect neurons, the signal produced by an external input moves only in one direction and the output of the network is just the output of the last group of neurons in the network. In this case the network behaves mathematically like a nonlinear function of the inputs. The second type of network behaviour is observed when there are feedback loops in the neuron connections. In this case the network behaves like a dynamical system, so the output of the neurons varies with time. The neuron output can oscillate, or settle down into steady state values, or, since the threshold function introduces nonlinearity into the system, become chaotic.

The modelling capability of the ANN can be ascribed to its ability to learn the mathematical function underlying the system operation. If the network is designed and trained properly, it can perform generalization rather than simple curve fitting. Any continuous function can be implemented by a three-layer feedforward neural network [Lorentz, 1976]. A combination of neural networks can describe a discontinuous function. The ANNs can be used as a preprocessor: If the ANN cannot map the input patterns into output patterns accurately, then some input variables are probably missing. If an input variable can be removed without hurting the network performance, then the variable is probably irrelevant. The ANNs can also work as postprocessor: From the ANN model new connections and relations may be found. A network can actually improve a theoretical model, by finding new relationships we had not thought about. When a neural network has been trained, it can be considered as a model. By changing the input values to the input layer the network's response and output can be studied. From that an understanding of what the network has learned can also be derived.

The rapid and successful proliferation of applications incorporating ANN technology in fields as diverse as commerce, science, industry and medicine depends on three characteristics of ANNs: the direct and straightforward manner ANNs acquire knowledge; the compact form in which knowledge is stored and the ease and speed with which this knowledge can be accessed; the robustness of an ANN in the presence of noise in input data. ANNs would have gained even higher acceptance if ANNs had an inherent ability to explain in a comprehensible form, the process by which a given decision or output generated by an ANN had been reached. However, there are many ways of coding knowledge into neural networks and also to extract rules from neural networks. Figure 1 illustrates how rules of type "if the premise then the conclusion" can be coded into a neural network.

In order to extract knowledge from a neural network we first need to understand how the knowledge acquired by the ANN is encoded. The knowledge of an ANN is encoded by a) the network architecture itself, b) the activation functions and c) the weights. The task of extracting explanations (or rules) from a trained ANN is therefore to interpret in a comprehensible form the collective effects of a), b) and c). Many algorithms have been developed based on a-c [Fu, 1994]. In a) we consider the ANN as a black box and extract rules that map inputs directly into outputs. In b-c we extract rules at the individual unit levels. An example of such an algorithm is "the SUBSET Algorithm" [Towell and Shavlik, 1993]:

for each hidden and output unit:
    extract up to S_p subsets of the positively-weighted incoming links for which the summed weight is greater than the bias on the unit;
    for each element p of the S_p subsets:
        search for a set S_N of a set negative-attributes so that the summed weights of p plus the summed weights of N - n (where N is the set of all negative-attributes and n is an element of S_N) exceed the threshold on the unit:
        with each element n of the S_N set, form
        a rule: 'if p and NOT n, then the concept designated by the unit'.

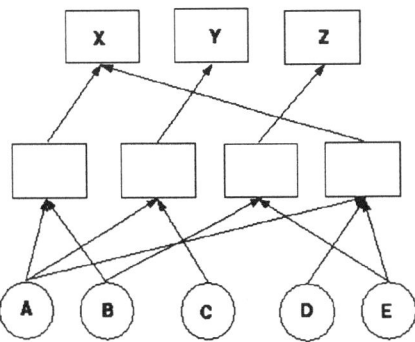

Figure 1. A rule-based inference system can be mapped into a neural network architecture as follows. Rules: If A and B then X, If A and C then Y, If B and E then Z, and If A and D and E then X.

An example of the black box approach is described by "The RULENEG Algorithm" [Pop et al., 1994]:

initialise the Rule-Holder to empty
for every pattern s from the training set
    find the class C for s by use of the ANN (C = ANN(s))
    if s is not classified by the existing rules
        initialise a new rule r for class C
        for every input i into the network
            make a copy ś of s
            negate the i-th entry in ś
            find the class Ć for ś by use of
            the ANN Ć = ANN(ś)
            if C is not equal to Ć
                add i-th input and its truth value to r
        end for every input
        add r to the Rule-Holder
end for every pattern

The real power of using AI methods therefore comes with the use of integrated AI-methods, so called Intelligent Hybrid Systems [Goonatilake and Khebbal, 1995].

Another major breakthrough for using ANNs and IHSs within science probably comes with the discovery by Wray and Green [1994], cited from Klimas et al., [1996].

> It is often said that neural networks yield no usable information on the physics of the system that they model. However, it appears that this prejudice may not be correct. Wray and Green [1994] have considered the situation in which an artificial neural network has been used to produce a black box model of a

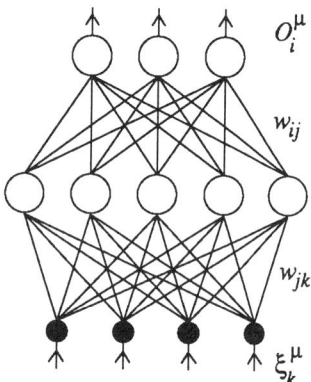

Figure 2. A multi-layer backpropagation network.

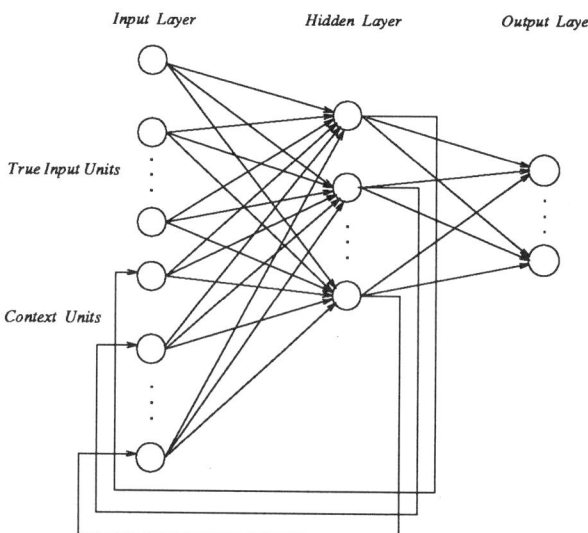

Figure 3. An Elman recurrent network.

nonlinear dynamical system. They have shown how such an artificial neural network is equivalent to a Volterra series and give the equation for the *n*th-order Volterra kernel in terms of the internal parameters of the network. The Volterra series can be shown to be a general solution of nonlinear differential equation models [Fliess et al., 1983]. While nothing of the sort appears to have been attempted in space physics applications, one wonders if the properties of the nonlinear differential equation model can be deduced from the Volterra series. If so, then a method for deducing the model that is equivalent to a trained neural network would be available. Even if the differential equation model cannot be deduced, the Volterra equation equivalent to the neural network can certainly be obtained and this may be sufficient.

In a very interesting article Vassiliadis et al., [1996] present a method that converts a type of nonlinear filter, the nonlinear autoregressive moving average (ARMA) model to a physical model, namely a nonlinear damped oscillator.

3. FOUR NEURAL NETWORK PARADIGMS

Let me now give a brief description of the main features of the neural networks paradigms I will show applied for predictions.

A feed-forward Multi-Layer Backpropagation network (MLBP) (Fig. 2) learns to map an input vector to an output vector from examples with known answers. The network most often consists of one input, one hidden layer and an output layer. The hidden layer creates a representation of the features in the input vector ξ. The output O_i^μ of a single hidden-layer neural network with an input pattern μ is given by

$$O_i^\mu = g_1(\sum_j w_{ij} g_2(\sum_k w_{jk} \xi_k^\mu)) \qquad (2)$$

where w_{ij} and w_{jk} are the weights between the input and hidden layer and between the hidden and output layer, respectively.

The transfer functions ($g_{1,2}$) are often sigmoid-shaped functions, e.g. tanh. If the transfer functions ($g_{1,2}$) were chosen to be linear, e.g. ($g_{1,2}(x) = x$), then the network would become identical to a linear filter (Iyemori et al., 1979). A MLBP, with non-linear transfer functions, could therefore be regarded as a non-linear generalization of a linear filter.

The weights are updated by the gradient descent algorithm according to

$$\Delta w(t+1) = -\eta \nabla_w E + \alpha \Delta w(t) \qquad (3)$$

where η the learning rate and α the momentum term are used to smooth or speed up the learning process and avoid local minima.

We used the following error function (E),

$$E = \frac{1}{2} \sum_\mu (O^\mu - T^\mu)^2 \qquad (4)$$

where O^μ is the actual output of the network and T^μ the corresponding target (e.g. D_{st}).

A very interesting extension to the MLBP network is the Elman network [Elman, 1990]. Elman networks are two-layer backpropagation networks (Fig. 3), with the addition of a feedback connection from the hidden layer to input layer. This feedback path allows Elman networks to learn to recognize and generate temporal patterns, as well as spatial patterns. If the number of input, hidden and output units are denoted respectively by R, S_1, and S_2, then at time t the hidden unit $V_j(t)$ is computed from

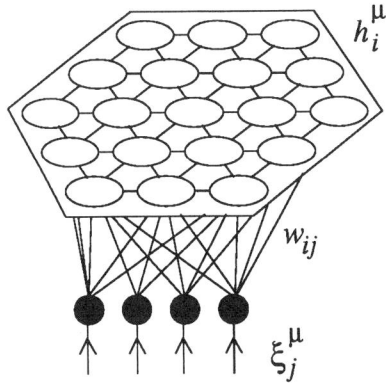

Figure 4. A self-organizing map network.

$$V_j(t) = tanh\left(\sum_{k=1}^{R} w_{jk}I_k(t) + \sum_{k=R+1}^{R+S_1} w_{jk}V_k(t-1)\right) \quad (5)$$

where $I_k(t)$ is the input to the kth input unit at time t, $V_j(t-1)$ the output of the jth hidden unit at time $t-1$, and w_{jk} the connection strength between the input unit k and hidden unit j. The linear activation function in the output layer results in the output

$$O_i(t) = \sum_{j=1}^{S_1} W_{ij}V_j(t) \quad (6)$$

where W_{ij} is the connection strength between the hidden and the output layer.

An unsupervised neural network, the self-organized map neural network (SOM) (Fig. 4) [Kohonen, 1984] clusters similar input patterns on a map. The net input to a node is

$$h_i^\mu = \sum_j w_{ij}\xi_j^\mu. \quad (7)$$

The index i refers to the output nodes (in the map), j refers to the input nodes. The winning node i^* is the node that has its weight vector closest to the input pattern, i.e.

$$\mathbf{w}_{i^*} \cdot \xi \geq \mathbf{w}_i \cdot \xi. \quad (8)$$

It is only the winning node that is active in the map. The winning node is then updated to make the weights come closer to the input. The nodes in the nearest neighbourhood of the winning node are also updated in the same direction. The Kohonen learning rule is

$$\Delta w_{ij} = \eta \Lambda(i, i^*)\left(\xi_j^\mu - w_{ij}\right) \quad (9)$$

where Λ is the neighbourhood function. All nodes in the map can be included at the start of the learning process, but the neighbourhood is then decreased as learning progresses. In

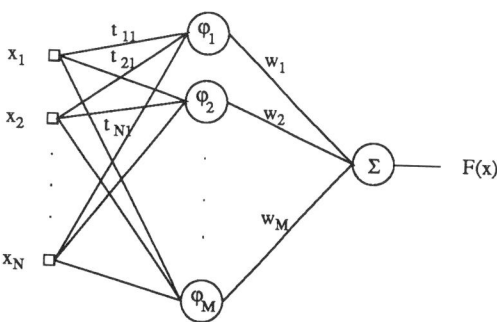

Figure 5. A radial basis function network.

this way the large-scale features are found first, and smaller details are discerned later.

Finally a radial-basis function network (RBF) (Fig. 5) is again a network consisting of an input, a hidden layer and an output layer. Here the learning is carried out in two steps, first unsupervised and then through supervised learning. Most often the transfer function in the hidden layer is a gaussian function, causing a local response φ_i, and then a linear function in the output layer $F(\mathbf{x})$.

$$\varphi_i = \exp\left[\frac{(\mathbf{x} - \mathbf{t}_i)^T(\mathbf{x} - \mathbf{t}_i)}{(2\sigma^2)}\right] \quad (10)$$

$$F(\mathbf{x}) = \sum_i w_i \varphi_i(\mathbf{x}) \quad (11)$$

where $\mathbf{x}$ is the input vector, $\mathbf{t}_i$ a weight vector of the hidden neuron i and w_i a weight in the output layer.

The RBF networks are very fast compared to MLBP networks. The training is often hundreds of times faster than that for a MLBP network. A RBF network constructs local approximations to nonlinear input-output mapping, whereas a MLBP network constructs a global mapping.

4. PREDICTION OF GEOMAGNETIC STORMS

To predict geomagnetic storms and activity 1-3 days or months ahead [Joselyn, 1995] we need to use solar data as input. We started to use AI-methods by using inductive expert systems to predict geomagnetic activity 1-3 days in advance, from input about occurrence of solar phenomena [Lundstedt, 1990]. From examples, the inductive expert system created rules for when different geomagnetic categories should occur depending on the history of solar phenomena. However, the expert system showed little success. By using a MLBP we had much more success in predicting geomagnetic activity 1-3 days ahead [Lundstedt, 1992b]. Since if-then-rules can be extracted from a neural network, we will extend our study

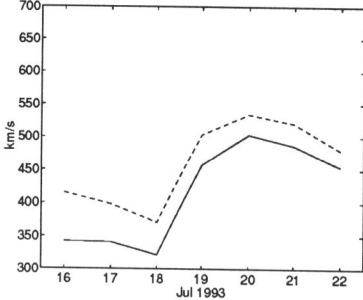
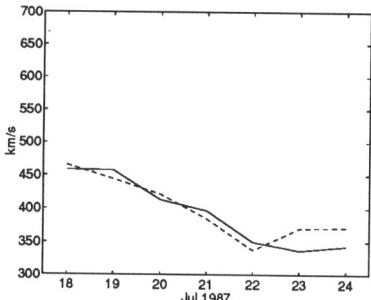

Figure 6. The solid line represents the observed solar wind velocity V and the dashed line the predicted V from solar magnetic field data, using a hybrid intelligent system. The left figure illustrates a prediction for a fast solar wind and the right figure illustrates a prediction for an event of a slow solar wind.

doing that. We also plan to retrain the network with more solar input data and further optimize the net.

Instead of using occurrence of solar phenomena as input, we could also use the solar wind variation derived from solar data or the occurrence of solar wind structures as input to networks predicting geomagnetic storms. A RBF network has been used to map the photospheric magnetic field to the computed coronal source surface magnetic field by Zhao and Hoeksema [1994; 1995]. The input data, the solar magnetic field, was observed at Wilcox Solar Observatory [Scherrer et al., 1977, Hoeksema, 1984] for the years 1976 to 1995. Predictions of the solar wind parameters from the coronal magnetic field have also been developed. Predictions of the daily average interplanetary magnetic field (IMF) at 1 AU from daily full disc solar magnetograms are discussed by Wintoft and Lundstedt, [1995]. The training set (cycle 21) included 2000 magnetograms and the test set, (cycle 22) 1374 magnetograms. In the test set the polarity was correctly predicted in 84% of the cases.

Today we are also able to predict the solar wind velocity (V) from only solar magnetic field data using a hybrid system of a RBF network and a MHD-model [Wintoft and Lundstedt, 1996] (Fig. 6). A potential field model [Hoeksema, 1984] was used to calculate the magnetic field strengths on the same field line at the photosphere $B_\odot = B(R_\odot)$ and the source surface ($= 2.5 R_\odot$) $B_s = B(R_s)$ from WSO magnetograms. The RMS magnetic field B_{RMS} was formed from daily WSO magnetograms. Defining a vector $x(t) = (B_\odot(t), B_s(t), B_{RMS}(t))$, the input to network was the time series $x(t-2), x(t-1), x(t)$ and the output $V(t+3)$ i.e. the velocity three days ahead. The RBF network was trained on magnetograms during solar cycle 21 and tested on solar cycle 22. A correlation coefficient of 0.58, a RMSE (root mean square error, km/s) of 90 km/s and an average relative variance of 0.68 were obtained. The IHS is doing a better job than the method presented by Wang and Sheeley [1994]. They reached a correlation coefficient of 0.4 for daily solar wind predictions. The next step will be to use SOHO solar magnetic fields measurements with better time and spatial resolution.

We have also used a MLBP to predict the geomagnetic activity 26-29 days in advance from the IMF computed at Wilcox Solar Observatory [Lundstedt and Hoeksema, 1992]. We trained a MLBP to associate patterns in the interplanetary magnetic field for several rotations with the resulting geomagnetic activity. The ANN predicted CHs and quiet times well but missed most of the activity caused by transient events such as CMEs.

Gothaskar and Khobragade, [1993] have used a MLBP to detect solar wind structures that might cause geomagnetic storms from studies of power spectra of interplanetary scintillations (IPS). A set of 840 IPS spectra, obtained from IPS observations at 327 MHz, was used. The efficiency of the MLBP network to recognize normal to disturbed spectra was ~ 74% to 78% when compared with classification by two experts. Snel and Lundstedt [1993] carried out a study of classifying structures in the solar wind at times of geomagnetic storms with the use of a SOM network. The input parameters that were used were the solar wind density (n), velocity (V), southward directed magnetic field (B_z), electric field (E) and pressure (p).

The most accurate predictions of geomagnetic activity are obtained by applying an ANN to hourly or minutes averaged satellite data. The hourly average solar wind parameters and interplanetary magnetic field (IMF) values used were compiled from data of several satellites by J.H.King (NASA GSFC).

Many groups [Freeman et al., 1993; Detman, 1994; Lundstedt 1991, 1992a; Lundstedt and Wintoft, 1994] have used the solar wind data to predict geomagnetic storms. Different solar wind parameters have been selected as inputs for the neural networks. Most often, the solar wind velocity (V), density (n), and the southward directed magnetic field (B_z), for a time history, have been used. However, the electric field (E_y) and dynamic pressure (p) and the other magnetic field components and standard deviations have also been used as

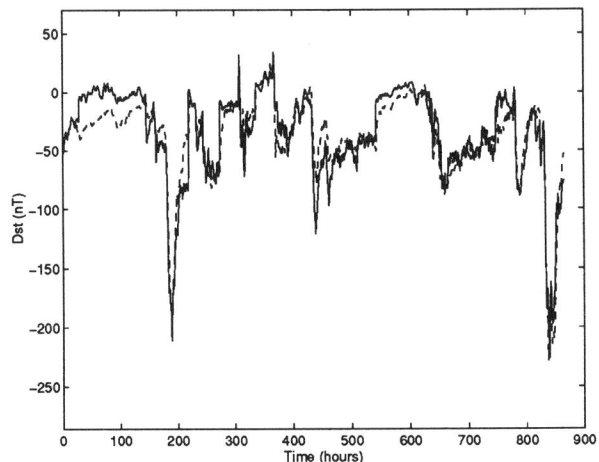

Figure 7. The solid line represents the observed D_{st} and the dashed line the predicted D_{st} one hour ahead with an Elman neural network.

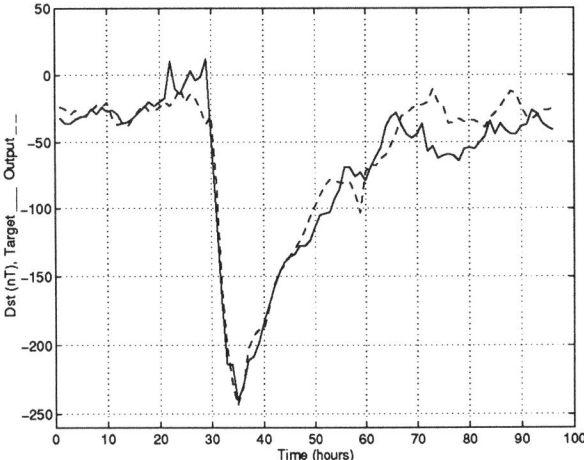

Figure 8. The solid line represents the observed D_{st} and the dashed line the predicted D_{st} two hours ahead with an Elman neural network.

inputs. Even the D_{st} index as input was used by Freeman et al. [1993]. Freeman et al. use as input, the interplanetary total magnetic field magnitude (B), the southward magnetic field component, dynamic pressure, and the D_{st}-index days 0, -1, -2. From that they predicted D_{st} one hour ahead. In their IHS Magnetospheric Specification and Forecast Model (MSFM) Freeman et al. used the predicted D_{st} as input parameter to the MSFM, which forecasted the charged-particle fluxes for different locations in the magnetosphere.

A MLBP network [Lundstedt and Wintoft, 1994] easily learns to predict the initial and main phase of a geomagnetic storm, but not the recovery phase. When on the other hand a longer time window between 18-24 hours is used, then even the recovery phase can be predicted [Gleisner et al., 1996a]. Solar wind and geomagnetic data including both quite and storm-time periods from the years 1963-1987, were used in training and testing the networks. The training set consisted of 62 periods covering 6600 hours and the test set consisted of 22 periods covering 2100 hours. We used hourly averages of n, V, and B_z as input to the networks, while the output was D_{st} one hour forward in time.

When we extended our study to Elman networks, i.e. recurrent backpropagation networks, then all phases of the geomagnetic storm and all intensities of storms were possible to predict one hour in advance (Fig. 7) [Wu and Lundstedt, 1996a].

As an average for the test data predictions one hours ahead the correlation coefficient between the observed and predicted D_{st} reached 0.92 and the corresponding prediction efficiency (1 - average relative variance) was 85 %.

This study has now also been extended to predictions of D_{st} two hours ahead, (Fig. 8) [Wu and Lundstedt, 1996b]. For predictions two hours ahead the correlation coefficient between the observed and predicted D_{st} reached 0.90 and the corresponding prediction efficiency was 82%.

A natural extension of studies of geomagnetic storms has been to predict geomagnetic disturbances on shorter timescales, 1-5 minutes. Hernandez et al., [1993] were first to predict the electrojet index AL with the use of neural networks. However a clipping problem is evident in their results.

We started by predicting the AE-index from solar wind data (Fig. 9) [Gleisner et al., 1996b]. We have not found any clipping problems in our results.

In an attempt to have our neural networks explain what they have learned, both various energy coupling functions and separate solar wind parameters were presented to the input layer. It was found that the ANNs find a better function by themselves describing the solar wind-magnetosphere/ionosphere coupling than the suggested energy coupling functions. To us it means that several non-linear processes are involved, and that the ANNs are capable of describing that but not the energy coupling functions. With the solar wind variables n, V, B_y, B_z as input to the network, 76% of the AE index variance is accounted for. The best coupling parameter in this study is $p^{\frac{1}{2}} V B_s$, and 71% of the AE index variance is accounted for. B_s is defined as $B_s = -B_z$ when $B_z < 0$ and $B_s = 0$ when $B_z > 0$.

To predict geomagnetic disturbances on all time-scales, we need to use solar and solar wind data as input to neural networks. To have the neural networks explain what they have learned, we need to use hybrid intelligent systems. We are therefore developing an intelligent hybrid system, Lund Space Weather Model (Fig. 10), which consists of modules each predicting or explaining the space weather at different time-scales. The availability of real-time solar wind data

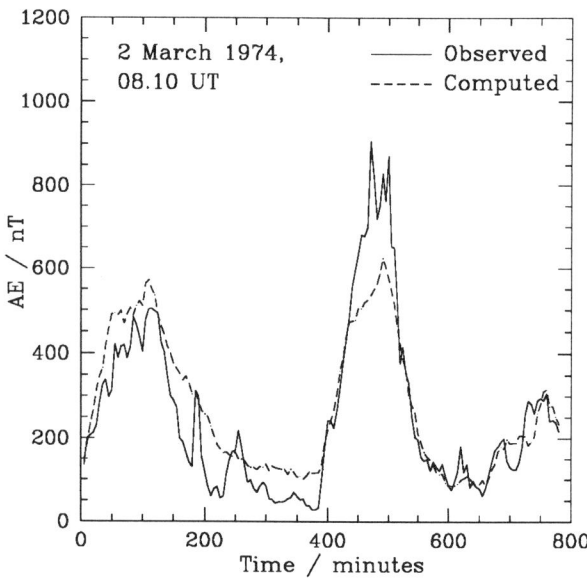

Figure 9. The solid line represents the observed AE and the dashed line the predicted AE with the use of a neural network.

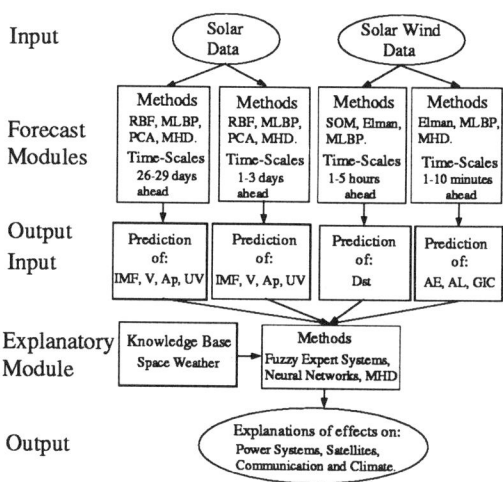

Figure 10. The Lund Space Weather Model is an intelligent hybrid system that predicts and models the space weather.

during 24 hours per day and high time and spatial resolution magnetic field images from SOHO would make it possible to present predictions on all time scales with very high accuracy.

5. CONCLUSIONS

During recent years, an immense interest in space weather and effects has developed. We believe AI methods are very powerful tools for studies of space weather. We have lately seen much success in using ANNs. Some of the most accurate predictions are now made with ANNs. A matter of great importance is therefore to have IHSs explaining what the ANNs have learned.

We also need multi-satellite observations of the space weather to be able to model a 3D and dynamic picture of the space weather. The ISTP program with recently launched spacecraft Geotail, WIND, SOHO, will facilitate part of that. However, we need more satellites to better cover the space weather. We also need satellites monitoring the solar wind closer to the magnetosphere than WIND as proposed by Sandahl et al., [1995] in order to understand the substorm/storm relationship.

Acknowledgements. We would like to thank Dr. Todd Hoeksema for making the WSO magnetic field data available and also Dr. J. King for the solar wind data. I would like to thank the two anonymous referees for valuable and interesting comments and suggestions.

REFERENCES

Baker, D.N., Statistical analysis in the study of solar wind-magnetosphere coupling, *Solar wind-magnetosphere coupling*, 17, Eds. Y.Kamide and J.A. Slavin, Terra Scientific Publishing, 17-38, 1986.

Baumjohann, W., and G. Haerendel, Entry and dissipation of energy in the earth's magnetosphere, *Space Astronomy and Solar System Exploration*, Proceedings of summer school held at Alpach, Austria, July 29 - August 8, 1986, ESA SP-268 May, 121-130, 1987.

Boteler, D., Geomagnetically Induced Currents: Present Knowledge and Future Research, *IEEE PES* Winter Meeting, Columbus, OH, January 31 - February 5, 1993.

Detman, T., Ring Current Predictions and Physics from Neural Networks, *Eos* Transactions, AGU, 1994 Fall Meeting, 75, No. 44 November 1, Supplement, 1994.

Elman, J.L., Finding structure in time, *Cognitive Science*, 14, 179-211, 1990.

Farrugia, C.J., M.P. Freeman, L.F. Burlaga, R.P. Lepping, and K. Takahashi, The Earth's magnetosphere under continued forcing: Substorm activity during the passage of an interplanetary magnetic cloud, *J. Geophys. Res.*, 98 ,7657-7671, 1993.

Fliess, M., M. Lamnabhi, and F. Lamnabhi-Lagarrigue, An algebraic approach to nonlinear functional expansions, *IEEE Trans. circuits Syst.*, 30, 554, 1983.

Freeman, J., A. Nagai, P. Reiff, W. Denig, S. Gussenhoven-Shea, M. Heinemann, F. Rich, and M. Hairston, The use of neural networks to predict magnetospheric parameters for input to a magnetospheric forecast model, in *Proceedings of Artificial Intelligence Applications in Solar-Terrestrial Physics Workshop* in Lund Sweden September 22-24, 1993, Eds. J.A. Joselyn, H. Lundstedt and J. Trollinger, 167-181, 1993.

Fu, L. M., Rule generation from neural networks, *IEEE Transactions on Systems, Man, and Cybernetics*, 28, no.8, 1114-1124, 1994.

Gleisner, H., H. Lundstedt, and P. Wintoft, Predicting geomagnetic storms from solar wind data using time-delay neural networks, *Ann. Geophys.*, in press, 1996a.

Gleisner, H., H. Lundstedt, and P. Wintoft, The response of the auroral electrojets to the solar wind modelled with neural networks, *J. Geophys. Res.*, in press, 1996b.

Goldberg, D., Genetic Algorithms in Search, Optimisation, and Machine Learning, *Addison-Wesley, Reading,MA*, 1989.

Gosling, J.T., D.J. McComas, J.T. Phillips, and S.J. Bame, Geomagnetic activity associated with earth passage of interplanetary shock disturbances and coronal mass ejections, *J. Geophys. Res.*, 96, 7831-7839, 1991.

Gonzalez, W.D., J.A. Joselyn, Y. Kamide, H.W. Kroehl, G. Rostocker, B.T. Tsurutani, and V.M. Vasyliunas, What is a geomagnetic storm?, *J. Geophys. Res.*, 99, 5771-5792, 1994.

Goonatilake, S., and S. Khebbal, Intelligent Hybrid Systems, *John Wiley and Sons*, 1995.

Gothaskar, P., and S. Khobragade, Classification of Interplanetary Scintillation Power Spectra Using Artificial Neural Network, in *Proceedings of Artificial Intelligence Applications in Solar-Terrestrial Physics Workshop* in Lund Sweden September 22-24, 1993, Eds. J.A. Joselyn, H. Lundstedt and J. Trollinger, 109-114, 1993.

Hernandez, J., V., T. Tajima, and W. Horton, Neural net forecasting for geomagnetic activity, *Geophys. Res. Lett.*, 20, 2707, 1993.

Hertz, J.K., A. Krogh, and R.G. Palmer, Introduction to the theory of neural computation, *A Lecture Notes Volume in the Santa Fe Institute Studies in the Sciences of Complexity*, Addison-Wesley, 1991.

Hoeksema, T.J., Structure and Evolution of the Large Scale Solar Heliospheric Magnetic Fields, *Ph.D dissertation, Stanford University*, 1984.

Iyemori, T., H. Maeda, and T. Kamei, Impulse response of geomagnetic indices to interplanetary magnetic field, *J. Geomagn. Geoelectr.*, 6, 577, 1979.

Jackson, P., Introduction to Expert Systems, *Addison-Wesley*, 1986.

Joselyn, J., Geomagnetic Activity Forecasting: The State of the Art, *Rev. of Geophys.*, 33, 383-401, 1995.

Klein, L.W., and L.F. Burlaga, Magnetic clouds at 1 AU, *J. Geophys. Res.*, 87, 613-624, 1982.

Klimas, A.J., D. Vassiliadis, D.N. Baker, and D.A. Roberts, The organized nonlinear dynamics of the magnetosphere, *J. Geophys. Res.*, 101, 13.098-13.113, 1996.

Kosko, B., Neural Networks and Fuzzy Systems, *Prentice Hall*, Englewood Cliffs, NJ, 1992.

Kohonen, T., Self-Organization and Associative Memory, *Springer Series in Information Science Springer-Verlag*, 1984.

Lindsay, G.M., C.T. Russell, and J.G. Luhmann, Coronal mass ejections and stream interaction region characteristics and their potential geomagnetic effectiveness, *J. Geophys. Res.*, 100, 16,999-17,013, 1995.

Lorentz, G.G., The 13th problem of Hilbert, in *Mathematical Developments Arising from Hilberts Problem*. American Mathematical Society, Province, RL, 1976.

Lundstedt, H., An inductive expert system for solar predictions, in *Proceedings of Solar-Terrestrial Prediction Workshop*, in Leura Australia October 16-20, 1989, Eds. Thompson, R.J., Cole, D.G., Wilkinson, P.J., Shea, M.A., Smart, D. and Heckman, G,. 125-129, 1990.

Lundstedt, H., Neural networks and predictions, in *IAGA programs and abstract*, XX General Assembly, IUGG, Vienna, 1991.

Lundstedt, H., A trained neural network, geomagnetic activity and solar wind variation, in *Proceeding of Solar-Terrestrial Workshop in Ottawa May 18-22*, Ed. M.A. Shea, NOAA, 607-610, 1992a.

Lundstedt, H., Neural networks and predictions of solar-terrestrial effects, *Planet. Space Sci.*, 40, 457-464, 1992b.

Lundstedt, H., Solar caused potential in gas-pipelines in southern Sweden. In *Proceeding of Solar-Terrestrial Workshop in Ottawa May 18-22*, Ed. M.A. Shea, NOAA, 607-610, 1992c.

Lundstedt, H., Solar origin of geomagnetic storms and predictions, *J. of Atmosph. and Terr. Phys.*, 58, no. 7, 821-830, 1996.

Lundstedt, H., and J.T. Hoeksema, Neural images of the interplanetary magnetic field and predictions of geomagnetic storms 27 days ahead, in *Eos*, October 27, 1992.

Lundstedt, H., and P. Wintoft, Prediction of geomagnetic storms from solar wind data with the use of a neural network, *Ann. Geophysicae*, 12, 19-24, 1994.

Lundstedt, H., P. Wintoft, J.-G. Wu, and H. Gleisner, AI methods and space weather forecasting, *Artificial Intelligence and Knowledge Based Systems for Space*, 5th Workshop, 10-11 Oct., ESTEC, ESA, 235, 1995.

McPherron, R.L., Possible applications of expert systems and fuzzy logic in solar terrestrial physics, in *Proceedings of Artificial Intelligence Applications in Solar-Terrestrial Physics Workshop*, in Lund Sweden September 22-24, 1993, Eds. J.A. Joselyn, H. Lundstedt and J. Trollinger, 1-13, 1993.

Pop, E., R. Hayward, and J. Diederich, RULENEG: extracting rules from a trained ANN by stepwise negation, *Queensland Univ. of Tech., Neurocomputing Res. Centre*, 1994.

Sandahl, I., H. Lundstedt, H. Koskinen, and K-H. Glassmeier, On the need for solar wind monitoring close to the magnetosphere, in *Proceedings of 16th NSO/Sacramento Peak International workshop, Solar Drivers of Interplanetary and Terrestrial Disturbances*, October 16-22, 1995.

Scherrer, P.H., J.M. Wilcox, L. Svalgaard, T.L. Duvall, Jr. P.H. Dittmer, and E.K. Gustafson, The mean magnetic field of the sun: observations at Stanford, *Solar Phys.*, 54, 353-361, 1977.

Snel, R., and H. Lundstedt, Selforganising maps of solar wind structures, in *Proceedings of Artificial Intelligence Applications in Solar-Terrestrial Physics Workshop*, in Lund Sweden September 22-24, 1993, Eds. J.A. Joselyn, H. Lundstedt and J. Trollinger, 97-107, 1993.

Towell, G., and J. Shavlik, The Extraction of Refined Rules From Knowledge Based Neural Networks, *Machine Learning*, 131, 71-101, 1993.

Tsurutani, B.T., and W.D. Gonzalez, Tweaking the magnetosphere, *Nature*, 358, 26, 1992.

Tsurutani, B.T., W.D. Gonzalez, F. Tang, S.I. Akasofu, and E.J. Smith, Origin of Interplanetary Southward Magnetic Fields Responsible for Major Magnetic Storms Near Solar Maximum (1978-1979), *J. Geophys. Res.*, 93, 8519-8531, 1988.

Tsurutani, B.T., W.D. Gonzalez, A.L.C. Gonzalez, F. Tang, J.K. Arballo, and M. Okada, Interplanetary origin of geomagnetic activity in the declining phase of the solar cycle, *J. Geophys. Res.*, 100, 21717, 1995.

Vassiliadis, D., A.J. Klimas, and D.N. Baker, D.N., Nonlinear ARMA Models for the D_{st} Index and Their Physical Interpretation, in Proceedings of the Third International Conference on Substorms (ICS-3), Versailles, France, May 13-17, 1996.

Viljanen, A. and R. Pirjola, Geomagnetically Induced Currents in the Finnish High-Voltage Power System, *Surveys in Geophys.*, 14, 383-408, 1994.

Wang, Y.-M., and N.R. Sheeley Jr., Global evolution of interplanetary sector structure, coronal holes, and solar wind streams during 1976-1993: Stackplot displays based on solar magnetic observations, *J. Geophys. Res.*, 99, 6597-6608, 1994.

Wintoft, P., and H. Lundstedt, IMF polarity prediction from WSO magnetograms using radial basis neural network, *IUGG XXI General Assembly Meeting, Boulder, July 2-14, 1995*, B140, 1995.

Wintoft, P., and H. Lundstedt, Space Weather Modeling with Intelligent Hybrid Systems: Predicting the Solar Wind Velocity, presented at the 31st Scientific Assembly, Birmingham 14-21 July 1996, U.K. and will be published in Advances in Space Research.

Wray, J., and G.G. R. Green, Calculation of the Volterra kernels of nonlinear dynamic systems using an artificial neural network, *Biol. Cybernet.*, 71(3), 187, 1994.

Wu, J.-G., and H. Lundstedt, Prediction of Geomagnetic Storms From Solar Wind Data Using Elman Recurrent Neural Networks, *Geophys. Res. Letters*, 23, 319-322, 1996a.

Wu, J.-G., and H. Lundstedt, Geomagnetic Storm Predictions From Solar Wind Data with the use of Dynamic Neural Networks, revised Sept., *J. Geophys., Res.*, 1996b.

Zhao, X., and J.T Hoeksema, A Magnetostatic Coronal Model with Horizontal Electric Currents: Modeling the Coronal Streamer Belt, in *Proceedings of the First SOHO Workshop, Annapolis, Maryland, USA, 25-28 August 1992*, ESA SP-348, November, 117-120, 1992.

Zhao, X., and J.T. Hoeksema, A coronal magnetic field model with horizontal volume and sheet currents, *Solar Physics*, 151, 91-105, 1994.

Zhao, X., and J.T. Hoeksema, Prediction of the interplanetary magnetic field strength, *J. Geophys., Res.*, 100, 19-33, 1995.

Review of Techniques for Magnetic Storm Forecasting

Thomas R. Detman

Space Environment Center, NOAA, Boulder, CO, 80303, USA.

Dimitris Vassiliadis

Universities Space Research Association, Greenbelt, MD 20771.

Today a wide variety of techniques are available for nowcasting and forecasting magnetic storm activity. A brief review of linear time series prediction techniques, with examples, is used to lay a foundation for the description of newer non-linear techniques based on state-space reconstruction. We illustrate the state-space prediction technique in application to predict *Dst* from ISEE-3 solar wind data. Upstream solar wind data, such as from ISEE-3 or WIND close to the L_1 libration point, provide a prediction lead time of 0.5–1.5 hours. To go beyond the L_1 prediction lead time some information about the solar wind between the L_1 point and the Sun is required. Remote sensing is the measurement of something from a distance, like solar magnetograms or X-ray images. Both empirical and physically based models, driven by remote sensing data, promise a way to make forecasts a few days into the future. A combination of the statistical time series prediction techniques operating on the output of physically based models, driven by remote sensing data, may offer the first capability of predicting magnetic storms a few days in advance. We illustrate this combination of techniques using the output of a potential field model [Wang and Sheeley, 1988] as input to a linear prediction filter to forecast the planetary geomagnetic index. Finally, practical forecasting requires verification. We describe some of the standard measures of forecast performance: skill score, prediction efficiency, and correlation coefficient. The value of cross validation testing is emphasized.

INTRODUCTION

Earlier papers in this monograph have described the damaging effects of magnetic storms [*Kappenman and Radasky, this volume; Boteler, this volume*]. The NOAA Space Environment Center (SEC) in Boulder, Colorado, in collaboration with the USAF, runs a full-time operational service 24 hours-a-day, every day. The Space Weather Operations (SWO) of SEC issues warnings, alerts, and forecasts in order to help customers (e.g. power companies or satellite operators) avoid damage and wasted expense that might be caused by magnetic storms. Before a customer can make an informed business decision based on a forecast, they need to balance the cost of taking action against the risk of not taking action [*Murphy and Daan*, 1985]. This is only possible if the expected accuracy of that forecast is known, thus verification must be part of operational forecasting. Verification of forecasts since 1988 [*Doggett*, 1993] show that improvement is needed in those forecasts.

In the sections below we describe the basic elements of linear prediction, nonlinear prediction based on state space reconstruction, and forecast verification. Finally, we show an end-to-end prediction scheme that is based on combining time series methods with solar observation data and physical models.

LINEAR TIME SERIES PREDICTION TECHNIQUES

Statistical time series analysis and prediction has a long history. Previous reviews of note in space physics are the reviews on statistical methods by Baker [1986] and Reiff [1983], and the review of linear prediction by Clauer [1986]. Parallel developments and descriptions exist in many different fields of study including: economics, communication and control theory, seismic geology, digital speech analysis and coding. Descriptions from different fields often have differing terminologies, approaches and emphasis. In this brief review we describe basic ideas, connect some of the differing terminology, and demonstrate basic techniques with examples.

Moving Average (MA) Linear Filters

A filter is a transformation that maps a sequence of inputs into a sequence of outputs; it can also serve as a model of the system that produced the output data. The term "moving average" refers to the fact that the output of the filter is a weighted average of preceding inputs. In the case where time is a continuous variable, the filter is a weighting function also known as the impulse response. The behavior of the filter is described by equation (1) which contains a convolution integral. In equation (1), $x(t)$ is the filter input, $f(t)$ is the weighting function, or impulse response, and $y(t)$ is the filter output.

$$y(t) = \int f(\tau) \, x(t - \tau) \, d\tau \quad (1)$$

$$y_t = \sum_{i=0}^{M-1} f_i \, x_{t-i} \quad (2)$$

The form of equation (1) is good for analytical work (or analogue computers); equation (2) is the sampled data, or digital, equivalent of equation (1). In (2) M is the number of filter coefficients; these coefficients determine the current output as a linear combination or weighted moving average of M previous inputs. The equations of (3) show the basic form, and equation (4) is the same thing written in matrix-vector form.

$$\begin{array}{c} x_2 f_0 + x_1 f_1 + x_0 f_2 = y_2 \\ x_3 f_0 + x_2 f_1 + x_1 f_2 = y_3 \\ x_4 f_0 + x_3 f_1 + x_2 f_2 = y_4 \\ \vdots \quad \vdots \quad \vdots \quad \vdots \end{array} \quad (3)$$

$$X f = y \quad (4)$$

The general set of coupled equations, described by equations (2)-(4), is extremely well studied. The number of input-output pairs, N_p, of the general type shown in (3), should be larger than the number of coefficients, M, to be found (more equations than unknowns); we only consider this case here. In this case there is generally no f which exactly satisfies (4); we say that the system (3) is overdetermined. We find instead the f which gives the least squared error as defined in (5); superscript T denotes transpose.

$$E^2 = (y - Xf)^T (y - Xf) \quad (5)$$

The minimum squared error in (5) is found by setting the partial derivative of E^2 with respect to f to be zero. This leads to the *normal equations*, shown in (6).

$$[X^T X] \, f = X^T y \quad (6)$$

The least squares technique originated independently with Gauss and Legendre in the early 19th century. Many methods to solve the normal equations (6) have since been developed; we will discuss solution techniques later.

A classic example of the application of MA filters to the study of geomagnetic activity is work done by Bargatze et al. [1985]. They considered the auroral geomagnetic index AL to be the output of a linear MA filter driven by the rectified solar wind dawn-to-dusk electric field, VBs, as input. VBs is computed as the solar wind velocity, V, times the southward component of the the interplanetary magnetic field (IMF), Bs. The Bargatze data set consists of 34 isolated activity intervals, each with a well-defined beginning and end, sorted by intensity. Each set of 5 consecutive intervals (beginning with 1-5, then 2-6, and so on) were used to solve (6) for the impulse response filter, f. Figure 1 is a re-creation of those results.

The obvious interpretation of Figure 1 is that the system impulse response depends on the level of activity, which implies that the system is nonlinear.

Autoregressive (AR) Linear Prediction Filters

The terms autoregressive, all-pole model, recursive, and autonomous, all generally refer to systems (or models, or filters) that are self-driving or lack an observable input. Sunspot numbers are usually modeled as the output of a autonomous system. Equation (7) describes a basic autoregressive

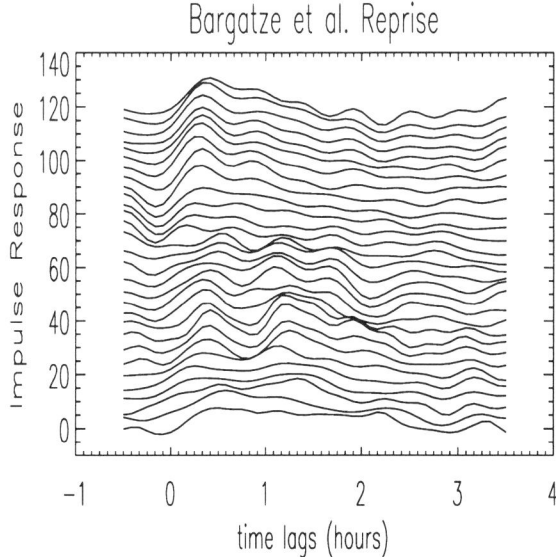

Figure 1. A stack plot of linear MA prediction filters relating VBs to AL. The filters are ordered by the level of activity of the data sets used to compute them, as explained in the text. The activity level increases from bottom to top. Successive filters are offset by 4 nT/(mV/m). (A re-creation of Figure 3 from Bargatze, et al. [1985].)

(AR) linear prediction filter.

$$y_t = \sum_{j=1}^{N} g_j \, y_{t-j} \qquad (7)$$

The current filter output is a linear combination of the previous N outputs. Feedback is the essential feature. Because of the feedback, AR filters are fundamentally different from MA filters in two important ways. For MA filters the response goes to zero M steps after an impulse (delta-function-like) input. In contrast to that an AR filter response can, at least conceptually, go on forever because of the feedback. Thus the phrase "infinite impulse response" (IIR) refers to the impulse response of an AR filter while the phrase "finite impulse response" (FIR) means a MA filter; an AR filter can also be unstable. If the filter contains sufficient "positive feedback" the magnitude of the output exhibits unbounded exponential growth (in jargon, it *blows up*). Simply put, AR and MA filters represent different kinds of dynamics.

The behavior of the AR filter becomes more interesting if either an input signal or a noise disturbance is added to the right side of (7). We examine the effect of an input signal in the next section on ARMA filters. The case of a noise disturbance on the right side of (7) is considered next.

In some applications, equation (7) will have an additional term on the right hand side. The additional term can be interpreted as either prediction error, or white (uncorrelated) gaussian noise driving the system (called *innovation*). Under *certain assumptions* these two interpretations are equivalent. (If one has found the optimum filter g that solves equation (7) with the minimum mean square error, subject to those "certain assumptions," then the resulting time series of prediction errors will be white Gaussian noise.) Equation (8), like (3) above, illustrates the form;

$$\begin{aligned}
y_2 g_0 + y_1 g_1 + y_0 g_2 &= y_3 \\
y_3 g_0 + y_2 g_1 + y_1 g_2 &= y_4 \\
y_4 g_0 + y_3 g_1 + y_2 g_2 &= y_5 \\
\vdots \qquad \vdots \qquad \vdots &\quad \vdots
\end{aligned} \qquad (8)$$

$$A \, g = y \qquad (9)$$

equation (9) is the matrix-vector form. Note that (9) has the same form as (4).

Feedback is very important in system dynamics. Note that (7) is a model of a system that has feedback; stability is an issue whenever there is feedback. There is an important distinction, regarding feedback, as to whether the actual use of the filter involves feedback or not. If the input to the filter is supplied entirely by measurements (observations) then the filter does not have feedback; it is running in feed-forward mode. This mode may be used, for example, to predict future observations. If, on the other hand, some or all of the filter inputs are previous filter outputs then the filter itself has feedback. So, although (7) models a system that has feedback, the filter described by it can be used in either a feedback mode or a pure feed-forward mode (input supplied by observations); the results and usually the purpose are different.

Notice that the vector of coefficients which compose g can be used in equation (7) to extrapolate the time series into the future, or to fill gaps (by blending forward and backward extrapolations across the gap). There is a justification for using (7) to extrapolate in this way over time; the power spectrum of the y_t time series computed by the maximum entropy method (MEM) is also based on those same coefficients [*Ulrych and Clayton*, 1976]. The spectrum of the extrapolated series will approximate that of the preceding data. In some applications this will be far superior to linear interpolation.

In contrast to this extremely brief and elementary description of AR filtering (or modeling), the history of this subject is long and rich. Our desire to exemplify a quality *ethos* [*Campbell*, 1995] in this short paper requires us to mention the comprehensive review by Kay and Marple [1981] for anyone interested in refinements and connecting relationships that we have omitted.

ARMA and MLR

In general a filter can have both AR and MA types of inputs. The defining equation (10) has two summations on the right hand side — the summations given in (2) and (7).

$$y_t = \sum_{j=1}^{N} g_j y_{t-j} + \sum_{i=0}^{M-1} f_i x_{t-i} \quad (10)$$

The matrices and vectors can be combined, however, so that the final form is identical to that of (4) and (9), and the resulting normal equations take the same form as (6).

The general linear least squares problem, also called multi-linear regression (MLR), or multi-channel filtering, shares the same basic forms as (4), (9) and (6). In the general problem the columns of the matrix X, or A, are taken to be basis vectors. Then g has the interpretation of specifying a particular linear combination of those basis vectors. The desired solution of (9) is the value of g for which Ag is the best possible approximation of y, again in the least squares sense, as with linear MA or AR filters. A common use of MLR is to fit a polynomial to a set of data.

Examination of (3) or (8) should convince one that the columns of A (or X) may be very similar to one another if the time series y_t has a broad autocorrelation function. If this is the case, the square symmetric matrix, e.g. $A^T A$, appearing in the normal equations may be ill conditioned or even singular. A flexible and robust method for solving the normal equations even if they are singular, called singular value decomposition (SVD) is described in *Numerical Recipes*, Press, et al., 1992.

For an example of AR linear prediction filtering (LPF) we have applied the technique to making predictions of the daily planetary A index, Ap. (This technique could also be applied to Dst; we will discuss Dst later.)

Before we can discuss any results, however, we first have to define how to measure success.

FORECAST VERIFICATION: MEASURING PERFORMANCE

Measures of prediction accuracy fall into two broad categories: those based on signal detection theory (we will not discuss these) and those based on the statistics: statistics of the prediction errors or joint statistics of forecasts and observations [*Brooks and Doswell*, 1996; *Murphy and Winkler*, 1987]. Measures based on statistics can be divided into three groups, absolute, relative, and normalized. The root-mean-squared (rms) error, for example, is a measure of the absolute or actual errors. (If the forecast is in units of nT, then the rms error has units of nT.) Relative measures compare two techniques or forecasters. The rms error divided by the standard deviation of the observations is an example of a normalized error measure; it is a dimensionless ratio.

Prediction Efficiency, Skill Score, and Correlation

Suppose we apply a prediction filter to a set of data. Let f_i ($i=1,Np$) represent the predictions made by the filter, and let x_i ($i=1,Np$) represent the corresponding observations; the prediction errors are then given by ($f_i - x_i$). The prediction efficiency (*PE*) of the filter, on that set of data, is given by (11), where $\bar{x}$ is the average of the Np sample observations.

$$PE = 1 - \left(\frac{MSE}{VAR}\right) \quad , \text{where} \quad (11)$$

$$MSE = \frac{1}{Np} \sum_{i=1}^{Np} (f_i - x_i)^2 \quad , \text{and}$$

$$VAR = \frac{1}{Np} \sum_{i=1}^{Np} (x_i - \bar{x})^2.$$

In terrestrial weather prediction, skill scores, which are relative measures, are often used to measure performance. A skill score describes the performance of one technique or forecaster relative to a reference forecast that may be another technique or forecaster, or a simple statistical forecast based on the data sample. The common practice, and the one we use, is to take the mean value of the sample observations, $\bar{x}$, as the reference forecast; in weather jargon this is called the *sample climatology*. The skill score with respect to sample climatology (*SS*) for our prediction filter is given by (12):

$$SS = 1 - \left(\frac{MSE}{MSE_{SC}}\right) \quad , \text{where} \quad (12)$$

$$MSE_{SC} = \frac{1}{Np} \sum_{i=1}^{Np} (x_i - \bar{x})^2 = VAR.$$

Note that the mean-squared prediction error produced by using the sample mean for every forecast (MSE_{SC}) is the sample variance (*VAR*), and therefore *PE* and *SS*, as we have defined them, are identical. We use these terms interchangeably.

The *PE* or *SS* is the fraction of the variance that was predicted. A *PE* equal to 1 indicates perfect predictions. The *PE* or *SS* obtained by always predicting the mean value of the observed quantity (assuming one knows it ahead of time) is exactly 0.

The correlation coefficient, Rc, defined in (13), is a conventional standard prediction quality measure.

$$Rc = \frac{\frac{1}{Np}\sum_{i=1}^{Np}(x_i - \bar{x})(f_i - \bar{f})}{\sigma_x \sigma_f} \quad , \text{where} \quad (13)$$

$$\sigma_x^2 = \frac{1}{Np}\sum_{i=1}^{Np}(x_i - \bar{x})^2 \quad , \text{and}$$

$$\sigma_f^2 = \frac{1}{Np}\sum_{i=1}^{Np}(f_i - \bar{f})^2 \quad .$$

It answers the question: "How much does the average forecast tell you about what will happen?" It is a relative measure; a weakness of the Rc is that it is insensitive to errors in magnitude. So for example, a forecast that is always off by a fixed factor, i.e. $f_i = 5\ x_i$, would give an Rc of 1; however the SS would be a large negative number.

In figure 2 the correlation coefficient is used to show the recurrence of geomagnetic activity with solar rotation.

Cross Validation

In the development of forecasting techniques there is some trial and error experimenting to see what works. There is a potential trap in this process, called over-fitting; it is tempting to use filters of increasingly higher order, since the in-sample fitting error always decreases with increasing order. If the order is too high, however one is fitting the noise as well as the signal in the regression data sample. Clearly it is necessary to test, or validate, the filter on a statistically independent dataset, a process called cross validation. (This is closely related to the problem of generalization in neural network training [*Lundstedt, this volume*].)

The Akaike Information theoretic Criterion (AIC), [*Akaike*, 1972, 1974; *Ulrych and Clayton*, 1976], can be used to *estimate* the number of model parameters or filter coefficients that would be optimal if the filter could be tested on a statistically identical, but independent data set. The AIC is based on the "maximum likelihood" principle of probability theory. If the amount of data available for fitting is too small to be split up for cross validation, the AIC may be used to avoid "over fitting". In that case, for a given data set consisting of Np input-output pairs, one should choose the filter that gives the smallest $AIC = Np\ log(MSE) + 2P$, where MSE is the mean squared prediction error and P is the total number of free parameters or coefficients, e.g. $P = N + M$ for the ARMA models discussed above.

If you use all the available data to construct a prediction filter, then you have no independent data left to test the filter. Cross validation is done by splitting the available data into 2 or more groups. One group is then used to construct (fit) a model or prediction filter, and then a different group (or groups) of data is used for testing.

If the amount of available data is too small for cross validation then use the AIC, but remember, it only gives an estimate of the optimum filter order. Cross validation is a real test on independent data.

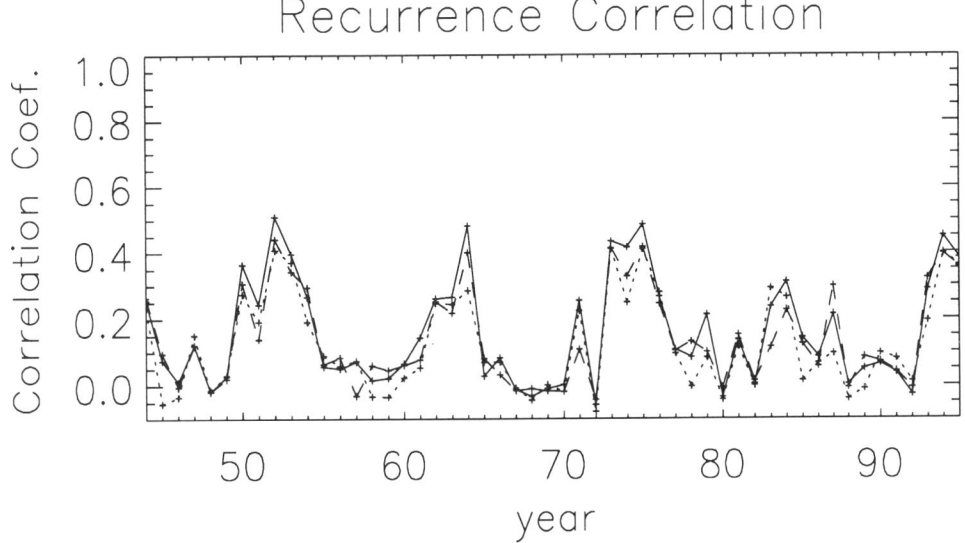

Figure 2. Recurrence of geomagnetic activity: As viewed from the Earth the sun rotates in about 27 days. Long-lived solar features can lead to increased activity on successive rotations. The strength of recurrence is shown by the correlation coefficient between two Ap observations made 26 days (dotted), 27 days (solid), or 28 days (dashed) apart.

Figure 3 shows the results of a sequence of cross validation tests. The historical record of observed daily Ap values for the years 1967 through 1976 were used to fit 29 different AR linear prediction filters, as described earlier. These filters differed only in length, the number of coefficients as given by N in equation (7). Each filter was then used to make one-step-ahead predictions through the test interval, 1989 through 1993, using only prior Ap observations as input. Figure 3 also indicates the skill score and correlation coefficient for published operational 1-day ahead Ap predictions during the same test period. Based on the given performance measures, on average, the linear prediction filters (based on input-free system models) would have outperformed the human forecasters during this time period. In fairness to the human forecasters however, we point out that these simple linear prediction filters, driven only by previous Ap observations, will not predict the large rare events that cause the most problems. A human forecaster can take linear (or nonlinear) predictions as numerical guidance and combine it intuitively with other sources of information.

From figure 2 it can be seen that 1989 through 1993 was a period of relatively low recurrence. Figure 4 shows how the same set of filters as in Figure 3 (except now extended up to length $N=40$) would have performed in 1994, a year of high activity recurrence. Figure 4 shows some interesting features. The Ap is more predictable when recurrence is strong, even for filters shorter than the recurrence period, and there is a significant improvement in performance when the filter length is increased from just under the recurrence period to just over it (note the sharp rise in SS from $N=24$ to $N=28$). There is little rise in SS between $N=12$ and $N=14$; this feature is frequently seen in cross validation tests based on database (fit) and test periods different from those represented here.

There are some important points to make about the old classical linear techniques summarized above; as shown by Figures 3 and 4 they can still make a contribution, and they can provide a bench-mark against which to compare more sophisticated nonlinear techniques, such as neural nets. Finally, linear techniques are employed within an important class of nonlinear prediction techniques, as discussed below.

NONLINEAR ARMA FILTERS AND STATE SPACE RECONSTRUCTION—LOCAL LINEAR PREDICTION

The AR, MA, and ARMA filters are readily extended to nonlinear cases where the theory of nonlinear dynamical systems [*Abarbanel et al.*, 1993; *Casdagli*, 1992; *Weigend and Gershenfeld*, 1994] becomes relevant. A significant idea is that of state space reconstruction for either linear or nonlinear systems. The introduction of the state space representation goes back at least to Yule [1927], who in fact used

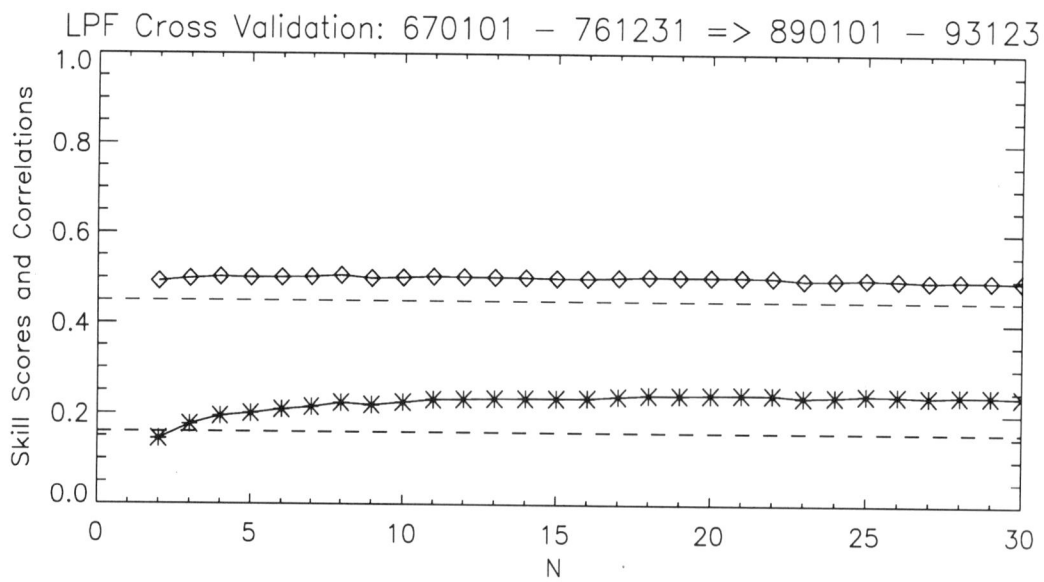

Figure 3. Performance of AR linear prediction filters as a function of filter length, N. Predictions are 1-step (1-day) ahead. Input is the N preceding observed values of Ap. Skill scores (asterisks, lower curve) and correlation coefficient (diamonds, upper curve) are shown as a function of filter length, N. Skill score and correlation for operational predictions during the same test period are indicated by dashed lines.

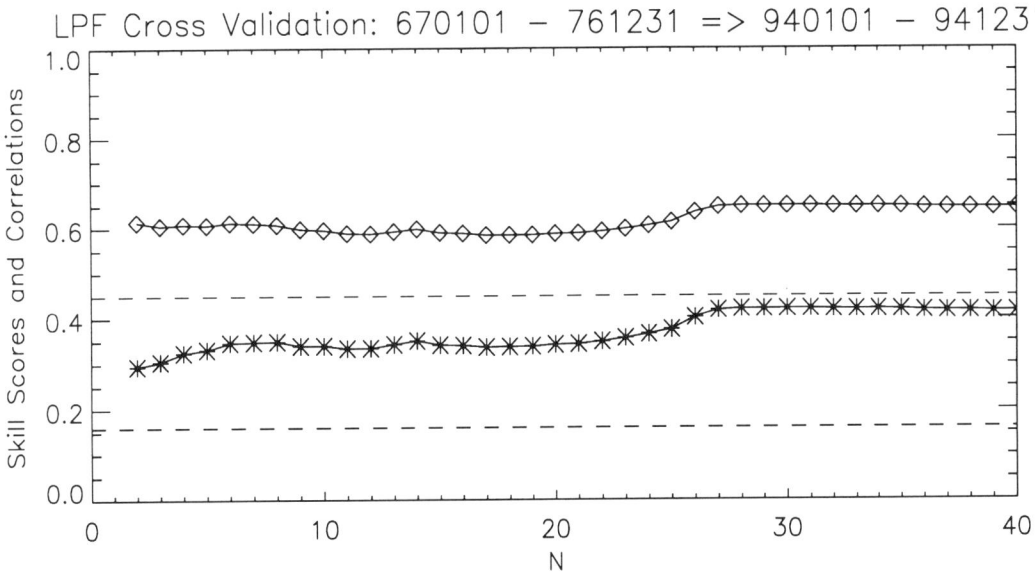

Figure 4. Same as Figure 3, except that the test period is 1994, and the filter length, N, is extended to 40. The top SS is 0.421 and the top Rc is 0.650, both achieved by the N=31 AR filter.

it for solar activity studies. The comparison of linear filters and nonlinear systems is explored in several papers e.g. [*Broomhead et al.*, 1992]. A neural network can be considered as a form of nonlinear dynamical system [*Lundstedt, this volume*].

Equation (7)(repeated here for convenience) describes a linear dynamical system in that the state of the system at time t–1, y_{t-1}, determines y_t, the next output.

$$y_t = \sum_{j=1}^{N} g_j \, y_{t-j} \qquad (7)$$

For the linear case the coefficients, g_j, are constant. In the corresponding nonlinear case the g_j depend on y_{t-1}, the current state of the system.

$$y_t = \sum_{j=1}^{N} g_j(y_{t-1}) \, y_{t-j} \qquad (14)$$

The activity-dependent response $g(y)$ is calculated from the regression (8) using the same numerical methods as for linear filters. The difference is that for a linear filter the whole dataset is used in the calculation of g. In the nonlinear calculation of $g(y)$, however, only a subset of the data is used, namely vectors similar enough to y, so that the response is also similar. In practice, the vectors that have the smallest distance from y are selected. Usually the distance function is the Euclidean, but other metrics have been proposed [*Garcia et al.*, 1996].

In practice one has many consecutive observations of the state of the system, y, that define trajectories or orbits of the system in state space. The crux of the technique is to answer the question: When the system was in this state (or a similar state) before, what did it do next? The answer is found from a (small) number of state vectors, those nearest to y in the state space and their respective outputs. The vectors y_{NN} are called the nearest neighbors of y in the state space. Given y, the choice of the number of the nearest neighbors determines the response, $g(y)$, calculated from a set of Equations (8) for the subset y_{NN} of the full database. The subset y_{NN} must be determined for each new prediction.

A system with nonlinear dynamics is then approximated by a sequence of local-linear filters, $g(y_{NN})$, one for each $y(t)$ in succession. Each region or "neighborhood" of the state space has different dynamics and is best approximated by a different number $NN(y)$ of nearest neighbors around y [*Casdagli*, 1991]. Other properties of the dynamics (periodicity, stability, etc.) also depend on the position of y in the state space.

For input-output data the filters are extended to nonlinear ARMA filters.

$$y_{t+1} = \sum_{j=0}^{N-1} g_j(y_t, x_t) \, y_{t-j} + \sum_{i=0}^{M-1} f_i(y_t, x_t) \, x_{t-i} \qquad (15)$$

The next output, y_{t+1} is determined by the state-input vector (y_t, x_t). The nearest neighbors are the vectors with the smallest distances to (y_t, x_t).

As an example of nonlinear ARMA we consider the *Dst* index. Numerous studies have applied linear ARMA filtering to prediction of *Dst* from solar wind data, e.g. [*Burton, et al.*, 1975; *Feldstein, et al.*, 1984; *Fay, et al.*, 1986; *McPherron, et al.*, 1986; *Murayama*, 1986; *Iyemori, et al.*, 1979; *Trattner and Rucker*, 1990]. The *Dst* is the defining index for magnetic storms, but it is not a good choice for a state variable because it represents a combination of both the magnetospheric ring current and magnetopause currents which fluctuate with the solar wind dynamic pressure. The preferred state variable is Dst^0, where Dst^0 stands for the solar wind-pressure-corrected geomagnetic disturbance, obtained from *Dst* and the solar wind dynamic pressure, P_{SW} by (16), where b and c are fixed constants.

$$Dst^0 = Dst - b\sqrt{P_{SW}} - c \qquad (16)$$

The standard dynamic model for the state variable Dst^0 is given by (17)

$$\frac{dDst^0}{dt} = aE_w(t) - \frac{Dst^0}{\tau} \qquad (17)$$

where the input is the rectified (westward) solar wind convection electric field $E_w = VB_{South}$, the product of the bulk speed and the southward component of the magnetic field. Linear studies based on different data sets report different values for the constants $a, b, c,$ and τ in (16) and (17). This standard linear model can generally explain (fit) about 70% to 75% of the variance in *Dst*, that is, a *PE* or *SS* of about 0.7 to 0.75.

Detman [1995] obtained an improved linear fit by considering Dst^0 to be a sum of two components each with dynamics as given by (17), but one with $\tau_1 = 3$ hours and the other with $\tau_2 = 24$ hours. The observed *Dst* was then found by using multi-linear regression (MLR) to solve (18)

$$Dst = a_1 Dst^{01} + a_2 Dst^{02} + b\sqrt{P_{SW}} + c \quad . \qquad (18)$$

By integrating (17) with fixed values of τ to get Dst^{01} and Dst^{02} then solving (18), Detman included the pressure correction as part of the least squares fitting process, and then iterated the entire solution process in a trial and error procedure to find his best combination of τ_1 and τ_2.

This approach was suggested (but not implemented) by Murayama [1986]. Murayama used a different method to find the values of two simultaneous decay time constants. His method in fact demonstrates and important principle—in practice, the representation of a given dynamic system in terms of AR, MA, or ARMA models is not unique. Murayama computed an $M=49$ MA filter for Dst^0 in terms of E_w (using 1–hour *Dst* and near Earth solar wind data), then showed that the MA filter was well represented by a double exponential decay with $\tau_1 = 2$ hours and $\tau_2 = 21$ hours. Considering the different data sets and procedures used the Detman and Murayama results are in good agreement.

Valdivia et al. [1996] have constructed several models for predicting *Dst* including higher order linear models, a nonlinear ARMA state-input space reconstruction model, and nonlinear AR state space model for predicting the severity and duration of a storm from *Dst* data alone, early in the main phase of the storm.

The solar wind-pressure, P_{SW}, may be used either to obtain Dst^0, using (16), or as a second input, or both, so that the state-input space may contain lags from all three variables: $(Dst_t, Dst_{t-1} ..., Ew_t, Ew_{t-1}, ..., P_t, P_{t-1}, ...)$, analogous to the multi-linear regression of the earlier sections. Figure 5 shows an example of this approach where four variables $(Ew_t, Ew_t\sqrt{P_{SW}}, \frac{d}{dt}\sqrt{P_{SW}}, Dst^0_{t-1})$ define the state-input space. The first three of these variables are treated as if they were obtained in real-time from a spacecraft in the solar wind, but the 4th variable, $Dst^0(t-1)$, is derived from the previous prediction. Actually, $dDst^0/dt$ at time t is the predicted quantity and $Dst^0(t)$ is obtained by doing a semi-implicit (trapezoid-rule) time step. (Note the filter has feedback, as discussed earlier.) This is a practical approach since observed *Dst* is not currently available in, or near, real-time. For the local linear state-input space predictions in Figure 5 the number of nearest neighbors was held constant at 256 from a data base of 1343, and the time resolution of the data was 1 hour. For the linear prediction in Figure 5, $PE=0.79$ and $Rc=0.97$; for the state space prediction, $PE=0.94$ and $Rc=0.98$. The choice of 256 for the number of nearest neighbors in Figure 5 was a compromise between high prediction efficiency (*PE*) and stability of the decay rate parameter $\tau(t)$, shown in the bottom panel.

Several studies have used the representation (15) to study the nonlinearity, stability, and predictability of auroral geomagnetic indices AE, AU, and AL from solar wind parameters [*Price and Prichard*, 1993; *Vassiliadis*, 1994; *Price et al.*, 1994; *Vassiliadis et al.*, 1995, *Vassiliadis et al.*, 1996]. Recently Klimas et al. [1996] reviewed the application of nonlinear methods of time series analysis in magnetospheric physics. The AE/AL studies showed a significantly higher prediction performance using nonlinear ARMA filters

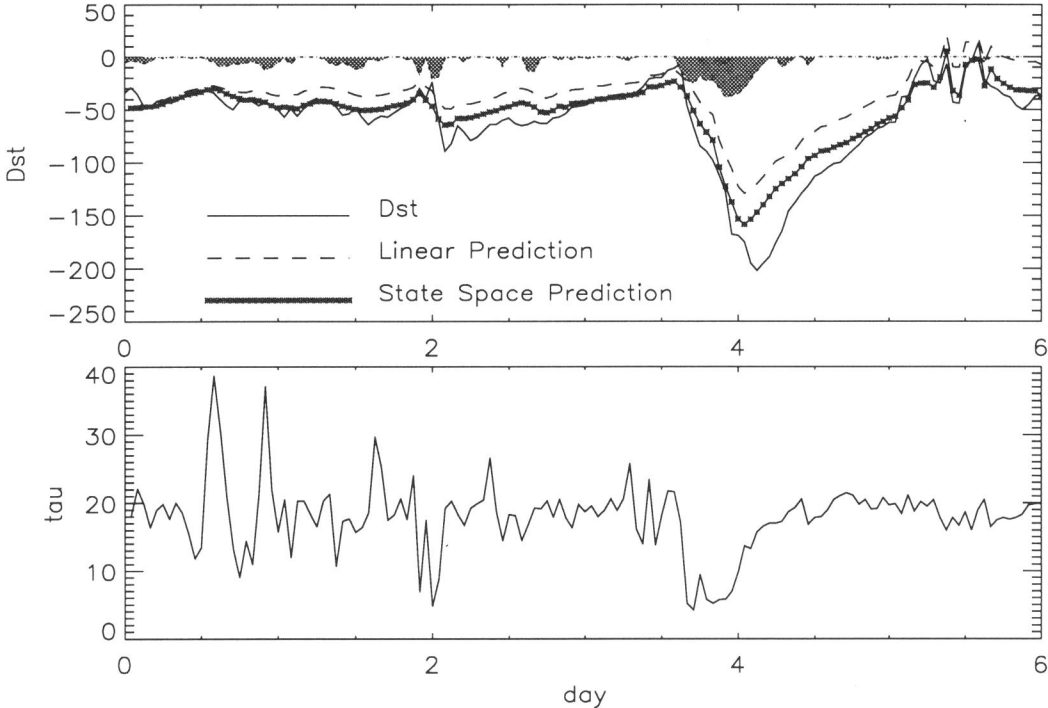

Figure 5. Local linear state-input space prediction of Dst. ISEE-3 solar wind data and Dst from 1978 08 17 through 1978 10 11 were used to create a state-input space database as described in the text. The test interval shown begins at 00 UT on 1979 03 31. Top panel shows Dst, and Ew (times 4, in gray), along with one-step ahead predictions by a linear filter fitted to the same database, and one-step ahead nonlinear predictions as described in the text. Bottom panel shows the decay rate parameter $\tau(t)$ obtained from the local linear filter coefficient of $Dst^0(t-1)$.

than when using linear prediction filters [*Vassiliadis*, 1994]. Given a continuous stream of solar wind input data, a continuous running prediction is possible (as with Dst). For test purposes using the same data set that was used to construct Figure 1 [*Bargatze, et al.*, 1985] (which has a 2.5 minute sample interval) the predictions can be started from an initial state known to be correct; in this case the prediction error reaches its asymptotic level in about 1 hour [*Price et al.*, 1994] to about 60 nT. Predictions from different initial conditions subject to the same sequence of solar wind inputs converge to the same uniquely determined evolution. The convergence is exponential, with a time scale of 30 minutes [*Vassiliadis et al.*, 1995].

There are differing results regarding the optimal number of nearest neighbors for short-term (5 min – 1 hour) predictions of AE indices. If the optimal number of nearest neighbors is a small fraction of the total database this suggests that the dynamics are highly nonlinear, however the optimum number also depends on the choice of state variables and state-space (embedding) dimension. This interesting issue is discussed in the Klimas et al. [1996] review.

Although nonlinear filters have a high predictive capability, their physical meaning is less obvious, than say, linear MA filters. Several theorems show that under general conditions the properties of the nonlinear filter (topology of state space, periodicities, stability time scales, and, in some cases, effective degrees of freedom) are identified with the real dynamics, which may not be directly measurable. These embedding theorems for autonomous [*Takens*, 1981; *Sauer et al.* 1991] and input-output systems [*Casdagli*, 1992] address generic properties of data embedding, or state-space construction.

In addition to properties of the dynamics, some physics can be learned from the filter coefficients g_j or (g_j, f_i), for example the (variable) Dst decay parameter, τ, in the Burton et al. [1975] model as shown in the bottom panel of Figure 5, or coupling constants, such as the *a* in Equation (17). For comparison, the linear MA filters time scales and coupling constants are obtained directly from the filter response (Fig. 1). More recent work on the state space technique [*Vassiliadis, et al.*, 1996] discusses further the compromise between extracting physics and maintaining low prediction error and describes new approaches to the problem.

TECHNIQUES FOR MEDIUM AND LONG TERM FORECASTING

Predicting magnetic storms or other geomagnetic activity more than about one hour in advance depends on conditions in the solar wind upstream of the L_1 Lagrange point.

An interesting new technique for making medium term predictions based on L_1 data was described earlier in this conference [*Chen, et al.*, 1996]. Chen's technique is based on early recognition of large scale patterns (e.g. flux ropes) in the solar wind, when they occur.

Chen et al. [1996] proposed characterizing the interplanetary magnetic field (IMF) by a set of features (amplitude and period of rotation). In feature space certain geoeffective events such as magnetic clouds are distinguishable from regular IMF activity. Moreover, because the cloud field changes in a sinusoidal pattern, detection of a cloud's positive Bz phase gives an extended lead time (from a several up to tens of hours) for the subsequent, and more geoeffective, negative Bz oscillation.

Techniques based on Remote Sensing and Models

For making reliable, scheduled, operational, long-term forecasts of geomagnetic activity, including magnetic storms, remote sensing is essential. The magnetic field of the Sun determines the structure of the corona and the solar wind, and it is the interaction of the solar wind with the magnetosphere of Earth that drives geomagnetic activity, including magnetic storms. Remote sensing is the ability to see or measure from a distance; solar images (of various wavelengths) represent remote sensing information.

The information produced by remote sensing techniques is generally incomplete or ambiguous in some way. Models based on physical principles offer a way to resolve the ambiguity or interpret the incomplete information. As a simple example, we humans automatically do this when we interpret a 2-dimensional image of the sun in terms of a mental model that is a 3-dimensional sphere. By using remote sensing information to set boundary conditions for physics-based models, those models can evolve that information in space and time.

Models exist for all regions of space, from the photosphere of the sun to the crust of the Earth. Eventually a causal chain of models will exist that propagate observations of the Sun all the way to geomagnetically induced currents in the ground. Construction of just such a system has been proposed, e.g. [*Akasofu and Fry*, 1986; *Dryer et al.*, 1986; *Heinemann et al.*, 1993], but since most of the necessary models are still in research or prototype stages, it may be some time before a complete chain can be assembled with all of its links.

Statistical (data based) models, such as those discussed in the preceding parts of this paper, can be constructed to bridge the gap between the outputs of existing physical models and terrestrial magnetic activity, as represented by geomagnetic indices such as Dst or Ap. This technique of combining data based models with physics based models will be illustrated by example test predictions of the Ap index, although it could just as easily be applied to the Dst index.

Background: Existing Operations and Models

The Wilcox Solar Observatory (WSO) combines daily magnetograms to produce global magnetic maps of the sun. The Naval Research Laboratory (NRL) in Washington, D.C. uses these maps as input boundary conditions to run models. One of these models is the flux transport code of DeVore, Sheeley, and Boris [1984]. The flux transport code evolves the global map one (or more) solar rotation(s) into the future by simulating the effects of differential rotation, diffusion, and meridional flow. The output of this model is another global magnetic map of the sun, but representing predicted conditions (e.g.) 27 days into the future.

Two other (potential field) models determine the coronal magnetic field configuration from the global magnetic map; this provides a basis to predict solar wind conditions at Earth. These models are the potential field source surface (PFSS) model and the current sheet (CS) model of Wang and Sheeley [1988]. Both models determine the coronal magnetic field configuration by solving Laplace's equation within a spherical shell ($R_S \leq r \leq R_{ss}$), where R_S is the solar surface and R_{ss} is called the source surface. Normally the source surface is placed at $r = 2.5\ R_S$ based on empirical results. Both models are time-independent in the frame rotating with the Sun; beyond the source surface the two models differ. The version used for the results reported here is called the current sheet (CS) model because heliospheric current sheets are included.

The output of the CS model is in the form of global maps showing the distribution of two variables, *Br* and *Ef*, at 1 AU (1 AU is the average distance from the Sun to the Earth). *Br* represents the magnitude and polarity of the magnetic field at 1 AU, and *Ef* is the flux-tube expansion factor, determined from the ratio of the flux tube area at 1 AU to its area on the photosphere; *Ef*=1 represents purely radial expansion.

The flux-tube expansion factor, *Ef*, is inversely correlated with the solar wind velocity at 1 AU [*Wang and Sheeley*, 1990]. Using this correlation the *Ef* can be interpreted as a solar wind speed at 1 AU, after allowing for transit time. If the flux transport code were applied to the global magnetic map before the CS model were run then model output would represent predicted conditions at 1 AU up to 27 days into the future. (The prediction lead time depends on the position of Earth's helio-longitude on the solar magnetic map.) Postscript plot files of model inputs and outputs (for both the estimate and prediction mode) are routinely available by anonymous ftp from NRL. (The address is maple.nrl.navy.mil; sub-directory pub/nrsfiles contains 4 sub-directories, for different solar observatories.)

Current Operational Forecasts

Forecasters at two centers cooperate to make operational predictions of Ap. The 50th Weather Squadron (50 WS), at Falcon Air Force Base near Colorado Springs, CO, predicts Ap for the next 45 days for the military. The Space Environment Center Space Weather Operations (SEC-SWO) in Boulder, CO, distributes daily predictions of Ap for the next 3 days, and publishes weekly predictions of Ap for the next 27 days.

Currently no operational forecasts of Dst are made.

The following table shows the correlation coefficient, *Rc*, and skill score, *SS*, of the operational Ap forecasts for the interval 1989 through 1993 as a function of forecast lead time. This table shows that, on average, there is little or no skill in the forecasts beyond 1 day.

Table 1

lead (days)	1	2	3	4	5	6	7
Rc	0.45	0.33	0.21	0.15	0.15	0.15	0.12
SS	0.16	0.05	−0.03	−0.08	−0.08	−0.09	−0.10

Beginning in 1993 a study was started to evaluate Ap forecasts based on the global plots produced by the Wang and Sheeley Models [*Gehred*, 1996]. Although these forecasts were subjective, based on a set of guiding principles, the conclusion of the study was that use of the global plots from the models produced a slight improvement. Could a linear prediction filter, as described above, make better use of the same data?

Linear ARMA Prediction of Ap from the Output of the Wang and Sheeley Current Sheet Model

A crucial step in this process (for which we are thankful to Y.-M. Wang) is to extract 1-dimensional data sets from the 2-dimensional output global maps from the CS model. These 1-dimensional data sets are determined by tracing those field lines which intersect Earth back to the source surface. For the results presented here the flux transport code was not used.

It is well known that magnetic storm activity is more frequent and stronger near spring and fall equinoxes than near the summer and winter solstices. This was explained [*Russell and McPherron*, 1973] to be a consequence of the fact that the rotation axis and magnetic dipole axis of Earth are tilted with respect to the plane of Earth's orbit about the Sun, the ecliptic plane. The effect of the tilt is that in Spring a heliospheric sector with magnetic polarity *toward* the Sun favors magnetic reconnection between the interplanetary magnetic field and the Earth's magnetic field. In the Fall, *away* polarity favors reconnection.

The CS model outputs are provided in solar (Carrington) coordinates and must be converted into a time series of daily *Ef* and *Br* at the sub-Earth point on the source surface. Next the Russell-McPherron (R-M) effect is taken into account in a simple way; we just multiply *Br* times the R-M tilt angle. For the daily average R-M tilt angle we simply use 23° times the cosine of the ecliptic longitude of Earth. We call the result *Be*, for B-effective.

Using just the time series of *Ef*, and *Be* we tried fitting multi-channel MA linear filters for predicting the Ap index. For example, the Ap index prediction is given as a linear combination of *log(Ef)* and *Be* at lags of 4, 5, and 6 days preceding the verification day. (The verification day is day of the Ap observation against which the prediction is verified.) The filter coefficients are found by the general linear least squares (MLR) technique described earlier. Although results were positive, this approach ignores the predictive value of previous Ap observations that is evident in Figures 3 and 4.

Figures 3 and 4 show 1-step (1-day) ahead predictions, but the same technique can be used for any lead time. In a separate set of experiments we calculated skill scores of AR linear prediction filters that predict Ap with a lead time of 25 days (almost one solar rotation). Specifically, using only Ap data from 1967 through 1976 (the same as for Figures 3 and 4) we constructed a filter to predict Ap with a lead time of 25 days by using as input observed Ap at lags of 25, 26, 27, 28, and 29 days before the day being predicted. Not too surprisingly, the results depend strongly on where in the solar cycle the test period is chosen. This is shown in the following table:

Table 2

Test Interval	Correlation Coef.	Skill Score
1989 – 1993	0.111	–0.196
1994	0.502	0.191
1995	0.494	0.202

The important question is, how much improvement can be obtained by adding the information supplied by the Wang and Sheeley model outputs?

Table 3

Test Interval	Correlation Coef.	Skill Score
1989 – 1993	0.162	0.009
1994	0.573	0.283
1995	0.539	0.136

The results shown in Table 3 were obtained using both model outputs and observed Ap from 1977 through 1988 as a database to fit a multi-channel ARMA prediction filter.

An example of the predictions made by this technique is shown in Figure 6. The R-M effect is evident from the fact that open diamonds are associated with increased Ap where the R-M angle is positive and filled diamonds are associated with increased Ap where the R-M angle is negative.

The database for the filter used to make Figure 6 was different from that used to make the above table; the database used for Figure 6 consisted of model outputs and observed Ap from 1993 and 1994. A comparison of the skill score for 1995 from Figure 6 (0.284) with the skill score for 1995 from the preceding table (0.136) suggests that performance of this technique is sensitive to the choice of the database used to fit the filter. Another example supporting that conclusion is a skill score of 0.304, for a filter constructed using 1983 and 1984 as a database and 1994 as the cross validation test interval, compared with skill score of only 0.283 for 1994 in the preceding table.

Discussion of the Combined Techniques

Figure 6 clearly shows that the strengths of physically based models can be combined with the strengths of statistical time series prediction techniques. Both parts of this hybrid technique are subject to improvement. The nonlinear state-space prediction techniques discussed above could replace the linear predictions used in this example. The sensi-

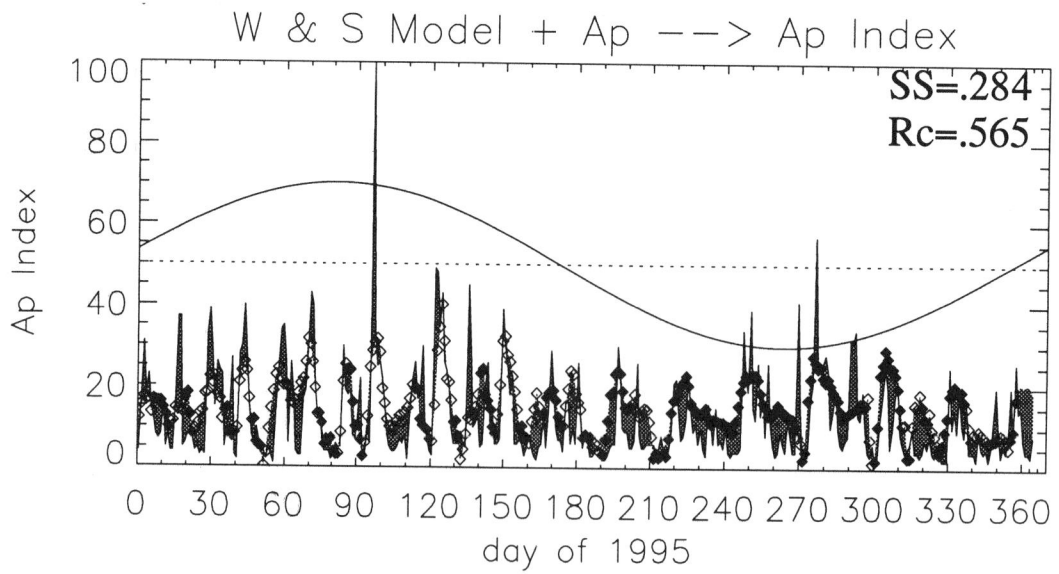

Figure 6. Wang and Sheeley Model based predictions of Ap for 1995. The observed Ap is plotted as a solid line. Predicted Ap is plotted as diamonds, open where model polarity is toward sun, solid where model polarity is away. The prediction error is filled in gray, and the sine curve, centered on Ap=50, represents the Russell-McPherron tilt angle.

tivity of the skill scores to the choice of the fitting database suggests that the local-linear prediction technique based on nearest neighbors in state-space might give significant further improvement. Further research and development of this hybrid technique is planned.

We think this hybrid technique is a good candidate of immediate operational development and beta testing because all of its key elements are either already operational or could be made so with relatively little effort.

Acknowledgments. Detman, wishes thank Walter Gonzalez for beginning his interest in the dynamics of Dst, in particular the variability of the decay rate. Both authors thank R. L. McPherron and another reviewer for helpful suggestions. Thanks also to J. Joselyn, L. Puga, and T. Onsager for valuable suggestions. D. Vassiliadis was supported by NASA grant NRA-94-SSM-SR&T-046.

REFERENCES

Abarbanel, H.D.I., R. Brown, and J.J. Sidorowhich, The analysis of observed chaotic data in physical systems, *Rev. Mod Phys., 65,* 1331, 1993.

Akaike, H., Use of an information theoretic quantity for statistical model identification, *Proc. 5th Hawaii Int. Conf on System Sciences,* 1972.

Akaike, H., A new look at statistical model identification. *IEEE Trans. Automat. Control AC-19,* 1974.

Akasofu, S.-I., and C.F Fry, A First Generation Numerical Geomagnetic Storm Prediction Scheme, *Planet. Space Sci., 34,* 77-92, 1986.

Baker, D.N., Statistical Analysis in the Study of Solar Wind-Magnetosphere Coupling, in *Solar Wind-Magnetosphere Coupling,* edited by Y. Kamide and J.A. Slavin, 17-38, Terra Scientific Publishing Co., 1986.

Bargatze, L.F., D.N. Baker, R.L. McPherron, and E.W Hones, Jr. Magnetospheric impulse response for many levels of geomagnetic activity, *J. Geophys. Res., 90,* 6387, 1985.

Boteler, D.H., Geomagnetic effects on power systems, this volume.

Brooks, H.E., and C.A. Doswell, A comparison between measures-oriented and distributions-oriented verification methods in forecast verification, *Weather and Forecasting, 11,* in press, 1996.

Broomhead, D.S., J.P. Huke, and M.R. Muldoon, Linear filters and Non-linear systems, *J. R. Statist. Soc. B, 54,* 373-382, 1992.

Burton, R.K., R.L. McPherron, and C.T. Russell, An empirical relationship between interplanetary conditions and Dst, *J. Geophys. Res., 80,* 4204-4214, 1975.

Campbell, Charles P., *Ethos:* Character and Ethics in Technical Writing, *IEEE Trans. on Professional Communication 38,* 132, 1995.

Casdagli, M., Chaos and deterministic versus stochastic non-linear modelling, *J. R. Statist. Soc. B, 54,* 203, 1991.

Casdagli, M., A dynamical systems approach to modeling input-output systems, in *Nonlinear Modeling and Forecasting,* edited by Casdagli, M. and S. Eubank, Addison-Wesley, 1992.

Chen, J., P.J. Cargill, and P. Palmadesso, Real-Time Prediction of Magnetic Storms, *J. Geophys. Res,* Special Section on Magnetic Storms (submitted) 1996.

Clauer, C.R., The Technique of Linear Prediction Filters Applied to Studies of Solar Wind-Magnetosphere Coupling in *Solar Wind-Magnetosphere Coupling,* edited by Y. Kamide and J. A. Slavin, 39-57, Terra Scientific Publishing Co., 1986.

Detman, T., D. Vassiliadis, and G. Burkhart, Application of the State-Space Method to Prediction of Dst from Solar Wind Data and Characterization of Ring Current Dynamics, *IUGG XXI General Assembly, Abstracts Week B,* GAB32E-11, 1995.

DeVore, C.R., N.R. Sheeley, Jr., and J.P. Boris, The concentration of the large-scale solar magnetic field by a meridional surface flow, *Sol. Phys., 92,* 1, 1984.

Dryer, M., S.-I. Akasofu, H.W. Kroehl, R. Sagalyn, S.T. Wu, T F. Tascione, and Y. Kamide, The solar/interplanetary/magnetosphere/ionosphere connection: a strategy for prediction of geomagnetic storms, *Adv. Astro. Sci., 58,* 1986.

Fay, R.A., C.R. Garrity, R.L. McPherron, and L.F. Bargatze, Prediction Filters for the Dst Index and Polar Cap Potential, in *Solar Wind-Magnetosphere Coupling,* edited by Y. Kamide and J. A. Slavin, 111-117, Terra Scientific Publishing Co., 1986.

Feldstein, Y.I., V. YU. Pisarsky, N.M. Rudneva, and A. Grafe, Ring Current Simulation in Connection with Interplanetary Space Conditions, *Planet. Space Sci,* 32, 975-984, 1984.

Garcia, P., J. Jimenez, A. Marcano, and F Moleiro, Local optimal metrics and nonlinear modeling of chaotic time series, *Phys. Rev. Lett., 76,* 9, 1449-1452, 1996.

Gehred, P. A., Wang and Sheeley medium-range planetary A index forecast verification statistics, *NOAA Technical Memorandum ERLSEL-91,* Space Environment Center, Boulder, CO, 1996.

Gonzalez, W. D., A unified view of solar wind-magnetosphere coupling functions, *Planet. Space Sci., 38,* 627-632, 1990.

Heinemann, M., N.C. Maynard, D.N. Anderson, and F. Marcos, Space Weather Forecasting System, in *Solar-Terrestrial Predictions-IV, Proceedings of a Workshop at Ottawa, Canada, May 18-22, 1992, Vol. 2,* National Oceanic and Atmospheric Administration, Environmental Research Laboratories, Boulder, CO. 1993.

Iyemori, Toshihiko, Hiroshi Maeda, and Toyohisa Kamei, Impulse Response of Geomagnetic Indices to Interplanetary Magnetic Field, *J. Geomag. Geoelectr, 31,* 1-9, 1979.

Kappenman, J.G., and W.A. Radasky, The impact of geomagnetic disturbances on electric power systems and geomagnetic forecast improvements needed of the electric utility industry, this volume.

Kanter, M., Lower bounds for nonlinear prediction error in moving-average processes, *Ann. Probability, 7,* 2, 128-138, 1979.

Kay, Steven M., and Stanley Lawrence Marple, Jr., Spectrum Analysis - A Modern Perspective, *Proc. IEEE, 69,* 1380-1419, 1981.

Klimas, A.J., D.N. Baker, and D. Vassiliadis, The organized nonlinear dynamics of the magnetosphere, *J. Geophys. Res., 101,* 13,089, 1996.

Lundstedt, H., AI techniques in geomagnetic storm forecasting, this volume.

Marshall, K.T., and R. M. Oliver, *Decision Making and Forecasting,* 407 pp., New York, NY, McGraw-Hill, 1995.

McPherron, R.L., D.N. Baker, and L.F, Bargatze, Linear Filters as a Method of Real Time Prediction of Geomagnetic Activity, in *Solar Wind Magnetosphere Coupling,* edited by Y. Kamide and J.A. Slavin, 85-92, Terra Scientific Publishing Co., 1986.

McPherron, R.L., Monitoring the solar wind with the Dst index, *Eos Trans. AGU, 76,* 17, S206, 1995.

Murayama, T., Coupling Function between the Solar Wind and the Dst Index, in *Solar Wind Magnetosphere Coupling,* edited by Y. Kamide and J.A. Slavin, 119-126, Terra Scientific Publishing Co., 1986.

Murphy, A.H., Forecast Verification, in *Economic Value of Weather and Climate Forecasts,* edited by R.W Katz and A.H. Murphy, Cambridge University Press, Cambridge, U.K., in press, 1996.

Murphy, A.H., and H. Daan, Forecast evaluation, in *Probability, Statistics and Decision Making in the Atmospheric Sciences,* edited by A. H. Murphy and R.W. Katz, Westview Press, Boulder, CO, 1985.

Murphy, A.H. and R.L. Winkler, A general framework for forecast verification, *Mon. Wea. Rev., 115,* 1330-1338, 1987.

Press, W. H., B.P. Flannery, S.A. Teukolsky, and W. T. Vettering, *Numerical recipes, the Art of Scientific Computing (FORTRAN Version),* Cambridge Univ. Press, Cambridge, UK., 52-64, 1992.

Price, C.P., and D. Prichard, The nonlinear response of the magnetosphere: 30 October 1978, *Geophys. Res. Lett, 20,* 771, 1993.

Price, C.P., D. Prichard, and J.E. Bischoff, Non-linear input-output analysis of the auroral electrojet index, *J. Geophys. Res., 99,* 13,227, 1994.

Reiff, P.H., The use and misuse of statistical analysis, in *Solar-Terrestrial Physics,* Eds. R.L. Caravillano and J.M. Forbes, D. Reidel, 1983.

Russell, C.T., and R.L. McPherron, Semiannual variation of geomagnetic activity, *J. Geophys. Res., 78,* 92, 1973.

Sauer, T., J.A. Yorke and M. Casdagli, Embedology, *J. Stat. Phys., 65,* 3-4,1991.

Shepp, L.A., D. Slepian, and A.D. Wyner, On prediction of moving-average processes, *The Bell System Technical Journal, 59,* 3, 367-415, 1980.

Takens, F., Detecting Strange Attractors in Turbulence, in *Dynamical Systems and Turbulence,* edited by A. Dold and B. Eckmann, *Lecture Notes in Mathematics 898,* 366- 381, Springer Verlag, Berlin, 1981.

Trattner, K. J., and H.O. Rucker, Linear Prediction Theory in Studies of Solar Wind-Magnetosphere Coupling, *Ann. Geophysicae 8,* 733-738, 1990.

Ulrych, T.J., and R.W. Clayton, Time series modeling and maximum entropy, *Phys. Earth Planet. Inter. 12,* 188-200, 1976.

Valdivia, J.A., A.S. Sharma, and K. Papadopoulos, Dynamics and prediction of magnetospheric storms, *Eos, Trans. AGU Supplement, 75,* 44, S544, 1994.

Valdivia, J.A., A.S. Sharma, and K Papadopoulos, Prediction of Magnetospheric Storms by Nonlinear Dynamical Models, *Geophys. Res. Lett.* (accepted) 1996.

Vassiliadis, D., The input-state space approach to the prediction of auroral geomagnetic activity from solar wind variables, in *Proceedings of the International Workshop on Artificial Intelligence Applications in Solar Terrestrial Physics,* edited by Joselyn, J. and H. Lundstedt and J. Trolinger, NOAA, Boulder, 145-151, 1994.

Vassiliadis, D., A.J. Klimas, D.N. Baker and D.A. Roberts, A description of solar wind magnetosphere coupling based on nonlinear filters, *J. Geophys. Res., 100,* 3, 3495, 1995.

Vassiliadis, D., A.J. Klimas, D.N. Baker and D.A. Roberts, Nonlinear predictor error for the vBz-AL coupling, *J. Geophys. Res., 100,* 3495, 1996.

Vassiliadis, D., A.J. Klimas, and D.N. Baker, Nonlinear ARMA Models for the Dst Index and their Physical Interpretation, in *Proceedings of the Third International Conference on Substorms (ICS-3),* Versailles, France, May 13-17, 1996.

Wang,Y.-M., and N.R. Sheeley, Jr.,The solar origin of long-term variations of the interplanetary magnetic field strength, *J. Geophys. Res., 93,* 11,227, 1988.

Wang,Y.-M., and N.R. Sheeley, Jr., Solar wind speed and coronal flux-tube expansion, *Astrophys. J., 355,* 726, 1990.

Weigend, A., and N. Gershenfeld, *Time Series Prediction,* Addison-Wesley, Reading, MA, 1994.

Yule, G., On a method of investigation periodicity in disturbed series with special reference to Wolfer's sunspot numbers, *Phil. Trans. Roy. Soc. London, A 226,* 1927.